Linear Continuous-Time Systems

Linear Continuous-Time Systems

Lyubomir T. Gruyitch

CRC Press
Taylor & Francis Group
Boca Raton London New York

CRC Press is an imprint of the
Taylor & Francis Group, an **informa** business

CRC Press
Taylor & Francis Group
6000 Broken Sound Parkway NW, Suite 300
Boca Raton, FL 33487-2742

First issued in hardback 2019

ISBN-13: 978-1-138-03950-6 (hbk)

Visit the Taylor & Francis Web site at
http://www.taylorandfrancis.com

and the CRC Press Web site at
http://www.crcpress.com

Dedication

Miodrag RAKITCH, D. Sci.

(Belgrade, January 1, 1923 - Belgrade, December 28, 1998)

Professor and Vice-Chairman of the Faculty of Electrical Engineering,
University of Belgrade, Serbia,
the son of a rich Serb who owned an industry before World War II,
which communists nationalized after the War,
the coauthor of the famous Belgrade artificial hand and
the father of robotics in Serbia,
the first President of the Association for
Systems, Control and Measurement of Serbia
and its first Distinguished and Honorary Member,
the Supervisor of my Master of Science thesis, and
the Member of the jury for the defence of my D.Sci. Dissertation.

For his
brave rejection of Hitler Nazis during World War II,
courage and integrity although it was not in line
with that of the Yugoslav Communist Party,
fully objective evaluations of students' and researchers' works
and deep devotion to support them
regardless of their membership or non-membership
in the Yugoslav Communist Party.
For his
permanent moral support since our first scientific discussion
after his postgraduate lecture on linear systems
up to the end of his rich and inspiring life.

Lyubomir T. GRUYITCH

Contents

List of Figures

Preface

On the state of the art

This book discovers and resolves fundamental lacunae of the theory and practice of the continuous-*time time*-invariant linear dynamical Multiple Input - Multiple Output ($MIMO$) systems, for short systems in the sequel.

Initial (input, state and output) conditions contain all relevant information on the system history, on the external actions and on their consequences upon the system until the initial moment. The initial conditions are unpredictable, unknown a priori and untouchable. It is well known that the Laplace transform of derivatives of a variable contains the corresponding initial conditions. They do exist in the complex domain. They appear in the Laplace transform of the system state and the system output. Their appearance for the systems of the order two or higher is under double sums that contain products of the system parameters and initial conditions. Such double sums seemed nontreatable effectively mathematically. In order to avoid the originated mathematical problem of how to study effectively the system described completely, in the forced regime under arbitrary initial conditions, the superposition principle has been exploited to justify the assumption on the zero initial conditions that has been unconditionally accepted and has governed the courses and the research on the systems.

The fact is that the system transfers simultaneously the influence of the input variables, i.e., the influence of the input vector, and the influence of the initial conditions on its state and its output.

The fact is that the system being dynamical possesses its internal dynamics that determines its state and state variables, i.e., its state vector. This fact has been ignored by the unjustified accepted condition that the concept of state has a sense if and only if the derivatives of the input vector do not influence the system internal dynamics. This condition is physically meaningless, but it has appeared as very useful to develop a mathematical machinery for the related studies. This explains the reason for which the theory and practice have been well developed only for one class of the physical systems. In order to treat other classes of the physical system their mathematical models have to be transformed mathematically with the complete loss of the physical sense of the new mathematical model.

A fundamental lacuna is that the concept of the state, state variables, state vector and state space is well defined and widely effectively used only in the framework of the systems described by the first order vector linear differential state equation and by the algebraic vector linear output equation, which are called *Input-State-Output* (*ISO*) (or *state-space*) systems. This is a fundamental lacuna due to the nonexistence of the clear, well defined, concept of the state also for the much less theoretically treated systems described by an arbitrary order vector linear input-output differential equation, which are called *Input-Output* (*IO*) systems, as well as for a new, more general class of the systems described by an arbitrary order vector linear differential state equation and by a vector linear differential output equation, which are called *Input-Internal and Output dynamical* (*IIO*) systems.

A fundamental lacuna is also the use of the system transfer function matrix $G(s)$ defined and valid only for all zero initial conditions and a nonzero input vector to test Lyapunov stability properties defined for arbitrary initial conditions and for the zero input vector.

Another fundamental lacuna is the restriction of the bounded input stability concept imposed by the condition that all initial conditions are equal to zero.

An additional fundamental lacuna is the nonexistence of a complex domain method for effective treatment based on the Laplace transform of the systems subjected simultaneously to both input actions and initial conditions. It is the consequence of the unreal and unjustifiable assumption on all zero initial conditions, which is accepted for the definition and for the determination of the system transfer function (transfer function matrix in general) $G(s)$ and for the system matrix $P(s)$. One of its consequences is the inconsistent link between the stability definitions and the stability criteria determined in terms of the system transfer function matrix $G(s)$. The inconsistence is due to the definition of the former exclusively for the free regime under arbitrary initial conditions and the definition of the latter exclusively for the forced regime under zero initial conditions. Another consequence is related to the unclear criterion for the pole-zero cancellation. The third consequence is the treatment of mainly the system behavior in the forced regime under zero initial conditions.

The conceptual drawback of the widely accepted existing theory is the nonexistence of the state space theory for the *IO* dynamical systems. Moreover, the *IIO* systems were not studied. These drawbacks reflect a deep scientific and engineering vagueness.

Among many excellent books on the linear dynamical systems theory, the author consulted in particular the books by the following writers (mentioned in alphabetical order) when writing this book: B. D. O. Anderson and J. B. Moore [3], P. J. Antsaklis and A. N. Michel [4], P. Borne et al. [5], W. L. Brogan [7], G. S. Brown and D. P. Campbell [8], F. M. Callier and C. A. Desoer [10] and [11], C.-T. Chen [13], H. Chestnut and R. W. Mayer [14], J. J. D'Azzo and C. H. Houpis [15], C. A. Desoer [17], C. A. Desoer and M. Vidyasagar [19], W. Hahn [76], T. Kailath [77], N. N. Krasovskii [83] and [84], B. C. Kuo [85] and [86], H. Kwakernaak and R. Sivan [87], A. M. Lyapunov [88], L. A. MacColl [89], J. M. Maciejowski [90], J. L. Melsa and D. G. Schultz [91], B. R. Milojković and Lj. T. Grujić* [93], K. Ogata [106] and [107], D. H. Owens [108], H. M. Power and R. J. Simpson [109], H. H. Rosenbrock [110], R. E. Skelton [114], M. Vidyasagar [118], J. C. West [119], D. M. Wiberg [120], W. A. Wolovich [121], and W. M. Wonham [122].

On the book

Linearized continuous-*time* mathematical descriptions of physical (biological, economical, technical) systems can be studied in the *time* domain or in the *complex* domain. In the former case, all system variables are expressed in terms of *time t*. In the latter case, all system variables are represented by their Laplace transforms in terms of *the complex variable s*. In this book, the linear *time*-invariant continuous-*time* mathematical models of physical (biological, economical, technical) dynamical systems will be called *the systems*. The book treats some general aspects of them. The discrete-*time* counterpart to this book, which is a refined version of the book [47], is the accompanying book [9]. These two books are completely literally mutually analogous and constitute an entity.

We treat herein the three classes of the continuous-*time* *time*-invariant dynamical systems-**systems**.

- *The Input-Output* (*IO*) *systems* described by the $\nu - th$ order linear vector differential equation in the output vector $\mathbf{Y} \in \mathfrak{R}^N$.

- *The (first order) Input-State-Output* (*ISO*) *systems* determined by the first-order linear vector differential equation in the state vector $\mathbf{X}$ - *the state equation*, $\mathbf{X} \in \mathfrak{R}^n$, and by the algebraic vector equation in the output $\mathbf{Y}$ - *the output equation.* They are well known as *the state-space systems.*

- *the Input-Internal and Output dynamical* (IIO) *systems* characterized by the α-*th* order linear vector differential equation in the internal dynamics vector $\mathbf{R}$ - *the internal dynamics equation*, $\mathbf{R} \in \mathfrak{R}^{\rho}$, and by the ν-*th* order linear vector (differential if $\nu > 0$, algebraic if $\nu = 0$) equation in the output vector $\mathbf{Y}$ - *the output equation.* This class of the systems has not been studied so far.

The *time* domain mathematical models and their studies enable the direct insight in the physical properties and in the dynamical behavior of the systems. The *time* domain is adequate for explaining physical phenomena of, and processes in, dynamical systems. It is appropriate also for defining dynamical properties of the systems. However, their direct mathematical treatment in the *time* domain is not very effective in the case of the analysis of their qualitative dynamical properties.

The complex domain mathematical descriptions of the systems enable the effective mathematical treatment of their qualitative properties. Its basic tools are the Laplace transform and, by it induced, the fundamental dynamic system characteristic that is well known, in general, as the transfer function matrix $G(s)$ of a *Multiple-Input Multiple-Output* ($MIMO$) system, or, in a simpler case, the transfer function $G(s)$ of a *Single-Input Single-Output* ($SISO$) system. The usage of $G(s)$ enables us to study effectively many system qualitative dynamical properties [e.g., the system response *under all zero initial conditions*, system controllability and observability, Lyapunov stability of a completely controllable and observable system, *Bounded-Input Bounded-Output* ($BIBO$) stability under *zero initial conditions*]. It represents the powerful mathematical system characteristic to treat dynamical properties of the system, by the definition, under *all zero initial conditions*. Its substantial feature is its full independence of external actions (i.e., of the system inputs), and, naturally, of initial conditions since it is defined *for all zero initial conditions*. It is completely determined exclusively by the system itself so that it describes in the complex domain how the system transfers in the course of *time* the influence of the external action on the system to its output variables (i.e., to its output vector). It is *only the system input-output transfer function (matrix), nothing more.* It is inapplicable in the cases when the initial conditions are not all equal to zero. It does not and cannot express in the complex domain how the initial conditions influence the system output behavior in the course of *time*.

This raises the following basic question:

Does a linear *time*-invariant system have a complex domain characteristic mathematically expressed in the form of a matrix such that it satisfies the following two conditions:

- it describes in the complex domain how the system transfers in the course of *time* the influences of *both* the input vector action and of *all initial conditions* on the system output response, and

- it is completely determined only by the system itself, meaning its full independence of *both the input vector and the vector of all initial conditions*?

The reply is affirmative.

Such system characteristic is its **full (complete) transfer function matrix** $F(s)$. We will define and determine it precisely for all three classes of the systems. Its usage permits us to refine studies of system dynamical properties (e.g., of the complete system response, of Lyapunov stability, Bounded Input (BI) stability properties, of the system minimal realization, of the system tracking and of its trackability [70]). It shows exactly when poles and zeros may (not) be cancelled.

Even the simplest first-order linear $SISO$ dynamical system induces this characteristic in a matrix form rather than in a scalar form.

It seems that the problem which represented the obstacle to the determination of the full transfer function matrix is mainly related to the application of the Laplace transform to the IO (*Input-Output*) (and to *the Input-Internal and Output dynamical,* i.e., IIO) systems descriptions.

The Laplace transform of the IO and IIO systems introduces double sums that contain the products of initial conditions and system parameters [(3.12), (3.13), (3.18) and (3.19) in Subsection 3.3.3]. We will show how those double sums can be represented, e.g., for the IO system, as a product of a matrix $G_{IOo}(s)$ and of the vector $\mathbf{C}_{IOo}$ composed of all (input and output) initial conditions,

$$\mathbf{C}_{IOo} = \begin{bmatrix} I_0 \\ I_0^{(1)} \\ \cdots \\ I_0^{(\mu-1)} \\ Y_0 \\ Y_0^{(1)} \\ \cdots \\ Y_0^{(\nu-1)} \end{bmatrix} \in \mathfrak{R}^{\mu M+\nu N},$$

The vector $\mathbf{C}_{IOo}$ is subvector of the vector $\mathbf{V}_{IO}(s)$ composed of the Laplace transform $\mathbf{I}(s)$ of the input vector $\mathbf{I}(t)$ and of all (input and output) initial conditions composing the vector $\mathbf{C}_{IOo}$,

$$\mathbf{V}_{IO}(s) = \begin{bmatrix} \mathbf{I}(s) \\ \mathbf{C}_{IOo} \end{bmatrix} \in \mathfrak{R}^{(\mu+1)M+\nu N}.$$

The complete list of the notation is in Appendix A.

The matrix $G_{IOo}(s)$ is determined exclusively by the system parameters. It is completely independent of the initial conditions and of the input vector. It is *the system transfer function matrix relative to the initial conditions*, which along with the system transfer function matrix $G_{IO}(s)$, compose the *full transfer function matrix* $F_{IO}(s)$ of the *Input-Output* (IO) *systems*,

$$F_{IO}(s) = \left[G_{IO}(s) \vdots G_{IOo}(s) \right] \Longleftrightarrow \mathbf{Y}(s) = F_{IO}(s)\mathbf{V}_{IO}(s).$$

The analogy holds also for the *Input-State-Output* (ISO) *systems*,

$$F_{ISO}(s) = \left[G_{ISO}(s) \vdots G_{ISOo}(s) \right] \Longleftrightarrow \mathbf{Y}(s) = F_{ISO}(s)\mathbf{V}_{ISO}(s),$$

and for the *Input-Internal dynamics-Output* (IIO) *systems*,

$$F_{IIO}(s) = \left[G_{IIO}(s) \vdots G_{IIOo}(s) \right] \Longleftrightarrow \mathbf{Y}(s) = F_{IIO}(s)\mathbf{V}_{IIO}(s),$$

The use of the system full transfer function matrix $F(s)$ requires the same knowledge of mathematics as for the application of the system transfer function matrix $G(s)$. Nothing more. However, it requires some new notation that is presented in the simple symbolic vector or matrix form analogous to the scalar form.

The introduction of both the system full transfer function matrix $F(s)$ and the Laplace transform $\mathbf{V}(s)$ of the complete action vector $\mathbf{v}(t)$ enables us to generalize the block diagram technique to the full block diagram technique.

The definition and the determination of the full transfer function matrix $F(s)$ is unified herein for all three classes of the systems. The new compact, simple, scalar-like, vector-matrix notation permits us to unify the treatment of the IO systems and of the IIO systems in order to study them in the same manner as the ISO systems. Lyapunov stability theory (definitions, method and theorems) becomes directly applicable to the IO systems and to the IIO systems in the same manner as to the ISO systems. We will discover herein that the properties of the system full transfer function matrix $F(s)$ and of the system transfer function matrix $G(s)$ show that $F(s)$ is adequate for Lyapunov stability study and for the investigations of BI stability properties under arbitrary bounded initial conditions while the application of $G(s)$ can yield essentially the wrong result (e.g., that the equilibrium state is asymptotically stable although it is really unstable). We achieve herein the crucial generalization of the BI stability concept. Besides, we refine inherently Lyapunov stability and BI stability criteria by proving new ones in the complex domain. This will be possible only due to the use of the system full transfer function matrix $F(s)$.

The purpose of this book is to contribute to the advancement of the linear dynamical systems theory, of the corresponding university courses, to open new directions for the research in this theory and for its engineering applications. It represents a further development of the existing linear systems theory, which will not be repeated herein.

The system full transfer function matrix $F(s)$ was introduced to seniors through the undergraduate courses on linear dynamical systems and on control systems at first at the Department of Electrical Engineering, University of Natal in Durban (UND), R. South Africa, 1993, and at the National Engineering School (Ecole Nationale d'Ingénieurs de Belfort - $ENIB$) in Belfort, France, 1994, and continued until 1999. This was a topic of lectures to freshmen or juniors of the new University of Technology Belfort - Montbéliard ($UTBM$), which was created as the union of the $ENIB$ and the Polytechnic Institute of Sevenans. The UND students had the lecture notes [29] available during the course. The lecture notes [25]-[28], [49], [51], [63], were immediately, after the classes, available to the students in the copy center of $ENIB$ / $UTBM$.

The hope is that *the XXI century linear systems courses and linear control courses:*

- **will incorporate**:

- *all three classes of the systems*: IO, ISO and IIO systems,

- *the system full (complete) transfer function matrix* $F(s)$ [68] as the basic system dynamical characteristic in the complex domain, as well as its applications to various issues of dynamical systems in general, and control systems in particular, e.g., to the system complete response, pole-zero cancellation, stability, observability, controllability, tracking, trackability ([25]-[34], [36]-[45], [47], [49], [51], [53], [56]-[63], [68], [70]-

[75], [96] - [105]), and optimality, which are the basic issues of the system theory in general, and the control theory in particular,

and

- *the tracking and trackability theory*, as the fundamentals of the control science and of control engineering, which express the primary control goal (presented in the accompanying books [58], [70]),

- **will refine the study in the complex domain** *of the qualitative system properties* ***by using the system full transfer function matrix F(s)*** *instead of the system transfer function matrix G(s),*

- **will devote more attention to the basic system characteristics** *such as*

- *system regimes,*
- *system stationary vectors,*
- *system equilibrium vectors,*

and

- **will pay attention to the differences between**

- *the transfer function matrix realization* and *the system realization,*
- *the irreducible complex rational matrix functions* and *the degenerate complex rational matrix functions.*

Note 1 ***On the proofs presentation***
Since the book provides the novel results, all the details of the derivations of formulae and proofs of the main results are preserved and exposed in the text, even those that might appear very simple (or even trivial) to some readers, but with which some students or young researchers might be unfamiliar. Such presentation gives the reader the full understanding and the easy verification of the statements and the results.

In gratitude

The author is indebted to the publisher, the editor and their team, in particular to:

Ms. Nora Konopka, Editorial Director, Engineering, for leading and organizing the publication process elegantly and effectively

Ms. Michele Dimont, Production Editor, for leading effectively the editing process

Ms. Kyra Lindholm, Senior Editorial Assistant – Engineering, for leading effectively administrative process

Ms. Michele Smith, Editorial Assistant – Engineering, for very useful assistance

The author is indebted to Professor Zoran M. Buchevats for discovering several typographical errors, which helped the author to correct them.

The author is grateful to Mr. George Pearson with MacKichan Company for his very kind and effective assistance to resolve various problems related to the SWP application among which are the problem of the adaptation of the trim size of the book typed by using Scientific Work Place and the problem of effective application of SWP to generate simultaneously the Author Index and Subject Index.

* The author's name is written Ljubomir T. Grujić in Serbo-Croatian using the Croatian Roman alphabet, Lyubomir T. Gruyitch in English and Lyoubomir T. Grouyitch in French [27]). Since it is not possible** to preserve the original Cyrillic writing of Serb names in other languages, and since it is important to preserve personal identification, in order to at least preserve the correct pronunciation, the adequate transcription should be applied, which, in this case, the author's name is transcribed in other languages as shown herein.

** It is common to transcribe the an Serb Cyrillic letter "ћ" in English into "ch". However, "ch" is always used as English transcription of another Slavic (Russian, Serb,...) Cyrillic letter that is transcribed in the Croatian Roman alphabet as "č" (this software does not have this Cyrillic letter). In order to avoid confusion, I have accepted the use of "tch" for "ћ" because the linguists have not implemented another solution that might be more adequate.

Lyubomir T. Gruyitch, the author
Belgrade, January 4, 2017

Part I

BASIC TOPICS OF LINEAR CONTINUOUS-*TIME* *TIME*-INVARIANT DYNAMICAL SYSTEMS

Chapter 1

Introduction

1.1 *Time*

All processes, motions and movements, all behaviors of systems and their responses, as well as all external actions on the systems, occur and propagate in *time*. It is natural from the physical point of view to study systems directly in the temporal domain. This requires us to be clear on how we understand what is *time* and what are its properties, which we explain in brief as follows (for the more complete analysis see: [54], [64], [65], [66]).

Definition 2 ***Time***

***Time** (i.e., **the temporal variable**) denoted by t or by τ is an independent scalar physical variable such that:*

- *its **value** called **instant** or **moment** determines uniquely **when** somebody or something started / interrupted to exist,*
- *its values determine uniquely **since when and until when** somebody or something existed / exists or will exist,*
- *its values determine uniquely **how long** somebody or something existed / exists or will exist,*
- *its values determine uniquely whether an event E_1 occurs **then when** another event E_2 has not yet happened, or the event E_1 takes place just **then when** the event E_2 happens, or the event E_1 occurs **then when** the event E_2 has already happened,*
- *its value **occupies (covers, encloses, imbues, impregnates, is over and in, penetrates) equally** everybody and everything (i.e., beings, objects, energy, matter, and space) **everywhere and always**, and*
- *its value has been, is, will be, **permanently changing smoothly, strictly monotonously continuously, equally** in all spatial directions*

and their senses, in and around everybody and everything, ***independently*** *of everybody and everything (i.e., independently of beings, objects, energy, matter, and space),* ***independently*** *of all other variables, independently of all happenings, movements and processes.*

Time is a basic constituent of the existence of everybody and everything. All human trials during millenniums have failed to explain or to express the nature or the phenomenon of *time* in terms of other well defined notions [64], [65, Axiom 25, p. 53], [66, Axiom 25, p. 52]. The nature of *time*, the physical content of it, cannot be explained in terms of other basic constituents of the existence (in terms of energy, matter and space) or in terms of other physical phenomena or variables. *Time* has its own original nature that we can only call it *the nature of time*, i.e., *the temporal nature* or *the time nature.*

The value of *time* t (τ), i.e., **instant** or **moment**, is denoted also by t (or by τ), respectively. It is an **instantaneous** (**momentous**) and **elementary *time* value.** It can happen exactly once and then it is the same everywhere for and in everybody and everything (i.e., for and in beings, energy, matter, objects, and space), for all other variables, happenings, movements, processes, biological, physical and social systems. It is not repeatable. Nobody and nothing can influence the flow of instants.

The physical dimension of *time* is denoted by $[T]$, where T stands for *time*. It cannot be expressed in terms of the physical dimension of another variable. Its physical dimension is one of the basic physical dimensions. It is used to express the physical dimensions of most of the physical variables. A selected unity 1_t of *time* can be arbitrarily chosen and then fixed. If it is *second* s then $1_t = s$, which we denote by $t \langle 1_t \rangle = t \langle s \rangle$.

There can be assigned *exactly one* (which is denoted by $\exists!$) real number to every moment (instant), and vice versa. The numerical value **num** t of the moment t is a real number, $num\ t \in \mathfrak{R}$, where $\mathfrak{R}$ is the set of all real numbers.

Theorem 3 ***Universal time speed law*** *[64], [65, Axiom 25, p. 53], [66, Axiom 25, p. 52]*

Time is the unique physical variable such that the speed v_t (v_τ) *of the evolution (the flow) of its values and its numerical values:*

a) is invariant with respect to a choice of a relative zero moment t_{zero}, *initial moment* t_0, *a time scale and a time unit* 1_t, *i.e., invariant relative to a choice of a time axis, invariant relative to a selection of spatial coordinates, invariant relative to everybody and everything,*

and

b) its value (its numerical value) is invariant and equals one arbitrary time unit per the same time unit (equals one), respectively,

$$v_t = 1[TT^{-1}]\left\langle 1_t 1_t^{-1}\right\rangle = 1[TT^{-1}]\left\langle 1_\tau 1_\tau^{-1}\right\rangle = v_\tau,\ numv_t = numv_\tau = 1, \tag{1.1}$$

relative to arbitrary time axes T *and* T_τ*, i.e., its numerical value equals 1 (one) with respect to all time axes (with respect to any accepted relative zero instant* t_{zero}*, chosen initial instant* t_0*, time scale and any selected time unit* 1_t*), with respect to all spatial coordinate systems, to all beings and all objects.*

Time set $\mathcal{T}$ is the set of all moments. It is an open, unbounded and connected set. It is in the biunivoque (one-to-one) correspondence with the set $\mathfrak{R}$ of all real numbers,

$$\begin{gathered}\mathcal{T} = \{t : num\ t \in \mathfrak{R},\ dt > 0,\ t^{(1)} \equiv 1\},\\ \forall t \in \mathcal{T},\ \exists! x \in \mathfrak{R} \Longrightarrow x = num\ t\\ and\\ \forall x \in \mathfrak{R},\ \exists! t \in \mathcal{T} \Longrightarrow num\ t = x,\\ num\ inf\mathcal{T} = num\ t_{\inf} = -\infty\ and\ num\ sup\mathcal{T} = num\ t_{\sup} = \infty.\end{gathered} \tag{1.2}$$

The rule of the correspondence determines an **accepted relative zero numerical *time* value** t_{zero}, a ***time* scale** and a ***time* unit** denoted by 1_t (or by 1_τ). The *time* unit can be a millisecond, second, minute, hour, day, etc.

We accept herein the relative zero moment t_{zero} to have the zero numerical value, $num\ t_{zero} = 0$. Besides, we adopt it to be also the initial moment t_0, $t_0 = t_{zero}$, $num\ t_0 = 0$, in view of the *time*-invariance of the systems to be studied. This determines the subset $\mathfrak{T}_0$ of $\mathfrak{T}$,

$$\mathfrak{T}_0 = \{t : t \in \mathfrak{T}, numt \in [0, \infty[\}.$$

Note 4 *It has been the common custom to use the letters t and* τ *to denote not only time itself, and an arbitrary instant, but also the numerical value of the arbitrary instant relative to the chosen zero instant, e.g.,* $t = 0$ *is used in the sense* $numt = 0$*. From the physical point of view this is incorrect. The numerical value num t of the instant t is a real number without a physical dimension, while the instant t is a temporal value that has the physical dimension - the temporal dimension T of time. This is overcome by using* *the normalized, dimensionless, mathematical temporal variable, denoted by* $\bar{t}$ *and defined by*

$$\bar{t} = \frac{t}{1_t}[-],$$

so that the time set $\mathcal{T}$ *is to be replaced by*

$$\overline{\mathcal{T}} = \{\bar{t}[-] : \bar{t} = num\bar{t} = num\ t \in \mathfrak{R},\ d\bar{t} > 0,\ \bar{t}^{(1)} \equiv 1\}.$$

With this in mind, ***we will use in the sequel the letter*** t ***also for*** $\bar{t}$*, and* $\mathcal{T}$ *also for* $\overline{\mathcal{T}}$. *Hence,*

$$t[-] = \bar{t} = numt[-].$$

Between any two different instants $t_1 \in \mathcal{T}$ and $t_2 \in \mathcal{T}$ there is a third instant $t_3 \in \mathcal{T}$, either $t_1 < t_3 < t_2$ or $t_2 < t_3 < t_1$. The *time* set $\mathcal{T}$ is **continuum**. It is called also **the continuous-*time* set**.

1.2 *Time*, physical principles, and systems

Being a physical variable, *time* possesses some general properties valid for all physical variables. They are summarized in the form of the following principles [64], [65, Axiom 25, p. 53], [66, Axiom 25, p. 52]:

Principle 5 ***Physical Continuity and Uniqueness Principle*-PCUP*: scalar form***

A physical variable can change its value from one value to another one only by passing through every intermediate value. It possesses a unique local instantaneous real value in any place (in any being or in any object) at any moment.

Principle 6 ***Physical Continuity and Uniqueness Principle*-PCUP*: matrix and vector form***

A vector physical variable or a matrix (vector) of physical variables can change, respectively, its vector or matrix (vector) value from one vector or matrix (vector) value to another only by passing elementwise through every intermediate vector or matrix (vector) value. It possesses a unique local instantaneous real vector or matrix (vector) value in any place [i.e., being or object] and at any moment, respectively.

Principle 7 ***Physical Continuity and Uniqueness Principle*-PCUP*: system form***

The system physical variables (including their derivatives or integrals, which are also physical variables) can change, respectively, their (scalar, vector or matrix) values from one (scalar, vector or matrix) value to another only by passing elementwise through every intermediate (scalar, vector or matrix) value. They possess unique local instantaneous real (scalar, vector or matrix) values in any place and at any moment.

The $PCUP$ appears important for an accurate modeling physical system.

Corollary 8 ***Mathematical model of a physical variable, mathematical model of a physical system and PCUP***

a) For a mathematical (scalar or vector) variable to be, respectively, an adequate description of a physical (scalar or vector) variable, it is necessary that it obeys the Physical Continuity and Uniqueness Principle.

b) For a mathematical model of a physical system to be an adequate description of the physical system, it is necessary that all its system variables obey the Physical Continuity and Uniqueness Principle; i.e., that the mathematical model obeys the Physical Continuity and Uniqueness Principle.

The properties of *time* and the common properties of the physical variables expressed by $PCUP$ (Principle 5 through Principle 7) lead to the following:

Principle 9 ***Time Continuity and Uniqueness Principle-TCUP***

Any (scalar or vector) physical variable (any vector / matrix of physical variables) can change, respectively, its (scalar / vector / matrix) value from one (scalar / vector / matrix) value to another only continuously in time by passing (elementwise) through every intermediate (scalar / vector / matrix) value. It possesses a unique local instantaneous real (scalar / vector / matrix) value in any place (being or object) and at any moment.

Definition 10 *The system form of the $TCUP$ means that all system variables satisfy the $TCUP$.*

The $TCUP$ is very useful for the stability study of dynamic systems and for their control synthesis. This was shown in [48], [50], [55], [60], [67], [68], [69].

Corollary 11 ***Mathematical representation of a physical variable, mathematical model of a physical system and TCUP***

a) For a mathematical (scalar or vector) variable to be, respectively, an adequate description of a physical (scalar or vector) variable, it is necessary that it obeys the Time Continuity and Uniqueness Principle.

b) For a mathematical model of a physical system to be an adequate description of the physical system it is necessary that its system variables obey the Time Continuity and Uniqueness Principle; or equivalently, that the mathematical model obeys the Time Continuity and Uniqueness Principle.

c) For a mathematical model of a physical system to be an adequate description of the physical system it is necessary that its solutions are unique and continuous in time.

For more on *time*, its consistent physical and mathematical relativity theory, and its relationship to systems, see [52], [64], [65, Axiom 25, p. 53], [66, Axiom 25, p. 52].

1.3 *Time* and system dynamics

Every physical system exists and operates in *time*. Any variation of the system behavior occurs in the course of *time*. The values of all system variables depend on *time*.

There exist three main characteristic groups of the variables related to or linked with the dynamical system. Their definitions follow [13, Definition 3-6, p. 83], [58], [85, p. 105], [93, 2. Definition, p. 380], [94, 2. Definition, p. 380], [106, p. 4], [107, p. 664].

Definition 12 ***Input variables, input vector and input space***
A variable that acts on the system and its influence is essential for the system behavior is the ***system input variable*** *denoted by* $I \in \mathfrak{R}$. *The system can be under the action of several mutually independent input variables* $I_1, I_2, ..., I_M$. *They compose* ***the system input vector*** *(for short,* ***input****)*

$$\mathbf{I} = [I_1 \;\; I_2 \;\; ... \;\; I_M]^T \in \mathfrak{R}^M, \tag{1.3}$$

which spans ***the input space*** $\mathfrak{R}^M$. *The number* M *is the dimension of the input vector and the input space.*

The capital letters I *and* $\mathbf{I}$ *denote the total (scalar, vector) values of the variable* I *and the vector* $\mathbf{I}$ *relative to their total zero (scalar, vector) value, if it exists, or relative to their accepted zero (scalar, vector) value, respectively.*

The crucial characteristics of dynamical systems is the existence of their *internal dynamics* and/or *output dynamics*. The system internal dynamics represents its *state,* which is determined by the values of the corresponding variables called *state variables,* and the output dynamics is determined by the *output variables and their derivatives,* in the sense of the following definitions:

Definition 13 ***Output variables, output vector, output space and response***

A variable $Y \in \mathfrak{R}$ *is an* ***output variable*** *of the system if and only if its values result from the system behavior, they are (directly or indirectly) measurable, and we are interested in them.*

The number N *is the maximal number of the mutually independent output variables* $Y_1,\ Y_2,\ \cdots,\ Y_N$ *of the system. They form the* ***output vector*** $\mathbf{Y}$ *of the system, which spans* ***the output space*** $\mathfrak{R}^N$*:*

$$\mathbf{Y} = [Y_1\ Y_2\ \cdots\ Y_N]^T \in \mathfrak{R}^N. \tag{1.4}$$

The number N *is the dimension of the output vector and the output space.*

The time variation of the system output vector $\mathbf{Y}$ *is the* ***system (output) response.***

The capital letters Y *and* $\mathbf{Y}$ *denote the total (scalar, vector) values of the variable* Y *and the vector* $\mathbf{Y}$ *relative to their total zero (scalar, vector) value, if it exists, or relative to their accepted zero (scalar, vector) value, respectively.*

Note 14 *There are systems, the output variable of which is fed back to the system input. Such output variable is also the system input variable, and such system has its own (local) feedback.*

In what follows, the term *mathematical system* denotes the accepted mathematical model (description) of the corresponding physical system.

Definition 15 ***State of a dynamical system***

The (internal, output, full) state of a dynamical physical system at a moment $\tau \in \mathfrak{T}$ *is the system (internal, output, internal and output) dynamical situation at the moment* τ*, which, together with the input vector and its derivatives acting on the system at every moment* $(t \geq \tau) \in \mathfrak{T}$*, determines uniquely the system (internal, output, internal and output) behavior (i.e., the system (internal, output, full) state and the system output response) for all* $(t \geq \tau) \in \mathfrak{T}$*, respectively.*

The (internal, output, full) state of a dynamical mathematical system at a moment $\tau \in \mathfrak{T}$ *is the minimal amount of information* *about the system (internal, output, internal and output) dynamics at the moment* τ*, which, together with information about the action on the system (the system input and its derivatives acting on the system) at every moment* $(t \geq \tau) \in \mathfrak{T}$*, determines uniquely the system (internal, output, internal and output) behavior [i.e., the system (internal, output, full) state and the output response] for all* $(t \geq \tau) \in \mathfrak{T}$*, respectively.*

The minimal number n_I*,* n_O *or* n *of mutually independent variables* S_j*,* $j = 1, 2, \cdots, K$*,* $K \in \{n_I, n_O, n\}$*, the values* $S_j(\tau)$ *of which at any moment* $\tau \in \mathfrak{T}$ *are in the biunivoque correspondence with the system (internal, output, full) state at the same moment* τ*, are* ***the (internal, output, full) state variables of the system, respectively.*** *They compose* ***the (internal, output or full) state vector*** $\mathbf{S}_I$*,* $\mathbf{S}_O$ ***or*** $\mathbf{S}$ ***of the system,***

$$\mathbf{S}_I = [S_{I1} \cdots S_{In_I}]^T \in \mathfrak{R}^{n_I}, \; \mathbf{S}_O = [S_{O1} \cdots S_{On_O}]^T \in \mathfrak{R}^{n_O}, \tag{1.5}$$

$$\mathbf{S} = [S_1 \;\; S_2 \cdots S_n]^T \in \mathfrak{R}^n, \; n \geq n_I + n_O, \tag{1.6}$$

respectively. Every number n_I, n_O *and* n *is the dimension of the corresponding state vector and space.*

Definition 16 *The space* $\mathfrak{R}^K$*,* $K \in \{n_I, n_O, n\}$*, is* ***the (internal, output, full) state space of the system, respectively.***

This definition broadens and generalizes the well-known and commonly accepted definition of the state of the ISO systems.

Definition 17 ***State and motion***

The system (internal, output, full) state vector $S_{(.)}(t)$ *at a moment* $t \in \mathfrak{T}$ *is the vector value of the system (internal, output, full) motion* $\mathcal{S}_{(.)}(.; t_0; \mathbf{S}_0; \mathbf{I})$ *at the same moment* t*:*

$$\mathbf{S}_{(.)}(t) \equiv \mathbf{S}_{(.)}(t; t_0; \mathbf{S}_0; \mathbf{I}) \equiv \mathcal{S}_{(.)}(t; t_0; \mathbf{S}_0; \mathbf{I}) \Longrightarrow$$
$$\mathbf{S}_{(.)}(t_0) \equiv \mathcal{S}(t_0; t_0; \mathbf{S}_0; \mathbf{I}) \equiv \mathbf{S}_{(.)0}, \tag{1.7}$$

$(.) \in \{I, O, \phi\}$*, where* ϕ *denotes the empty space.*

Note 18 *Notice that:*

$$\mathbf{S}_I(t) \equiv \mathbf{S}_I(t; t_0; \mathbf{S}_0; \mathbf{I}) \equiv \mathbf{S}_I(t; t_0; \mathbf{S}_{I0}; \mathbf{I}).$$

1.4 Systems and complex domain

Considerations of dynamical properties of continuous-*time* *time*-invariant linear dynamical systems, called for short ***systems*** in the sequel, directly in the temporal domain can be mathematically more difficult and ineffective than in the complex domain.

The effectiveness of analysis and synthesis of *time*-invariant linear systems in the complex domain relies largely on the notion and on the properties of *the system transfer function for the SISO system*, and on its generalization – *the transfer function matrix for the MIMO system*. They represent input-output dynamical characteristics of the systems in the domain of the complex variable s,

$$s = (\sigma + j\omega) \in \mathfrak{C},\ \sigma \in \mathfrak{R},\ \omega \in \mathfrak{R}, \tag{1.8}$$

where $\mathbb{C}$ is the set of all complex numbers s, σ and ω are real numbers, or real valued scalar variables, and $j = \sqrt{-1}$ is the imaginary unit. They describe in the complex domain $\mathfrak{C}$ how the system transfers, in course of *time*, influences of the input variables $I_{(.)}$ (of the input vector $\mathbf{I}$) on the output variables $Y_{(.)}$ (on the output vector $\mathbf{Y}$) *under all zero initial conditions.*

For the $SISO$ systems, we will write I instead of I_1, $\mathbf{I} = (I_1) = (I) \in \mathfrak{R}^1$, and Y instead of Y_1, $\mathbf{Y} = (Y_1) = (Y) \in \mathfrak{R}^1$, or simply $I \in \mathfrak{R}$ and $Y \in \mathfrak{R}$, respectively.

Here, $\mathfrak{R}^1$ is the one-dimensional real vector space, the elements of which are one-dimensional real-valued vectors, while the elements of $\mathfrak{R}$ are scalars (real numbers). Division is not defined in $\mathfrak{R}^1$, but it is defined in $\mathfrak{R}$.

The superposition principle, i.e., the system linearity, permits us to treat, mathematically, separately the influence of the input vector and the influences of all initial conditions on the system dynamical behavior. We have treated the problems of the pole-zero cancellation, the system realization, $BIBO$ stability, and, most often, tracking by assuming all zero initial conditions; hence, by using the related system transfer function (matrix) and the block diagram technique. We use it also to investigate Lyapunov stability properties of the systems although they are defined exclusively for zero input vector and nonzero initial conditions. Besides, we study often the system response to the input vector action under all zero initial conditions.

The past (i.e., the history) and the present of a dynamic system influence its future behavior. If the system is without a memory and without a *time* delay, then the initial conditions express and transfer, in the very condensed form, the permanent influence of the system past on its future behavior.

Evidently, the past is untouchable; the initial conditions cannot be chosen, they are even most often unpredictable. The influence of the system's past is unavoidable; the initial conditions are imposed by the system history regardless of forms and intensities of current actions of the input variables.

When we wish to study the real system dynamical behavior, we may not avoid considering the influence of the initial conditions. O. I. Elgerd [20], H. M. Power and R. J. Simpson [109], and R. E. Skelton [114] observed this well by introducing various system transfer function matrices. Skelton defined the transfer function matrix relative to the initial state of the *ISO* system. Besides, he introduced the block diagram of the state-space system description with the initial state vector [114]. But, returning to the main stream of the system and control theories, he continued to use only the system transfer function matrix and the classical block diagram technique valid exclusively under all zero initial conditions.

The following examples of trivial *Single-Input Single-Output (SISO)* systems explain simply the crucial difference between the system transfer function $G(s)$ and the system *full (complete) transfer function matrix* $F(s)$. They show that the use of the latter is indispensable.

Example 19 *Let us consider the simplest systems,*

$$a)\ a\frac{dY}{dt} = bI,$$

$$\text{and}$$

$$b)\ a\frac{dY}{dt} = b\frac{dI}{dt},\quad (ab \neq 0) \in \mathfrak{R},$$

Y is the output variable, I is the input variable.

Their values are measured in their total scales.

Left (-), or right (+) Laplace transform $\mathcal{L}^{\mp}\{.\}$, or just Laplace transform $\mathcal{L}\{.\}$, (Appendix B.1), of the preceding equations read, respectively,

$$a)\ Y^{\mp}(s) = \frac{b}{as}I^{\mp}(s) + \frac{1}{s}Y(0^{\mp}),$$

$$b)\ Y^{\mp}(s) = \frac{bs}{as}I^{\mp}(s) - \frac{b}{as}I(0^{\mp}) + \frac{1}{s}Y(0^{\mp}).$$

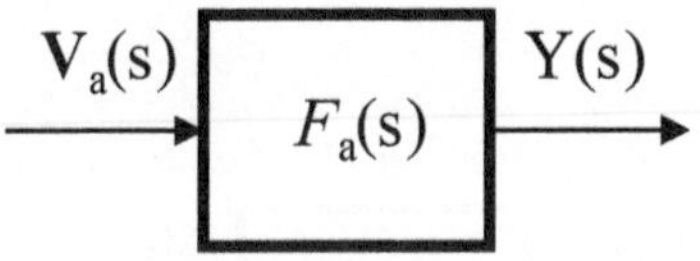

Figure 1.1: The full block of the system under a) in general for the nonzero initial output, $Y_0 \neq 0$.

We can set these equations in the more compact vector-matrix form,

$$a)\ Y^{\mp}(s) = \underbrace{\begin{bmatrix} \overbrace{\frac{b}{as}}^{G_a(s)} \vdots \overbrace{\frac{1}{s}}^{G_{ay0}(s)=G_{a0}(s)} \\ \end{bmatrix}}_{F_a(s)} \underbrace{\begin{bmatrix} I^{\mp}(s) \\ Y(0^{\mp}) \end{bmatrix}}_{\mathbf{V}_a(s)} = F_a(s)\mathbf{V}_a(s),$$

$$F_a(s) = \left[\frac{b}{as} \vdots \frac{1}{s}\right] = \left[G_a(s) \vdots G_{a0}(s)\right],\ \mathbf{V}_a(s) = \begin{bmatrix} I^{\mp}(s) \\ C_{0a}^{\mp} \end{bmatrix},\ C_{0a}^{\mp} = Y(0^{\mp}),$$

$$b)\ Y^{\mp}(s) = \underbrace{\begin{bmatrix} \overbrace{\frac{bs}{as}}^{G_b(s)} \vdots \underbrace{\overbrace{-\frac{b}{as}}^{G_{bi0}(s)} \vdots \overbrace{\frac{1}{s}}^{G_{by0}(s)}}_{G_{b0}(s)} \\ \end{bmatrix}}_{F_b(s)} \underbrace{\begin{bmatrix} I^{\mp}(s) \\ I(0^{\mp}) \\ Y(0^{\mp}) \end{bmatrix}}_{\mathbf{V}_b(s)} = F_b(s)\mathbf{V}_b(s),$$

$$F_b(s) = \left[\frac{bs}{as} \vdots -\frac{b}{as} \vdots \frac{1}{s}\right] = \left[G_b(s) \vdots G_{bi0}(s) \vdots G_{by0}(s)\right] =$$

$$= \left[G_b(s) \vdots G_{b0}(s)\right],\ \mathbf{V}_b(s) = \begin{bmatrix} I^{\mp}(s) \\ \mathbf{C}_{0b}^{\mp} \end{bmatrix},\ \mathbf{C}_{0b}^{\mp} = \begin{bmatrix} I(0^{\mp}) \\ Y(0^{\mp}) \end{bmatrix}.$$

The complex function $F_{(..)}(.) : \mathfrak{C} \longrightarrow \mathfrak{C}^{1xq}$ describes fully (completely) in the complex domain how the system transfers in the temporal domain influences of all actions: of the history through the initial conditions and the input variable, on its output behavior, where $q = 2$ in case a), and $q = 3$ in case b). The function $F_{(..)}(.)$ is a matrix function despite the fact that the dynamic systems are the simplest ones, scalar of the first order. We will call the function $F_{(..)}(.)$ ***the full (complete) matrix transfer function*** *of the system. We can use it by following Skelton [114] to extend the notion of the system block Fig. 1.1, Fig. 1.2, Fig. 1.3, and Fig. 1.4, respectively.*

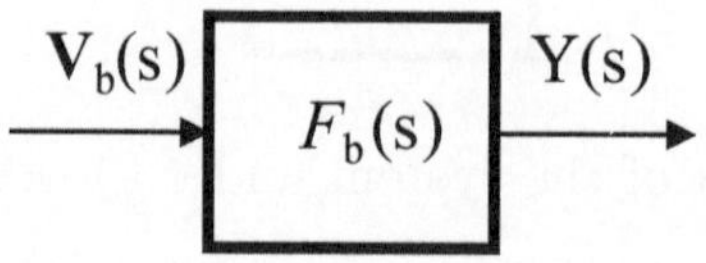

Figure 1.2: The full block of the system under b) in general for the nonzero initial input, $I_0 \neq 0$, and the initial output, $Y_0 \neq 0$.

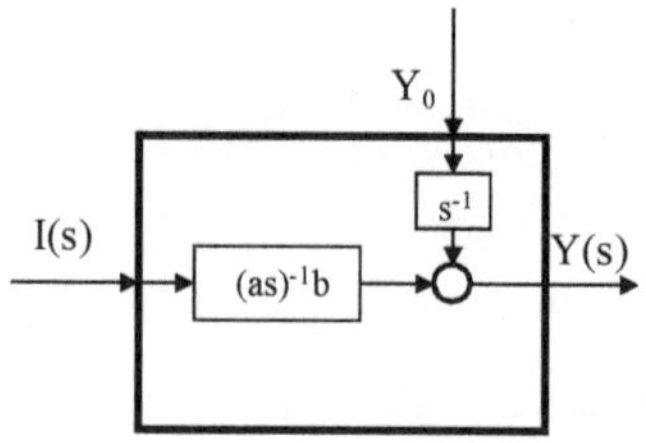

Figure 1.3: The full block diagram of the system under a).

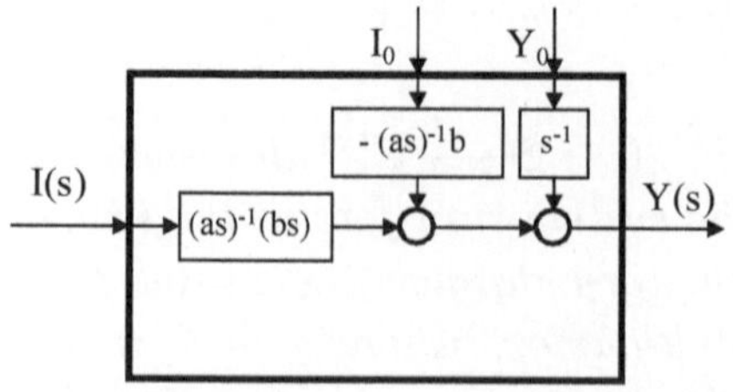

Figure 1.4: The full block diagram of the system under b).

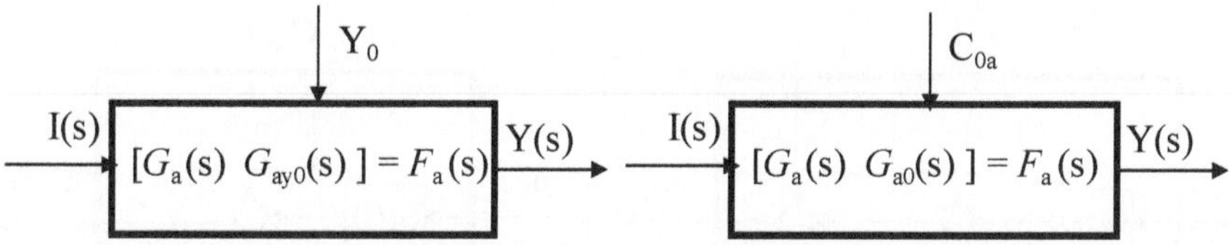

Figure 1.5: The system full block with the submatrices of the full system transfer function matrix $F(s)$ of the system under a).

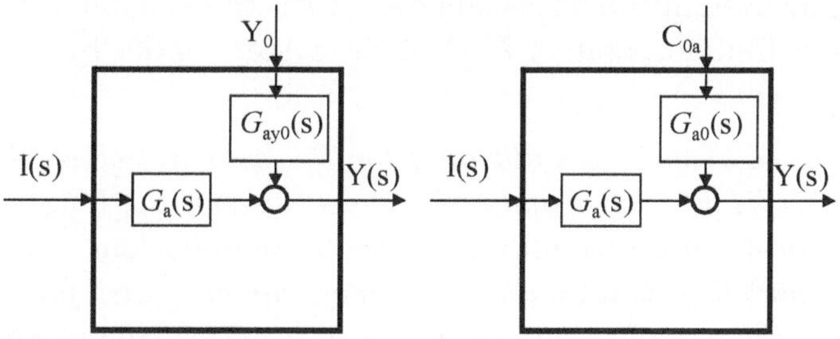

Figure 1.6: The system full block diagram with the submatrices of the full system transfer function matrix $F(s)$ of the system under a).

We can recognize transmissions and transformations of different influences through the system to its output. They are expressed and described in the complex domain by the corresponding transfer functions $G_{(..)}(.) : \mathfrak{C} \longrightarrow \mathfrak{C}$ *that are scalar entries in these examples of* $F_{(..)}(s)$*, Fig. 1.3 through Fig. 1.8. In the case of the system under b) above, we find its irreducible transfer function to be constant,* $G_b(s) = a^{-1}b = k$*,* $(k \neq 0) \in \mathfrak{R}$*. It yields the minimal system realization* $Y(t) = kI(t)$ *under zero initial condition, i.e., the transfer function realization. The same result for the system output response follows from* $Y(s) = G_b(s)I(s)$ *regardless of the form of* $G_b(s)$ *(either reducible,* $G_b(s) = (as)^{-1}bs$*, or irreducible,* $G_b(s) = k$*). We know that this is incor-*

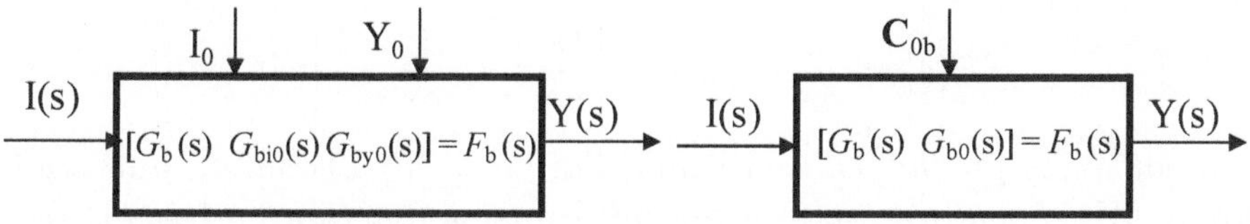

Figure 1.7: The system full block with the submatrices of the full system transfer function matrix $F(s)$ of the system under b).

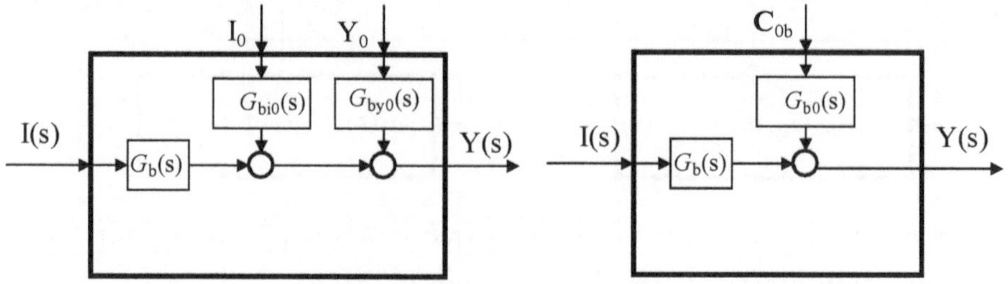

Figure 1.8: The system full block diagram with the submatrices of the full system transfer function matrix $F(s)$ of the system under b).

rect in general because it is valid only for the zero initial conditions. The equation $Y(t) = kI(t)$ corresponds to a static system, the behavior of which does not depend on initial conditions. Hence, dynamic (e.g., controllability, observability, stability, trackability) problems do not exist for such a system. However, the correct relationship between output and input in general is $Y(t) = k\left[I(t) - I(0^-)\right] + Y(0^-)$. It results from the system IO differential equation after its integration. It describes the output response of a dynamic system. Its equilibrium state $X = Y = 0$ is stable, but not attractive (hence, not asymptotically stable), i.e., the system is critically stable, not stable. The same result follows if we use the full transfer function matrix $F_b(s)$ and the vector $\mathbf{C}_{0b}^-$ of all initial conditions, or simply if we use $\mathbf{V}_b(s)$, in the expression for $Y^-(s)$,

$$Y^-(s) = F_b(s)\mathbf{V}_b(s),\ \mathbf{V}_b(s) = \begin{bmatrix} I^{\mp}(s) \\ \mathbf{C}_{0b}^- \end{bmatrix},\ \mathbf{C}_{0b}^- = \begin{bmatrix} I(0^{\mp}) \\ Y(0^{\mp}) \end{bmatrix},$$

and when we apply the inverse of the left Laplace transform to this equation. The denominator polynomial of $F_b(s)$ is its characteristic polynomial $\Delta(s)$ and, in this case, its minimal polynomial $m(s)$, $\Delta(s) = m(s) = s$. The cancellation of the zero s_1^0 and the pole s_1^* at the origin, $s_1^0 = s_1^* = 0$, of $G_b(s)$ is not possible in $F_b(s)$ although it is possible in the transfer function $G_b(s)$,

$$G_b(s) = \frac{bs}{as},\ F_b(s) = \left[\frac{bs}{as} \ \vdots\ -\frac{b}{as} \ \vdots\ \frac{1}{s}\right].$$

The use of $G_b(s)$ for the pole-zero cancellation or for Lyapunov stability test is wrong. We should use instead the full transfer function matrix $F_b(s)$.

Example 20 *The ignorance of the initial conditions in the complex domain becomes more severe if the reducible form of the system transfer function* *(matrix) has an unstable pole that is cancelled with the corresponding zero. Let*

$$Y^{(2)} + Y^{(1)} - 2Y = -I + I^{(1)},$$

so that

$$Y^{\mp}(s) = \underbrace{\left[\overbrace{\frac{s-1}{s^2+s-2}}^{G(s)} \vdots \overbrace{\frac{-1}{s^2+s-2}}^{G_{I_0}(s)} \vdots \overbrace{\frac{s+1}{s^2+s-2}}^{G_{Y_0}(s)} \vdots \overbrace{\frac{1}{s^2+s-2}}^{G_{Y_0^{(1)}}(s)} \right]}_{F(s)} \underbrace{\begin{bmatrix} I^{\mp}(s) \\ I(0^{\mp}) \\ Y(0^{\mp}) \\ Y^{(1)}(0^{\mp}) \end{bmatrix}}_{\mathbf{V}(s)},$$

$$Y^{\mp}(s) = F(s)\mathbf{V}(s), \ \mathbf{V}(s) = \begin{bmatrix} I^{\mp}(s) \\ \mathbf{C}_0 \end{bmatrix}, \ \mathbf{C}_0 = \begin{bmatrix} I(0^{\mp}) \\ Y(0^{\mp}) \\ Y^{(1)}(0^{\mp}) \end{bmatrix}.$$

The system transfer function $G(s) = \left(s^2+s-2\right)^{-1}(s-1)$ is reducible. Its irreducible form $G(s) = (s+2)^{-1}$ yields its minimal realization $Y^{(1)} + 2Y = I$, which is not the system minimal realization. Obviously, the irreducible form $(s+2)^{-1}$ of $G(s)$ may not be used to test Lyapunov stability properties of the system, to test system BIBO stability under bounded nonzero initial conditions, or to determine the system output response under nonzero initial conditions. The cancellation of the zero $s^0 = 1$ and the unstable pole $s^* = 1$ in $G(s) = \left(s^2+s-2\right)^{-1}(s-1)$ is not possible in $F(s)$.

The use of $F(s)$ enables us to get all correct results on the general validity of the pole-zero cancellation, the system minimal polynomial, the (minimal) system realization, on the system complete output response, Lyapunov stability properties, *BIBO* stability under bounded nonzero initial conditions (also under zero initial conditions), and on tracking (under nonzero or zero initial conditions [58], [70]). Moreover, the properties of $F(s)$ lead to the generalization of the block diagram technique; i.e., they imply the full (complete) block diagram technique (established in Chapter 9).

The full transfer function matrix $F(s)$ was introduced, defined and determined for *time*-invariant continuous-*time* linear systems in [29], and for *time*-invariant discrete-*time* linear systems in [51]. It was used in these references, as well as in [25]-[28], [47], [49], [51], [63], [68], for the analysis of the complete system output response. We will show in the sequel how it enables us to obtain new results (in the complex domain) on system minimal realization, the zero-pole cancellation, and stability analysis. It can be effectively exploited also for stabilizing, tracking [70], and/or optimal stabilizing control synthesis [35], [46], [111], [113]. However, these issues exceed the scope of this work that establishes a new basis for them.

1.5 Notational preliminaries

Lowercase ordinary letters denote scalars, bold (lowercase and capital, Greek and Roman) letters signify vectors, capital italic letters stand for matrices, and capital Fraktur letters indicate sets and spaces.

Note 21 ***On the new notation***

In order to define and use effectively the system full transfer function matrix $F(s)$, we need new, simple and elegant notation. For example, instead of using

$$\mathbf{Y}^{\mp}(s) = F(s)\begin{bmatrix} I^{\mp}(s) \\ I(0^{\mp}) \\ Y(0^{\mp}) \\ Y^{(1)}(0^{\mp}) \end{bmatrix}$$

we can use, [29],

$$\mathbf{Y}^{\mp}(s) = F(s)\mathbf{V}(s),\ \mathbf{V}(s) = \begin{bmatrix} I^{\mp}(s) \\ \mathbf{C}_0^{\mp} \end{bmatrix},$$

$$\mathbf{C}_0^{\mp} = \begin{bmatrix} I(0^{\mp}) \\ \mathbf{Y}^1(0^{\mp}) \end{bmatrix},\ \mathbf{Y}^1(0^{\mp}) = \begin{bmatrix} Y(0^{\mp}) \\ Y^{(1)}(0^{\mp}) \end{bmatrix},$$

by introducing the general compact vector notation

$$\mathbf{Y}^k = \begin{bmatrix} \mathbf{Y} \\ \mathbf{Y}^{(1)} \\ \dots \\ \mathbf{Y}^{(k)} \end{bmatrix} = \begin{bmatrix} \mathbf{Y}^{(0)} \\ \mathbf{Y}^{(1)} \\ \dots \\ \mathbf{Y}^{(k)} \end{bmatrix} = \begin{bmatrix} \mathbf{Y}^{(0)^T} \vdots \mathbf{Y}^{(1)^T} \dots \vdots \mathbf{Y}^{(k)^T} \end{bmatrix}^T \in \mathfrak{R}^{(k+1)N},$$

$$k \in \{0, 1, \dots\},\ \mathbf{Y}^0 = \mathbf{Y},$$

which is different from

$$\mathbf{Y}^{(k)} = \frac{d^k\mathbf{Y}}{dt^k} \in \mathfrak{R}^N,\ k \in \{1, \dots\}.$$

The system matrices $A_i \in \mathfrak{R}^{N\times N}$, $i \in \{0, 1, \dots, \nu\}$, induce the extended system matrix $A^{(\nu)}$,

$$A^{(\nu)} = \begin{bmatrix} A_0 \vdots\ A_1 \vdots \dots \vdots\ A_\nu \end{bmatrix} \in \mathfrak{R}^{N\times(\nu+1)N},\ A^{(\nu)} \neq A^\nu = \underbrace{AA\dots A}_{\nu-times} \in \mathfrak{R}^{N\times N}.$$

We use the matrix function $S_i^{(k)}(.) : \mathbb{C} \longrightarrow \mathbb{C}^{i(k+1)xi}$ *of* s,

$$S_i^{(k)}(s) = \left[s^0 I_i \vdots s^1 I_i \vdots s^2 I_i \vdots ... \vdots s^k I_i \right]^T \in \mathfrak{C}^{\ i(k+1)xi},$$
$$(k,i) \in \{(\mu, M)\,,\ (\nu, N)\}\,, \tag{1.9}$$

in order to set

$$\sum_{i=0}^{i=\nu} A_i s^i$$

into the compact form $A^{(\nu)} S_N^{(\nu)}(s)$,

$$\sum_{i=0}^{i=\nu} A_i s^i = A^{(\nu)} S_N^{(\nu)}(s) \in \mathbb{C}^{NxN}.$$

Note 22 *The higher order and/or higher dimensional system, the more advantageous new notation.*

We will use the symbolic vector notation and operations in the elementwise sense as follows:

- *the zero and unit vectors,*

$$\mathbf{0}_N = [0\ \ 0\ ...0]^T \in \mathfrak{R}^N,\ \mathbf{1}_N = [1\ \ 1\ ...1]^T \in \mathfrak{R}^N,$$

- *the matrix* E *associated elementwise with the vector* ε,

$$\varepsilon = [\varepsilon_1\ \ \varepsilon_2\ \ ...\ \varepsilon_N]^T \in \mathfrak{R}^N \Longrightarrow$$
$$E = diag\{\varepsilon_1\ \ \varepsilon_2\ \ ...\ \varepsilon_N\} \in \mathfrak{R}^{NxN},$$

- *the vector and matrix absolute values,*

$$|\varepsilon| = [|\varepsilon_1|\ \ |\varepsilon_2|\ \ ...\ |\varepsilon_N|]^T,\ |E| = diag\{|\varepsilon_1|\ \ |\varepsilon_2|\ \ ...\ |\varepsilon_N|\},$$

- *the elementwise vector inequality,*

$$\mathbf{w}\ = [w_1\ \ w_2\ \ ...\ w_N]^T,$$
$$\mathbf{w} \neq \varepsilon \Longleftrightarrow w_i \neq \varepsilon_i,\ \forall i = 1, 2, ..., N,$$

We define the scalar sign function:

- $sign(.) : \mathfrak{R} \rightarrow \{-1, 0, 1\}$ *the signum scalar function.*

Other notation is defined at its first use and in Appendix A.

Chapter 2

Classes of systems

2.1 *IO* systems

The original linear (usually linearized) mathematical models of physical dynamical systems result often directly in the form of *time*-invariant linear vector *Input-Output* (*IO*) differential equation of the form (2.1),

$$\sum_{k=0}^{k=\nu} A_k \mathbf{Y}^{(k)}(t) = \sum_{k=0}^{k=\mu} B_k \mathbf{I}^{(k)}(t),\ det A_\nu \neq 0,\ \forall t \in \mathfrak{T}_0,\ \nu \geq 1,\ 0 \leq \mu \leq \nu,$$
$$\mathbf{Y}^{(k)}(t) = \frac{d^k \mathbf{Y}(t)}{dt^k},\ A_k \in \mathfrak{R}^{N \times N},\ B_k \in \mathfrak{R}^{N \times M},\ k = 0, 1, .., \nu,$$
$$\mu < \nu \Longrightarrow B_i = O_{N,M},\ i = \mu + 1,\ \mu + 2, ..., \nu. \tag{2.1}$$

$\mathfrak{R}^k$ is the k-dimensional real vector space, $k \in \{M, N, n\}$. $\mathbb{C}^k$ denotes the k-dimensional complex vector space. $O_{M \times N}$ is the zero matrix in $\mathfrak{R}^{M \times N}$, and O_N is the zero matrix in $\mathfrak{R}^{N \times N}$, $O_N = O_{N \times N}$. Analogously, $\mathbf{0}_k \in \mathfrak{R}^k$ is the zero vector in $\mathfrak{R}^k$. The total *input vector* $\mathbf{I}$ (1.3) (Definition 12),

$$\mathbf{I} = [I_1\ I_2\ ...\ I_M]^T \in \mathfrak{R}^M,$$

and the total *output vector* $\mathbf{Y}$ (1.4) (Definition 13),

$$\mathbf{Y} = [Y_1\ Y_2\ ...\ Y_N]^T \in \mathfrak{R}^N,$$

where the values I_i and Y_j are the total values of the input and the output variables (Definition 13), respectively. *The total value* of a variable means that its value is measured with respect to its total zero, if it has the total zero value; and if it does not have the total zero value, then an appropriate value is accepted to play the role of the total zero value. For example, temperature has the total zero value Kelvin zero. Position does not have a total zero value. We accept a point for the origin with respect to which we measure the positions of all other points, things and objects.

We introduce the following compact notation for the extended matrices, which was proposed in [29] (see Note 21 in *Notational preliminaries,* Section 1.5):

$$A^{(\nu)} = \left[A_0 \vdots\ A_1 \vdots ... \vdots A_\nu\right] \in \mathfrak{R}^{N\times(\nu+1)N},$$

$$B^{(\mu)} = \left[B_0 \vdots B_1 \vdots ... \vdots B_\mu\right] \in \mathfrak{R}^{N\times(\mu+1)M}, \tag{2.2}$$

$$\mathbf{I}^{\mu}(t) = \left[\mathbf{I}^T(t) \vdots \mathbf{I}^{(1)^T}(t) \vdots ... \vdots \mathbf{I}^{(\mu)^T}(t)\right]^T \in \mathfrak{R}^{(\mu+1)M}, \tag{2.3}$$

$$\mathbf{Y}^{\nu}(t) = \left[\mathbf{Y}^T(t) \vdots \mathbf{Y}^{(1)^T}(t) \vdots ... \vdots \mathbf{Y}^{(\nu)^T}(t)\right]^T \in \mathfrak{R}^{(\nu+1)N}, \tag{2.4}$$

so that the corresponding initial vectors are

$$\mathbf{I}_{0\mp}^{\mu-1} = \mathbf{I}^{\mu-1}(0^{\mp}) = \left[\mathbf{I}_{0(\mp)}^T \vdots \mathbf{I}_{0(\mp)}^{(1)^T} \vdots ... \vdots \mathbf{I}_{0(\mp)}^{(\mu-1)^T}\right]^T \in \mathfrak{R}^{\mu M}, \tag{2.5}$$

$$\mathbf{Y}_{0\mp}^{\nu-1} = \mathbf{Y}^{\nu-1}(0^{\mp}) = \left[\mathbf{Y}_{0(\mp)}^T \vdots \mathbf{Y}_{0(\mp)}^{(1)^T} \vdots ... \vdots \mathbf{Y}_{0(\mp)}^{(\nu-1)^T}\right]^T \in \mathfrak{R}^{\nu N}. \tag{2.6}$$

We set ν in the parentheses in $A^{(\nu)}$ in order to distinguish $A^{(\nu)}$ from the $\nu - th$ power A^{ν} of A,

$$A^{(\nu)} = \left[A_0 \vdots\ A_1 \vdots ... \vdots A_\nu\right] \neq A^{\nu} = \underbrace{AA....A}_{\nu\ times}.$$

However, the superscript μ is not in the parentheses in $\mathbf{I}^{\mu}(t)$ because $\mathbf{I}^{(\mu)}(t)$ denotes the $\mu - th$ derivative $d^{\mu}\mathbf{I}(t)/dt^{\mu}$ of $\mathbf{I}(t)$,

$$\mathbf{I}^{\mu}(t) = \left[\mathbf{I}^T(t) \vdots \mathbf{I}^{(1)^T}(t) \vdots ... \vdots \mathbf{I}^{(\mu)^T}(t)\right]^T \neq \mathbf{I}^{(\mu)}(t) = \frac{d^{\mu}\mathbf{I}(t)}{dt^{\mu}}.$$

The above compact notation permits us to set the *IO* vector differential Equation (2.1) into the following simple, elegant and compact form:

$$A^{(\nu)}\mathbf{Y}^{\nu}(t) = B^{(\mu)}\mathbf{I}^{\mu}(t),\ \forall t \in \mathfrak{T}_0. \tag{2.7}$$

We will continue to use mainly (2.7) instead of (2.1) in the sequel.

The left-hand side of the first Equation (2.1) expresses simultaneously the internal dynamics, i.e., *the internal state* $\mathbf{S}_I$, the output dynamics, i.e., *the output state* $\mathbf{S}_O$, and the full dynamics, i.e., *the state* $\mathbf{S}$, of the *IO* system, where

$$\mathbf{S} = \mathbf{S}_I = \mathbf{S}_O = \mathbf{Y}^{\nu-1}. \tag{2.8}$$

The output vector $\mathbf{Y}(t)$ and its derivatives $\mathbf{Y}^{(1)}(t)$, $\mathbf{Y}^{(2)}(t)$, ... , $\mathbf{Y}^{(\alpha-1)}(t)$ determine them at any $t \in \mathfrak{T}$. The extended output vector $\mathbf{Y}^{\nu-1}$ composed of the output vector $\mathbf{Y}(t)$ and its derivatives $\mathbf{Y}^{(1)}(t)$, $\mathbf{Y}^{(2)}(t)$, ... $\mathbf{Y}^{(\nu-1)}(t)$,

$$\begin{gathered}\mathbf{S}(t) = [S_1(t) \;\; S_2(t) \;\cdots\; S_{\nu N}(t)]^T = \\ = \mathbf{Y}^{\nu-1}(t) = \left[\mathbf{Y}^T(t) \vdots \mathbf{Y}^{(1)^T}(t) \vdots \ldots \vdots \mathbf{Y}^{(\nu-1)^T}(t)\right]^T = \\ = \left[\mathbf{Y}^{(0)^T}(t) \vdots \mathbf{Y}^{(1)^T}(t) \vdots \ldots \vdots \mathbf{Y}^{(\nu-1)^T}(t)\right]^T \in \mathfrak{R}^{\nu N},\end{gathered} \tag{2.9}$$

is *the state vector at the moment* t of the *IO* system (Definition 15), $\mathbf{S} = \mathbf{Y}^{\nu-1}$. The system motion $\mathrm{S}(.;t_0;S_0;\mathbf{I})$ is $\mathcal{Y}^{\nu-1}(.;t_0;Y_0^{\nu-1};\mathbf{I})$,

$$\begin{gathered}\mathcal{S}(.;t_0;\mathbf{S}_0;\mathbf{I}) = \mathcal{Y}^{\nu-1}(.;t_0;\mathbf{Y}_0^{\nu-1};\mathbf{I}) \Longrightarrow \\ \mathbf{S}(t;t_0;\mathbf{S}_0;\mathbf{I}) = \mathbf{Y}^{\nu-1}(t;t_0;\mathbf{Y}_0^{\nu-1};\mathbf{I}) \Longrightarrow \\ \mathbf{S}(t_0;t_0;\mathbf{S}_0;\mathbf{I}) = \mathbf{Y}^{\nu-1}(t_0;t_0;\mathbf{Y}_0^{\nu-1};\mathbf{I}) \equiv \mathbf{Y}_0^{\nu-1}.\end{gathered} \tag{2.10}$$

The system output response $\mathcal{Y}(.;t_0;Y_0^{\nu-1};\mathbf{I})$ is the first component of the system motion $\mathcal{Y}^{\nu-1}(.;t_0;\mathbf{Y}_0^{\nu-1};\mathbf{I})$,

$$\mathbf{Y}(t_0;t_0;\mathbf{Y}_0^{\nu-1};\mathbf{I}) \equiv \mathbf{Y}_{0.} \tag{2.11}$$

$\mathfrak{R}^+$ is the set of all positive real numbers. We will prove the stability conditions for input vector functions $\mathbf{I}(.) : \mathfrak{T}_0 \rightarrow \mathfrak{R}^M$, which belong to the class $\mathfrak{L}$ of *time* dependent bounded functions such that their Laplace transforms are strictly proper real rational vector functions of the complex variable s,

$$\mathfrak{L} =$$
$$= \left\{ \mathbf{I}(.) : \left(\begin{array}{c} \exists\gamma(\mathbf{I}) \in \mathfrak{R}^{+} \Longrightarrow \|\mathbf{I}(t)\| < \gamma(\mathbf{I}),\ \forall t \in \mathfrak{T}_0, \\ \mathcal{L}^{\mp}\{\mathbf{I}(t)\} = \mathbf{I}^{\mp}(s) = \left[I_1^{\mp}(s)\ \ I_2^{\mp}(s)\ \ \ldots\ I_M^{\mp}(s)\right]^T \Longrightarrow \\ \mathbf{I}^{\mp}(s) = \underbrace{blockdiag\left\{I_1^{\mp}(s)\ \ I_2^{\mp}(s)\ \ \ldots\ I_M^{\mp}(s)\right\}}_{I(s)}\mathbf{1}_M \Longrightarrow \\ \mathbf{1}_M = \left[1 \vdots 1 \vdots \ldots 1\right]^T \in \mathfrak{R}^M \Longrightarrow \mathbf{I}^{\mp}(s) = I(s)\mathbf{1}_M, \\ I_k^{\mp}(s) = \dfrac{\sum_{j=0}^{j=\zeta_k} a_{kj}s^j}{\sum_{j=0}^{j=\psi_k} b_{kj}s^j}, 0 \leq \zeta_k < \psi_k,\ \forall k = 1, 2, ..., M, \end{array} \right) \right\}. \tag{2.12}$$

The left Laplace transform $\mathbf{I}^{-}(s)$, the right Laplace transform $\mathbf{I}^{+}(s)$, or just the Laplace transform $\mathbf{I}(s)$ of the input vector function $\mathbf{I}(.) \in \mathfrak{L}$ should be strictly proper in order to ensure that the original $\mathbf{I}(t)$ does not contain an impulse component because such component is unbounded.

Notice that the zero input vector function $\mathbf{I}(.)$, $\mathbf{I}(t) \equiv \mathbf{0}_M$, belongs to $\mathfrak{L}$.

$\mathfrak{C}^{ki} = \mathfrak{C}^{k}(\mathfrak{R}^{i})$ is *the family of all functions defined and k-times continuously differentiable on* $\mathfrak{R}^{i}$, and

$\mathfrak{C}^{k} = \mathfrak{C}^{k}(\mathfrak{T}_0)$ is *the family of all functions defined, continuous and k-times continuously differentiable on* $\mathfrak{T}_0$, $\mathfrak{C} = \mathfrak{C}^{0}(\mathfrak{T}_0)$,

$\mathfrak{I}^{k}$ is a given, or to be determined, family of all bounded and k-*times* continuously differentiable permitted input vector functions $\mathbf{I}(.) \in \mathfrak{C}^{k} \cap \mathfrak{L}$,

$$\mathfrak{I}^{k} \subset \mathfrak{C}^{k} \cap \mathfrak{L}, \tag{2.13}$$

$\mathfrak{I}^{0} = \mathfrak{I}$ is the family of all bounded continuous permitted input vector functions $\mathbf{I}(.) \in \mathfrak{C} \cap \mathfrak{L}$,

$$\mathfrak{I} \subset \mathfrak{C} \cap \mathfrak{L}. \tag{2.14}$$

$\mathfrak{I}_{-}^{k}$ is a subfamily of $\mathfrak{I}^{k}$, $\mathfrak{I}_{-}^{k} \subset \mathfrak{I}^{k}$, such that the real part of every pole of the Laplace transform $\mathbf{I}(s)$ of every $\mathbf{I}(.) \in \mathfrak{I}_{-}^{k}$ is negative, $\mathfrak{I}_{-} = \mathfrak{I}_{-}^{0}$.

Note 23 *The condition* $\det A_\nu \neq 0$ *is a sufficient condition, but not necessary, for all the output variables of system (2.1) to have the same order* ν *of their highest derivatives. If the order* k *of the highest derivative of one* *output variable Y_i were lower than the highest derivative order ν of some other output variable Y_j, then all entries of the i-th column of $A_{k+1}, \dots, A_\nu$ would be equal to zero implying their singularity: $detA_m = 0,\ m \geq k+1$.*

Example 24 *Let*

$$M = 1,\ N = 2,\ \nu = 1,\ \mu = 0,$$

$$\underbrace{\begin{bmatrix} 2 & 2 \\ 4 & 4 \end{bmatrix}}_{A_1} \mathbf{Y}^{(1)}(t) + \underbrace{\begin{bmatrix} \frac{1}{2} & 1 \\ 1 & 2 \end{bmatrix}}_{A_0} \mathbf{Y}(t) = \begin{bmatrix} 4 \\ 2 \end{bmatrix} I(t) \Longrightarrow$$

$$detA_1 = \begin{vmatrix} 2 & 2 \\ 4 & 4 \end{vmatrix} = 0,\ det\left(A_1 s + A_0\right) = \begin{vmatrix} 2s+\frac{1}{2} & 2s+1 \\ 4s+1 & 4s+2 \end{vmatrix} = 0.$$

Both Y_1 and Y_2 have the same highest derivative that is the first order, but A_1 is singular. This illustrates that the condition $detA_\nu \neq 0$ is not necessary (although it is sufficient) for all the output variables of system (2.1) to have the same order of their highest derivatives. There does not exist a solution to this vector differential equation because $det\left(A_1 s + A_0\right) \equiv 0$. We explain this first by considering the scalar form of the mathematical model of the system,

$$2Y_1^{(1)}(t) + 2Y_2^{(1)}(t) + \frac{1}{2}Y_1(t) + Y_2(t) = 4I(t),$$
$$4Y_1^{(1)}(t) + 4Y_2^{(1)}(t) + Y_1(t) + 2Y_2(t) = 8I(t).$$

We multiply the first equation by 2. The result is

$$4Y_1^{(1)}(t) + 4Y_2^{(1)}(t) + Y_1(t) + 2Y_2(t) = 4I(t),$$
$$4Y_1^{(1)}(t) + 4Y_2^{(1)}(t) + Y_1(t) + 2Y_2(t) = 8I(t).$$

The left-hand sides of these equations are the same. Their right-hand sides are different. They do not have a solution for $I(t) \neq 0$.

The Laplace transform $\mathcal{L}\{.\}$ of the system mathematical model for all zero initial conditions reads

$$\left(A_1 s + A_0\right)\mathbf{Y}(s) = \begin{bmatrix} 2s+\frac{1}{2} & 2s+1 \\ 4s+1 & 4s+2 \end{bmatrix} \mathbf{Y}(s) = \begin{bmatrix} 4 \\ 2 \end{bmatrix} I(s),$$

where $\mathbf{Y}(s)$ and $I(s)$ are the Laplace transforms $\mathcal{L}\{.\}$ of $\mathbf{Y}(t)$ and of $I(t)$, $\mathbf{Y}(s) = \mathcal{L}\{\mathbf{Y}(t)\}$ and $I(s) = \mathcal{L}\{I(t)\}$. Since $det\left(A_1 s + A_0\right) \equiv 0$, then the preceding vector equation is not solvable in $\mathbf{Y}(s)$.

Condition 25 *The matrix A_ν of the IO system (2.1) obeys*

$$detA_\nu \neq 0. \tag{2.15}$$

The condition (2.15), imposed already in (2.1), ensures

$$\exists s \in \mathfrak{C} \Longrightarrow \det\left(\sum_{k=0}^{k=\nu} A_k s^k\right) \neq 0,$$

and it permits the solvability of the Laplace transform of (2.1) in $\mathbf{Y}(s)$; see (D.4) in Appendix D.

Note 26 ***Throughout this book we accept the validity of Condition 25.***

Besides, $A_k \in \mathfrak{R}^{N\times N}$, $B_k \in \mathfrak{R}^{N\times M}$, $k = 0, 1, .., \nu$, and Condition 25 imply

$$deg\left(\sum_{k=0}^{k=\nu} A_k s^k\right) = \nu,$$
$$deg\left[det\left(\sum_{k=0}^{k=\nu} A_k s^k\right)\right] = \eta,\ \eta = \nu N,$$
$$deg\left[adj\left(\sum_{k=0}^{k=\nu} A_k s^k\right)\right] = \sigma,\ \sigma = \nu\left(N-1\right),$$
$$deg\left(\sum_{k=0}^{k=\mu} B_k s^k\right) = \mu, \tag{2.16}$$

where

$$deg\left(\sum_{k=0}^{k=\nu} A_k s^k\right),\ deg\left[adj\left(\sum_{k=0}^{k=\nu} A_k s^k\right)\right]\ and\ deg\left(\sum_{k=0}^{k=\mu} B_k s^k\right)$$

denote the greatest power of s over all elements of

$$\sum_{k=0}^{k=\nu} A_k s^k,\ adj\left(\sum_{k=0}^{k=\nu} A_k s^k\right)\ and\ \sum_{k=0}^{k=\mu} B_k s^k,$$

respectively.

However, in general

$$deg\left[det\left(\sum_{k=0}^{k=\nu} A_k s^k\right)\right] = \eta,\ 0 \leq \eta \leq \nu N,$$

$$deg\left[adj\left(\sum_{k=0}^{k=\nu} A_k s^k\right)\right] = \sigma,\ 0 \leq \sigma \leq \nu\left(N-1\right),$$

Definition 27 *The quadruple* $(\nu,\ \mu,\ A^{(\nu)},\ B^{(\mu)})$ *is* ***a realization of the IO system (2.1); i.e., of (2.7), for an arbitrary input vector function and for arbitrary input and output initial conditions.***

Comment 28 *The realization* $(\nu,\ \mu,\ A^{(\nu)},\ B^{(\mu)})$ *of the IO system (2.1), i.e., of (2.7), is* ***its IO realization.*** *For the ISO realization of the IO system (2.1), i.e., of (2.7), see Appendix C.1.*

We will distinguish the order of a system, from its dimension and its dynamical dimension (see in the sequel Definition 220 in Section 12.1).

Definition 29 ***The order of the IO system (2.1)*** *is the number* ν *of the highest derivative of the output vector function* $\mathbf{Y}(.)$ *in (2.1).*

Definition 30 ***The dimension,*** *denoted by* dim_{IO}, ***of the IO system (2.1)*** *is the dimension* N *of its output vector* $\mathbf{Y}$: $dim_{IO} = dim\mathbf{Y} = N$.

Example 31 *The IO system*

$$\underbrace{\begin{bmatrix} 3 \vdots 0 \vdots 0 \\ 0 \vdots 0 \vdots 0 \\ 0 \vdots 0 \vdots 0 \end{bmatrix}}_{A_2}\mathbf{Y}^{(2)}(t) + \underbrace{\begin{bmatrix} 0 \vdots 0 \vdots 0 \\ 0 \vdots 1 \vdots 0 \\ 0 \vdots 0 \vdots 0 \end{bmatrix}}_{A_1}\mathbf{Y}^{(1)}(t) + \underbrace{\begin{bmatrix} 0 \vdots 0 \vdots 0 \\ 0 \vdots 0 \vdots 0 \\ 0 \vdots 0 \vdots 1 \end{bmatrix}}_{A_0}\mathbf{Y}(t) =$$

$$= \begin{bmatrix} 2 \vdots 0 \\ 0 \vdots 1 \\ 1 \vdots 0 \end{bmatrix}\mathbf{I}(t) + \begin{bmatrix} 1 \vdots 0 \\ 0 \vdots 1 \\ 1 \vdots 1 \end{bmatrix}\mathbf{I}^{(2)}(t)$$

yields

$$\nu = 2,\ \mu = 2,\ N = 3,\ M = 2,\ detA_\nu = detA_2 = det\begin{bmatrix} 3 \vdots 0 \vdots 0 \\ 0 \vdots 0 \vdots 0 \\ 0 \vdots 0 \vdots 0 \end{bmatrix} = 0,$$

$$deg\left[det\left(\sum_{k=0}^{k=\nu=2} A_k s^k\right)\right] = deg\begin{vmatrix} 3s^2 & 0 & 0 \\ 0 & s & 0 \\ 0 & 0 & 1 \end{vmatrix} = deg\left(3s^3\right) = 3 = \eta > \nu = 2,$$

$$deg\left[adj\left(\sum_{k=0}^{k=\nu=2} A_k s^k\right)\right] = deg\begin{bmatrix} s & 0 & 0 \\ 0 & 3s^2 & 0 \\ 0 & 0 & 3s^3 \end{bmatrix} = deg\left(3s^3\right) = 3 = \sigma,$$

$$deg\left(\sum_{k=0}^{k=\mu=2} B_k s^k\right) = deg\begin{bmatrix} 2+s^2 & 0 \\ 0 & 1+s^2 \\ 1+s^2 & s^2 \end{bmatrix} = deg\left(2+s^2\right) = 2 = \mu,$$

$$\sigma = \ \eta = 3 > \nu = 2,\ \eta < \nu N = 2\text{x}3 = 6,\ \sigma = \ 3 < \nu\left(N-1\right) = 2\text{x}2 = 4.$$

This is the second order system, $\nu = 2$. Its dimension equals $N = 3$, $dim\mathbf{Y} = N = 3$. Their product is bigger than the degree η of the system characteristic polynomial, $\nu N = 6 \geq 3 = \eta$. The nonzero entries of the matrices A_2 through A_0 show that the second derivative exists only of the first output variable Y_1, the first derivative exists only of the second output variable Y_2, and there is not any derivative of the third output variable Y_3 in the system mathematical model.

Example 32 *The IO system*

$$\begin{bmatrix} 3 \vdots 0 \vdots 1 \\ 2 \vdots 0 \vdots 0 \\ 0 \vdots 1 \vdots 1 \end{bmatrix} \mathbf{Y}^{(2)}(t) + \begin{bmatrix} 0 \vdots 0 \vdots 0 \\ 0 \vdots 1 \vdots 0 \\ 0 \vdots 0 \vdots 0 \end{bmatrix} \mathbf{Y}^{(1)}(t) + \begin{bmatrix} 0 \vdots 0 \vdots 0 \\ 0 \vdots 0 \vdots 0 \\ 0 \vdots 0 \vdots 1 \end{bmatrix} \mathbf{Y}(t) =$$

$$= \begin{bmatrix} 2 \vdots 0 \\ 0 \vdots 1 \\ 1 \vdots 0 \end{bmatrix} \mathbf{I}(t) + \begin{bmatrix} 1 \vdots 0 \\ 0 \vdots 1 \\ 1 \vdots 1 \end{bmatrix} \mathbf{I}^{(2)}(t)$$

induces

$$\nu = 2,\ \mu = 2,\ N = 3,\ \ M = 2,$$

$$detA_\nu = detA_2 = det\begin{bmatrix} 3 \vdots 0 \vdots 1 \\ 2 \vdots 0 \vdots 0 \\ 0 \vdots 1 \vdots 1 \end{bmatrix} = 2 \neq 0,$$

$$deg\left(\sum_{k=0}^{k=\nu=2} A_k s^k\right) = deg \begin{bmatrix} 3s^2 & 0 & s^2 \\ 2s^2 & s & 0 \\ 0 & s^2 & s^2+1 \end{bmatrix} = 2 = \nu,$$

$$deg\left[det\left(\sum_{k=0}^{k=\nu=2} A_k s^k\right)\right] = deg \begin{vmatrix} 3s^2 & 0 & s^2 \\ 2s^2 & s & 0 \\ 0 & s^2 & s^2+1 \end{vmatrix} =$$
$$= deg\left(2s^6 + 3s^5 + 3s^3\right) = 6 = \eta = 2\text{x}3 = \nu N,$$

$$deg\left[adj\left(\sum_{k=0}^{k=\nu=2} A_k s^k\right)\right] =$$
$$= deg \begin{bmatrix} s^3+s & s^4 & -s^3 \\ -2s^4-2s^2 & 3s^4+3s^2 & 2s^4 \\ 2s^4 & -3s^4 & 3s^3 \end{bmatrix} =$$
$$= deg\left(s^4\right) = 4 = \sigma = \nu\left(N-1\right) = 2\left(3-1\right),$$

$$deg\left(\sum_{k=0}^{k=\mu=2} B_k s^k\right) = deg \begin{bmatrix} 2+s^2 & 0 \\ 0 & 1+s^2 \\ 1+s^2 & s^2 \end{bmatrix} = deg\left(2+s^2\right) = 2 = \mu,$$
$$\sigma = 4 = 2\text{x}2 = 2\text{x}\left(3-1\right) = \nu\left(N-1\right) < \eta = \nu N = 6.$$

This is also the second order system of dimension three. In this case, the product of the system order ($\nu = 2$) and of the system dimension ($N = 3$) equals the degree ($\eta = 6$) of the system characteristic polynomial, $\nu N = 6 = \eta$.

2.2 *ISO* systems

Many fundamental system and control concepts, and the related complete results, have enriched the theory of linear systems by studying the *Input-State-Output* (*ISO*) *systems* described by *the state Equation* (2.17) and by *the output Equation* (2.18),

$$\frac{d\mathbf{X}(t)}{dt} = A\mathbf{X}(t) + B\mathbf{I}(t),\ \forall t \in \mathfrak{T}_0, \tag{2.17}$$

$$\mathbf{Y}(t) = C\mathbf{X}(t) + D\mathbf{I}(t),\ \forall t \in \mathfrak{T}_0, \tag{2.18}$$

$$\mathbf{X} \in \mathfrak{R}^n,\ A \in \mathfrak{R}^{n\text{x}n},\ B \in \mathfrak{R}^{n\text{x}M},\ \mathbf{I} \in \mathfrak{R}^M,$$
$$\mathbf{Y} \in \mathfrak{R}^N,\ C \in \mathfrak{R}^{N\text{x}n},\ C \neq O_{N,n},\ D \in \mathfrak{R}^{N\text{x}M},\ n \geq N.$$

The vector $\mathbf{X}$ is simultaneously the system *internal state vector* $\mathbf{S}_I$, and the system *state vector* $\mathbf{S}$,

$$\mathbf{S} = \mathbf{S}_I = \mathbf{X} \in \mathfrak{R}^n. \tag{2.19}$$

(Definition 15). The system motion $\mathrm{S}(.;t_0;S_0;\mathbf{I})$ is $\mathcal{X}(.;t_0;X_0;\mathbf{I})$,

$$\begin{gathered}\mathcal{S}(.;t_0;\mathbf{S}_0;\mathbf{I}) = \mathcal{X}(.;t_0;\mathbf{X}_0;\mathbf{I}) \Longrightarrow \\ \mathbf{S}(t;t_0;\mathbf{S}_0;\mathbf{I}) = \mathbf{X}(t;t_0;\mathbf{X}_0;\mathbf{I}), \\ \mathbf{S}(t_0;t_0;\mathbf{S}_0;\mathbf{I}) = \mathbf{X}(t_0;t_0;\mathbf{X}_0;\mathbf{I}) \equiv \mathbf{X}_0. \end{gathered} \tag{2.20}$$

The system output response $\mathcal{Y}(.;t_0;X_0;\mathbf{I})$ obeys

$$\mathbf{Y}(t_0;t_0;\mathbf{X}_0;\mathbf{I}) \equiv \mathbf{Y}_{0.} \tag{2.21}$$

We commonly call the ISO mathematical model (2.17) and (2.18) *the state-space system (description)*.

Note 33 ***Equivalence between the*** IO ***system and the*** ISO ***system***
Section C.1 in Appendix C presents the transformation of the IO system (2.1), Section 2.1, into the equivalent ISO system (2.17) and (2.18). Section C.2 in Appendix C shows how the ISO system (2.17) and (2.18) can be transformed into the IO system (2.1).

The well-known definition of the realization of the transfer function matrix of the ISO system is valid only under zero initial conditions. The following is its generalization to the systems with arbitrary input vector function $\mathbf{I}(.)$ and initial state vector $\mathbf{X}_{0\mp}$:

Definition 34 *The quadruple $(A,\ B,\ C,\ D)$ is* ***a realization of the ISO system (2.17) and (2.18)*** *for an arbitrary input vector function $\mathbf{I}(.)$ and for an arbitrary initial state vector $\mathbf{X}_{0\mp}$.*

Comment 35 *The realization $(A,\ B,\ C,\ D)$ of the ISO system (2.17) and (2.18) is* ***its ISO realization.*** *For its IO realization, see Appendix C.2.*

Definition 36 ***The order of the ISO system (realization) (2.17) and (2.18)*** *is the order of the highest derivative of the state vector function $\mathbf{X}(.)$ in (2.17); i.e., it is one (1).*

This definition characterizes all ISO systems as the first order systems. Their state Equation (2.17) is in *the Cauchy form*, also called *the normal form*. It contains only the first derivative $\mathbf{X}^{(1)}(t)$ of the state vector $\mathbf{X}(t)$, and there is not any derivative or integral of the input vector $\mathbf{I}(t)$.

Definition 37 ***The dimension****, denoted by* dim_{ISO}*,* ***of the ISO system (2.17) and (2.18)*** *is the dimension* n *of its state vector* $\mathbf{X}$*,* $dim_{ISO} = dim\mathbf{X} = n$*.*

2.3 *IIO* systems

The general description, in terms of the total vector coordinates, of time-invariant continuous-time linear *Input-Internal and Output dynamical systems*, for short *IIO systems*, without a delay, has the following general form:

$$\sum_{k=0}^{k=\alpha} Q_k \mathbf{R}^{(k)}(t) = \sum_{k=0}^{k=\beta} P_k \mathbf{I}^{(k)}(t),\ \alpha \geq 1, \alpha \geq \beta \geq 0,$$

$$\sum_{k=0}^{k=\nu} E_k \mathbf{Y}^{(k)}(t) = \sum_{k=0}^{k=\alpha} R_k \mathbf{R}^{(k)}(t) + \sum_{k=0}^{k=\mu\leq\nu} T_k \mathbf{I}^{(k)}(t),\ \nu, \mu \in \mathfrak{R}_+,$$

$$\forall t \in \mathfrak{T}_0, \tag{2.22}$$

where $Q_k \in \mathfrak{R}^{\rho x \rho}$, $\mathfrak{R}_+$ is the set of all nonnegative real numbers, $\mathbf{R} \in \mathfrak{R}^{\rho}$, $P_k \in \mathfrak{R}^{\rho x M}$, $E_k \in \mathfrak{R}^{N x N}$, $R_k \in \mathfrak{R}^{N x \rho}$, $R_\alpha = O_{N,\rho}$, and $T_k \in \mathfrak{R}^{N x M}$.

Condition 38 *The matrices* Q_k *and* E_k *obey:*

$$detQ_\alpha \neq 0,\ which\ implies\ \exists s \in \mathfrak{C} \Longrightarrow det\left[\sum_{k=0}^{k=\alpha} s^k Q_k\right] \neq 0,$$

$$detE_\nu \neq 0,\ which\ implies\ \exists s \in \mathfrak{C} \Longrightarrow det\left[\sum_{k=0}^{k=\nu} s^k E_k\right] \neq 0. \tag{2.23}$$

Note 39 *We accept the validity of Condition 38 in the sequel. The reasons are the same for which we accepted Condition 25 (for details see Example 24), (Section 2.1).*

We define $Q^{(\alpha)}$, $P^{(\beta)}$, $E^{(\nu)}$, $R^{(\alpha)}$, and $T^{(\mu)}$ by (2.24) in the sense of (2.2)

(Subsection 2.1),

$$Q^{(\alpha)} = \left[Q_0 \vdots\ Q_1 \vdots ... \vdots Q_\alpha \right] \in \mathfrak{R}^{\rho x(\alpha+1)\rho},$$

$$P^{(\beta)} = \left[P_0 \vdots\ P_1 \vdots ... \vdots P_\beta \right] \in \mathfrak{R}^{\rho x(\beta+1)M},$$

$$E^{(\nu)} = \left[E_0 \vdots\ E_1 \vdots ... \vdots E_\nu \right] \in \mathfrak{R}^{N x(\nu+1)N},$$

$$R^{(\alpha)} = \left[R_0 \vdots\ R_1 \vdots ... \vdots R_{\alpha-1} \vdots O_{N,\rho} \right] \in \mathfrak{R}^{N x(\alpha+1)\rho},$$

$$T^{(\mu)} = \left[T_0 \vdots T_1 \vdots ... \vdots T_\mu \right] \in \mathfrak{R}^{N x(\mu+1)M}, \tag{2.24}$$

and $\mathbf{R}^{\alpha-1}$ in the sense of (2.3) and (2.4) (Section 2.1):

$$\mathbf{S} = \mathbf{R}^{\alpha-1}(t) = \left[\mathbf{R}^T(t) \vdots \mathbf{R}^{(1)^T}(t) \vdots ... \vdots \mathbf{R}^{(\alpha-1)^T}(t) \right]^T \in \mathfrak{R}^{\alpha\rho}. \tag{2.25}$$

The left-hand side of the first Equation (2.22) describes *the internal dynamics of the system*, i.e., *the internal state* $\mathbf{S}_I$ *of the system (Definition 15)*, the left-hand side of the second Equation (2.22) describes *the output state* $\mathbf{S}_O$ of the system if and only if $\nu > 0$,

$$\mathbf{S}_I = \mathbf{R}^{\alpha-1}. \tag{2.26}$$

The output vector $\mathbf{Y}(t)$ and its derivatives $\mathbf{Y}^{(1)}(t)$, $\mathbf{Y}^{(2)}(t)$,..., $\mathbf{Y}^{(\nu-1)}(t)$ determine *the output dynamics, i.e., the output state* of the *IIO* system at any $t \in \mathfrak{T}$. We call the extended output vector $\mathbf{Y}^{\nu-1}$,

$$\mathbf{Y}^{\nu-1} = \left[\mathbf{Y}^T(t) \vdots \mathbf{Y}^{(1)^T}(t) \vdots ... \vdots \mathbf{Y}^{(\nu-1)^T}(t) \right]^T \in \mathfrak{R}^{\nu N}, \tag{2.27}$$

also *the output state vector* $\mathbf{S}_O$ of the *IIO* system if and only if $\nu \geq 1$,

$$\mathbf{S}_O = \mathbf{Y}^{\nu-1},\ \nu \geq 1. \tag{2.28}$$

The left-hand sides of both Equations (2.22) determine *the state* $\mathbf{S}$ *of the IIO system* (2.22), which is also called *the system full state,*

$$\mathbf{S} = \left\{ \begin{array}{c} \left[\begin{array}{c} \mathbf{S}_I \\ \mathbf{S}_O \end{array} \right] = \left[\begin{array}{c} \mathbf{R}^{\alpha-1} \\ \mathbf{Y}^{\nu-1} \end{array} \right],\ \nu \geq 1, \\ \mathbf{S}_I = \mathbf{R}^{\alpha-1} = \mathbf{S}_I,\ \nu = 0. \end{array} \right\} \tag{2.29}$$

The system internal motion $\mathrm{S}_I(.;t_0;S_0;\mathbf{I})$ is $\mathcal{R}^{\alpha-1}(.;t_0;R_0^{\alpha-1};\mathbf{I})$,

$$\begin{gathered}\mathcal{S}_I(.;t_0;\mathbf{S}_{I0};\mathbf{I})=\mathcal{R}^{\alpha-1}(.;t_0;\mathbf{R}_0^{\alpha-1};\mathbf{I})\Longrightarrow\\ \mathbf{S}_I(t;t_0;\mathbf{S}_{I0};\mathbf{I})=\mathcal{R}^{\alpha-1}(t;t_0;\mathbf{R}_0^{\alpha-1};\mathbf{I})\Longrightarrow\\ \mathbf{S}_I(t_0;t_0;\mathbf{S}_{I0};\mathbf{I})=\mathbf{R}_0^{\alpha-1}=\mathbf{S}_{I0}.\end{gathered}\tag{2.30}$$

The system extended output response is *the system output motion* denoted by $\mathcal{Y}^{\nu-1}(.;t_0;\mathbf{Y}_0^{\nu-1};\mathbf{I})$,

$$\begin{gathered}\mathcal{S}_O(.;t_0;\mathbf{S}_{O0};\mathbf{I})=\mathbf{Y}^{\nu-1}(.;t_0;\mathbf{Y}_0^{\nu-1};\mathbf{I})\Longrightarrow\\ \mathbf{S}_O(t;t_0;\mathbf{S}_{O0};\mathbf{I})=\mathbf{Y}^{\nu-1}(t;t_0;\mathbf{Y}_0^{\nu-1};\mathbf{I})\Longrightarrow\\ \mathbf{S}_O(t_0;t_0;\mathbf{S}_{O0};\mathbf{I})=\mathbf{Y}_0^{\nu-1}=\mathbf{S}_{O0}.\end{gathered}\tag{2.31}$$

The IIO system output response $\mathcal{Y}(.;t_0;Y_0^{\nu-1};\mathbf{I})$ is the first component of the system full output response $\mathcal{Y}^{\nu-1}(.;t_0;\mathbf{Y}_0^{\nu-1};\mathbf{I})$,

$$\mathcal{Y}(t_0;t_0;\mathbf{Y}_0^{\nu-1};\mathbf{I})\equiv\mathbf{Y}(t_0;t_0;\mathbf{Y}_0^{\nu-1};\mathbf{I})\equiv\mathbf{Y}_{0.}\tag{2.32}$$

The *time* evolution $\mathcal{S}(.;t_0;\mathbf{S}_0;\mathbf{I})$ of the *IIO* system *state vector* $\mathbf{S}$,

$$\mathbf{S}=\left\{\begin{array}{c}\left[\begin{array}{c}\mathbf{S}_I\\ \mathbf{S}_O\end{array}\right]\in\mathfrak{R}^{\alpha\rho+\nu N},\ \nu\geq1,\\ \mathbf{S}_I\in\mathfrak{R}^{\alpha\rho},\ \nu=0.\end{array}\right\}\tag{2.33}$$

is the system *(full) motion*. It is the vector function

$$\mathcal{S}(.;t_0;\mathbf{S}_0;\mathbf{I})=\left\{\begin{array}{c}\left[\begin{array}{c}\mathcal{S}_I(.;t_0;\mathbf{S}_{I0};\mathbf{I})\\ \mathcal{S}_O(.;t_0;\mathbf{S}_{O0};\mathbf{I})\end{array}\right],\ \nu\geq1,\\ \mathcal{S}_I(.;t_0;\mathbf{S}_{I0};\mathbf{I}),\ \nu=0.\end{array}\right\}.\tag{2.34}$$

The preceding simple vector notation, (2.25) and (2.28), permits us to set (2.22) into its compact form

$$\begin{gathered}Q^{(\alpha)}\mathbf{R}^{\alpha}(t)=P^{(\beta)}\mathbf{I}^{\beta}(t),\ \forall t\in\mathfrak{T}_0,\\ E^{(\nu)}\mathbf{Y}^{\nu}(t)=R^{(\alpha)}\mathbf{R}^{\alpha}(t)+T^{(\mu)}\mathbf{I}^{\mu}(t),\ \forall t\in\mathfrak{T}_0.\end{gathered}\tag{2.35}$$

Definition 40 ***The order of the IIO system (2.22)*** *is the number equal to the sum $\alpha+\nu$ of the highest derivatives of the substate vector function* $\mathbf{R}(.)$ *and of the output vector function* $\mathbf{Y}(.)$ *in (2.22).*

Definition 41 ***The dimension,*** *denoted by* dim_{IIO}, ***of the IIO system (2.22)*** *is the sum of the dimension* ρ *of its substate vector* $\mathbf{R}$ *and of the dimension* N *of its output vector* $\mathbf{Y}$*; i.e.,* $\rho+N$, $dim_{IIO} = dim\left[\mathbf{R}^T \;\; \mathbf{Y}^T\right]^T = \rho + N$.

The results valid for the *IIO* systems are valid also for the *IO* systems, the *ISO* systems, and the systems described by (2.36),

$$Q^{(\alpha)}\mathbf{R}^{\alpha}(t) = P^{(\beta)}\mathbf{I}^{\beta}(t),\ \forall t \in \mathfrak{T}_0,$$
$$\mathbf{Y}(t) = R^{(\alpha)}\mathbf{R}^{\alpha}(t) + T^{(\mu)}\mathbf{I}^{\mu}(t),\ \forall t \in \mathfrak{T}_0. \tag{2.36}$$

P. J. Antsaklis and A. N. Michel [4, p. 553] called these systems *the Polynomial Matrix Description* of the systems (for short: *the PMD systems*).

We will call the *IIO* systems determined by $\alpha = 1$, $\beta = 0$, $Q_1 = I_{\rho,\rho}$, $Q_0 = -A$, $P_0 = B$, $R_0 = R = C$, $R_1 = O_{N,\rho}$, $\mathbf{R} = \mathbf{X}$, $\nu = 0$, and $\mu \geq 1$, i.e.,

$$\mathbf{X}^{(1)}(t) = A\mathbf{X}(t) + B\mathbf{I}(t),\ \forall t \in \mathfrak{T}_0,$$
$$\mathbf{Y}(t) = C\mathbf{X}(t) + T^{(\mu)}\mathbf{I}^{\mu}(t),\ \forall t \in \mathfrak{T}_0. \tag{2.37}$$

Rosenbrock systems (*RS*) after H. H. Rosenbrock [110].

The characteristics of another subclass of the *IIO* systems are $\nu = 0$, $E_0 = I_N$, $R_\alpha = O_{N,\rho}$, and $T_0 = T$, $T_k = O_{N\times M}$, $k = 1, 2, ..., \mu$, $\mu = 0$,

$$\sum_{k=0}^{k=\alpha} Q_k\mathbf{R}^{(k)}(t) = \sum_{k=0}^{k=\beta} P_k\mathbf{I}^{(k)}(t),\ \forall t \in \mathfrak{T}_0.$$
$$\mathbf{Y}(t) = \sum_{k=0}^{k=\alpha} R_k\mathbf{R}^{(k)}(t) + T\mathbf{I}(t),\ \forall t \in \mathfrak{T}_0, \tag{2.38}$$

or, equivalently, in the compact form:

$$Q^{(\alpha)}\mathbf{R}^{\alpha}(t) = P^{(\beta)}\mathbf{I}^{\beta}(t),\ t \in \mathfrak{T},\ \alpha \geq 1, \alpha \geq \beta \geq 0,$$
$$\mathbf{Y}(t) = R^{(\alpha)}\mathbf{R}^{\alpha}(t) + T\mathbf{I}(t) = \mathbf{Y}(t) = R^{(\alpha-1)}\mathbf{R}^{\alpha-1}(t) + T\mathbf{I}(t),$$
$$R^{(\alpha)}\in\mathfrak{R}^{N\times\rho(\alpha+1)},\ R^{(\alpha-1)}\in\mathfrak{R}^{N\times\rho\alpha},\ \mathbf{R}^{\alpha-1}\in\mathfrak{R}^{\rho\alpha},\ T\in\mathfrak{R}^{N\times M},\ \forall t \in \mathfrak{T}_0. \tag{2.39}$$

We will call the systems described by (2.39) *General Input-State-Output Systems* (for short: *GISO systems*). They have the state, but they do not possess the output dynamics. Their special case represents the first order *ISO* systems.

For the relationships among different classes of systems, see Section C.3 in Appendix C. Let in the sequel

$$\gamma = \max\{\beta, \mu\}. \tag{2.40}$$

Definition 42 *The quintuple* $(E^{(\nu)}, P^{(\beta)}, Q^{(\alpha)}, R^{(\alpha-1)}, T^{(\mu)})$ *is* ***an IIO realization of the IIO system (2.22), equivalently of (2.35), for an arbitrary input vector function and for arbitrary input, internal (dynamics) and output initial conditions.***

Chapter 3

System Regimes

3.1 System regime meaning

The dynamical behavior (*the behavior*) of a system expresses the manner of the temporal evolution of a process, work, movement of the system, or the manner of the temporal evolution of the system response. The system behavior depends on:

a) *the system properties*, and

b) *the actions upon the system.*

The system properties can be *quantitative* or *qualitative*. *A quantitative system property* is, for example, the settling *time* of the system response. Quantitative system properties characterize the system behavior under specific external and internal conditions. *Qualitative properties of the system* are, for example, its controllability, observability, stability, and trackability ([30], [31]-[34], [37]-[43], [53]-[60], [56], [70]-[75], [96]-[104]). They characterize the system behavior for a set of external and/or internal conditions. Their set can be finite or infinite, bounded or unbounded, but not a singleton that is typically tied with a quantitative system property.

There exist two different principal *actions upon the system*:

∘ *Actions on the system, which are completely determined by the system history and by past (external and/or internal) influences on the system.* These actions are expressed by ***initial conditions*** (initial conditions of input variables, state, and output variables). In view of the accepted system linearity, we will permit *arbitrary initial conditions*.

∘ *Actions influencing the system since the initial moment $t_0 = 0$ on.* These actions are usually ***the external actions***. They are called ***the input variables*** if and only if their influence on the system is essential for its behavior.

The behavior of the system can be:

∘ *independent of time* (invariable in *time*, i.e., *time-invariant*, *time-independent*), which is a constant behavior also called a *stationary behavior*,

or

∘ *dependent on time* (variable in *time*, i.e., *time-dependent*, *time-varying*). In this case, the system behavior is either *periodic* or *aperiodic* (called also: *transient*). The system behavior is periodic if and only if it repeats itself every σ seconds, where σ is a positive real number. Such a minimal number σ is *the period* T of the behavior. Otherwise, the behavior is *aperiodic (transient)*.

Definition 43 ***A system regime*** *represents the set of all (initial and exterior) conditions under which the system operates, and the type of its behavior (i.e. the type of the temporal evolution: process, work, movement of the system and/or its response).*

This definition shows that system regimes can be different and that they can be classified with respect to the following criteria:

- *The existence (the nonexistence) of the initial conditions.* Their existence (nonexistence) signifies that their values are different from (equal to) zero, respectively.
- *The existence (the nonexistence) of the exterior actions.*
- *The realization of the system demanded/required behavior*. If the system is an object/a plant, its demanded behavior is called its *desired behavior* and it is defined by its *desired response* (or, more precisely, by its *desired output response*) denoted by $\mathbf{Y}_d(.\,;\,.\,;\,.) = \mathbf{Y}_d(.)$.
- *The type of the system behavior*.

We distinguish different system regimes relative to these criteria.

3.2 System regimes and initial conditions

Relative to *the existence of the initial conditions* a system can be in a regime during which:

∘ all initial conditions are *absent*, i.e., *equal to zero*,

or

◦ there are *nonzero initial conditions.*

These regimes do not have special names.

Some qualitative dynamical properties (e.g., Lyapunov stability properties, controllability, trackability and observability) concern the system behavior under nonzero initial conditions. However, Bounded-Input Bounded-Output ($BIBO$) stability has assumed so far all initial conditions to be equal to zero.

Initial conditions, being unpredictable, are most often different from zero.

3.3 Forced and free regimes

3.3.1 Introduction

Let

$$(\sigma, \infty[\in \{]\sigma, \infty[, \ [\sigma, \infty[\}.$$

We classify in this case the system regimes relative to the following criterion: *the existence of the exterior actions.*

Definition 44 *(a) A system is in* ***a forced regime*** *on* $(\sigma, \infty[$ *if and only if there exists a moment* τ *in* $(\sigma, \infty[$*, when the input vector different from the zero vector acts on the system:*

$$\exists \tau \in (\sigma, \infty[\Longrightarrow \mathbf{I}(\tau) \neq \mathbf{0}_M.$$

(b) A system is in ***a free regime*** *on* $(\sigma, \infty[$ *if and only if its input vector is equal to the zero vector always on* $(\sigma, \infty[$ *:*

$$\mathbf{I}(t) = \mathbf{0}_M, \, \forall t \in (\sigma, \infty[.$$

The expression "on $(\sigma, \infty[$*" is to be omitted if and only if* $\sigma = 0$*, i.e.,* $(\sigma, \infty[= [\sigma, \infty[= \mathfrak{T}_0$.

Definition 45 *A system behavior is* ***trivial*** *if and only if the system movement is equal to the zero vector all the time. Otherwise, it is* ***nontrivial.***

For a system behavior to be trivial in a free regime, it is necessary and sufficient that all initial conditions are equal to zero. From the physical viewpoint, this signifies that the physical behavior of the system in a free regime can be nontrivial if and only if there is an accumulated energy in the system at the initial moment.

Lyapunov stability properties, and essentially linear system observability, concern the system behavior in a free regime. *BIBO* stability, practical stability, controllability, trackability, and tracking concern the system behavior in a forced regime.

We shall use the following descriptions of an arbitrary system, the forms of which depend on the domain of the independent variable that can be *time* $t \in \mathfrak{T}_0$, or a complex variable $s \in \mathfrak{C}$.

3.3.2 The temporal domain descriptions. The independent variable is *time* $t \in \mathfrak{T}_0$

The description of the IO system in terms of the total coordinates

- in a *forced regime* is determined by (2.1), (in Section 2.1),

$$\sum_{k=0}^{k=\nu} A_k \mathbf{Y}^{(k)}(t) = \sum_{k=0}^{k=\mu} B_k \mathbf{I}^{(k)}(t), \ \forall t \in \mathfrak{T}_0, \tag{3.1}$$

or, equivalently by (2.7), (Section 2.1),

$$A^{(\nu)} \mathbf{Y}^{\nu}(t) = B^{(\mu)} \mathbf{I}^{\mu}(t), \ \forall t \in \mathfrak{T}_0, \tag{3.2}$$

- in a *free regime* has the form

$$\sum_{k=0}^{k=\nu} A_k \mathbf{Y}^{(k)}(t) = \mathbf{0}_N, \ \forall t \in \mathfrak{T}_0. \tag{3.3}$$

or, equivalently,

$$A^{(\nu)} \mathbf{Y}^{\nu}(t) = \mathbf{0}_N, \ \forall t \in \mathfrak{T}_0. \tag{3.4}$$

The description of the ISO system in terms of the total coordinates

- in a *forced regime* is determined by (2.17) and (2.18), (in Section 2.2),

$$\frac{d\mathbf{X}(t)}{dt} = A\mathbf{X}(t) + B\mathbf{I}(t), \ \forall t \in \mathfrak{T}_0, \tag{3.5}$$

$$\mathbf{Y}(t) = C\mathbf{X}(t) + D\mathbf{I}(t), \ \forall t \in \mathfrak{T}_0, \tag{3.6}$$

- in a *free regime* takes the following form, respectively:

$$\frac{d\mathbf{X}(t)}{dt} = A\mathbf{X}(t), \ \forall t \in \mathfrak{T}_0, \tag{3.7}$$

$$\mathbf{Y}(t) = C\mathbf{X}(t), \ \forall t \in \mathfrak{T}_0. \tag{3.8}$$

The description of the IIO system in terms of the total coordinates

- in a *forced regime* is determined by (2.22), (Section 2.3),

$$\sum_{k=0}^{k=\alpha} Q_k \mathbf{R}^{(k)}(t) = \sum_{k=0}^{k=\beta\leq\alpha} P_k \mathbf{I}^{(k)}(t), \ \forall t \in \mathfrak{T}_0,$$
$$\sum_{k=0}^{k=\nu} E_k \mathbf{Y}^{(k)}(t) = \sum_{k=0}^{k=\alpha} R_k \mathbf{R}^{(k)}(t) + \sum_{k=0}^{k=\mu} T_k \mathbf{I}^{(k)}(t), \ \forall t \in \mathfrak{T}_0, \tag{3.9}$$

or, equivalently by (2.35), (in Section 2.3),

$$Q^{(\alpha)} \mathbf{R}^{\alpha}(t) = P^{(\beta)} \mathbf{I}^{\beta}(t), \ \forall t \in \mathfrak{T}_0,$$
$$E^{(\nu)} \mathbf{Y}^{\nu}(t) = R^{(\alpha)} \mathbf{R}^{\alpha}(t) + T^{(\mu)} \mathbf{I}^{\mu}(t), \ \forall t \in \mathfrak{T}_0, \tag{3.10}$$

- in a *free regime* becomes:

$$Q^{(\alpha)} \mathbf{R}^{\alpha}(t) = \mathbf{0}_{\rho}, \ \forall t \in \mathfrak{T}_0,$$
$$E^{(\nu)} \mathbf{Y}^{\nu}(t) = R^{(\alpha)} \mathbf{R}^{\alpha}(t), \ \forall t \in \mathfrak{T}_0. \tag{3.11}$$

The mathematical models simplify essentially if the system is in a free regime.

The temporal domain mathematical descriptions of the systems do not show explicitly the influence of initial conditions. It appears only in the solutions of the mathematical models.

3.3.3 The complex domain system descriptions. The independent variable is the complex variable $s \in \mathbb{C}$

Laplace transform (Appendix B.1) is the basic tool to obtain the complex domain description of the system.

The complex domain description of the IO system

is determined by Laplace transform of (3.1), when we apply the properties of Laplace transform,

- in a *forced regime* by

$$\sum_{k=0}^{k=\nu} A_k \left[s^k \mathbf{Y}(s) - \sum_{i=1}^{i=k} s^{k-i} \mathbf{Y}^{(i-1)}(0^{\mp}) \right] =$$
$$= \sum_{k=0}^{k=\mu} B_k \left[s^k \mathbf{I}(s) - \sum_{i=1}^{i=k} s^{k-i} \mathbf{I}^{(i-1)}(0^{\mp}) \right], \tag{3.12}$$

- in a *free regime* by

$$\sum_{k=0}^{k=\nu} A_k \left[s^k \mathbf{Y}(s) - \sum_{i=1}^{i=k} s^{k-i} \mathbf{Y}^{(i-1)}(0^{\mp}) \right] = \mathbf{0}_N. \tag{3.13}$$

The complex domain description of the ISO system

is determined by Laplace transform of (3.5) and (3.6), when we apply the properties of Laplace transform,

- in a *forced regime* by

$$(sI - A)\,\mathbf{X}(s) - \mathbf{X}(0^{\mp}) = B\mathbf{I}(s), \tag{3.14}$$
$$\mathbf{Y}(s) = C\mathbf{X}(s) + D\mathbf{I}(s), \tag{3.15}$$

- which in a *free regime* becomes

$$(sI - A)\,\mathbf{X}(s) - \mathbf{X}(0^{\mp}) = \mathbf{0}_N, \tag{3.16}$$
$$\mathbf{Y}(s) = C\mathbf{X}(s). \tag{3.17}$$

The complex domain description of the IIO system

follows from the application of Laplace transform and its properties to (3.9) in a *forced regime*

$$\begin{aligned}
&\sum_{k=0}^{k=\alpha} Q_k \left[s^k \mathbf{R}(s) - \sum_{i=1}^{i=k} s^{k-i} \mathbf{R}^{(i-1)}(0^{\mp}) \right] = \\
&= \sum_{k=0}^{k=\beta\leq\alpha} P_k \left[s^k \mathbf{I}(s) - \sum_{i=1}^{i=k} s^{k-i} \mathbf{I}^{(i-1)}(0^{\mp}) \right], \\
&\sum_{k=0}^{k=\nu} E_k \left[s^k \mathbf{Y}(s) - \sum_{i=1}^{i=k} s^{k-i} \mathbf{Y}^{(i-1)}(0^{\mp}) \right] = \\
&= \sum_{k=0}^{k=\alpha} R_k \left[s^k \mathbf{R}(s) - \sum_{i=1}^{i=k} s^{k-i} \mathbf{R}^{(i-1)}(0^{\mp}) \right] + \\
&+ \sum_{k=0}^{k=\mu} T_k \left[s^k \mathbf{I}(s) - \sum_{i=1}^{i=k} s^{k-i} \mathbf{I}^{(i-1)}(0^{\mp}) \right], \qquad (3.18)
\end{aligned}$$

in a *free regime*

$$\begin{aligned}
&\sum_{k=0}^{k=\alpha} Q_k \left[s^k \mathbf{R}(s) - \sum_{i=1}^{i=k} s^{k-i} \mathbf{R}^{(i-1)}(0^{\mp}) \right] = \mathbf{0}_\rho, \\
&\sum_{k=0}^{k=\nu} E_k \left[s^k \mathbf{Y}(s) - \sum_{i=1}^{i=k} s^{k-i} \mathbf{Y}^{(i-1)}(0^{\mp}) \right] = \\
&= \sum_{k=0}^{k=\alpha} R_k \left[s^k \mathbf{R}(s) - \sum_{i=1}^{i=k} s^{k-i} \mathbf{R}^{(i-1)}(0^{\mp}) \right]. \qquad (3.19)
\end{aligned}$$

The complex domain descriptions in free regimes are essentially simpler from those related to forced regimes. However, both contain explicitly initial conditions. We resolve them in terms of Laplace transforms of unknown system variables.

The solution of the complex domain description* (3.12) *of the IO system

in a *forced regime* under **nonzero initial conditions**

$$\mathbf{Y}(s) = \left(\sum_{k=0}^{k=\nu} A_k s^k\right)^{-1} \bullet$$
$$\bullet \left\{ \begin{array}{c} \sum_{k=0}^{k=\nu} A_k \left[\sum_{i=1}^{i=k} s^{k-i}\mathbf{Y}^{(i-1)}(0^{\mp})\right] + \\ +\sum_{k=0}^{k=\mu} B_k \left[s^k \mathbf{I}(s) - \sum_{i=1}^{i=k} s^{k-i}\mathbf{I}^{(i-1)}(0^{\mp})\right] \end{array} \right\}, \tag{3.20}$$

in a *free regime* under **nonzero initial conditions**

$$\mathbf{Y}(s) = \left(\sum_{k=0}^{k=\nu} A_k s^k\right)^{-1} \sum_{k=0}^{k=\mu} A_k \left[\sum_{i=1}^{i=k} s^{k-i}\mathbf{Y}^{(i-1)}(0^{\mp})\right]. \tag{3.21}$$

The solution of the complex domain description of the ISO system

follows from (3.14) and (3.15)

in a *forced regime* under **nonzero initial conditions**

$$\mathbf{X}(s) = (sI - A)^{-1}\left[B\mathbf{I}(s) + \mathbf{X}(0^{\mp})\right], \tag{3.22}$$
$$\mathbf{Y}(s) = C\mathbf{X}(s) + D\mathbf{I}(s), \tag{3.23}$$

in a *free regime* under **nonzero initial conditions**

$$\mathbf{X}(s) = (sI - A)^{-1}\mathbf{X}(0^{\mp}), \tag{3.24}$$
$$\mathbf{Y}(s) = C\mathbf{X}(s). \tag{3.25}$$

The solution of the complex domain description of the IIO system

results from (3.18)

in a *forced regime* under **nonzero initial conditions**

$$\mathbf{R}(s) = \left(\sum_{k=0}^{k=\alpha} Q_k s^k\right)^{-1} \bullet$$
$$\bullet \left\{ \begin{array}{c} \sum_{k=0}^{k=\alpha} Q_k \left[\sum_{i=1}^{i=k} s^{k-i}\mathbf{R}^{(i-1)}(0^{\mp})\right] + \\ +\sum_{k=0}^{k=\beta\leq\alpha} P_k \left[s^k \mathbf{I}(s) - \sum_{i=1}^{i=k} s^{k-i}\mathbf{I}^{(i-1)}(0^{\mp})\right] \end{array} \right\}, \tag{3.26}$$

$$\mathbf{Y}(s) = \left(\sum_{k=0}^{k=\nu} E_k s^k \right)^{-1} \bullet$$

$$\bullet \left\{ \begin{array}{c} \sum_{k=0}^{k=\nu} E_k \left[\sum_{i=1}^{i=k} s^{k-i} \mathbf{Y}^{(i-1)}(0^{\mp}) \right] + \\ + \sum_{k=0}^{k=\alpha} R_k \left[s^k \mathbf{R}(s) - \sum_{i=1}^{i=k} s^{k-i} \mathbf{R}^{(i-1)}(0^{\mp}) \right] + \\ + \sum_{k=0}^{k=\mu} T_k \left[s^k \mathbf{I}(s) - \sum_{i=1}^{i=k} s^{k-i} \mathbf{I}^{(i-1)}(0^{\mp}) \right] \end{array} \right\} \quad (3.27)$$

in a *free regime* under **nonzero initial conditions**

$$\mathbf{R}(s) = \left(\sum_{k=0}^{k=\alpha} Q_k s^k \right)^{-1} \left\{ \sum_{k=0}^{k=\alpha} Q_k \left[\sum_{i=1}^{i=k} s^{k-i} \mathbf{R}^{(i-1)}(0^{\mp}) \right] \right\}, \quad (3.28)$$

$$\mathbf{Y}(s) = \left(\sum_{k=0}^{k=\nu} E_k s^k \right)^{-1} \bullet$$

$$\bullet \left\{ \begin{array}{c} \sum_{k=0}^{k=\nu} E_k \left[\sum_{i=1}^{i=k} s^{k-i} \mathbf{Y}^{(i-1)}(0^{\mp}) \right] + \\ \sum_{k=0}^{k=\alpha} R_k \left[s^k \mathbf{R}(s) - \sum_{i=1}^{i=k} s^{k-i} \mathbf{R}^{(i-1)}(0^{\mp}) \right] \end{array} \right\}. \quad (3.29)$$

Although the complex domain mathematical models of the systems simplify in free regimes, they still have complex forms because of double sums of the terms containing initial conditions.

Let us accept that **all initial conditions are equal to zero**.

The complex domain description for zero initial conditions

from (3.12) of the *IO* system

in a *forced regime* under **zero initial conditions**

$$\mathbf{Y}(s) = \left[\left(\sum_{k=0}^{k=\nu} A_k s^k \right)^{-1} \left(\sum_{k=0}^{k=\mu} B_k s^k \right) \right] \mathbf{I}(s), \quad (3.30)$$

in a *free regime* under **zero initial conditions**

$$\mathbf{Y}(s) \equiv \mathbf{0}, \quad (3.31)$$

from (3.14) and (3.15) of the *ISO* system

in a *forced regime* under **zero initial conditions**

$$\mathbf{X}(s) = \left[(sI - A)^{-1} B\right] \mathbf{I}(s), \tag{3.32}$$

$$\mathbf{Y}(s) = C\mathbf{X}(s) + D\mathbf{I}(s), \tag{3.33}$$

in a *free regime* under **zero initial conditions**

$$\mathbf{X}(s) \equiv \mathbf{0}, \tag{3.34}$$

$$\mathbf{Y}(s) \equiv \mathbf{0}, \tag{3.35}$$

from (3.18) of the *IIO* system

in a *forced regime* under **zero initial conditions**

$$\mathbf{R}(s) = \left[\left(\sum_{k=0}^{k=\alpha} Q_k s^k\right)^{-1} \left(\sum_{k=0}^{k=\beta\leq\alpha} P_k s^k\right)\right] \mathbf{I}(s), \tag{3.36}$$

$$\mathbf{Y}(s) = \left(\sum_{k=0}^{k=\nu} E_k s^k\right)^{-1} \left[\left(\sum_{k=0}^{k=\alpha} R_k s^k\right) \mathbf{R}(s) + \left(\sum_{k=0}^{k=\mu} T_k s^k\right) \mathbf{I}(s)\right], \tag{3.37}$$

in a *free regime* under **zero initial conditions**

$$\mathbf{R}(s) \equiv \mathbf{0}, \tag{3.38}$$

$$\mathbf{Y}(s) \equiv \mathbf{0}. \tag{3.39}$$

3.3.4 Basic problem

In all the preceding cases, in the forced regime, the relationships between Laplace transforms of the output and input vectors are determined only by system parameters, whatever is the form of the input vector function.

This raises the following basic problem to be solved in this book:

Problem 46 ***The basic problem***

Find the complex domain description of the system such that the relationship between the Laplace transforms of the output and input vectors is determined only by system parameters, whatever is the form of the input vector function, and for arbitrary initial conditions.

The solutions to this problem for different classes of the systems will be presented in the sequel. They discover the existence of the dynamic system characteristic that generalizes the system transfer function matrix $G(s)$. We will call it ***the system full (complete) transfer function matrix*** denoted by $F(s)$ ([27], [29], [49], [63]-[68]). Its use enables us to treat fully and correctly many qualitative dynamical properties as well as quantitative dynamical characteristics of the systems in the complex domain.

3.4 Desired regime

3.4.1 Introduction

We shall analyze system regimes relative to the following criterion: *the realization of the demanded system behavior.* This criterion has the full sense for a system that is an object O (a plant P), which is to be controlled or which is controlled. A desired regime of a system (object/plant) is defined by its demanded (output) response $\mathbf{Y}_d(.)$.

Definition 47 ***Desired regime***

A system is in ***a desired*** *(also called:* ***nominal*** *or* ***nonperturbed)*** ***regime on*** $\mathfrak{T}_0$ *(for short: in* ***a desired regime****) if and only if it realizes its desired (output) response* $\mathbf{Y}_d(t)$ *all the time,*

$$\mathbf{Y}(t) = \mathbf{Y}_d(t), \ \forall t \in \mathfrak{T}_0. \tag{3.40}$$

This definition implies directly a necessary (but not a sufficient) condition for a system to be in a desired (nominal, nonperturbed) regime.

Proposition 48 *In order for the plant to be in a desired (nominal, nonperturbed) regime, i.e.,*

$$\mathbf{Y}(t) = \mathbf{Y}_d(t), \ \forall t \in \mathfrak{T}_0 \Longrightarrow \mathbf{Y}_0 = \mathbf{Y}_{d0},$$

it is necessary that the initial real output vector is equal to the initial desired output vector,

$$\mathbf{Y}_0 = \mathbf{Y}_{d0}.$$

Hence, the system cannot be in a nominal regime (on $\mathfrak{T}_0$) if its initial real output vector is different from the initial desired output vector:

$$\mathbf{Y}_0 \neq \mathbf{Y}_{d0} \Longrightarrow \exists \sigma \in \mathfrak{T}_0 \Longrightarrow \mathbf{Y}(\sigma) \neq \mathbf{Y}_d(\sigma).$$

Since the real initial output vector $\mathbf{Y}(0) = \mathbf{Y}_0$ is most often different from the desired initial output vector $\mathbf{Y}_d(0) = \mathbf{Y}_{d0}$, then the system is most often in a *nondesired* (*nonnominal, perturbed, disturbed*) regime.

Definition 49 ***Nominal input***
*An input vector function $\mathbf{I}^*(.)$ of a system is **nominal relative to the desired response** $\mathbf{Y}_d(.)$, which is denoted by $\mathbf{I}_N(.)$, if and only if $\mathbf{I}(.) = \mathbf{I}^*(.)$ ensures that the corresponding real response $\mathbf{Y}(.) = \mathbf{Y}^*(.)$ of the system obeys $\mathbf{Y}^*(t) = \mathbf{Y}_d(t)$ all the time as soon as all the internal and the output system initial conditions are desired (nominal, nonperturbed).*

This definition is general. It will be specified for different classes of systems with respect to the internal and output system initial conditions.

Note 50 *An input vector function $\mathbf{I}^*(.)$ can be nominal relative to the desired response $\mathbf{Y}_{d1}(.)$ of a system, but it need not be nominal with respect to another desired response $\mathbf{Y}_{d2}(.)$ of the system. This explains the relative sense of the notion "nominal relative to the desired response $\mathbf{Y}_d(.)$".*

Definition 51 $\mathbf{Y}_d(.)$ ***realizable in*** $\mathfrak{I}^i$
*The desired response $\mathbf{Y}_d(.)$ of the system is **realizable in** $\mathfrak{I}^i$ if and only if there exist $\mathbf{I}^*(.) \in \mathfrak{I}^i$ and the initial conditions $\mathbf{Y}_0^*$, $\mathbf{Y}_0^{*(1)}$, ... of the output vector function $\mathbf{Y}(.)$ and of its derivatives such that $\mathbf{I}^*(.)$ is nominal relative to $\mathbf{Y}_d(.)$,*

$$\exists \mathbf{I}^*(.) \in \mathfrak{I}^i, \ \exists \mathbf{Y}_0^*, \mathbf{Y}_0^{*(1)}, .. \in \mathfrak{R}^N \Longrightarrow \mathbf{Y}(t; \mathbf{I}^*; \mathbf{Y}_0^*, \mathbf{Y}_0^{*(1)}, ..) = \mathbf{Y}_d(t), \ \forall t \in \mathfrak{T}_0.$$

Definition 52 $\mathbf{Y}_d(.)$ ***realizable on*** $\mathfrak{I}^i$
*The desired response $\mathbf{Y}_d(.)$ of the system is **realizable on** $\mathfrak{I}^i$ if and only if for every $\mathbf{I}(.) \in \mathfrak{I}^i$ there exist the initial conditions $\mathbf{Y}_0^*$, $\mathbf{Y}_0^{*(1)}$, ... of the output vector function $\mathbf{Y}(.)$ such that $\mathbf{Y}(.; \mathbf{I}; \mathbf{Y}_0^*, \mathbf{Y}_0^{*(1)}, ...)$ is equal to the desired output vector function $\mathbf{Y}_d(.)$,*

$$\forall \mathbf{I}^*(.) \in \mathfrak{I}^i, \ \exists \mathbf{Y}_0^*, \mathbf{Y}_0^{*(1)}, .. \in \mathfrak{R}^N \Longrightarrow \mathbf{Y}(t; \mathbf{I}^*; \mathbf{Y}_0^*, \mathbf{Y}_0^{*(1)}, ..) = \mathbf{Y}_d(t), \ \forall t \in \mathfrak{T}_0.$$

Comment 53 *The realizability of $\mathbf{Y}_d(.)$ in $\mathfrak{I}^i$ is necessary, but not sufficient, for the realizability of $\mathbf{Y}_d(.)$ on $\mathfrak{I}^i$. The realizability of $\mathbf{Y}_d(.)$ on $\mathfrak{I}^i$ is sufficient, but not necessary, for the realizability of $\mathbf{Y}_d(.)$ in $\mathfrak{I}^i$.*

Problem 54 *Under what conditions there exists a nominal vector function $\mathbf{I}_N(.)$ relative to the system desired (nominal) output response? Equivalently. Under what conditions is the system desired output response realizable in $\mathfrak{I}^i$ and/or realizable on $\mathfrak{I}^i$?*

Some qualitative system properties [e.g., Lyapunov stability properties of the desired motion when the internal dynamical behavior is also (or, is considered formally as) the output behavior] have a sense if and only if there exists an affirmative solution to the preceding problem.

3.4.2 *IO* systems

We follow Definition 47 and Definition 49, (in Subsection 3.4.1), in order to apply them to the *IO* object (2.1), (in Section 2.1).

We need the following complex matrix functions, the first one is $S_i^{(k)}(.) : \mathfrak{C} \longrightarrow \mathfrak{C}^{\ i(k+1)xi}$, for the presentation of the complex domain condition for an input vector function to be nominal for the system relative to its desired output vector response,

$$S_i^{(k)}(s) = \left[s^0 I_i \vdots s^1 I_i \vdots s^2 I_i \vdots ... \vdots s^k I_i\right]^T \in \mathfrak{C}^{\ i(k+1)xi},$$
$$(k,i) \in \{(\mu, M),\ (\nu, N)\}, \tag{3.41}$$

in which I_i is the *i-th* order identity matrix, $I_i \in \mathfrak{R}^{ixi}$, and the second one is $Z_k^{(\varsigma-1)}(.) : \mathfrak{C} \rightarrow \mathfrak{C}^{(\varsigma+1)kx\varsigma k}$,

$$Z_k^{(\varsigma-1)}(s) = \begin{bmatrix} O_k & O_k & O_k & ... & O_k \\ s^0 I_k & O_k & O_k & ... & O_k \\ ... & ... & ... & ... & ... \\ s^{\varsigma-1} I_k & s^{\varsigma-2} I_k & s^{\varsigma-3} I_k & ... & s^0 I_k \end{bmatrix}, \ \varsigma \geq 1,$$
$$Z_k^{(\varsigma-1)}(s) \in \mathfrak{C}^{(\varsigma+1)kx\varsigma k},\ (\varsigma, k) \in \{(\mu, M),\ (\nu, N)\} \tag{3.42}$$

They will enable us also to resolve effectively the basic Problem 46 (in Subsection 3.3.4), and the equivalent main task to solve Problem 112 (Chapter 5).

Note 55 *If $\varsigma = 0$, then the matrix $Z_k^{(\varsigma-1)}(s) = Z_k^{(-1)}(s)$ should be completely omitted rather than replaced by the zero matrix. The matrix $Z_k^{(\zeta-1)}(s)$ is not defined for $\zeta \leq 0$ and should be treated as a nonexisting one. Derivatives exist only for natural numbers, i.e., $\mathbf{Y}^{(\varsigma)}(t)$ can exist only for $\varsigma \geq 1$. Matrix function $Z_k^{(\zeta-1)}(.)$ is related to Laplace transform of derivatives only.*

Theorem 56 *In order for an input vector function $\mathbf{I}^*(.)$ to be nominal for the IO system (2.1), i.e., for (2.7), relative to its desired response $\mathbf{Y}_d(.)$: $\mathbf{I}^*(.) = \mathbf{I}_N(.)$, it is necessary and sufficient that 1) and 2) hold:*

1) rank $B^{(\mu)} = N \leq M$*, equivalently rank*$\sum_{k=0}^{k=\mu} B_k s^k = rank B^{(\mu)} S_M^{(\mu)}(s)$ $= N \leq M$*, and*

2) the following equation is valid in the time domain:

$$\sum_{k=0}^{k=\mu} B_k \mathbf{I}^{*(k)}(t) = \sum_{k=0}^{k=\nu} A_k \mathbf{Y}_d^{(k)}(t), \ \forall t \in \mathfrak{T}_0, \tag{3.43}$$

i.e., its following compact form

$$B^{(\mu)} \mathbf{I}^{*^{\mu}}(t) = A^{(\nu)} \mathbf{Y}_d^{\nu}(t), \ \forall t \in \mathfrak{T}_0, \tag{3.44}$$

or equivalently in the complex domain:

$$\mathbf{I}^*(s) = \left(\sum_{k=0}^{k=\mu} B_k s^k \right)^T \left[\left(\sum_{k=0}^{k=\mu} B_k s^k \right) \left(\sum_{k=0}^{k=\mu} B_k s^k \right)^T \right]^{-1} \bullet$$
$$\bullet \left\langle \sum_{k=0}^{k=\mu} B_k \left[\sum_{i=1}^{i=k} s^{k-i} \mathbf{I}^{*^{(i-1)}}(0) \right] + \sum_{k=0}^{k=\nu} A_k \left[s^k \mathbf{Y}_d(s) - \sum_{i=1}^{i=k} s^{k-i} \mathbf{Y}_d^{(i-1)}(0) \right] \right\rangle, \tag{3.45}$$

i.e.,

$$\mathbf{I}^*(s) = \left(B^{(\mu)} S_M^{(\mu)}(s) \right)^T \left[\left(B^{(\mu)} S_M^{(\mu)}(s) \right) \left(B^{(\mu)} S_M^{(\mu)}(s) \right)^T \right]^{-1} \bullet$$
$$\bullet \left\langle B^{(\mu)} Z_M^{(\mu-1)}(s) \mathbf{I}^{*^{\mu-1}}(0) + A^{(\nu)} \left[S_N^{(\nu)}(s) \mathbf{Y}_d(s) - Z_N^{(\nu-1)}(s) \mathbf{Y}_d^{\nu-1}(0) \right] \right\rangle, \tag{3.46}$$

Proof. *Necessity.* Let a vector function $\mathbf{I}^*(.)$ be nominal for the *IO* system (2.1), i.e., for (2.7), relative to its desired response $\mathbf{Y}_d(.)$. Definition 49 (Subsection 3.4.1) holds. It and (2.1), i.e., (2.7), imply

$$\sum_{k=0}^{k=\nu} A_k \mathbf{Y}_d^{(k)}(t) = \sum_{k=0}^{k=\mu} B_k \mathbf{I}^{*(k)}(t), \ \forall t \in \mathfrak{T}_0,$$
$$A^{(\nu)} \mathbf{Y}_d^{\nu}(t) = B^{(\mu)} \mathbf{I}^{*\mu}(t), \ \forall t \in \mathfrak{T}_0.$$

These equations are (3.43) and (3.44) in another forms, respectively. Their Laplace transforms solved in $\mathbf{I}^*(s)$ are given in (3.45) and (3.46), respectively. Since they are solvable in $\mathbf{I}^*(.)$ it follows that the conditions 1) and 2) hold.

Sufficiency. Let the conditions 1) and 2) hold. The input vector function $\mathbf{I}^*(.)$ to the IO system (2.1), i.e., (2.7),

$$A^{(\nu)}\mathbf{Y}^{\nu}(t) = B^{(\mu)}\mathbf{I}^{*\mu}(t),\ \forall t \in \mathfrak{T}_0,$$

satisfies (3.43), hence (3.44):

$$A^{(\nu)}\mathbf{Y}_d^{\nu}(t) = B^{(\mu)}\mathbf{I}^{*\mu}(t),\ \forall t \in \mathfrak{T}_0. \tag{3.47}$$

These equations and

$$\varepsilon = \mathbf{Y}_d - \mathbf{Y} \tag{3.48}$$

yield

$$A^{(\nu)}\varepsilon^{\nu}(t) = \mathbf{0}_N,\ \forall t \in \mathfrak{T}_0, \tag{3.49}$$

Definition 49 requires $\varepsilon^{\nu}(0) = \mathbf{0}_{N(\nu+1)}$, which implies the trivial solution $\varepsilon(t) = \mathbf{0}_N$, $\forall t \in \mathfrak{T}_0$, of (3.49). This and $\varepsilon = \mathbf{Y}_d - \mathbf{Y}$ prove $\mathbf{Y}(t) = \mathbf{Y}_d(t)$, $\forall t \in \mathfrak{T}_0$.

Let the input vector function $\mathbf{I}^*(.)$ to the IO system (2.1), i.e., to (2.7), obeys (3.45), equivalently (3.46). Laplace transforms of (2.1) and (2.7) read for the input vector function $\mathbf{I}(.) = \mathbf{I}^*(.)$:

$$\mathbf{I}^*(s) = \left(\sum_{k=0}^{k=\mu} B_k s^k\right)^T \left[\left(\sum_{k=0}^{k=\mu} B_k s^k\right)\left(\sum_{k=0}^{k=\mu} B_k s^k\right)^T\right]^{-1} \bullet$$

$$\bullet \left\langle \sum_{k=0}^{k=\mu} B_k \left[\sum_{i=1}^{i=k} s^{k-i} \mathbf{I}^{*(i-1)}(0)\right] + \sum_{k=0}^{k=\nu} A_k \left[s^k \mathbf{Y}(s) - \sum_{i=1}^{i=k} s^{k-i} \mathbf{Y}^{(i-1)}(0)\right]\right\rangle,$$

i.e.,

$$\mathbf{I}^*(s) = \left(B^{(\mu)} S_M^{(\mu)}(s)\right)^T \left[\left(B^{(\mu)} S_M^{(\mu)}(s)\right)\left(B^{(\mu)} S_M^{(\mu)}(s)\right)^T\right]^{-1} \bullet$$

$$\bullet \left\langle B^{(\mu)} Z_M^{(\mu-1)}(s) \mathbf{I}^{*^{\mu-1}}(0) + A^{(\nu)}\left[S_N^{(\nu)}(s)\mathbf{Y}(s) - Z_N^{(\nu-1)}(s)\mathbf{Y}^{\nu-1}(0)\right]\right\rangle.$$

These equations multiplied on the left by $\left(\sum_{k=0}^{k=\mu} B_k s^k\right)$, i.e., by $B^{(\mu)} S_M^{(\mu)}(s)$, respectively, and (3.45) and (3.46) imply, respectively,

$$\sum_{k=0}^{k=\nu} A_k \left[s^k \mathbf{Y}_d(s) - \sum_{i=1}^{i=k} s^{k-i} \mathbf{Y}_d^{(i-1)}(0)\right] =$$

$$= \sum_{k=0}^{k=\nu} A_k \left[s^k \mathbf{Y}(s) - \sum_{i=1}^{i=k} s^{k-i} \mathbf{Y}^{(i-1)}(0)\right]$$

and

$$A^{(\nu)}\left[S_N^{(\nu)}(s)\mathbf{Y}_d(s)-Z_N^{(\nu-1)}(s)\mathbf{Y}_d^{\nu-1}(0)\right]=$$
$$=A^{(\nu)}\left[S_N^{(\nu)}(s)\mathbf{Y}(s)-Z_N^{(\nu-1)}(s)\mathbf{Y}^{\nu-1}(0)\right].$$

Definition 49 requires $\mathbf{Y}^{\nu-1}(0)=\mathbf{Y}_d^{\nu-1}(0)$ that reduces the preceding equations to

$$\sum_{k=0}^{k=\nu}A_ks^k\left(\mathbf{Y}_d(s)-\mathbf{Y}(s)\right)=\sum_{k=0}^{k=\nu}A_ks^k\varepsilon\left(s\right)=\mathbf{0}_N,$$

$$A^{(\nu)}S_N^{(\nu)}(s)\left(\mathbf{Y}_d(s)-\mathbf{Y}(s)\right)=A^{(\nu)}S_N^{(\nu)}(s)\varepsilon\left(s\right)=\mathbf{0}_N.$$

These equations imply $\varepsilon\left(s\right)=\mathbf{0}_N$ due to Condition 25 (in Section 2.1), hence $\varepsilon\left(s\right)=\mathbf{0}_N$, $\forall s\in\mathfrak{C}$, which is equivalent to $\varepsilon\left(t\right)=\mathbf{0}_N$, $\forall t\in\mathfrak{T}_0$, i.e., $\mathbf{Y}(t)=\mathbf{Y}_d(t)$, $\forall t\in\mathfrak{T}_0$ ■

This theorem is general. It holds for all *IO* systems (2.1), i.e., (2.7).

Note 57 *We will treat in the sequel only IO systems that satisfy the condition $N\leq M$ for the realizability of their desired output vector functions $\mathbf{Y}_d(.)$.*

3.4.3 *ISO* systems

We specify now Definition 49 (in Subsection 3.4.1) in the framework of the *ISO* systems (2.17) and (2.18) (Section 2.2).

Definition 58 *A functional vector pair $[\mathbf{I}^*(.),\mathbf{X}^*(.)]$ is* ***nominal for the ISO system (2.17) and (2.18) relative to its desired response*** $\mathbf{Y}_d(.)$, *which is denoted by $[\mathbf{I}_N(.),\mathbf{X}_N(.)]$, if and only if $[\mathbf{I}(.),\mathbf{X}(.)]=[\mathbf{I}^*(.),\mathbf{X}^*(.)]$ ensures that the corresponding real response $\mathbf{Y}(.)=\mathbf{Y}^*(.)$ of the system obeys $\mathbf{Y}^*(t)=\mathbf{Y}_d(t)$ all the time,*

$$[\mathbf{I}^*(.),\mathbf{X}^*(.)]=[\mathbf{I}_N(.),\mathbf{X}_N(.)]\Longleftrightarrow\langle\mathbf{Y}^*(t)=\mathbf{Y}_d(t),\ \forall t\in\mathfrak{T}_0\rangle.$$

The time evolution $\mathbf{X}_N(t;\mathbf{X}_{N0};\mathbf{I}_N)$, $\mathbf{X}_N(0;\mathbf{X}_{N0};\mathbf{I}_N)\equiv\mathbf{X}_{N0}$, of the nominal state vector $\mathbf{X}_N$ is ***the desired motion*** $\mathbf{X}_d(.;\mathbf{X}_{d0};\mathbf{I}_N)$ ***of the ISO system (2.17) and (2.18) relative to its desired response*** $\mathbf{Y}_d(.)$, *for short,* ***the desired motion,***

$$\mathbf{X}_d(t;\mathbf{X}_{d0};\mathbf{I}_N)\equiv\mathbf{X}_N(t;\mathbf{X}_{N0};\mathbf{I}_N),\ \mathbf{X}_d(0;\mathbf{X}_{d0};\mathbf{I}_N)\equiv\mathbf{X}_{d0}\equiv\mathbf{X}_{N0}.\quad(3.50)$$

I is the identity matrix of the dimension n: $I_n = I$. Notice that

$$\begin{bmatrix} -B & sI - A \\ D & C \end{bmatrix} \in \mathfrak{C}^{(N+n)\mathsf{x}(M+n)}.$$

Theorem 59 *In order for a functional vector pair* $[\mathbf{I}^*(.), \mathbf{X}^*(.)]$ *to be nominal for the ISO system (2.17) and (2.18) relative to its desired response* $\mathbf{Y}_d(.)$, $[\mathbf{I}^*(.), \mathbf{X}^*(.)] = [\mathbf{I}_N(.), \mathbf{X}_N(.)]$, *it is necessary and sufficient that it obeys the following equations:*

$$-B\mathbf{I}^*(t) + \frac{d\mathbf{X}^*(t)}{dt} - A\mathbf{X}^*(t) = \mathbf{0}_n, \ \forall t \in \mathfrak{T}_0, \tag{3.51}$$

$$D\mathbf{I}^*(t) + C\mathbf{X}^*(t) = \mathbf{Y}_d(t), \ \forall t \in \mathfrak{T}_0, \tag{3.52}$$

or equivalently,

$$\begin{bmatrix} -B & sI - A \\ D & C \end{bmatrix} \begin{bmatrix} \mathbf{I}^*(s) \\ \mathbf{X}^*(s) \end{bmatrix} = \begin{bmatrix} \mathbf{X}_0{}^* \\ \mathbf{Y}_d(s) \end{bmatrix}. \tag{3.53}$$

Proof. *Necessity.* Let $[\mathbf{I}^*(.), \mathbf{X}^*(.)]$ be a nominal functional (input and state) vector pair for the ISO system (2.17) and (2.18) relative to its desired response $\mathbf{Y}_d(.)$. Hence, the system is in its desired regime relative to $\mathbf{Y}_d(.)$. Definition 58 shows that $[\mathbf{I}(.), \mathbf{X}(.)] = [\mathbf{I}^*(.), \mathbf{X}^*(.)]$ implies $\mathbf{Y}(.) = \mathbf{Y}_d(.)$. This and the ISO model (2.17) and (2.18) yield the following equations:

$$\frac{d\mathbf{X}^*(t)}{dt} = A\mathbf{X}^*(t) + B\mathbf{I}^*(t), \ \forall t \in \mathfrak{T}_0,$$

$$\mathbf{Y}(t) = \mathbf{Y}_d(t) = C\mathbf{X}^*(t) + D\mathbf{I}^*(t), \ \forall t \in \mathfrak{T}_0, \tag{3.54}$$

which can be easily set in the form of Equations (3.51) and (3.52). Application of Laplace transform together with its properties transforms Equations (3.51) and (3.52) into (3.53).

Sufficiency. We accept that all the conditions of the theorem are valid. We chose $[\mathbf{I}(.), \mathbf{X}(.)] = [\mathbf{I}^*(.), \mathbf{X}^*(.)]$. Equation (3.51) written in the normal state form,

$$\frac{d\mathbf{X}^*(t)}{dt} = A\mathbf{X}^*(t) + B\mathbf{I}^*(t), \ \forall t \in \mathfrak{T}_0,$$

shows that the pair $[\mathbf{I}^*(.), \mathbf{X}^*(.)]$ satisfies (2.17). Furthermore, the equation (2.18) takes the following form:

$$\mathbf{Y}(t) = C\mathbf{X}^*(t) + D\mathbf{I}^*(t), \ \forall t \in \mathfrak{T}_0.$$

It, subtracted from (3.52), yields

$$\mathbf{Y}(t) - \mathbf{Y}_d(t) = \mathbf{0}, \ \forall t \in \mathfrak{T}_0,$$

i.e.,

$$\mathbf{Y}(t) = \mathbf{Y}_d(t), \ \forall t \in \mathfrak{T}_0.$$

Laplace transform of Equations (3.51) and (3.52) yield (3.53). This completes the proof ■

The initial state vector $\mathbf{X}^*(0)$ rests free for the choice. Let us consider the existence of the solutions of Equations (3.51) and (3.52), or equivalently (3.53). There are $(M+n)$ unknown variables and $(N+n)$ equations. The unknown variables are the entries of $\mathbf{I}^*(s) \in \mathfrak{C}^M$ and $\mathbf{X}^*(s) \in \mathfrak{C}^n$.

Case 60 $N > M$

If $N > M$, then Equations (3.51) and (3.52), or equivalently (3.53), do not have a solution. The number of the unknown variables is less than the number of the available equations.

Let us consider the case that seems contrary to this assertion : $rankD = M$, which implies $\det\left(D^T D\right) \neq 0$. This creates an impression that we can find $\mathbf{I}^*(s)$ and $\mathbf{X}^*(s)$ from (3.53):

$$\mathbf{I}^*(s) = \left(D^T D\right)^{-1}\left[-D^T C\mathbf{X}^*(s) + D^T\mathbf{Y}_d(s)\right],$$
$$\left[sI - A + B\left(D^T D\right)^{-1} D^T C\right]\mathbf{X}^*(s) = B\left(D^T D\right)^{-1} D^T\mathbf{Y}_d(s) + \mathbf{X}_0{}^*.$$

In order for these solutions to satisfy

$$\mathbf{Y}^*(s) = C\mathbf{X}^*(s) + D\mathbf{I}^*(s) = \mathbf{Y}_d(s)$$

the following condition should hold:

$$D\left(D^T D\right)^{-1} D^T = I_N.$$

In view of $rankD = M < N$, we accept $D \in \mathfrak{R}^{N\times M}$ in the following form:

$$D = \begin{bmatrix} \widetilde{D} \\ O \end{bmatrix} \in \mathfrak{R}^{N\times M}, \widetilde{D} \in \mathfrak{R}^{M\times M}, \ rank\widetilde{D} = M, \ O = O_{(N-M),M},$$

so that the condition $D\left(D^T D\right)^{-1} D^T = I_N$ reads

$$D\left(D^T D\right)^{-1} D^T = \begin{bmatrix} \widetilde{D} \\ O \end{bmatrix}\left\{\left(\begin{bmatrix} \widetilde{D} \\ O \end{bmatrix}^T \begin{bmatrix} \widetilde{D} \\ O \end{bmatrix}\right)^{-1}\right\}\begin{bmatrix} \widetilde{D} \\ O \end{bmatrix}^T =$$

$$= \begin{bmatrix} \widetilde{D} \\ O \end{bmatrix}\left(\widetilde{D}^T\widetilde{D}\right)^{-1}\begin{bmatrix} \widetilde{D} \\ O \end{bmatrix}^T = \begin{bmatrix} \widetilde{D}\left(\widetilde{D}^T\widetilde{D}\right)^{-1}\widetilde{D}^T & O_{M,(N-M)} \\ O_{(N-M),M} & O_{(N-M),(N-M)} \end{bmatrix} \neq I_N.$$

The condition $D\left(D^TD\right)^{-1}D^T = I_N$ *is unrealizable. Hence, if* $N > M$ *then Equations (3.51) and (3.52), or equivalently (3.53), do not have a solution. If the number* N *of the output variables, i.e., the dimension* N *of the output vector, is bigger than the number* M *of the input variables, i.e., the dimension* M *of the input vector, then there is not a nominal functional vector pair* $[\mathbf{I}^*(.), \mathbf{X}^*(.)]$ *for the ISO system (2.17) and (2.18) relative to its desired response* $\mathbf{Y}_d(.)$.

In general, the following should hold if $N \geq M$:

$$\underbrace{\begin{bmatrix} -B & sI-A \\ D & C \end{bmatrix}}_{(N+n)\text{x}(M+n)} \underbrace{\left\{\begin{bmatrix} -B & sI-A \\ D & C \end{bmatrix}^T \begin{bmatrix} -B & sI-A \\ D & C \end{bmatrix}\right\}^{-1}}_{(M+n)\text{x}(M+n)} \bullet$$

$$\bullet \underbrace{\begin{bmatrix} -B & sI-A \\ D & C \end{bmatrix}^T}_{(M+n)\text{x}(N+n)} = I_{N+n},\ \forall s \in \mathfrak{C},.$$

for the existence of a solution $[\mathbf{I}^*(t), \mathbf{X}^*(t)]$ *to (3.51) and (3.52), i.e., to (3.53), which is impossible.*

Case 61 $N \leq M$

If $N \leq M$ *and*

$$rank\begin{bmatrix} -B & sI-A \\ D & C \end{bmatrix} = N + n \leq M + n$$

for all complex numbers s for which $\det(sI - A) \neq 0$,

then

$$\det\left\{\begin{bmatrix} -B & sI-A \\ D & C \end{bmatrix}\begin{bmatrix} -B & sI-A \\ D & C \end{bmatrix}^T\right\} \neq 0.$$

Equations (3.51) and (3.52), or equivalently (3.53), have the solution determined by

$$\begin{bmatrix} \mathbf{I}^*(s) \\ \mathbf{X}^*(s) \end{bmatrix} = \begin{bmatrix} -B & sI-A \\ D & C \end{bmatrix}^T \bullet$$

$$\bullet \left\{\begin{bmatrix} -B & sI-A \\ D & C \end{bmatrix}\begin{bmatrix} -B & sI-A \\ D & C \end{bmatrix}^T\right\}^{-1}\begin{bmatrix} \mathbf{X}_0{}^* \\ \mathbf{Y}_d(s) \end{bmatrix}.$$

Case 62 $N = M$
If $N = M$ *and*

$$rank\begin{bmatrix} -B & sI-A \\ D & C \end{bmatrix} = N + n = M + n$$

$$\textit{for all complex numbers } s \textit{ for which } \det(sI - A) \neq 0,$$

then Equations (3.51) and (3.52), or equivalently (3.53), have the unique solution determined by

$$\begin{bmatrix} \mathbf{I}^*(s) \\ \mathbf{X}^*(s) \end{bmatrix} = \begin{bmatrix} -B & sI-A \\ D & C \end{bmatrix}^{-1} \begin{bmatrix} \mathbf{X}_0{}^* \\ \mathbf{Y}_d(s) \end{bmatrix}.$$

Conclusion 63 *In order to exist a nominal functional vector pair*

$$[\mathbf{I}_N(.), \mathbf{X}_N(.)]$$

for system (2.17) and (2.18) relative to its desired response $\mathbf{Y}_d(.)$ *it is necessary and sufficient that the conditions of Case 61 hold (see [4, Equations (2.73)-(2.76) on p. 115]). Then, the functional vector pair* $[\mathbf{I}_N(.), \mathbf{X}_N(.)]$ *is nominal relative to the desired response* $\mathbf{Y}_d(.)$ *of system (2.17) and (2.18).*

This resolves completely both the problem of the existence of a nominal functional vector pair $[\mathbf{I}_N(.), \mathbf{X}_N(.)]$ for the *ISO* system (2.17) and (2.18) relative to its desired response $\mathbf{Y}_d(.)$, and the problem of the realizability of $\mathbf{Y}_d(.)$. The conditions show that the realizability of $\mathbf{Y}_d(.)$ is independent of the characteristics of $\mathbf{Y}_d(.)$, and that it is determined only by the system parameters, i.e., by the system realization (A, B, C, D).

Note 64 *We will treat in the sequel only ISO systems that satisfy the conditions of Case 61 for the realizability of their desired output vector functions* $\mathbf{Y}_d(.)$.

3.4.4 *IIO* systems

The application of Definition 49 (Subsection 3.4.1) to the *IIO* system (2.22) (in Section 2.3) leads to the following:

Definition 65 *A functional vector pair* $[\mathbf{I}^*(.), \mathbf{R}^*(.)]$ *is* ***nominal for the IIO system (2.22) relative to its desired response*** $\mathbf{Y}_d(.)$, *which is denoted by* $[\mathbf{I}_N(.), \mathbf{R}_N(.)]$, *if and only if* $[\mathbf{I}(.), \mathbf{R}(.)] = [\mathbf{I}^*(.), \mathbf{R}^*(.)]$ *ensures* *that the corresponding real response* $\mathbf{Y}(.) = \mathbf{Y}^*(.)$ *of the system obeys* $\mathbf{Y}^*(t) = \mathbf{Y}_d(t)$ *all the time as soon as* $\mathbf{Y}_0^{\nu-1} = \mathbf{Y}_{d0}^{\nu-1}$,

$$[\mathbf{I}^*(.), \mathbf{R}^*(.)] = [\mathbf{I}_N(.), \mathbf{R}_N(.)] \Longleftrightarrow$$
$$\Longleftrightarrow \left\langle \mathbf{Y}_0^{\nu-1} = \mathbf{Y}_{d0}^{\nu-1} \Longrightarrow \mathbf{Y}^*(t) = \mathbf{Y}_d(t),\ \forall t \in \mathfrak{T}_0 \right\rangle.$$

Let

$$\mathbf{w}_1(s) = \sum_{k=0}^{k=\beta\leq\alpha} P_k \left[\sum_{i=1}^{i=k} s^{k-i} \mathbf{I}_0^{*(i-1)} \right] - \sum_{k=0}^{k=\alpha} Q_k \left[\sum_{i=1}^{i=k} s^{k-i} \mathbf{R}_0^{*(i-1)} \right], \tag{3.55}$$

$$\mathbf{w}_2(s) = \left\{ \begin{array}{c} \sum_{k=0}^{k=\nu} E_k \left[s^k \mathbf{Y}_d(s) - \sum_{i=1}^{i=k} s^{k-i} \mathbf{Y}_{d0}^{(i-1)} \right] + \\ + \sum_{k=0}^{k=\alpha} R_k \left[\sum_{i=1}^{i=k} s^{k-i} \mathbf{R}_0^{*(i-1)} \right] + \\ + \sum_{k=0}^{k=\mu} T_k \left[\sum_{i=1}^{i=k} s^{k-i} \mathbf{I}_0^{*(i-1)} \right] \end{array} \right\}. \tag{3.56}$$

Theorem 66 *In order for a functional vector pair* $[\mathbf{I}^*(.), \mathbf{R}^*(.)]$ *to be nominal for the IIO system (2.22) relative to its desired response* $\mathbf{Y}_d(.)$,

$$[\mathbf{I}^*(.), \mathbf{R}^*(.)] = [\mathbf{I}_N(.), \mathbf{R}_N(.)],$$

it is necessary and sufficient that it obeys the following equations:

$$\sum_{k=0}^{k=\beta\leq\alpha} P_k \mathbf{I}^{*(k)}(t) - \sum_{k=0}^{k=\alpha} Q_k \mathbf{R}^{*(k)}(t) = P^{(\beta)} \mathbf{I}^{*\beta}(t) - Q^{(\alpha)} \mathbf{R}^{*\alpha}(t) = \mathbf{0},\ \forall t \in \mathfrak{T}_0, \tag{3.57}$$

$$\sum_{k=0}^{k=\alpha} R_k \mathbf{R}^{*(k)}(t) + \sum_{k=0}^{k=\mu} T_k \mathbf{I}^{*(k)}(t) = R^{(\alpha)} \mathbf{R}^{*\alpha}(t) + T^{(\mu)} \mathbf{I}^{*\mu}(t) =$$
$$= \sum_{k=0}^{k=\nu} E_k \mathbf{Y}_d^{(k)}(t) = E^{(\nu)} \mathbf{Y}_d^{\nu}(t),\ \forall t \in \mathfrak{T}_0, \tag{3.58}$$

or equivalently,

$$\left[\begin{array}{cc} P^{(\beta)} S_M^{(\beta)}(s) & -Q^{(\alpha)} S_\rho^{(\alpha)}(s) \\ T^{(\mu)} S_M^{(\mu)}(s) & R^{(\alpha)} S_\rho^{(\alpha)}(s) \end{array} \right] \left[\begin{array}{c} \mathbf{I}^*(s) \\ \mathbf{R}^*(s) \end{array} \right] = \left[\begin{array}{c} \mathbf{w}_1(s) \\ \mathbf{w}_2(s) \end{array} \right]. \tag{3.59}$$

Proof. *Necessity*, Let $[\mathbf{I}^*(.), \mathbf{R}^*(.)]$ be a nominal functional (input and substate) vector pair for the *IIO* system (2.22) relative to its desired response $\mathbf{Y}_d(.)$. Hence, the system is in its desired regime relative to $\mathbf{Y}_d(.)$. Definition 65 shows that $[\mathbf{I}(.), \mathbf{R}(.)] = [\mathbf{I}^*(.), \mathbf{R}^*(.)]$ implies $\mathbf{Y}(.) = \mathbf{Y}_d(.)$. This and the *IIO* system model (2.22) yield the following equations:

$$Q^{(\alpha)}\mathbf{R}^{*^{\alpha}}(t) = P^{(\beta)}\mathbf{I}^{*^{\beta}}(t),\ \forall t \in \mathfrak{T}_0, \tag{3.60}$$

$$E^{(\nu)}\mathbf{Y}_d^{\nu}(t) = R^{(\alpha)}\mathbf{R}^{*^{\alpha}}(t) + T^{(\mu)}\mathbf{I}^{*^{\mu}}(t),\ \forall t \in \mathfrak{T}_0, \tag{3.61}$$

which can be easily set in the form of Equations (3.57) and (3.58). When we apply Laplace transform together with its properties to Equations (3.57) and (3.58) then they become

$$\begin{gathered}\left[P^{(\beta)}S_M^{(\beta)}(s)\right]\mathbf{I}^*(s) - \left[Q^{(\alpha)}S_\rho^{(\alpha)}(s)\right]\mathbf{R}^*(s) = \\ = \underbrace{\sum_{k=0}^{k=\beta\leq\alpha} P_k\left[\sum_{i=1}^{i=k} s^{k-i}\mathbf{I}_0^{*(i-1)}\right] - \sum_{k=0}^{k=\alpha} Q_k\left[\sum_{i=1}^{i=k} s^{k-i}\mathbf{R}_0^{*(i-1)}\right]}_{\mathbf{w}_1(s)\ (3.55)},\end{gathered} \tag{3.62}$$

$$\begin{gathered}\left[T^{(\mu)}S_M^{(\mu)}(s)\right]\mathbf{I}^*(s) + \left[R^{(\alpha)}S_\rho^{(\alpha)}(s)\right]\mathbf{R}^*(s) = \\ = \underbrace{\left\{\begin{array}{c} \sum_{k=0}^{k=\nu} E_k\left[s^k\mathbf{Y}_d(s) - \sum_{i=1}^{i=k} s^{k-i}\mathbf{Y}_{d0}^{(i-1)}\right] + \\ + \sum_{k=0}^{k=\alpha} R_k\left[\sum_{i=1}^{i=k} s^{k-i}\mathbf{R}_0^{*(i-1)}\right] + \\ + \sum_{k=0}^{k=\mu} T_k\left[\sum_{i=1}^{i=k} s^{k-i}\mathbf{I}_0^{*(i-1)}\right] \end{array}\right\}}_{\mathbf{w}_2(s)\ (3.56)},\end{gathered} \tag{3.63}$$

or equivalently (3.59).

Sufficiency. We accept that all the conditions of the theorem hold. We choose $[\mathbf{I}(.), \mathbf{R}(.)] = [\mathbf{I}^*(.), \mathbf{R}^*(.)]$. Equation (3.57) written in the form (3.60) shows that the pair $[\mathbf{I}^*(.), \mathbf{R}^*(.)]$ satisfies the first Equation (2.22). Besides, the second Equation (2.22) takes the following form:

$$E^{(\nu)}\mathbf{Y}^{\nu}(t) = R^{(\alpha)}\mathbf{R}^{*^{\alpha}}(t) + T^{(\mu)}\mathbf{I}^{*^{\mu}}(t),\ \forall t \in \mathfrak{T}_0.$$

It, subtracted from (3.58), implies

$$E^{(\nu)}\left[\mathbf{Y}_d^{\nu}(t) - \mathbf{Y}^{\nu}(t)\right] = \mathbf{0}_N,\ \forall t \in \mathfrak{T}_0,$$

i.e.,

$$E^{(\nu)}\varepsilon^{\nu}(t) = \mathbf{0}_N, \ \forall t \in \mathfrak{T}_0.$$

The solution of this time-invariant linear differential Equation is trivial, $\varepsilon^{\nu}(t) \equiv \mathbf{Y}_d^{\nu}(t) - \mathbf{Y}^{\nu}(t) \equiv \mathbf{0}_{(\nu+1)N}$, because all initial conditions are equal to zero, $\varepsilon_0^{\nu} = \mathbf{Y}_{d0}^{\nu-1} - \mathbf{Y}_0^{\nu-1} = \mathbf{0}_{(\nu+1)N}$, due to the conditions of the theorem and Definition 65, $\mathbf{Y}_0^{\nu-1} = \mathbf{Y}_{d0}^{\nu-1}$. This completes the proof ■

Let us consider the existence of the solutions of Equations (3.62) and (3.63), i.e., of (3.59). There are $(M+\rho)$ unknown variables and $(N+\rho)$ Equations. The unknown variables are the entries of $\mathbf{I}^*(s) \in \mathfrak{C}^M$ and of $\mathbf{R}^*(s) \in \mathfrak{C}^{\rho}$.

Case 67 $N > M$

If $N > M$, then Equations (3.62) and (3.63), or equivalently (3.59), do not have a solution for the same reasons explained in Case 60.

Case 68 $N \leq M$

If $N \leq M$ and

$$rank \begin{bmatrix} P^{(\beta)}S_M^{(\beta)}(s) & -Q^{(\alpha)}S_\rho^{(\alpha)}(s) \\ T^{(\mu)}S_M^{(\mu)}(s) & R^{(\alpha)}S_\rho^{(\alpha)}(s) \end{bmatrix} = N + \rho \leq M + \rho$$

for almost all complex numbers s, then

$$\det \left\{ \begin{array}{c} \begin{matrix} P^{(\beta)}S_M^{(\beta)}(s) & -Q^{(\alpha)}S_\rho^{(\alpha)}(s) \\ T^{(\mu)}S_M^{(\mu)}(s) & R^{(\alpha)}S_\rho^{(\alpha)}(s) \end{matrix} \bullet \\ \bullet \begin{bmatrix} P^{(\beta)}S_M^{(\beta)}(s) & -Q^{(\alpha)}S_\rho^{(\alpha)}(s) \\ T^{(\mu)}S_M^{(\mu)}(s) & R^{(\alpha)}S_\rho^{(\alpha)}(s) \end{bmatrix}^T \end{array} \right\} \neq 0$$

for almost all complex numbers s.

There is a solution determined by

$$\begin{bmatrix} \mathbf{I}^*(s) \\ \mathbf{R}^*(s) \end{bmatrix} = \begin{bmatrix} P^{(\beta)}S_M^{(\beta)}(s) & -Q^{(\alpha)}S_\rho^{(\alpha)}(s) \\ T^{(\mu)}S_M^{(\mu)}(s) & R^{(\alpha)}S_\rho^{(\alpha)}(s) \end{bmatrix}^T \bullet$$

$$\bullet \left\{ \begin{array}{c} \begin{bmatrix} P^{(\beta)}S_M^{(\beta)}(s) & -Q^{(\alpha)}S_\rho^{(\alpha)}(s) \\ T^{(\mu)}S_M^{(\mu)}(s) & R^{(\alpha)}S_\rho^{(\alpha)}(s) \end{bmatrix} \bullet \\ \bullet \begin{bmatrix} P^{(\beta)}S_M^{(\beta)}(s) & -Q^{(\alpha)}S_\rho^{(\alpha)}(s) \\ T^{(\mu)}S_M^{(\mu)}(s) & R^{(\alpha)}S_\rho^{(\alpha)}(s) \end{bmatrix}^T \end{array} \right\}^{-1} \begin{bmatrix} \mathbf{w}_1(s) \\ \mathbf{w}_2(s) \end{bmatrix},$$

Case 69 $N = M$

If $N = M$ *and*

$$rank \begin{bmatrix} P^{(\beta)} S_M^{(\beta)}(s) & -Q^{(\alpha)} S_\rho^{(\alpha)}(s) \\ T^{(\mu)} S_M^{(\mu)}(s) & R^{(\alpha)} S_\rho^{(\alpha)}(s) \end{bmatrix} = N + \rho = M + \rho$$

for almost all complex numbers s.

then Equations (3.62) and (3.63), or equivalently (3.59), have the unique solution determined by

$$\begin{bmatrix} \mathbf{I}^*(s) \\ \mathbf{R}^*(s) \end{bmatrix} = \begin{bmatrix} P^{(\beta)} S_M^{(\beta)}(s) & -Q^{(\alpha)} S_\rho^{(\alpha)}(s) \\ T^{(\mu)} S_M^{(\mu)}(s) & R^{(\alpha)} S_\rho^{(\alpha)}(s) \end{bmatrix}^{-1} \begin{bmatrix} \mathbf{w}_1(s) \\ \mathbf{w}_2(s) \end{bmatrix}.$$

We have solved the problem of the existence of a nominal functional vector pair $[\mathbf{I}_N(.), \mathbf{R}_N(.)]$ for the *IIO* system (2.22) relative to its desired response $\mathbf{Y}_d(.)$, $[\mathbf{I}^*(.), \mathbf{R}^*(.)] = [\mathbf{I}_N(.), \mathbf{R}_N(.)]$, and the problem of the realizability of $\mathbf{Y}_d(.)$. The realizability of $\mathbf{Y}_d(.)$ is independent of the characteristics of $\mathbf{Y}_d(.)$. It is determined only by the system parameters.

Note 70 *We will treat in the sequel only IIO systems that satisfy the conditions of Case 68 for the realizability of their desired output vector functions* $\mathbf{Y}_d(.)$.

3.5 Deviations and mathematical models

3.5.1 Introduction

A nominal regime can take place only under the special conditions. Otherwise, the system is in *a nonnominal regime* called also *a perturbed regime* that is more probable than the nominal one. This permits us to call a nonnominal regime also *a real regime*. It poses the problem of analysis of the relationship between behaviors of the system in real regimes and its nominal behavior (its behavior in a nominal regime).

We use *the output deviation (vector)* $\mathbf{y}$, (3.64), and *the output error (vector)* ε (3.65) [already defined in (3.48), Subsection 3.4.2], of the real behavior from the nominal one in order to measure their distance at every moment,

$$\textit{the output deviation vector: } \mathbf{y} = \mathbf{Y} - \mathbf{Y}_d, \tag{3.64}$$

$$\textit{the output error vector: } \varepsilon = \mathbf{Y_d} - \mathbf{Y}, \tag{3.65}$$

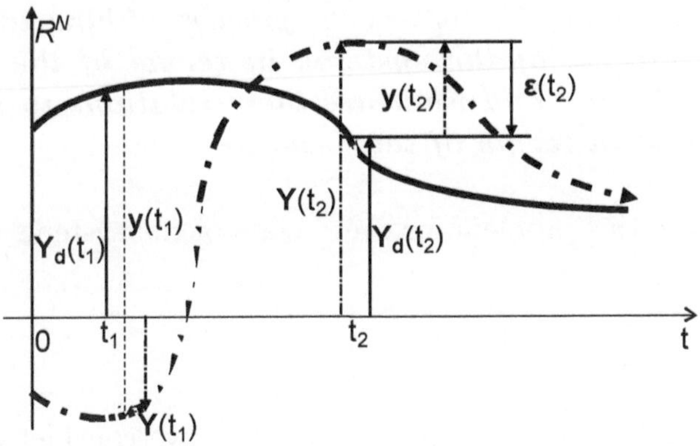

Figure 3.1: Lyapunov coordinate transformations. In this figure e stands for ε.

so that

$$\varepsilon = -\mathbf{y}. \tag{3.66}$$

They are *Lyapunov coordinate transformations* because A. M. Lyapunov [88] was the first to relate the coordinates of a real *motion* to the coordinates of the nominal (i.e., desired) *motion* of the system; Fig. 3.1.

Comment 71 *The preceding equations show the equivalence among*

$$\varepsilon = \mathbf{0}_N, \ \mathbf{y} = \mathbf{0}_N \ \text{ and } \mathbf{Y} = \mathbf{Y}_d, \tag{3.67}$$

$$\mathbf{Y} = \mathbf{Y}_d \Longleftrightarrow \varepsilon = \mathbf{0}_N \Longleftrightarrow \mathbf{y} = \mathbf{0}_N. \tag{3.68}$$

The zero output error vector $\varepsilon = \mathbf{0}_N$ *and the zero output deviation vector* $\mathbf{y} = \mathbf{0}_N$ *correspond to the total desired output vector* $\mathbf{Y}_d$.

Conclusion 72 *We can reduce the study of the properties of the desired (nominal) total response [of the desired (nominal) total movement] to the study of the same properties of the zero deviation (of the zero error) vector.*

Lyapunov coordinate transformations hold also for other variables,

$$\mathbf{d} = \mathbf{D} - \mathbf{D}_N, \tag{3.69}$$

$$\mathbf{i} = \mathbf{I} - \mathbf{I}_N, \tag{3.70}$$

$$\mathbf{r} = \mathbf{R} - \mathbf{R}_N, \tag{3.71}$$

$$\mathbf{x} = \mathbf{X} - \mathbf{X}_N. \tag{3.72}$$

These equations define them in general.

Problem 73 *Conclusion 72 imposes the problem of* ***how to determine mathematical models of the systems in terms of the deviations, what are their forms, and what are their relations to the original models expressed in terms of total coordinates.***

We will resolve this problem for every class of the systems separately in what follows.

3.5.2 *IO* systems

As emphasized in Note 57 (in Subsection 3.4.2), we consider only well designed IO systems so that the dimension M of the input vector $\mathbf{I}$ is not less than the dimension N of the output vector $\mathbf{Y}$.

The IO system model expressed in total values of the variables has the form (2.1) (in Section 2.1) in an arbitrary regime, i.e., in a real regime,

$$\sum_{k=0}^{k=\nu} A_k \mathbf{Y}^{(k)}(t) = \sum_{k=0}^{k=\mu} B_k \mathbf{I}^{(k)}(t),\ det A_\nu \neq 0,\ \forall t \in \mathfrak{T}_0,\ \nu \geq 1,\ 0 \leq \mu \leq \nu. \tag{3.73}$$

If the system is in its desired (nominal) regime, then (3.73) becomes (3.74) due to (3.43) (Theorem 56 in Subsection 3.4.2):

$$\sum_{k=0}^{k=\nu} A_k \mathbf{Y}_d^{(k)}(t) = \sum_{k=0}^{k=\mu} B_k \mathbf{I}_N^{(k)}(t),\ \forall t \in \mathfrak{T}_0. \tag{3.74}$$

Assumption 74 *The desired output response of the IO system (2.1), i.e., (3.73), is realizable.*

We accept this assumption and we subtract (3.74) from (3.73),

$$\sum_{k=0}^{k=\nu} A_k \left[\mathbf{Y}^{(k)}(t) - \mathbf{Y}_d^{(k)}(t)\right] = \sum_{k=0}^{k=\mu} B_k \left[\mathbf{I}^{(k)}(t) - \mathbf{I}_N^{(k)}(t)\right],\ \forall t \in \mathfrak{T}_0. \tag{3.75}$$

We apply the deviations $\mathbf{y} = \mathbf{Y} - \mathbf{Y}_d$ (3.64) and $\mathbf{i} = \mathbf{I} - \mathbf{I}_N$ (3.70), (in Section 3.5.1), to (3.75):

$$\sum_{k=0}^{k=\nu} A_k \mathbf{y}^{(k)}(t) = \sum_{k=0}^{k=\mu} B_k \mathbf{i}^{(k)}(t),\ \forall t \in \mathfrak{T}_0,\ \nu \geq 1,\ 0 \leq \mu \leq \nu. \tag{3.76}$$

Note 75 *This is the IO system model expressed in terms of the deviations of all variables. It has exactly the same form, the same order, and the same matrices as the system model expressed in total values of the variables (2.1), i.e., (3.73). They possess the same characteristics and properties by noting once more that* $\mathbf{y} = \mathbf{0}$ *represents* $\mathbf{Y} = \mathbf{Y}_d$. *For example, they have the same transfer function matrices, and the stability properties of* $\mathbf{y} = \mathbf{0}$ *of (3.76) are simultaneously the same stability properties of* $\mathbf{Y}_d(t)$ *of (2.1), i.e., of (3.73). Therefore, we will continue with the IO system description in terms of the deviations (3.76).*

The compact form of (3.76) reads, in view of (2.2), (2.3), and (2.4) (in Section 2.1), as follows:

$$A^{(\nu)}\mathbf{y}^{\nu}(t) = B^{(\mu)}\mathbf{i}^{\mu}(t),\ \forall t \in \mathfrak{T}_0,\ 0 \leq \mu \leq \nu. \tag{3.77}$$

This is the compact *IO* system description in terms of the deviations, i.e., it is the deviation form of (2.7) (in Section 2.1). We will use mainly (3.77).

3.5.3 *ISO* systems

We continue to consider only the *ISO* systems having the number M of input variables (i.e., the dimension of the input vector $\mathbf{I}$) not less than the number N of the output variables (i.e., the dimension of the output vector $\mathbf{Y}$), $N \leq M$, Note 64 (in Subsection 3.4.3).

Equations **(2.17)**, **(2.18)**, (Section 2.2), i.e., **(3.78)** (3.79),

$$\frac{d\mathbf{X}(t)}{dt} = A\mathbf{X}(t) + B\mathbf{I}(t),\ \forall t \in \mathfrak{T}_0, \tag{3.78}$$

$$\mathbf{Y}(t) = C\mathbf{X}(t) + D\mathbf{I}(t),\ \forall t \in \mathfrak{T}_0, \tag{3.79}$$

represent the mathematical description of the *ISO* system in an arbitrary regime, i.e., in a real regime. Its description in a nominal regime relative to $\mathbf{Y}_d(.)$ follows from Theorem 59, Equations (3.51), (3.52), (in Subsection 3.4.3),

$$\frac{d\mathbf{X}_N(t)}{dt} = A\mathbf{X}_N(t) + B\mathbf{I}_N(t),\ \forall t \in \mathfrak{T}_0, \tag{3.80}$$

$$\mathbf{Y}_d(t) = C\mathbf{X}_N(t) + D\mathbf{I}_N(t),\ \forall t \in \mathfrak{T}_0. \tag{3.81}$$

Assumption 76 *The desired output response of the ISO system (2.17) and (2.18), i.e., (3.78) and (3.79), is realizable.*

We adopt this assumption to hold. It permits us to use (3.80) and (3.81), which we subtract from (3.78) and (3.79),

$$\frac{d\left[\mathbf{X}(t)-\mathbf{X}_N(t)\right]}{dt}=A\left[\mathbf{X}(t)-\mathbf{X}_N(t)\right]+B\left[\mathbf{I}(t)-\mathbf{I}_N(t)\right], \tag{3.82}$$

$$\mathbf{Y}(t)-\mathbf{Y}_d(t)=C\left[\mathbf{X}(t)-\mathbf{X}_N(t)\right]+D\left[\mathbf{I}(t)-\mathbf{I}_N(t)\right],\ \forall t\in\mathfrak{T}_0. \tag{3.83}$$

The application of $\mathbf{x}=\mathbf{X}-\mathbf{X}_N$ (3.72), $\mathbf{y}=\mathbf{Y}-\mathbf{Y}_d$ (3.64) and $\mathbf{i}=\mathbf{I}-\mathbf{I}_N$ (3.70), (in Subsection 3.5.1), to (3.82) and (3.83) yields the mathematical model in terms of the deviations of the *ISO* system (2.17) and (2.18), i.e., (3.78) and (3.79),

$$\frac{d\mathbf{x}(t)}{dt}=A\mathbf{x}(t)+B\mathbf{i}(t),\ \forall t\in\mathfrak{T}_0, \tag{3.84}$$

$$\mathbf{y}(t)=C\mathbf{x}(t)+D\mathbf{i}(t),\ \forall t\in\mathfrak{T}_0. \tag{3.85}$$

Note 77 *This mathematical model in terms of the deviations of all variables of the ISO system has exactly the same form, the same order, and the same matrices as the system model expressed in total values of the variables (2.17) and (2.18), i.e., (3.78) and (3.79). They possess the same characteristics and properties by noting once more that* $\mathbf{x}=\mathbf{0}$ *replaces* $\mathbf{X}=\mathbf{X}_N$ *and* $\mathbf{y}=\mathbf{0}$ *stands for* $\mathbf{Y}=\mathbf{Y}_d$. *For example, they have the same transfer function matrices; the stability properties of* $\mathbf{x}=\mathbf{0}$ *of (3.84) are the same as of* $\mathbf{X}=\mathbf{X}_N$ *of (2.17), i.e., of (3.78), and the tracking properties of* $\mathbf{y}=\mathbf{0}$ *of (3.84) and (3.85) are simultaneously the same tracking properties of* $\mathbf{Y}_d(t)$ *of (2.17) and (2.18), i.e., of (3.78) and (3.79). Therefore, we will continue with the ISO system description in terms of the deviations (3.84) and (3.85).*

3.5.4 *IIO* systems

We assume in the sequel (Note 70, Subsection 3.4.4) that the *IIO* system (2.22), (in Section 2.3), is well designed so that the dimension N of the output vector $\mathbf{Y}$ does not exceed the dimension M of the input vector $\mathbf{I}$, $N\leq M$, and that Condition 38 holds (Section 2.3).

The mathematical model (2.22), which is rewritten here as (3.86), of the *IIO* system in terms of the total values of the variables holds for an arbitrary regime,

$$\sum_{k=0}^{k=\alpha} Q_k \mathbf{R}^{(k)}(t) = \sum_{k=0}^{k=\beta} P_k \mathbf{I}^{(k)}(t),\ detQ_\alpha \neq 0,\ \forall t \in \mathfrak{T}_0,\ \alpha \geq 1, \alpha \geq \beta \geq 0,$$
$$\sum_{k=0}^{k=\nu} E_k \mathbf{Y}^{(k)}(t) = \sum_{k=0}^{k=\alpha} R_k \mathbf{R}^{(k)}(t) + \sum_{k=0}^{k=\mu} T_k \mathbf{I}^{(k)}(t),\ detE_\nu \neq 0,$$
$$\forall t \in \mathfrak{T}_0,\ \nu \geq 0,\ \mu \geq 0. \tag{3.86}$$

Its form in the nominal regime reads [Theorem 66, Equations (3.57) and (3.58) in Subsection 3.4.4]:

$$\sum_{k=0}^{k=\alpha} Q_k \mathbf{R}_N^{(k)}(t) = \sum_{k=0}^{k=\beta} P_k \mathbf{I}_N^{(k)}(t),\ \forall t \in \mathfrak{T}_0, \tag{3.87}$$

$$\sum_{k=0}^{k=\nu} E_k \mathbf{Y}_d^{(k)}(t) = \sum_{k=0}^{k=\alpha} R_k \mathbf{R}_N^{(k)}(t) + \sum_{k=0}^{k=\mu} T_k \mathbf{I}_N^{(k)}(t),\ \forall t \in \mathfrak{T}_0. \tag{3.88}$$

Assumption 78 *The desired output response of the IIO system (2.22), i.e., (3.86), is realizable.*

This assumption enables us to reduce the mathematical model of the *IIO* system (2.22), i.e., (3.86), given in terms of the total coordinates to that in terms of the deviations. The latter follows from the former when we subtract (3.87) and (3.88) from (3.86),

$$\sum_{k=0}^{k=\alpha} Q_k \left[\mathbf{R}^{(k)}(t) - \mathbf{R}_N^{(k)}(t)\right] = \sum_{k=0}^{k=\beta} P_k \left[\mathbf{I}^{(k)}(t) - \mathbf{I}_N^{(k)}(t)\right],$$
$$\sum_{k=0}^{k=\nu} E_k \left[\mathbf{Y}^{(k)}(t) - \mathbf{Y}_d^{(k)}(t)\right] = \sum_{k=0}^{k=\alpha} R_k \left[\mathbf{R}^{(k)}(t) - \mathbf{R}_N^{(k)}(t)\right] +$$
$$+ \sum_{k=0}^{k=\mu} T_k \left[\mathbf{I}^{(k)}(t) - \mathbf{I}_N^{(k)}(t)\right],\ \forall t \in \mathfrak{T}_0,$$

and when we apply $\mathbf{r} = \mathbf{R} - \mathbf{R}_N$ (3.71), $\mathbf{y} = \mathbf{Y} - \mathbf{Y}_d$ (3.64) and $\mathbf{i} = \mathbf{I} - \mathbf{I}_N$ (3.70), (in Section 3.5.1),

$$\sum_{k=0}^{k=\alpha} Q_k \mathbf{r}^{(k)}(t) = \sum_{k=0}^{k=\beta\leq\alpha} P_k \mathbf{i}^{(k)}(t),\ detQ_\alpha \neq 0,\ \forall t \in \mathfrak{T}_0,\ \alpha \geq 1,\ \alpha \geq \beta \geq 0,$$
$$\sum_{k=0}^{k=\nu} E_k \mathbf{y}^{(k)}(t) = \sum_{k=0}^{k=\alpha} R_k \mathbf{r}^{(k)}(t) + \sum_{k=0}^{k=\mu\leq\nu} T_k \mathbf{i}^{(k)}(t),\ \forall t \in \mathfrak{T}_0,\ \nu \geq 0,\ \mu \geq 0. \tag{3.89}$$

Note 79 *This is the IIO system model in terms of the deviations of all variables. It has exactly the same form, order, and matrices as the system model expressed in total values of the variables (2.22), i.e., (3.86). All of them possess the same characteristics and properties by noting again that* $\mathbf{y} = \mathbf{0}$ *represents* $\mathbf{Y} = \mathbf{Y}_d$. *They have the same transfer function matrices; the stability properties of* $\mathbf{r} = \mathbf{0}$ *of (3.89) are the same as the stability properties of* $\mathbf{R} = \mathbf{R}_N$ *of (3.87), and the tracking properties of* $\mathbf{y} = \mathbf{0}$ *of (3.89) are simultaneously the same tracking properties of* $\mathbf{Y}_d(t)$ *of (2.22), i.e., (3.86). Therefore, we will continue with the IIO system description in terms of the deviations (3.89).*

The compact form of (3.89) reads, in view of (2.24), (in Section 2.3), as follows:

$$Q^{(\alpha)}\mathbf{r}^{\alpha}(t) = P^{(\beta)}\mathbf{i}^{\beta}(t),\ detQ_\alpha \neq 0,\ \forall t \in \mathfrak{T}_0,\ \alpha \geq 1,\ \alpha \geq \beta \geq 0,$$
$$E^{(\nu)}\mathbf{y}^{\nu}(t) = R^{(\alpha)}\mathbf{r}^{\alpha}(t) + T^{(\mu)}\mathbf{i}^{\mu}(t),\ detE_\nu \neq 0,$$
$$\forall t \in \mathfrak{T}_0,\ \nu \geq 0,\ \mu \geq 0. \tag{3.90}$$

This follows also from (2.35), (in Section 2.3), when it is written for the nominal regime, due to $\mathbf{r} = \mathbf{R} - \mathbf{R}_N$ (3.71), $\mathbf{y} = \mathbf{Y} - \mathbf{Y}_d$ (3.64) and $\mathbf{i} = \mathbf{I} - \mathbf{I}_N$ (3.70).

Note 80 *The time evolution of* $\mathbf{r}^{\alpha-1}(t;\mathbf{r}_0^{\alpha-1};\mathbf{i})$ *expresses explicitly the state evolution of the IIO system (3.90). The time evolution of both* $\mathbf{r}^{\alpha-1}(t;\mathbf{r}_0^{\alpha-1};\mathbf{i})$ *and* $\mathbf{y}^{\nu-1}(t;\mathbf{r}_0^{\alpha-1};\mathbf{y}_0^{\nu-1};\mathbf{i})$, *i.e., of*

$$\left[\mathbf{r}^{\alpha-1^T}(t;\mathbf{r}_0^{\alpha-1};\mathbf{i}) \vdots \mathbf{y}^{\nu-1^T}(t;\mathbf{r}_0^{\alpha-1};\mathbf{y}_0^{\nu-1};\mathbf{i})\right]^T,$$

expresses explicitly the complete (full) internal and output dynamics - IOD (for short: the full dynamics, or shortest: dynamics - D) of the IIO system (3.90). The time evolution of $\mathbf{y}^{\nu-1}(t;\mathbf{r}_0^{\alpha-1};\mathbf{y}_0^{\nu-1};\mathbf{i})$ *expresses explicitly the output dynamics (OD) of the system. The time evolution of* $\mathbf{y}(t;\mathbf{r}_0^{\alpha-1};\mathbf{y}_0^{\nu-1};\mathbf{i})$ *is the output response (OR) of the system.*

From (2.22), or from (3.89):

- together with (2.35), (in Section 2.3), we find the *PMD* system form in terms of deviations:

$$Q^{(\alpha)}\mathbf{r}^{\alpha}(t) = P^{(\beta)}\mathbf{i}^{\beta}(t),\ \forall t \in \mathfrak{T}_0,$$
$$\mathbf{y}(t) = R^{(\alpha)}\mathbf{r}^{\alpha}(t) + T^{(\mu)}\mathbf{i}^{\mu}(t),\ \forall t \in \mathfrak{T}_0. \tag{3.91}$$

- together with (2.37), (in Section 2.3), we get directly the compact description in terms of the deviations of the Rosenbrock system (2.37),

$$\mathbf{x}^{(1)}(t) = A\mathbf{x}(t) + B\mathbf{i}(t),\ \forall t \in \mathfrak{T}_0,$$
$$\mathbf{y}(t) = C\mathbf{x}(t) + T^{(\mu)}\mathbf{i}^{\mu}(t),\ \forall t \in \mathfrak{T}_0, \tag{3.92}$$

- together with (2.39), (in Section 2.3), we find the description in terms of the deviations of the *IIDO* system,

$$\sum_{k=0}^{k=\alpha} Q_k\mathbf{r}^{(k)}(t) = \sum_{k=0}^{k=\beta} P_k\mathbf{i}^{(k)}(t),\ \forall t \in \mathfrak{T}_0.$$
$$\mathbf{y}(t) = R\mathbf{r}(t) + T\mathbf{i}(t),\ \forall t \in \mathfrak{T}_0. \tag{3.93}$$

or in the compact form,

$$Q^{(\alpha)}\mathbf{r}^{\alpha}(t) = P^{(\beta)}\mathbf{i}^{\beta}(t),\ \forall t \in \mathfrak{T}_0,$$
$$\mathbf{y}(t) = R\mathbf{r}(t) + T\mathbf{i}(t),\ \forall t \in \mathfrak{T}_0. \tag{3.94}$$

3.6 Stationary and nonstationary regimes

3.6.1 Introduction

Let us consider the system regimes relative to the following criterion: *the type of the system behavior.* The system internal dynamical (i.e., the state) behavior and its output dynamical behavior can be:

- after some moment constant, unchangeable, in *time*, which is its *steady*, or *stationary*, behavior,

- occasionally, or permanently, variable, nonconstant, which can be repeatable after some *time* interval, or nonrepeatable. If it is repeatable, then it is *periodic*. If it is not repeatable, it is *aperiodic* or also called *transient.* They are *nonstationary* system behaviors.

Lyapunov stability properties concern a closeness of system motions to a desired motion in general, or to a stationary vector in particular, which represents the system stationary behavior.

We will define more precisely stationary and nonstationary system behaviors for every class of the linear systems treated herein.

A singular point characterizes a stationary regime. The singular point can be a **stationary point** in a general case, or an **equilibrium point** in a special case. We are interested in their existence and in system behaviors around them.

3.6.2 *IO* systems

Definition 81 *The IO system (3.76), i.e., (3.77), (Subsection 3.5.2),*

a) *is in* ***a stationary (steady) regime since*** $\sigma \in \mathfrak{T}_0$ *(i.e.,* ***on*** $[\sigma, \infty[$*)* ***relative to*** $\mathbf{i}(.)$ *if and only if its response is constant all the time on* $[\sigma, \infty[$*, that is that*

$$\mathbf{y}(t;\mathbf{y}_0^{\nu-1};\mathbf{i}) = \mathbf{y}(\sigma;\mathbf{y}_0^{\nu-1};\mathbf{i}) = \mathbf{y}_\sigma = \mathbf{const.}, \forall t \in [\sigma, \infty[. \tag{3.95}$$

If and only if this holds for $\sigma = 0$ *then the system is in* ***a stationary (steady) regime relative to*** $\mathbf{i}(.)$.

b) *Otherwise, the system is in* ***a nonstationary regime relative to*** $\mathbf{i}(.)$*, i.e.,*

$$\forall \sigma \in \mathfrak{T}_0,\ \exists \tau \in]\sigma, \infty[\Longrightarrow \mathbf{y}(\tau;\mathbf{y}_\sigma^{\nu-1};\mathbf{i}) \neq \mathbf{y}_\sigma. \tag{3.96}$$

In a nonstationary regime, the system is in

- ***a periodic regime relative to*** $\mathbf{i}(.)$ *if and only if there is such* $\sigma \in \mathfrak{T}_0 \cap \mathfrak{R}^+$ *that*

$$\mathbf{y}(t+\sigma;\mathbf{y}_0^{\nu-1};\mathbf{i}) = \mathbf{y}(t;\mathbf{y}_0^{\nu-1};\mathbf{i}),\ \forall t \in \mathfrak{T}_0, \tag{3.97}$$

where such minimal σ *is called* ***the period relative to*** $\mathbf{i}(.)$ *of the periodic regime and it is denoted by* T,

$$T = min\left\{\sigma \in \mathfrak{T}_0 \cap \mathfrak{R}^+ :\ \mathbf{y}(t+\sigma;\mathbf{y}_0^{\nu-1};\mathbf{i}) = \mathbf{y}(t;\mathbf{y}_0^{\nu-1};\mathbf{i}),\ \forall t \in \mathfrak{T}_0\right\}, \tag{3.98}$$

- ***a transient (aperiodic) regime relative to*** $\mathbf{i}(.)$ *if and only if it is not in a periodic regime, i.e.,*

$$\exists t \in \mathfrak{T}_0,\ \forall \sigma \in \mathfrak{T}_0 \cap \mathfrak{R}^+,\ \exists k \in \{1, 2, ..., n, ..\} \Longrightarrow$$
$$\mathbf{y}(t+k\sigma;\mathbf{y}_0^{\nu-1};\mathbf{i}) \neq \mathbf{y}(t;\mathbf{y}_0^{\nu-1};\mathbf{i}). \tag{3.99}$$

The system can be in a stationary or in a nonstationary (either in a periodic or in a transient) regime either in a free regime, or in a forced regime.

Definition 82 *A vector* $\mathbf{y}^{*^{\nu-1}} \in \mathfrak{R}^{\nu N}$ *is* ***a stationary vector of the IO system (3.76) relative to*** $\mathbf{i}(.)$, $\mathbf{y}^{*^{\nu-1}} = \mathbf{y}^{*^{\nu-1}}(\mathbf{i})$, *if and only if:*

$$\mathbf{y}^{\nu-1}(t;\mathbf{y}^{*^{\nu-1}};\mathbf{i}) = \mathbf{y}^{*^{\nu-1}}(\mathbf{i}),\ \forall t \in \mathfrak{T}_0. \tag{3.100}$$

It is denoted by the subscript « s » :

$$\mathbf{y}^{*^{\nu-1}}(\mathbf{i}) = \mathbf{y}_s^{\nu-1}(\mathbf{i}).$$

This definition permits for a vector $\mathbf{y}^{*^{\nu-1}}(\mathbf{i})$ to be a stationary vector of the *IO* system (3.76) relative to one input vector function $\mathbf{i}(.) = \mathbf{i}_1(.)$, $\mathbf{y}^{*^{\nu-1}}(\mathbf{i}_1) = \mathbf{y}_s^{\nu-1}(\mathbf{i}_1)$, but not relative to another input vector function $\mathbf{i}(.) = \mathbf{i}_2(.)$, $\mathbf{y}_s^{\nu-1}(\mathbf{i}_1) \neq \mathbf{y}_s^{\nu-1}(\mathbf{i}_2)$.

Theorem 83 *In order for a vector* $\mathbf{y}^{*^{\nu-1}} \in \mathfrak{R}^{N\nu}$ *to be a stationary vector of the IO system (3.76) relative to* $\mathbf{i}(.)$, $\mathbf{y}^{*^{\nu-1}} = \mathbf{y}_s^{\nu-1}(\mathbf{i}) \in \mathfrak{R}^{N\nu}$, *it is necessary and sufficient that*

$$\mathbf{y}^{\nu-1}(0;\mathbf{y}^{*^{\nu-1}};\mathbf{i}) = \mathbf{y}_0^{\nu-1} = \mathbf{y}^{*^{\nu-1}} = \begin{bmatrix} \mathbf{y}^{*T} & \mathbf{0}_N^T & \mathbf{0}_N^T & \ldots & \mathbf{0}_N^T \end{bmatrix}^T \in \mathfrak{R}^{N\nu}, \tag{3.101}$$

and

$$A_0\mathbf{y}^* = B^\mu \mathbf{i}^\mu(t) = \sum_{k=0}^{k=\mu} B_k \mathbf{i}^{(k)}(t),\ \forall t \in \mathfrak{T}_0. \tag{3.102}$$

Proof. *Necessity.* Let $\mathbf{y}^{*^{\nu-1}} = \mathbf{y}_s^{\nu-1}(\mathbf{i}) \in \mathfrak{R}^{N\nu}$ be a stationary vector of the *IO* system (3.76) relative to $\mathbf{i}(.)$. Hence, (3.100) holds. By differentiating (3.100) $\nu-$times we verify

$$\mathbf{y}^{(k)}(t;\mathbf{y}^{*^{\nu-1}};\mathbf{i}) = \mathbf{0}_N,\ \forall k = 1,2,...,\nu,\ \forall t \in \mathfrak{T}_0 \Longrightarrow \tag{3.103}$$

$$\mathbf{y}^{(k)}(0;\mathbf{y}^{*^{\nu-1}};\mathbf{i}) = \mathbf{y}^{*(k)} = \mathbf{0}_N,\ \forall k = 1,2,...,\nu. \tag{3.104}$$

The result (3.104) and (3.100) imply (3.101). Besides, (3.103) and (3.76), i.e., (3.77), imply (3.102).

Sufficiency, Let (3.101) and (3.102) be valid. We subtract (3.102) from (3.76), and use

$$\Delta\mathbf{y}^{(k)}(t;\mathbf{y}^{*^{\nu-1}};\mathbf{i}) = \mathbf{y}^{(k)}(t;\mathbf{y}^{*^{\nu-1}};\mathbf{i}) - \mathbf{y}^{*(k)}(\mathbf{i}) =$$

$$= \left\{ \begin{array}{c} \mathbf{y}(t;\mathbf{y}^{*^{\nu-1}};\mathbf{i}) - \mathbf{y}^*(\mathbf{i}),\ k = 0, \\ \mathbf{y}^{(k)}(t;\mathbf{y}^{*^{\nu-1}};\mathbf{i}),\ \forall k = 1,2,...,\nu \end{array} \right\},\ \forall t \in \mathfrak{T}_0, \tag{3.105}$$

to obtain

$$\sum_{k=0}^{k=\nu} A_k \Delta \mathbf{y}^{(k)}(t) = \mathbf{0}.$$

This is the homogenous linear differential equation with the zero initial conditions due to (3.101) and (3.105). Its solution is trivial,

$$\Delta \mathbf{y}^{(k)}(t; \mathbf{y}^{*^{\nu-1}}; \mathbf{i}) = \mathbf{y}^{(k)}(t; \mathbf{y}^{*^{\nu-1}}; \mathbf{i}) - \mathbf{y}^{*(k)}(\mathbf{i}) = \mathbf{0},\ \forall t \in \mathfrak{T}_0,\ \forall k = 0, 1, .., \nu-1,$$

i.e.,

$$\mathbf{y}^{\nu-1}(t; \mathbf{y}^{*^{\nu-1}}; \mathbf{i}) = \mathbf{y}^{*^{\nu-1}}(\mathbf{i}),\ \forall t \in \mathfrak{T}_0.$$

This is (3.100), which completes the proof. ■

Note 84 *The condition (3.102) shows that the IO system (3.76), i.e., (3.77), can be in a stationary regime even in the case when the input vector is time-varying. For example:*

$$\sum_{k=0}^{k=\nu} A_k \mathbf{y}^{(k)}(t) = B_0 \mathbf{i}(t) + B_1 \mathbf{i}^{(1)}(t),\ B_1 = -B_0,$$

for $\mathbf{i}(t) = (e^t + 1)\, \mathbf{1}_M$, *where the vector* $\mathbf{1}_M = [1 \ \ 1 \ \ ... \ \ 1]^T \in \mathfrak{R}^M$, *the preceding differential equation reduces to*

$$\sum_{k=0}^{k=\nu} A_k \mathbf{y}^{(k)}(t) = (e^t + 1)\, B_0 \mathbf{1}_M - B_0 e^t \mathbf{1}_M = B_0 \mathbf{1}_M,$$

so that the condition (3.102) takes the following form:

$$A_0 \mathbf{y}^*(\mathbf{i}) = B_0 \mathbf{1}_M.$$

If A_0 *is nonsingular, then the system has a unique stationary vector determined by*

$$\mathbf{y}^*(\mathbf{i}) = \mathbf{y}_s(\mathbf{i}) = A_0^{-1} B_0 \mathbf{1}_M,$$
$$\mathbf{y}^{*^{\nu-1}}(\mathbf{i}) = \mathbf{y}_s^{\nu-1}(\mathbf{i}) = \left[\left(A_0^{-1} B_0 \mathbf{1}\right)^T \ \ \mathbf{0}_N^T \ \ \mathbf{0}_N^T \ \ ... \ \mathbf{0}_N^T \right]^T = \mathbf{const.}$$

in view of (3.101). However, if the matrix A_0 *is the zero matrix* O_N, $A_0 = O_N$, *then the system is in a stationary regime if and only if*

$$B_0 \mathbf{1}_M = \mathbf{0}_M.$$

Then, every vector $\mathbf{y}^{*^{\nu-1}} = \begin{bmatrix} \mathbf{y}^{*^T} & \mathbf{0}_N^T & \mathbf{0}_N^T & ... & \mathbf{0}_N^T \end{bmatrix}^T \in \mathfrak{R}^{N\nu}$ *is the stationary vector relative to the input vector* $\mathbf{i}(t)$. *The system rests in the initial output vector* $\mathbf{y}_0^{\nu-1} = \begin{bmatrix} \mathbf{y}_0^T & \mathbf{0}_N^T & \mathbf{0}_N^T & ... & \mathbf{0}_N^T \end{bmatrix}^T \in \mathfrak{R}^{N\nu}$ *all the time, i.e., the initial output vector* $\mathbf{y}_0^{\nu-1}$ *is the stationary vector relative to the input vector* $\mathbf{i}$,

$$A_0 = O_N \text{ and } B_0 \mathbf{1}^M = \mathbf{0}_M \Longrightarrow$$
$$\mathbf{y}_s^{\nu-1}(\mathbf{i}) = \mathbf{y}_0^{\nu-1} = \begin{bmatrix} \mathbf{y}_0^T & \mathbf{0}_N^T & \mathbf{0}_N^T & ... & \mathbf{0}_N^T \end{bmatrix}^T \in \mathfrak{R}^{N\nu}.$$

To be specific let $N = 3$, $M = 4$ *and*

$$A_0 = O_3 \text{ and } B_0 = -B_1 = \begin{bmatrix} 3 & 1 & -6 & 2 \\ 5 & -7 & -2 & 4 \\ 11 & 9 & -24 & 4 \end{bmatrix} \in \mathfrak{R}^{3\times 4} \Longrightarrow$$

$$B_0 \mathbf{1}_4 = \begin{bmatrix} 3 & 1 & -6 & 2 \\ 5 & -7 & -2 & 4 \\ 11 & 9 & -24 & 4 \end{bmatrix} \begin{bmatrix} 1 \\ 1 \\ 1 \\ 1 \end{bmatrix} = \begin{bmatrix} 0 \\ 0 \\ 0 \\ 0 \end{bmatrix} = \mathbf{0}_4.$$

The stationary vector relative to the input vector $\mathbf{i}(t)$ *is the initial output vector*

$$\mathbf{y}^{\nu-1}(\mathbf{i}) = \mathbf{y}_0^{\nu-1} = \begin{bmatrix} \mathbf{y}_0^T & \mathbf{0}_3^T & \mathbf{0}_3^T & ... & \mathbf{0}_3^T \end{bmatrix}^T, \ \forall \mathbf{y}_0 \in \mathfrak{R}^3, \ \mathbf{i}(t) = \left(e^t + 1\right) \mathbf{1}^4,$$

whatever is the initial output vector

$$\mathbf{y}_0^{\nu-1} = \begin{bmatrix} \mathbf{y}_0^T & \mathbf{0}_3^T & \mathbf{0}_3^T & ... & \mathbf{0}_3^T \end{bmatrix}^T \in \mathfrak{R}^{3\nu}.$$

Note 85 *If* $M \geq N$, $rank B_0 = N$ *and* $B_k = O$, $\forall k = 1, 2, ..., \mu$, *then* $\det(B_0 B_0^T) \neq 0$ *so that (3.102) becomes*

$$A_0 \mathbf{y}_s = B_0 \mathbf{i}(t), \ \forall t \in \mathfrak{T}_0. \tag{3.106}$$

It is easy to verify that

$$\mathbf{i}(t) = B_0^T (B_0 B_0^T)^{-1} A_0 \mathbf{y}_s = \mathbf{const.}, \ \forall t \in \mathfrak{T}_0,$$

obeys Equation (3.106). In this case, only a constant input vector can force the system to stay in a stationary regime.

Note 86 *The condition (3.102) shows that for the IO system (3.76), i.e., (3.77), to have the unique stationary vector* $\mathbf{y}_s(\mathbf{i})$ *relative to* $\mathbf{i}$, *it is necessary and sufficient that* A_0 *is nonsingular and*

$$A_0^{-1}B^{\mu}\mathbf{i}^{\mu}(t) = A_0^{-1}\sum_{k=0}^{k=\mu} B_k\mathbf{i}^{(k)}(t) = \mathbf{const}, \ \forall t \in \mathfrak{T}_0.$$

Then, the unique stationary vector relative to $\mathbf{i}$ *is given by*

$$\mathbf{y}_s(\mathbf{i}) = A_0^{-1}B^{\mu}\mathbf{i}^{\mu}(t) = A_0^{-1}\sum_{k=0}^{k=\mu} B_k\mathbf{i}^{(k)}(t).$$

3.6.3 *ISO* systems

Definition 87 *The ISO system (3.84) and (3.85), (in Subsection 3.5.3),*

a) *is in* ***a stationary (steady) regime since*** $\sigma \in \mathfrak{T}_0$ *(or,* ***on*** $[\sigma, \infty[$*)* ***relative to*** $\mathbf{i}(.)$ *if and only if the following two conditions, (i) and (ii), are valid:*

(i) *its state vector (its motion) is constant all the time on* $[\sigma, \infty[$*:*

$$\mathbf{x}(t) = \mathbf{x}(t;\mathbf{x}_0;\mathbf{i}) = \mathbf{const.} = \mathbf{x}(\sigma;\mathbf{x}_0;\mathbf{i}) = \mathbf{x}_{\sigma}, \forall t \in [\sigma, \infty[, \tag{3.107}$$

(ii) *its response is constant all the time on* $[\sigma, \infty[$*:*

$$\mathbf{y}(t;\mathbf{x}_0;\mathbf{i}) = \mathbf{const.} = \mathbf{y}(\sigma;\mathbf{x}_0;\mathbf{i}) = \mathbf{y}_{\sigma}, \forall t \in [\sigma, \infty[. \tag{3.108}$$

If and only if this holds for $\sigma = 0$*, then the system is in* ***a stationary (steady) regime relative to*** $\mathbf{i}(.)$.

b) *Otherwise, the system is in* ***a nonstationary regime relative to*** $\mathbf{i}(.)$*, i.e.,*

$$\forall \sigma \in \mathfrak{T}_0, \ \exists \tau \in]\sigma, \infty[\Longrightarrow \mathbf{x}(\tau;\mathbf{x}_{\sigma};\mathbf{i}) \neq \mathbf{x}_{\sigma}, \\ and/or \ \mathbf{y}(\tau;\mathbf{x}_{\sigma};\mathbf{i}) \neq \mathbf{y}_{\sigma}. \tag{3.109}$$

In a nonstationary regime the system is in

- ***a periodic regime relative to*** $\mathbf{i}(.)$ *if and only if there is such* $\sigma \in \mathfrak{T}_0 \cap \mathfrak{R}^+$ *that both (3.110) and (3.111) hold,*

$$\mathbf{x}(t;\mathbf{x}_0;\mathbf{i}) = \mathbf{x}(t+\sigma;\mathbf{x}_0;\mathbf{i}), \ \forall t \in \mathfrak{T}_0, \tag{3.110}$$

$$\mathbf{y}(t;\mathbf{x}_0;\mathbf{i}) = \mathbf{y}(t+\sigma;\mathbf{x}_0;\mathbf{i}), \ \forall t \in \mathfrak{T}_0, \tag{3.111}$$

where such minimal σ is called ***the period relative to*** $\mathbf{i}(.)$ *of the periodic regime and it is denoted by* T,

$$T = min \left\{ \begin{array}{c} \sigma \in \mathfrak{T}_0 \cap \mathfrak{R}^+ : \ \mathbf{x}(t; \mathbf{x}_0; \mathbf{i}) = \mathbf{x}(t+\sigma; \mathbf{x}_0; \mathbf{i}) \ and \\ \mathbf{y}(t; \mathbf{x}_0; \mathbf{i}) = \mathbf{y}(t+\sigma; \mathbf{x}_0; \mathbf{i}), \ \forall t \in \mathfrak{T}_0 \end{array} \right\}, \quad (3.112)$$

- ***a transient (aperiodic) regime relative to*** $\mathbf{i}(.)$ *if and only if it is not in a periodic regime, i.e.,*

$$\exists t \in \mathfrak{T}_0, \ \forall \sigma \in \mathfrak{T}_0 \cap \mathfrak{R}^+, \ \exists k \in \{1, 2, ..., n, ..\} \Longrightarrow$$
$$\mathbf{x}(t; \mathbf{x}_0; \mathbf{i}) \neq \mathbf{x}(t + k\sigma; \mathbf{x}_0; \mathbf{i}) \ and/or \ \mathbf{y}(t; \mathbf{x}_0; \mathbf{i}) \neq \mathbf{y}(t + k\sigma; \mathbf{x}_0; \mathbf{i}). \quad (3.113)$$

The system can achieve a stationary, or nonstationary, regime either in a free regime, or in a forced regime.

Definition 88 *A vector* $\mathbf{x}^* \in \mathfrak{R}^n$ *is* ***a stationary vector (a stationary state) of the ISO system (3.84) and (3.85) relative to*** $\mathbf{i}(.)$, $\mathbf{x}^* = \mathbf{x}^*(\mathbf{i})$, *if and only if both (i) and (ii) hold:*

$$(i) \ \ \mathbf{x}(t; \mathbf{x}^*; \mathbf{i}) = \mathbf{x}(0; \mathbf{x}^*; \mathbf{i}) = \mathbf{x}_0 = \mathbf{x}^* = \mathbf{const.}, \forall t \in \mathfrak{T}_0, \quad (3.114)$$
$$(ii) \ \ \mathbf{y}(t; \mathbf{x}^*; \mathbf{i}) = \mathbf{y}(0; \mathbf{x}^*; \mathbf{i}) = \mathbf{y}_0(\mathbf{x}^*; \mathbf{i}) = \mathbf{const.}, \ \forall t \in \mathfrak{T}_0. \quad (3.115)$$

It is denoted by the subscript « s » :

$$\mathbf{x}^*(\mathbf{i}) = \mathbf{x}_s(\mathbf{i}).$$

This definition permits for a vector $\mathbf{x}^*(\mathbf{i})$ to be a stationary vector of the *ISO* system (3.84) and (3.85) relative to an input vector function $\mathbf{i}(.) = \mathbf{i}_1(.)$, $\mathbf{x}^*(\mathbf{i}_1) = \mathbf{x}_s(\mathbf{i}_1)$, but not relative to another input vector function $\mathbf{i}(.) = \mathbf{i}_2(.)$, $\mathbf{x}_s(\mathbf{i}_1) \neq \mathbf{x}_s(\mathbf{i}_2)$.

Theorem 89 *In order for a vector* $\mathbf{x}^* = \mathbf{x}^*(\mathbf{i}) \in \mathfrak{R}^n$ *to be a stationary vector of the ISO system (3.84) and (3.85) relative to* $\mathbf{i}(.)$ *it is necessary and sufficient that*

$$A\mathbf{x}^* + B\mathbf{i}(t) = \mathbf{0}_n, \ \forall t \in \mathfrak{T}_0, \quad (3.116)$$

and

$$C\mathbf{x}^* + D\mathbf{i}(t) = \mathbf{const.}, \ \forall t \in \mathfrak{T}_0. \quad (3.117)$$

Proof. *Necessity.* Let $\mathbf{x}^* = \mathbf{x}^*(\mathbf{i}) \in \mathfrak{R}^n$ be a stationary vector of the *ISO* system (3.84) and (3.85) relative to $\mathbf{i}(.)$. Hence, (3.107) holds. By differentiating it once and applying (3.114) we verify

$$\frac{d\mathbf{x}(t; \mathbf{x}^*; \mathbf{i})}{dt} = \frac{d\mathbf{x}^*}{dt} = \mathbf{0} = A\mathbf{x}(t; \mathbf{x}^*; \mathbf{i}) + B\mathbf{i}(t) = A\mathbf{x}^* + B\mathbf{i}(t), \ \forall t \in \mathfrak{T}_0,$$

which implies (3.116). This, Equations (3.85), (3.114) and (3.115) yield (3.117).

Sufficiency, Let (3.116) and (3.117) hold. We subtract (3.116) from (3.84), and we use

$$\Delta\mathbf{x}(t;\mathbf{x}^*;\mathbf{i}) = \mathbf{x}(t;\mathbf{x}^*;\mathbf{i}) - \mathbf{x}^*(\mathbf{i}),\ \forall t \in \mathfrak{T}_0, \Longrightarrow \\ \Delta\mathbf{x}(0;\mathbf{x}^*;\mathbf{i}) = \mathbf{x}^*(\mathbf{i}) - \mathbf{x}^*(\mathbf{i}) = \mathbf{0}_n, \tag{3.118}$$

to derive

$$\frac{d\left[\Delta\mathbf{x}(t;\mathbf{x}^*;\mathbf{i})\right]}{dt} = A\Delta\mathbf{x}(t;\mathbf{x}^*;\mathbf{i}).$$

This is the homogenous linear differential equation with the zero initial conditions due to (3.118). Its solution is trivial,

$$\Delta\mathbf{x}(t;\mathbf{x}^*;\mathbf{i}) = \mathbf{x}(t;\mathbf{x}^*;\mathbf{i}) - \mathbf{x}^*(\mathbf{i}) = \mathbf{0},\ \forall t \in \mathfrak{T}_0,$$

i.e.,

$$\mathbf{x}(t;\mathbf{x}^*;\mathbf{i}) = \mathbf{x}^*(\mathbf{i}),\ \forall t \in \mathfrak{T}_0.$$

This is (3.114), which implies (3.115) due to (3.85), and completes the proof. ■

Note 90 *The conditions (3.116) and (3.117) permit the existence of a stationary vector of the ISO system (3.84) and (3.85) relative to* $\mathbf{i}(.)$ *in spite* $\mathbf{i}(t)$ *is time varying. For example,*

$$A = \begin{bmatrix} 1 & 2 \\ 3 & 4 \end{bmatrix},\ B = \begin{bmatrix} 2 & -2 \\ 1 & -1 \end{bmatrix},\ C = \begin{bmatrix} 2 & 6 \\ 7 & 4 \end{bmatrix}, \\ D = \begin{bmatrix} 5 & -5 \\ -1 & 1 \end{bmatrix},\ \mathbf{i}(t) = 2\begin{bmatrix} te^t + 1 \\ te^t \end{bmatrix},$$

and (3.116) imply

$$A\mathbf{x}^* + B\mathbf{i}(t) = \begin{bmatrix} 1 & 2 \\ 3 & 4 \end{bmatrix}\mathbf{x}^* + \begin{bmatrix} 2 & -2 \\ 1 & -1 \end{bmatrix} 2 \begin{bmatrix} te^t + 1 \\ te^t \end{bmatrix} = \mathbf{0}_n,\ \forall t \in \mathfrak{T}_0,$$

i.e.,

$$\mathbf{x}^* = -\frac{1}{2}\begin{bmatrix} 4 & -2 \\ -3 & 1 \end{bmatrix}\begin{bmatrix} 2 & -2 \\ 1 & -1 \end{bmatrix} 2 \begin{bmatrix} te^t + 1 \\ te^t \end{bmatrix} = \\ = -\begin{bmatrix} 6 & -6 \\ -5 & 5 \end{bmatrix}\begin{bmatrix} te^t + 1 \\ te^t \end{bmatrix} = \begin{bmatrix} -6 \\ 5 \end{bmatrix},\ \forall t \in \mathfrak{T}_0.$$

This solution $\mathbf{x}^*$ *is constant. In order to be a stationary vector relative to* $\mathbf{i}(t)$, *it should fulfill (3.117). Let us verify,*

$$C\mathbf{x}^* + D\mathbf{i}(t) = \begin{bmatrix} 2 & 6 \\ 7 & 4 \end{bmatrix} \begin{bmatrix} -6 \\ 5 \end{bmatrix} + \begin{bmatrix} 5 & -5 \\ -1 & 1 \end{bmatrix} 2 \begin{bmatrix} te^t + 1 \\ te^t \end{bmatrix} =$$

$$= \begin{bmatrix} 18 \\ -22 \end{bmatrix} + 2 \begin{bmatrix} 5 \\ -1 \end{bmatrix} = \begin{bmatrix} 28 \\ -24 \end{bmatrix} = \mathbf{const.}, \forall t \in \mathfrak{T}_0.$$

The constant vector $\mathbf{x}^* = \begin{bmatrix} -6 & 5 \end{bmatrix}^T$ *satisfies (3.116) and (3.117). It is the stationary vector of the system relative to time-varying input vector* $\mathbf{i}(t) = 2\begin{bmatrix} te^t + 1 & te^t \end{bmatrix}^T$, $\mathbf{x}^* = \begin{bmatrix} -6 & 5 \end{bmatrix}^T = \mathbf{x}_s(\mathbf{i})$.

3.6.4 *IIO* systems

Definition 91 *The IIO system (3.89), i.e., (3.90), (Subsection 3.5.4),*

a) *is in* ***a stationary (steady) regime since*** $\sigma \in \mathfrak{T}_0$ *(i.e.,* ***on*** $[\sigma, \infty[$*)* ***relative to*** $\mathbf{i}(.)$ *if and only if the following two conditions, (i) and (ii), are valid:*

(i) *its substate vector (its internal dynamical behavior) is constant all the time on* $[\sigma, \infty[$*:*

$$\mathbf{r}(t; \mathbf{r}_0^{\alpha-1}; \mathbf{i}) = \mathbf{const.} = \mathbf{r}(\sigma; \mathbf{r}_0^{\alpha-1}; \mathbf{i}) = \mathbf{r}_\sigma, \forall t \in [\sigma, \infty[, \tag{3.119}$$

(ii) *its response is constant all the time on* $[\sigma, \infty[$*:*

$$\mathbf{y}(t; \mathbf{r}_0^{\alpha-1}; \mathbf{y}_0^{\nu-1}; \mathbf{i}) = \mathbf{const.} = \mathbf{y}(\sigma; \mathbf{r}_0^{\alpha-1}; \mathbf{y}_0^{\nu-1}; \mathbf{i}) = \mathbf{y}_\sigma, \forall t \in [\sigma, \infty[. \tag{3.120}$$

if and only if this holds for $\sigma = 0$, *then the system is in* ***a stationary (steady) regime relative to*** $\mathbf{i}(.)$.

b) *Otherwise, the system is in* ***a nonstationary regime relative to*** $\mathbf{i}(.)$, *i.e.,*

$$\forall \sigma \in \mathfrak{T}_0, \ \exists \tau \in]\sigma, \infty[\Longrightarrow \mathbf{r}(\tau; \mathbf{r}_0^{\alpha-1}; \mathbf{i}) \neq \mathbf{r}_\sigma, \\ \text{and/or } \mathbf{y}(\tau; \mathbf{r}_\sigma^{\alpha-1}; \mathbf{y}_\sigma^{\nu-1}; \mathbf{i}) \neq \mathbf{y}_\sigma. \tag{3.121}$$

In a nonstationary regime the system is in

- ***a periodic regime relative to*** $\mathbf{i}(.)$ *if and only if there is such* $\sigma \in \mathfrak{T}_0 \cap \mathfrak{R}^+$ *that both*

$$\mathbf{r}(t; \mathbf{r}_0^{\alpha-1}; \mathbf{i}) = \mathbf{r}(t+\sigma; \mathbf{r}_0^{\alpha-1}; \mathbf{i}), \ \forall t \in \mathfrak{T}_0, \text{ and} \\ \mathbf{y}(t; \mathbf{r}_0^{\alpha-1}; \mathbf{y}_0^{\nu-1}; \mathbf{i}) = \mathbf{y}(t+\sigma; \mathbf{r}_0^{\alpha-1}; \mathbf{y}_0^{\nu-1}; \mathbf{i}), \ \forall t \in \mathfrak{T}_0, \tag{3.122}$$

hold, where such minimal σ *is called* ***the period relative to*** $\mathbf{i}(.)$ *of the periodic regime and it is denoted by* T,

$$T = min \left\{ \begin{array}{c} \sigma \in \mathfrak{T}_0 \cap \mathfrak{R}^+ : \mathbf{r}(t; \mathbf{r}_0^{\alpha-1}; \mathbf{i}) = \mathbf{r}(t+\sigma; \mathbf{r}_0^{\alpha-1}; \mathbf{i}), \ \forall t \in \mathfrak{T}_0, \ and \\ \mathbf{y}(t; \mathbf{r}_0^{\alpha-1}; \mathbf{y}_0^{\nu-1}; \mathbf{i}) = \mathbf{y}(t+\sigma; \mathbf{r}_0^{\alpha-1}; \mathbf{y}_0^{\nu-1}; \mathbf{i}), \ \forall t \in \mathfrak{T}_0 \end{array} \right\}, \tag{3.123}$$

- ***a transient (aperiodic) regime relative to*** $\mathbf{i}(.)$ *if and only if it is not in a periodic regime, i.e.,*

$$\begin{gathered} \forall t \in \mathfrak{T}_0, \ \forall \sigma \in \mathfrak{T}_0 \cap \mathfrak{R}^+, \ \exists k \in \{1, 2, ..., n, ..\} \Longrightarrow \\ \mathbf{r}(t; \mathbf{r}_0^{\alpha-1}; \mathbf{i}) \neq \mathbf{r}(t+k\sigma; \mathbf{r}_0^{\alpha-1}; \mathbf{i}), \ and/or \\ \mathbf{y}(t; \mathbf{r}_0^{\alpha-1}; \mathbf{y}_0^{\nu-1}; \mathbf{i}) \neq \mathbf{y}(t+k\sigma; \mathbf{r}_0^{\alpha-1}; \mathbf{y}_0^{\nu-1}; \mathbf{i}). \end{gathered} \tag{3.124}$$

The stationary and nonstationary regimes are possible in free regimes and in forced regimes of the system.

Definition 92 *A vector* $\left[\left(\mathbf{r}^{*^{\alpha-1}}\right)^T \ \left(\mathbf{y}^{*^{\nu-1}}\right)^T \right]^T \in \mathfrak{R}^{\alpha\rho+\nu N}$ *is* ***a stationary vector (a stationary point) of the IIO system (3.89), i.e., (3.90), relative to*** $\mathbf{i}(.)$,

$$\left[\left(\mathbf{r}^{*^{\alpha-1}}\right)^T \ \left(\mathbf{y}^{*^{\nu-1}}\right)^T \right]^T = \left[\left(\mathbf{r}^{*^{\alpha-1}}(\mathbf{i})\right)^T \ \vdots \ \left(\mathbf{y}^{*^{\nu-1}}(\mathbf{i})\right)^T \right]^T,$$

if and only if both (i) and (ii) hold:

$$(i) \ \ \mathbf{r}^{\alpha-1}(t; \mathbf{r}_0^{*^{\alpha-1}}; \mathbf{i}) = \mathbf{const.} = \mathbf{r}_0^{*^{\alpha-1}}, \forall t \in \mathfrak{T}_0, \ and \tag{3.125}$$

$$(ii) \ \ \mathbf{y}^{\nu-1}(t; \mathbf{r}_0^{*^{\alpha-1}}; \mathbf{y}_0^{*^{\nu-1}}; \mathbf{i}) = \mathbf{const.} = \mathbf{y}_0^{*^{\nu-1}}, \ \forall t \in \mathfrak{T}_0. \tag{3.126}$$

It is denoted by the subscript « s » :

$$\left[\left(\mathbf{r}^{*^{\alpha-1}}(\mathbf{i})\right)^T \ \vdots \ \left(\mathbf{y}^{*^{\nu-1}}(\mathbf{i})\right)^T \right]^T = \left[\left(\mathbf{r}_s^{\alpha-1}(\mathbf{i})\right)^T \ \vdots \ \left(\mathbf{y}_s^{\nu-1}(\mathbf{i})\right)^T \right]^T.$$

The stationary vector $\left[\left(\mathbf{r}_s^{\alpha-1}(\mathbf{i})\right)^T \ \vdots \ \left(\mathbf{y}_s^{\nu-1}(\mathbf{i})\right)^T \right]^T$ is relative to the input vector function $\mathbf{i}(.)$. It can be stationary relative to $\mathbf{i}_1(.)$, but not relative to $\mathbf{i}_2(.) \neq \mathbf{i}_1(.)$.

Theorem 93 *In order for a vector* $\left[\left(\mathbf{r}^{*^{\alpha-1}}\right)^T \vdots \left(\mathbf{y}^{*^{\nu-1}}\right)^T\right]^T \in \mathfrak{R}^{\alpha\rho+\nu N}$ *to be a stationary vector of the IIO system (3.89), i.e., (3.90), relative to* $\mathbf{i}(.)$,

$$\left[\left(\mathbf{r}^{*^{\alpha-1}}\right)^T \vdots \left(\mathbf{y}^{*^{\nu-1}}\right)^T\right]^T = \left[\left(\mathbf{r}_s^{\alpha-1}(\mathbf{i})\right)^T \vdots \left(\mathbf{y}_s^{\nu-1}(\mathbf{i})\right)^T\right]^T,$$

it is necessary and sufficient that

$$\mathbf{r}^{\alpha-1}(0;\mathbf{r}_0^{*^{\alpha-1}};\mathbf{i}) = \mathbf{r}_0^{\alpha-1} = \mathbf{r}_0^{*^{\alpha-1}} = \left[\mathbf{r}^{*^T} \vdots\ \mathbf{0}_\rho^T \vdots\ \mathbf{0}_\rho^T \vdots \dots \vdots\ \mathbf{0}_\rho^T\right]^T \in \mathfrak{R}^{\alpha\rho},$$

$$\mathbf{y}^{\nu-1}(0;\mathbf{r}^{*^{\alpha-1}};\mathbf{y}_0^{*^{\nu-1}};\mathbf{i}) = \mathbf{y}_0^{\nu-1} =$$
$$= \mathbf{y}_0^{*^{\nu-1}} = \left[\mathbf{y}^{*^T} \vdots\ \mathbf{0}_N^T \vdots\ \mathbf{0}_N^T \vdots \dots \vdots\ \mathbf{0}_N^T\right]^T \in \mathfrak{R}^{N\nu}, \tag{3.127}$$

$$Q_0\mathbf{r}^* = \sum_{k=0}^{k=\beta\le\alpha} P_k\mathbf{i}^{(k)}(t) = P^\beta\mathbf{i}^\beta(t),\ \forall t \in \mathfrak{T}_0, \tag{3.128}$$

and

$$-R_0\mathbf{r}^* + E_0\mathbf{y}^* = \sum_{k=0}^{k=\mu} T_k\mathbf{i}^{(k)}(t) = T^{(\mu)}\mathbf{i}^\mu(t),\ \forall t \in \mathfrak{T}_0, \tag{3.129}$$

hold.

Proof. *Necessity.* Let $\left[\left(\mathbf{r}^{*^{\alpha-1}}\right)^T \ \left(\mathbf{y}^{*^{\nu-1}}\right)^T\right]^T \in \mathfrak{R}^{\alpha\rho+\nu N}$ be a stationary vector of the *IIO* system (3.89), i.e., (3.90), relative to $\mathbf{i}(.)$. Hence, (3.125) and (3.126) hold. By differentiating them we verify

$$\mathbf{r}^{(k)}(t;\mathbf{r}_0^{*^{\alpha-1}};\mathbf{i}) = \mathbf{0},\ \forall k = 1,2,...,\alpha,\ \forall t \in \mathfrak{T}_0 \Longrightarrow \tag{3.130}$$
$$\mathbf{r}^{(k)}(0;\mathbf{r}_0^{*^{\alpha-1}};\mathbf{i}) = \mathbf{r}^{*(k)} = \mathbf{0},\ \forall k = 1,2,...,\alpha, \tag{3.131}$$
$$\mathbf{y}^{(k)}(t;\mathbf{y}_0^{*^{\nu-1}};\mathbf{i}) = \mathbf{0},\ \forall k = 1,2,...,\nu,\ \forall t \in \mathfrak{T}_0 \Longrightarrow \tag{3.132}$$
$$\mathbf{y}^{(k)}(0;\mathbf{y}_0^{*^{\nu-1}};\mathbf{i}) = \mathbf{y}^{*(k)} = \mathbf{0},\ \forall k = 1,2,...,\nu. \tag{3.133}$$

The results (3.131) and (3.133), together with (3.125) and (3.126), imply (3.127). Besides, (3.130), (3.132) and (3.89), i.e., (3.90), imply (3.128) and (3.129).

Sufficiency, Let (3.127), (3.128) and (3.129) hold. We subtract (3.128) and (3.129) from (3.89), i.e., from (3.90), and we use

$$\Delta\mathbf{r}^{(k)}(t;\mathbf{r}_0^{*\alpha-1};\mathbf{i}) = \mathbf{r}^{(k)}(t;\mathbf{r}_0^{*\alpha-1};\mathbf{i}) - \mathbf{r}^{*(k)}(\mathbf{i}) =$$

$$= \left\{ \begin{array}{c} \mathbf{r}(t;\mathbf{r}_0^{*\alpha-1};\mathbf{i}) - \mathbf{r}^*(\mathbf{i}),\ k=0, \\ \mathbf{r}^{(k)}(t;\mathbf{r}_0^{*\alpha-1};\mathbf{i}),\ \forall k = 1,2,...,\alpha \end{array} \right\},\ \forall t \in \mathfrak{T}_0, \tag{3.134}$$

$$\Delta\mathbf{y}^{(k)}(t;\mathbf{r}_0^{*\alpha-1};\mathbf{y}_0^{*\nu-1};\mathbf{i}) = \mathbf{y}^{(k)}(t;\mathbf{r}_0^{*\alpha-1};\mathbf{y}_0^{*\nu-1};\mathbf{i}) - \mathbf{y}^{*(k)}(\mathbf{i}) =$$

$$= \left\{ \begin{array}{c} \mathbf{y}(t;\mathbf{r}_0^{*\alpha-1};\mathbf{y}_0^{*\nu-1};\mathbf{i}) - \mathbf{y}^*(\mathbf{i}),\ k=0, \\ \mathbf{y}^{(k)}(t;\mathbf{r}_0^{*\alpha-1};\mathbf{y}_0^{*\nu-1};\mathbf{i}),\ \forall k = 1,2,...,\nu \end{array} \right\},\ \forall t \in \mathfrak{T}_0, \tag{3.135}$$

to derive

$$\sum_{k=0}^{k=\alpha} Q_k \Delta\mathbf{r}^{(k)}(t) = \mathbf{0}_\rho,\ \forall t \in \mathfrak{T}_0,$$

$$-\sum_{k=0}^{k=\alpha} R_k \Delta\mathbf{r}^{(k)}(t) + \sum_{k=0}^{k=\nu} E_k \Delta\mathbf{y}^{(k)}(t) = \mathbf{0}_N,\ \forall t \in \mathfrak{T}_{0o}.$$

This is the system of the homogenous linear differential equations with the zero initial conditions due to (3.134) and (3.135). Their solutions are trivial,

$$\Delta\mathbf{r}^{(k)}(t;\mathbf{r}_0^{*\alpha-1};\mathbf{i}) = \mathbf{r}^{(k)}(t;\mathbf{r}_0^{*\alpha-1};\mathbf{i}) - \mathbf{r}^{*(k)}(\mathbf{i}) = \mathbf{0}_\rho,\ \forall t \in \mathfrak{T}_0,\ \forall k = 0,1,...,\alpha-1,$$

$$\Delta\mathbf{y}^{(k)}(t;\mathbf{r}_0^{*\alpha-1};\mathbf{y}_0^{*\nu-1};\mathbf{i}) = \mathbf{y}^{(k)}(t;\mathbf{r}_0^{*\alpha-1};\mathbf{y}_0^{*\nu-1};\mathbf{i}) - \mathbf{y}^{*(k)}(\mathbf{i}) = \mathbf{0}_N,\ \forall t \in \mathfrak{T}_0,$$
$$\forall k = 0,1,...,\nu-1,$$

i.e.,

$$\mathbf{r}^{\alpha-1}(t;\mathbf{r}_0^{*\alpha-1};\mathbf{i}) = \mathbf{r}_0^{*\alpha-1}(\mathbf{i}),\ \forall t \in \mathfrak{T}_0.$$

$$\mathbf{y}^{\nu-1}(t;\mathbf{r}_0^{*\alpha-1};\mathbf{y}_0^{*\nu-1};\mathbf{i}) = \mathbf{y}^{*\nu-1}(\mathbf{i}),\ \forall t \in \mathfrak{T}_0.$$

These are (3.125) and (3.126). They complete the proof. ■

3.7 Equilibrium regime

3.7.1 Introduction

An equilibrium regime can be ***static*** or ***dynamic***. It depends on the temporal characteristic of the total desired (nominal) behavior $\mathbf{Y}_d(.)$ of the system. If $\mathbf{Y}_d(t)$ is constant, then the equilibrium regime is **static**, and **the equilibrium point** in the deviation coordinates frame represents **a static equilibrium point**. If the desired (nominal) behavior $\mathbf{Y}_d(t)$ of the system is *time*-varying, then the equilibrium regime is **dynamic**, and the equilibrium point in the deviation coordinates frame represents a **dynamic equilibrium point**.

An equilibrium regime concerns the system behavior under the total nominal input vector $\mathbf{I}_N(.)$ and under all nominal total initial conditions in the total coordinates system.

However, if we deal with the system description in terms of the deviations, then an equilibrium regime concerns the system behavior in a free regime. Since such a mathematical model is *time*-invariant and linear (in the framework of this book), then its solution in the free regime under all zero initial conditions is trivial, i.e., it is identically equal to the zero deviation vector.

Conclusion 94 *The zero deviation vector represents an equilibrium vector (an equilibrium point) of every linear time-invariant mathematical model in terms of the deviations.*

Problem 95 *Does there exist another (nontrivial) equilibrium vector of a given linear time-invariant system described in terms of the deviations?*

The reply is affirmative.

Problem 96 *What are conditions for the existence of two or more nontrivial equilibrium points of a given linear time-invariant system described in terms of the deviations?*

We will present complete solutions to these problems for every class of the systems treated herein.

3.7.2 *IO* systems

The *IO* system (3.76), i.e., (3.77), (Subsection 3.5.2),

$$A^{(\nu)}\mathbf{y}^{\nu}(t) = B^{(\mu)}\mathbf{i}^{\mu}(t),\ \forall t \in \mathfrak{T}_0, \tag{3.136}$$

can have a single, unique equilibrium vector, but need not.

Definition 97 *The IO system (3.136) is in* ***an equilibrium regime*** *if and only if it is in a free regime and its response is constant all the time on* $\mathfrak{T}_0$*, that is that*

$$\mathbf{y}^*(t;\mathbf{y}^{*\nu-1};\mathbf{0}_M) = \mathbf{const.} = \mathbf{y}(0;\mathbf{y}^{*\nu-1};\mathbf{0}_M) = \mathbf{y}^*, \forall t \in \mathfrak{T}_0. \tag{3.137}$$

if and only if this holds then the vector $\mathbf{y}^{*\nu-1} = \left[\mathbf{y}^{*T}\vdots\ \mathbf{0}_N^T\vdots\ \mathbf{0}_N^T\vdots ...\vdots\ \mathbf{0}_N^T\right]^T \in \mathfrak{R}^{N\nu}$ ***is an equilibrium vector (an equilibrium point) of the system,*** *which is denoted by the subscript "e",*

$$\mathbf{y}^{*\nu-1} = \left[\mathbf{y}^{*T}\vdots\ \mathbf{0}_N^T\vdots\ \mathbf{0}_N^T\vdots ...\vdots\ \mathbf{0}_N^T\right]^T = \mathbf{y}_e^{\nu-1} = \left[\mathbf{y}_e^T\vdots\ \mathbf{0}_N^T\vdots\ \mathbf{0}_N^T\vdots ...\vdots\ \mathbf{0}_N^T\right]^T.$$

This definition and Definitions 81 and 82, (in Subsection 3.6.2), show that an equilibrium regime is a stationary regime relative to the zero deviation input vector $\mathbf{i}(t)\equiv \mathbf{0}_M$, and that an equilibrium vector is a stationary vector relative to the zero deviation input vector $\mathbf{i}(t)\equiv \mathbf{0}_M$.

The equilibrium vector of the system is independent of the real input vector. It depends only on the system itself. More precisely:

Theorem 98 *In order for a vector* $\mathbf{y}^{*\nu-1} = \left[\mathbf{y}^{*T}\vdots\ \mathbf{0}_N^T\vdots\ \mathbf{0}_N^T\vdots ...\vdots\ \mathbf{0}_N^T\right]^T \in \mathfrak{R}^{N\nu}$ *to be an equilibrium vector of the IO system (3.136), it is necessary and sufficient that*

$$A_0\mathbf{y}^* = \mathbf{0}_N. \tag{3.138}$$

Proof. This theorem results directly from Theorem 83 (in Subsection 3.6.2), in view of Definitions 81 and 97 ■

Theorem 99 *a) In order for a vector* $\mathbf{y}^{*\nu-1} = \left[\mathbf{y}^{*T}\vdots\ \mathbf{0}_N^T\vdots\ \mathbf{0}_N^T\vdots ...\vdots\ \mathbf{0}_N^T\right]^T \in \mathfrak{R}^{N\nu}$ *to be the unique equilibrium vector of the IO system (3.136), it is necessary and sufficient that the matrix* $A_0 \in \mathfrak{R}^{N\times N}$ *is nonsingular,*

$$\det A_0 \neq 0. \tag{3.139}$$

b) In order for the system (3.136) to have several different equilibrium vectors, it is necessary and sufficient that the matrix $A_0 \in \mathfrak{R}^{N\times N}$ *is singular,*

$$\det A_0 = 0. \tag{3.140}$$

Then, and only then, the system has infinitely many different equilibrium vectors that constitute the hyperplane $S_{eIO} \subset \mathfrak{R}^{\nu N}$ of all the equilibrium vectors, which passes through the origin $\mathbf{y}^{\nu-1} = \mathbf{0}_{\nu N}$,

$$S_{eIO} = \left\{ \mathbf{y}^{\nu-1} : \mathbf{y}^{\nu-1} = \left[\mathbf{y}^T \vdots\ \mathbf{0}_N^T \vdots\ \mathbf{0}_N^T \vdots \ldots \vdots\ \mathbf{0}_N^T \right]^T \in \mathfrak{R}^{\nu N},\ A_0 \mathbf{y} = \mathbf{0}_N \right\}. \tag{3.141}$$

Proof. The statement of the theorem follows directly from Theorem 98 and the well-known theorem on the number of the solutions of the homogeneous linear vector algebraic Equation (3.138). ■

The preceding Theorems 98 and 99 show that we can easily determine the equilibrium vector(s) of the system (3.136) by solving Equation 3.138.

Note 100 *Equation (3.141) holds also in the case (3.139). Then, and only then, the set of all the system equilibrium states is singleton,*

$$S_{eIO} = \left\{ \mathbf{y}^{\nu-1} : \mathbf{y}^{\nu-1} = \left[\mathbf{y}^T \vdots\ \mathbf{0}_N^T \vdots\ \mathbf{0}_N^T \vdots \ldots \vdots\ \mathbf{0}_N^T \right]^T \in \mathfrak{R}^{\nu N},\ A_0 \mathbf{y} = \mathbf{0}_N \right\} =$$
$$= \left\{ \mathbf{y}^{\nu-1} : \mathbf{y}^{\nu-1} = \mathbf{0}_{\nu N} \right\} = \left\{ \mathbf{0}_{\nu N} \right\} \Longleftrightarrow \det A_0 \neq 0.$$

Note 101 *If the matrix A_0 is the zero matrix O_N, $A_0 = O_N$, then every output vector $\mathbf{y}^{\nu-1} = \left[\mathbf{y}^T \vdots\ \mathbf{0}_N^T \vdots\ \mathbf{0}_N^T \vdots \ldots \vdots\ \mathbf{0}_N^T \right]^T \in \mathfrak{R}^{\nu N}$ is the equilibrium vector $\mathbf{y}_e^{\nu-1} \in \mathfrak{R}^{\nu N}$. The system rests in the initial output vector*

$$\mathbf{y}_0^{\nu-1} = \left[\mathbf{y}_0^T \vdots\ \mathbf{0}_N^T \vdots\ \mathbf{0}_N^T \vdots \ldots \vdots\ \mathbf{0}_N^T \right]^T,$$

i.e.,

$$A_0 = O_N \Longrightarrow \mathbf{y}_e^{\nu-1} = \mathbf{y}_0^{\nu-1} = \left[\mathbf{y}_0^T \vdots\ \mathbf{0}_N^T \vdots\ \mathbf{0}_N^T \vdots \ldots \vdots\ \mathbf{0}_N^T \right]^T,\ \forall \mathbf{y}_0 \in \mathfrak{R}^N,$$

whatever is the initial output vector $\mathbf{y}_0 \in \mathfrak{R}^N$.

3.7.3 *ISO* systems

The *ISO* system (3.84) and (3.85) (in Subsection 3.5.3),

$$\frac{d\mathbf{x}(t)}{dt} = A\mathbf{x}(t) + B\mathbf{i}(t),\ \forall t \in \mathfrak{T}_0, \tag{3.142}$$

$$\mathbf{y}(t) = C\mathbf{x}(t) + D\mathbf{i}(t),\ \forall t \in \mathfrak{T}_0. \tag{3.143}$$

can be in a single, unique, equilibrium regime, but need not.

Definition 102 *The ISO system (3.142) and (3.143), is in* ***an equilibrium regime*** *if and only if its state vector (its motion) is constant all the time on* $\mathfrak{T}_0$ *in a free regime, that is that*

$$\mathbf{x}(t;\mathbf{x}^*;\mathbf{0}_M) = \mathbf{const.} = \mathbf{x}(0;\mathbf{x}^*;\mathbf{0}_M) = \mathbf{x}^*, \forall t \in \mathfrak{T}_0. \tag{3.144}$$

if and only if this holds then the vector $\mathbf{x}^* \in \mathfrak{R}^n$ *is* ***an equilibrium vector (an equilibrium state) of the system****, which is denoted by the subscript "e",*

$$\mathbf{x}^* = \mathbf{x}_e.$$

A stationary regime in a free regime and a stationary state relative to the zero input vector $\mathbf{i}(t) \equiv \mathbf{0}_M$, i.e., in a free regime, represent an equilibrium regime and an equilibrium state, respectively, of the *ISO* system (3.142) and (3.143) in view of the preceding definition, and Definitions 87 and 88 (in Subsection 3.6.3). This fact and Theorem 89 (in Subsection 3.6.3) imply directly the following:

Theorem 103 *In order for a vector* $\mathbf{x}^* \in \mathfrak{R}^n$ *to be an equilibrium vector of the ISO system (3.142) and (3.143) it is necessary and sufficient that*

$$A\mathbf{x}^* = \mathbf{0}_n. \tag{3.145}$$

This theorem and the theorem on the number of the solutions of the linear homogeneous vector algebraic equation prove:

Theorem 104 *a) In order for a vector* $\mathbf{x}^* \in \mathfrak{R}^n$ *to be the unique equilibrium state of the ISO system (3.142) and (3.143), it is necessary and sufficient that the matrix* $A \in \mathfrak{R}^{nxn}$ *is nonsingular,*

$$\det A \neq 0. \tag{3.146}$$

Theorem 105 *b) In order for the ISO system (3.142) and (3.143) to have several different equilibrium vectors, it is necessary and sufficient that the matrix* $A \in \mathfrak{R}^{nxn}$ *is singular,*

$$\det A = 0. \tag{3.147}$$

Then, and only then the system has infinitely many different equilibrium vectors that constitute the hyperplane $S_{eISO} \subset \mathfrak{R}^n$ *of all the equilibrium vectors, which passes through the origin* $\mathbf{x} = \mathbf{0}_n$,

$$S_{eISO} = \{\mathbf{x} : \ \mathbf{x} \in \mathfrak{R}^n, \ A\mathbf{x} = \mathbf{0}_n\}. \tag{3.148}$$

Note 106 *Equation (3.148) of* S_{eISO} *holds regardless of the singularity* ($\det A = 0$) *or the regularity* ($\det A \neq 0$) *of the matrix* A. *If and only if the matrix* A *is nonsingular, i.e.,* $\det A \neq 0$, *then* S_{eISO} *is singleton,*

$$S_{eISO} = \{\mathbf{x} : \ \mathbf{x} \in \mathfrak{R}^n, \ \mathbf{x} = \mathbf{0}_n\} = \{\mathbf{0}_n\} \Longleftrightarrow \det A \neq 0.$$

Note 107 *If the matrix* A *is the zero matrix,* $A = O_n$, *then* S_{eISO} *is the whole state space,*

$$A = O_n \Longrightarrow S_{eISO} = \mathfrak{R}^n.$$

For example, the matrix A *of the system*

$$\frac{d\mathbf{x}(t)}{dt} = B\mathbf{i}(t), \ \mathbf{y}(t) = C\mathbf{x}(t) + D\mathbf{i}(t)$$

is the zero matrix, $A = O_n$. *The set* S_{eISO} *of all the system equilibrium states is the whole state space* $\mathfrak{R}^n$, $S_{eISO} = \mathfrak{R}^n$. *The system rests in its initial state* $\mathbf{x}_0$ *in the free regime, whatever is the initial state* $\mathbf{x}_0 \in \mathfrak{R}^n$,

$$\mathbf{x}_e = \mathbf{x}_0 \in \mathfrak{R}^n.$$

Notice also that this system is in the stationary regime if and only if

$$B\mathbf{i}(t) = \mathbf{0}_n, \ \forall t \in \mathfrak{T}_0.$$

Then, every state can be a stationary state of this system, which implies that then the system rests in its initial state for ever whatever is the initial state. Notice that the preceding condition does not imply the free regime, i.e., it is possible to hold for $\mathbf{i}(t) \neq \mathbf{0}_n$, $t \in \mathfrak{T}_0$. *To be specific, let*

$$A = O_2, \ B = \begin{bmatrix} -2 & 2 \\ 1 & -1 \end{bmatrix}, \ C = \begin{bmatrix} 2 & 6 \\ 7 & 4 \end{bmatrix},$$

$$D = \begin{bmatrix} 5 & -5 \\ -1 & 1 \end{bmatrix}, \ \mathbf{i}(t) = \begin{bmatrix} te^t + 1 \\ te^t + 1 \end{bmatrix} \neq \mathbf{0}_n, \ \forall t \in \mathfrak{T}_0.$$

This and (3.116) (Subsection 3.6.3) imply

$$A\mathbf{x}^* + B\mathbf{i}(t) = O_2\mathbf{x}^* + B\mathbf{i}(t) = B\mathbf{i}(t) = \begin{bmatrix} -2 & 2 \\ 1 & -1 \end{bmatrix} \begin{bmatrix} te^t + 1 \\ te^t + 1 \end{bmatrix} =$$

$$= \begin{bmatrix} -2te^t - 2 + 2te^t + 2 \\ te^t + 1 - te^t - 1 \end{bmatrix} = \begin{bmatrix} 0 \\ 0 \end{bmatrix} = \mathbf{0}_n, \ \forall t \in \mathfrak{T}_0.$$

The system is in the forced regime and every state can be its stationary state $\mathbf{x}_s$. *Hence, the system rests in its initial state* $\mathbf{x}_0 \in \mathfrak{R}^2$ *although it is in the forced regime under the action of the time varying input*

$$\mathbf{i}(t) = \begin{bmatrix} te^t + 1 \\ te^t + 1 \end{bmatrix},$$

whatever is its initial state $\mathbf{x}_0 \in \mathfrak{R}^2$,

$$\mathbf{x}_s(\mathbf{i}) = \mathbf{x}_0, \ \forall \mathbf{x}_0 \in \mathfrak{R}^n.$$

3.7.4 *IIO* systems

We define now an equilibrium vector of the *IIO* system (3.89), i.e., (3.90), (in Subsection 3.5.4),

$$\begin{gathered} Q^{(\alpha)}\mathbf{r}^\alpha(t) = P^{(\beta)}\mathbf{i}^\beta(t), \ detQ_\alpha \neq 0, \ \forall t \in \mathfrak{T}_0, \\ E^{(\nu)}\mathbf{y}^\nu(t) = R^{(\alpha)}\mathbf{r}^\alpha(t) + T^{(\mu)}\mathbf{i}^\mu(t), \ detE_\nu \neq 0, \ \forall t \in \mathfrak{T}_0. \end{gathered} \tag{3.149}$$

Definition 108 *The IIO system (3.149) is in* ***an equilibrium regime*** *if and only if its complete dynamics vector (its complete dynamic behavior) is constant in a free regime all the time on* $\mathfrak{T}_0$*:*

$$\begin{bmatrix} \mathbf{r}^{\alpha-1}(t; \mathbf{r}^{*\alpha-1}; \mathbf{0}_M) \\ \mathbf{y}^{\nu-1}(t; \mathbf{r}^{*\alpha-1}; \mathbf{y}^{*\nu-1}; \mathbf{0}_M) \end{bmatrix} = \mathbf{const.} = \begin{bmatrix} \mathbf{r}^{\alpha-1}(0; \mathbf{r}^{*\alpha-1}; \mathbf{0}_M) \\ \mathbf{y}^{\nu-1}(0; \mathbf{r}^{*\alpha-1}; \mathbf{y}^{*\nu-1}; \mathbf{0}_M) \end{bmatrix} =$$
$$= \begin{bmatrix} \mathbf{r}^{*\alpha-1} \\ \mathbf{y}^{*\nu-1} \end{bmatrix}, \forall t \in \mathfrak{T}_0. \tag{3.150}$$

If and only if this holds

- *and if* $\nu \geq 1$ *then the vector* $\left[\left(\mathbf{r}^{*\alpha-1}\right)^T \vdots \left(\mathbf{y}^{*\nu-1}\right)^T\right]^T$,

$$\begin{bmatrix} \mathbf{r}^{*\alpha-1} \\ \mathbf{y}^{*\nu-1} \end{bmatrix} = \begin{bmatrix} \left[\mathbf{r}^{*T} \vdots \mathbf{0}_\rho^T \vdots \mathbf{0}_\rho^T \vdots \ldots \vdots \mathbf{0}_\rho^T\right]^T \\ \left[\mathbf{y}^{*T} \vdots \mathbf{0}_N^T \vdots \mathbf{0}_N^T \vdots \ldots \vdots \mathbf{0}_N^T\right]^T \end{bmatrix} \in \mathfrak{R}^{\alpha\rho + N\nu},$$

is ***an equilibrium vector (an equilibrium point) of the system*** *(3.149), which is denoted by the subscript "e",*

$$\begin{bmatrix} \mathbf{r}^{*\alpha-1} \\ \mathbf{y}^{*\nu-1} \end{bmatrix} = \begin{bmatrix} \mathbf{r}_e^{\alpha-1} \\ \mathbf{y}_e^{\nu-1} \end{bmatrix} = \begin{bmatrix} \left[\mathbf{r}_e^T \vdots \mathbf{0}_\rho^T \vdots \mathbf{0}_\rho^T \vdots \ldots \vdots \mathbf{0}_\rho^T\right]^T \\ \left[\mathbf{y}_e^T \vdots \mathbf{0}_N^T \vdots \mathbf{0}_N^T \vdots \ldots \vdots \mathbf{0}_N^T\right]^T \end{bmatrix},$$

- *and if $\nu = 0$ then the vector $\mathbf{r}^{*\alpha-1}$,*

$$\mathbf{r}^{*\alpha-1} = \left[\mathbf{r}^{*T} \vdots\ \mathbf{0}_\rho^T \vdots\ \mathbf{0}_\rho^T \vdots \ldots \vdots\ \mathbf{0}_\rho^T\right]^T \in \mathfrak{R}^{\alpha\rho},$$

*is **an equilibrium vector (an equilibrium point) of the system** (3.149), hence of the Rosenbrock system (3.92), and of the IIO system (3.94), (in Subsection 3.5.4), which is denoted by the subscript "e",*

$$\mathbf{r}^{*\alpha-1} = \mathbf{r}_e^{\alpha-1} = \left[\mathbf{r}_e^T \vdots\ \mathbf{0}_\rho^T \vdots\ \mathbf{0}_\rho^T \vdots \ldots \vdots\ \mathbf{0}_\rho^T\right]^T.$$

A stationary regime and a stationary vector relative to the zero input $\mathbf{i}(t) \equiv \mathbf{0}_M$ represent an equilibrium regime and an equilibrium vector of the system, respectively. This follows from the preceding definition and Definitions 91 and 92 (in Subsection 3.6.4).

Theorem 109 *a) Let $\nu \geq 1$. In order for vector $\left[\left(\mathbf{r}^{*\alpha-1}\right)^T \vdots\ \left(\mathbf{y}^{*\nu-1}\right)^T\right]^T$,*

$$\left[\left(\mathbf{r}^{*\alpha-1}\right)^T \vdots\ \left(\mathbf{y}^{*\nu-1}\right)^T\right]^T =$$

$$= \left[\left[\mathbf{r}^{*T} \vdots\ \mathbf{0}_\rho^T \vdots\ \mathbf{0}_\rho^T \vdots \ldots \vdots\ \mathbf{0}_\rho^T\right]^T \vdots\ \left[\mathbf{y}^{*T} \vdots\ \mathbf{0}_N^T \vdots\ \mathbf{0}_N^T \vdots \ldots \vdots\ \mathbf{0}_N^T\right]^T\right]^T \in \mathfrak{R}^{\alpha\rho+\nu N},$$

to be an equilibrium vector of the IIO system (3.149), it is necessary and sufficient that

$$\begin{bmatrix} Q_0 & \vdots\ O_{\rho N} \\ -R_0 & \vdots\ E_0 \end{bmatrix} \begin{bmatrix} \mathbf{r}^* \\ \mathbf{y}^* \end{bmatrix} = \begin{bmatrix} \mathbf{0}_\rho \\ \mathbf{0}_N \end{bmatrix}. \tag{3.151}$$

b) Let $\nu = 0$. In order for a vector $\mathbf{r}^{*\alpha-1}$,

$$\mathbf{r}^{*\alpha-1} = \left[\mathbf{r}^{*T} \vdots\ \mathbf{0}_\rho^T \vdots\ \mathbf{0}_\rho^T \vdots \ldots \vdots\ \mathbf{0}_\rho^T\right]^T \in \mathfrak{R}^{\alpha\rho},$$

to be an equilibrium vector of the IIO system (3.149), hence of the Rosenbrock system (3.92), and of the system (3.94), it is necessary and sufficient that

$$Q_0\mathbf{r}^* = \mathbf{0}_\rho. \tag{3.152}$$

This theorem is a direct result of Definitions 91, 92 and 108, and Theorem 93 (in Subsection 3.6.4).

In order to determine an equilibrium vector of the *IIO* system (3.149), or to verify whether a vector

$$\begin{bmatrix} \mathbf{r}^{*^{\alpha-1}} \\ \mathbf{y}^{*^{\nu-1}} \end{bmatrix} = \begin{bmatrix} \left[\mathbf{r}^{*^T} \vdots\, \mathbf{0}_\rho^T \vdots\, \mathbf{0}_\rho^T \vdots ... \vdots\, \mathbf{0}_\rho^T\right]^T \\ \left[\mathbf{y}^{*^T} \vdots\, \mathbf{0}_N^T \vdots\, \mathbf{0}_N^T \vdots ... \vdots\, \mathbf{0}_N^T\right]^T \end{bmatrix}, \ \nu \geq 1,$$

$$\mathbf{r}^{*^{\alpha-1}} = \left[\mathbf{r}^{*^T} \vdots\, \mathbf{0}_\rho^T \vdots\, \mathbf{0}_\rho^T \vdots ... \vdots\, \mathbf{0}_\rho^T\right]^T, \ \nu = 0,$$

is its equilibrium vector, we should solve Equation (3.151) or Equation (3.152), respectively.

Theorem 110 *a) In order for:*

- a vector $\left[\left(\mathbf{r}^{*^{\alpha-1}}\right)^T \vdots \left(\mathbf{y}^{*^{\nu-1}}\right)^T\right]^T$ *if* $\nu \geq 1$, *where*

$$\left[\left(\mathbf{r}^{*^{\alpha-1}}\right)^T \vdots \left(\mathbf{y}^{*^{\nu-1}}\right)^T\right]^T =$$

$$= \left[\left[\mathbf{r}^{*^T} \vdots\, \mathbf{0}_\rho^T \vdots\, \mathbf{0}_\rho^T \vdots ... \vdots\, \mathbf{0}_\rho^T\right]^T \left[\mathbf{y}^{*^T} \vdots\, \mathbf{0}_N^T \vdots\, \mathbf{0}_N^T \vdots ... \vdots\, \mathbf{0}_N^T\right]^T\right]^T \in \mathfrak{R}^{\alpha\rho+\nu N},$$

to be the unique equilibrium vector of the IIO system (3.149), it is necessary and sufficient that the matrix

$$\begin{bmatrix} Q_0 & \vdots\, O_{\rho N} \\ -R_0 & \vdots\, E_0 \end{bmatrix} \in \mathfrak{R}^{(\rho+N)x(\rho+N)}$$

is nonsingular,

$$det \begin{bmatrix} Q_0 & O_{\rho N} \\ -R_0 & E_0 \end{bmatrix} \neq 0, \tag{3.153}$$

- a vector $\mathbf{r}^{*^{\alpha-1}}$ *if* $\nu = 0$, *where*

$$\mathbf{r}^{*^{\alpha-1}} = \left[\mathbf{r}^{*^T} \vdots\, \mathbf{0}_\rho^T \vdots\, \mathbf{0}_\rho^T \vdots ... \vdots\, \mathbf{0}_\rho^T\right]^T \in \mathfrak{R}^{\alpha\rho},$$

to be the unique equilibrium vector of the IIO system (3.149), hence of the Rosenbrock system (3.92), and of the system (3.94), it is necessary and sufficient that the matrix

$$Q_0 \in \mathfrak{R}^{\rho x \rho}$$

is nonsingular,

$$detQ_0 \neq 0, \tag{3.154}$$

b) In order for the system (3.149) to have several different equilibrium vectors, it is necessary and sufficient that

- the matrix

$$\begin{bmatrix} Q_0 & \vdots O_{\rho N} \\ -R_0 & \vdots \ E_0 \end{bmatrix} \in \mathfrak{R}^{(\rho+N)x(\rho+N)}$$

is singular if $\nu \geq 1$,

$$det \begin{bmatrix} Q_0 & \vdots O_{\rho N} \\ -R_0 & \vdots \ E_0 \end{bmatrix} = 0, \tag{3.155}$$

- the matrix

$$Q_0 \in \mathfrak{R}^{\rho x \rho}$$

is singular if $\nu = 0$,

$$detQ_0 = 0. \tag{3.156}$$

Then, and only then, the system has infinitely many different equilibrium vectors that constitute the hyperplane $S_{eIIO} \subset \mathfrak{R}^{\alpha\rho+\nu N}$ *of all the equilibrium vectors, which passes through:*

- the origin $\left[\left(\mathbf{r}^{\alpha-1}\right)^T \quad \left(\mathbf{y}^{\nu-1}\right)^T\right]^T = \mathbf{0}_{\alpha\rho+\nu N}$ *if* $\nu \geq 1$,

$$S_{eIIO} = \left\{ \begin{bmatrix} \mathbf{r}^{\alpha-1} \\ \mathbf{y}^{\nu-1} \end{bmatrix} : \left(\begin{array}{c} \begin{bmatrix} \mathbf{r}^{\alpha-1} \\ \mathbf{y}^{\nu-1} \end{bmatrix} = \begin{bmatrix} \left[\mathbf{r}^T \vdots \mathbf{0}_\rho^T \vdots \mathbf{0}_\rho^T \vdots ... \vdots \mathbf{0}_\rho^T\right]^T \\ \left[\mathbf{y}^T \vdots \mathbf{0}_N^T \vdots \mathbf{0}_N^T \vdots ... \vdots \mathbf{0}_N^T\right]^T \end{bmatrix}, \\ \begin{bmatrix} Q_0 & \vdots O_{\rho N} \\ -R_0 & \vdots \ E_0 \end{bmatrix} \begin{bmatrix} \mathbf{r} \\ \mathbf{y} \end{bmatrix} = \begin{bmatrix} \mathbf{0}_\rho \\ \mathbf{0}_N \end{bmatrix} \end{array} \right) \right\},$$

$$S_{eIIO} \subseteq \mathfrak{R}^{\alpha\rho+\nu N}, \tag{3.157}$$

- the origin $\mathbf{r}^{\alpha-1} = \mathbf{0}_{\alpha\rho}$ *if* $\nu = 0$,

$$S_{eIIO} = \left\{ \mathbf{r}^{\alpha-1} : \begin{array}{c} \mathbf{r}^{\alpha-1} = \left[\mathbf{r}^T \vdots \mathbf{0}_\rho^T \vdots \mathbf{0}_\rho^T \vdots ... \vdots \mathbf{0}_\rho^T\right]^T \in \mathfrak{R}^{\alpha\rho}, \\ Q_0\mathbf{r} = \mathbf{0}_\rho \end{array} \right\}. \tag{3.158}$$

Proof. This theorem follows directly from Theorem 109 and the theorem on the number of the solutions of the homogeneous linear vector algebraic equation. ∎

Note 111 *The set S_{eIIO} of the equilibrium states of the IIO system (3.149) becomes the singleton if and only if*

$$\det\begin{bmatrix} Q_0 & \vdots & O_{\rho N} \\ -R_0 & \vdots & E_0 \end{bmatrix} \neq 0,\ \nu \geq 1,$$
$$\det Q_0 \neq 0,\ \nu = 0.$$

The IIO system (3.149) then, and only then, has the unique equilibrium vector,

$$S_{eIIO} = \left\{ \begin{bmatrix} \mathbf{r}^{\alpha-1} \\ \mathbf{y}^{\nu-1} \end{bmatrix} : \left(\begin{array}{l} \begin{bmatrix} \mathbf{r}^{\alpha-1} \\ \mathbf{y}^{\nu-1} \end{bmatrix} = \begin{bmatrix} \left[\mathbf{r}^T \vdots \mathbf{0}_\rho^T \vdots \mathbf{0}_\rho^T \vdots ... \vdots \mathbf{0}_\rho^T\right]^T \\ \left[\mathbf{y}^T \vdots \mathbf{0}_N^T \vdots \mathbf{0}_N^T \vdots ... \vdots \mathbf{0}_N^T\right]^T \end{bmatrix} = \\ = \begin{bmatrix} \mathbf{0}_{\alpha\rho} \\ \mathbf{0}_{\nu N} \end{bmatrix} \in \mathfrak{R}^{\alpha\rho+\nu N} \end{array} \right) \right\} =$$
$$= \left\{ \begin{bmatrix} \mathbf{0}_{\alpha\rho} \\ \mathbf{0}_{\nu N} \end{bmatrix} \right\} = \{\mathbf{0}_{\alpha\rho+\nu N}\} \subset \mathfrak{R}^{\alpha\rho+\nu N},\ \nu \geq 1,$$
$$S_{eIIO} = \left\{ \mathbf{r}^{\alpha-1} : \mathbf{r}^{\alpha-1} = \left[\mathbf{r}^T \vdots \mathbf{0}_\rho^T \vdots \mathbf{0}_\rho^T \vdots ... \vdots \mathbf{0}_\rho^T\right]^T \right\} = \{\mathbf{0}_{\alpha\rho}\} \subset \mathfrak{R}^{\alpha\rho},\ \nu = 0.$$

Chapter 4

Transfer function matrix $G(s)$

The classical definition of the $SISO$ system transfer function determines it as the ratio of the Laplace transform of the system output to the Laplace transform of the system input *under all zero initial conditions*. It is then explained that the transfer function is the Laplace transform of the unit impulse response of the system *under all zero initial conditions.* For the $MIMO$ system, its transfer function matrix is consequently defined as the matrix composed of the system transfer functions, which relates the Laplace transform of the output vector to the Laplace transform of the input vector *under all zero initial conditions* [7], [15], [22], [85], [86], [87], [90], [106], [107], [108], [109], [114].

The transfer function matrix is defined, perhaps due to Desoer [17], in another literature on the linear systems as the Laplace transform of the corresponding matrix of the system unit impulse responses *under all zero initial conditions* [4], [11], [13], [77], [120], [121], [122].

These two definitions are equivalent. The former expresses the physical sense of the transfer function matrix from the point of view of the transmission and the transformation of the input vector onto and into the output system response, respectively. The latter contains the physical interpretation of the transfer function matrix from the point of view of the output system response to the specific (unit impulse) input vector. Both assume *all zero initial conditions.* They cannot relate the Laplace transform of the system output to the Laplace transform of its input as soon as any initial condition is not equal to zero.

The validity of the transfer function matrix $G(s)$ exclusively for all zero initial conditions restricts sharply its applications, validity and significance. Its application to the problems induced by nonzero initial conditions gives sometimes the wrong result. So wrong that is contrary to the exact result. For example, its application to the Lyapunov stability test gives the wrong result if there is the same zero and pole with non-negative real part(s), which are cancelled, and the real parts of all other poles are negative. Besides, the transfer function matrix $G(s)$ is inapplicable to the determination of the complete system response due to the influence of the input vector and nonzero initial conditions. The real initial conditions are rarely equal to zero. We devote the following part of the book to show how we can very effectively overcome this severe restriction.

Part II

FULL TRANSFER FUNCTION MATRIX $F(S)$ AND SYSTEM REALIZATION

Chapter 5

Problem statement

We consider usually the linear systems by separating the study of their reactions and properties under zero initial conditions from the study of their reactions and properties under zero input variables. This has been advantageous for attacking many problems among which are the zero-state system equivalency and the zero-input system equivalency [4]. P. J. Antsaklis and A. N. Michel noted that the zero-state system equivalency and the zero-input system equivalency do not necessarily imply the system equivalence ([4, p. 171]).

Another important note by Antsaklis and Michel [4, p. 387] is that different state-space realizations of the system transfer function matrix yield the same zero-state system response, but the corresponding zero-input (hence, also complete) system response can be quite different.

These observations by Antsaklis and Michel correspond to the reality of the system environment and its history, which require us to study the simultaneous influence of nonzero initial conditions and nonzero inputs. This appears important for the system overall (complete) both motion and (output) response, for the system equivalence, for the system realization and the system minimal realization, and for various system dynamic properties (such as $BIBO$ and L-stability when the system was not at rest at the initial moment, system total stability, system tracking and system optimality). This imposed the *Basic problem* 46 (in Subsection 3.3.4), which we pose now in another, more specific, form:

Problem 112 ***Main problem***

What is a compact mathematical description, in the complex domain $\mathfrak{C}$*, of that how the system transfers, in the course of time, the influence of all* *inputs and of arbitrary initial conditions on its state and/or on its output response?*

Can such a mathematical description be invariant relative to the system input and to all initial conditions, i.e., can it be completely independent of the system input and of all initial conditions?

What follows will show the solutions to this complex problem (which contains two subproblems) for the above described systems (Section 2.1 through Section 2.3).

The problem was solved by introducing, defining and determining ***the complete (full) matrix transfer function*** $F(.)$, and its complex matrix value that is ***the complete (full) transfer function matrix*** $F(s)$, of the *IO*, *ISO*, and *IIO* system in [29]. The solution was presented to senior undergraduates in the linear system and control courses at the University of Natal in Durban, RSA, [29] (1993), in the National Engineering School of Belfort (*Ecole Nationale d'Ingénieurs de Belfort - ENIBe*), France, [27] (1994-1996), and at the University of Technology Belfort-Montbéliard, Belfort, (*UTBM*), [26], [28], [49], [51], [63] (1999- 2003). It was exposed also in [68] (2003) and in [47] (2013). Besides, the full transfer function matrix of discrete-time systems was introduced to the students of the *UTBM* through the course on *Control of industrial processes* (2002, 2003) [51]. We will show in the sequel, additionally, by following [47], how the system full transfer matrix function $F(.)$ can be used to study various dynamic properties of the system and to solve the related problems.

Before we start defining various system transfer function matrices, we should clarify the notion and properties of *degenerate* and *nondegenerate matrix functions*, and their differences from the well-known reducible and irreducible matrices.

Chapter 6

Nondegenerate matrices

For the definitions and properties of the greatest common (left, right) divisors of the matrix polynomials, of the unimodular matrix polynomials and for their (left, right) coprimness see, for example, the books by P. J. Antsaklis and A. N. Michel [4, pp. 526-528, 535-540], C.-T. Chen [13, pp. 591-599] and T. Kailath [77, pp. 373-382].

A rational matrix function $M(.) = M_D^{-1}(.)M_N(.)$ $[M(.) = M_N(.)M_D^{-1}(.)]$ is called in [13, p. 605], [77, p. 370] *irreducible* if and only if its polynomial matrices $M_D(.)$ and $M_N(.)$ are (left and/or right) coprime. Such a definition of the irreducibility is adequate for a scalar case, i.e., when $M(.)$ is a rational function; $M_D(.)$ and $M_N(.)$ are then scalar polynomials. However, the generalization to rational matrix functions is not appropriate because the greatest common (left $L(.)$, and right $R(.)$) divisor of $M_D(.)$ and of $M_N(.)$ cancels itself in $M(.)$, in spite $L(.)$ and $R(.)$ are unimodular polynomial matrices,

$$M_D(s) = L(s)D(s),\ M_N(s) = L(s)N(s) \Longrightarrow$$
$$M(.) = M_D^{-1}(.)M_N(.) = D^{-1}(s)L^{-1}(s)L(s)N(s) = D^{-1}(s)N(s),$$

$$M_D(s) = D(s)R(s),\ M_N(s) = N(s)R(s) \Longrightarrow$$
$$M(.) = M_N(.)M_D^{-1}(.) = N(s)R(s)R^{-1}(s)D^{-1}(s) = N(s)D^{-1}(s).$$

In order to avoid any confusion, we will define *nondegenerate* and *degenerate* rational matrix functions [29].

Definition 113 *A rational matrix function $M(.) = M_D^{-1}(.)M_N(.)$ [respectively, $M(.) = M_N(.)M_D^{-1}(.)$] is*

a) ***row nondegenerate*** *if and only if respectively:*

(i) the greatest common left [right] divisor of $M_D(.)$ *and of* $M_N(.)$ *is a nonzero constant matrix, and*

(ii) the greatest common scalar factors of $\det M_D(s)$ *and of all elements of every row of* $(\operatorname{adj} M_D(s))\, M_N(s)$ *[respectively, of all elements of every row of* $M_N(s)\,(\operatorname{adj} M_D(s))$*] are nonzero constants.*

Otherwise, $M(.)$ *is* ***row degenerate.***

b) ***column nondegenerate*** *if and only if respectively:*

(i) the greatest common left [right] divisor of $M_D(.)$ *and* $M_N(.)$ *is a nonzero constant matrix, and*

(ii) the greatest common scalar factors of $\det M_D(s)$ *and of all elements of every column of* $(\operatorname{adj} M_D(s))\, M_N(s)$ *[respectively, of all elements of every column of* $M_N(s)\,(\operatorname{adj} M_D(s))$*] are nonzero constants.*

Otherwise, $M(.)$ *is* ***column degenerate.***

c) ***nondegenerate*** *if and only if respectively:*

(i) the greatest common left [right] divisor of $M_D(.)$ *and* $M_N(.)$ *is a nonzero constant matrix, and*

(ii) the greatest common scalar factor of $\det M_D(s)$ *and of all elements of* $(\operatorname{adj} M_D(s))\, M_N(s)$ *[respectively, of all elements of the product* $M_N(s)\,(\operatorname{adj} M_D(s))$*] is a nonzero constant.*

Otherwise, $M(.)$ *is* ***degenerate.***

This definition implies the following:

Note 114 *If a rational matrix function* $M(.) = M_D^{-1}(.)M_N(.)$ *[respectively,* $M(.) = M_N(.)M_D^{-1}(.)$*] is either row nondegenerate or column nondegenerate, or both, then it is also nondegenerate.*

Example 115 *Let*

$$M(s) = M_D^{-1}(s)M_N(s) = [(s+1)(s+2)]^{-1} \begin{bmatrix} s+1 & s+1 \\ s+2 & s+2 \end{bmatrix}.$$

It is both column nondegenerate and nondegenerate despite the fact that it is row degenerate. The greatest common factor of $\det M_D(s) = (s+1)(s+2)$ *and of all elements of the first row of* $M_N(s)$*,*

$$M_N(s) = \begin{bmatrix} s+1 & s+1 \\ s+2 & s+2 \end{bmatrix},$$

is $s+1$*. The greatest common factor of* $\det M_D(s) = (s+1)(s+2)$ *and of all elements of the second row of* $M_N(s)$ *is* $s+2$ *that is different from* $s+1$*.*

The greatest common factor of $\det M_D(s) = (s+1)(s+2)$ *and of all elements of the first column of* $M_N(s)$ *is* 1. *The same holds for the greatest common factor of* $\det M_D(s)$ *and of all elements of the second column of* $M_N(s)$, *as well as for the greatest common factor of* $\det M_D(s)$ *and of all elements of* $M_N(s)$. *The given* $M(s) = M_D^{-1}(s)M_N(s)$ *is both column nondegenerate and nondegenerate despite the fact that it is row degenerate.*

Example 116 *Let*

$$M(s) = \frac{1}{(s+1)(s+2)} \begin{bmatrix} s+1 & s+2 \\ s+1 & s+2 \end{bmatrix}.$$

It is both row nondegenerate and nondegenerate despite the fact that it is column degenerate.

Note 117 *If a rational matrix function* $M(.) = M_D^{-1}(.)M_N(.)$ *[respectively,* $M(.) = M_N(.)M_D^{-1}(.)$*] is degenerate, then it is both row degenerate and column degenerate.*

Example 118 *Let*

$$M(s) = \frac{1}{(s+1)(s+2)} \begin{bmatrix} s\,(s+1) & (s+3)\,(s+1) \\ 4\,(s+1) & (s+1)\,(s+2) \end{bmatrix}.$$

It is degenerate, and both column degenerate and row degenerate. Its nondegenerate form $M_{nd}(s)$,

$$M_{nd}(s) = \frac{1}{s+2} \begin{bmatrix} s & s+3 \\ 4 & s+2 \end{bmatrix}$$

is in this case also both column and row nondegenerate.

Note 119 *If a rational matrix function* $M(.) = M_D^{-1}(.)M_N(.)$ *[respectively,* $M(.) = M_N(.)M_D^{-1}(.)$*] is nondegenerate, then it is also irreducible, but the opposite does not hold in general (i.e., it can be irreducible but need not be nondegenerate).*

Example 120 *Let* $M(.) = M_D^{-1}(.)M_N(.)$,

$$M_D(s) = \begin{bmatrix} s^2 + s - 8 & 2s^2 + 3s - 2 \\ 2s^2 - 2s - 12 & 4s^2 - 2s \end{bmatrix} =$$

$$= \begin{bmatrix} s+2 & 6 \\ 2s & 12 \end{bmatrix} \begin{bmatrix} s-1 & 2s-1 \\ -1 & 0 \end{bmatrix},$$

$$M_N(s) = \begin{bmatrix} 2s^2+3\ s-2 & 6s^2+9s-6 & 4s^2+6s-4 \\ 4s^2-2s & 12s^2-6s & 8s^2-4s \end{bmatrix} =$$
$$= \begin{bmatrix} s+2 & 6 \\ 2s & 12 \end{bmatrix} \begin{bmatrix} 2s-1 & 6s-3 & 4s-2 \\ 0 & 0 & 0 \end{bmatrix}.$$

The polynomial matrices $M_D(.)$ *and* $M_N(.)$ *are left coprime. Their greatest left common divisor* $L(.)$,

$$L(s) = \begin{bmatrix} s+2 & 6 \\ 2s & 12 \end{bmatrix},\ L^{-1}(s) = \frac{1}{24}\begin{bmatrix} 12 & -6 \\ -2s & s+2 \end{bmatrix},\ detL(s) = 24,$$

is unimodular and cancels itself in $M(.)$. *The rational matrix function* $M(.)$ *is irreducible in the sense of the definition in [13, p. 605], [77, p. 370]. However, it is really further reducible, i.e., it is degenerate. The reduced form of* $M(s)$ *obtained after the cancellation of* $L(s)$ *reads*

$$M_{irr}(s) = \frac{1}{2s-1}\begin{bmatrix} 0 & -2s+1 \\ 1 & s-1 \end{bmatrix}\begin{bmatrix} 2s-1 & 6s-3 & 4s-2 \\ 0 & 0 & 0 \end{bmatrix} =$$
$$= \frac{1}{2s-1}\begin{bmatrix} 0 & 0 & 0 \\ 2s-1 & 3\,(2s-1) & 2\,(2s-1) \end{bmatrix}.$$

It is degenerate because the polynomial $2s-1$ *is common to* $detM_{irrD}(s) = 2s-1$ *and to all elements of* $(adjM_{irrD}(s))\,M_{irrN}(s)$:

$$(adjM_{irrD}(s))\,M_{irrN}(s) = \begin{bmatrix} 0 & 0 & 0 \\ 2s-1 & 3\,(2s-1) & 2\,(2s-1) \end{bmatrix}.$$

Evidently, the polynomial $2s-1$ *is not constant. We can cancel it in the denominator and in all entries of the matrix* $M_{irr}(s)$,

$$\begin{aligned} M_{irr}(s) &= \frac{1}{2s-1}\begin{bmatrix} 0 & \vdots & 0 & \vdots & 0 \\ 2s-1 & \vdots & 3\,(2s-1) & \vdots & 2\,(2s-1) \end{bmatrix} = \\ &= \frac{1}{2s-1}(2s-1)\begin{bmatrix} 0 \vdots 0 \vdots 0 \\ 1 \vdots 3 \vdots 2 \end{bmatrix}. \end{aligned}$$

The final, completely reduced form, hence fully irreducible form, i.e., the nondegenerate form $M_{nd}(s)$, *of* $M(s)$ *reads*

$$M_{nd}(s) = \begin{bmatrix} 0 \vdots 0 \vdots 0 \\ 1 \vdots 3 \vdots 2 \end{bmatrix}.$$

It is different from the irreducible form $M_{irr}(s)$,

$$M_{nd}(s) = \begin{bmatrix} 0\vdots 0\vdots 0 \\ 1\vdots 3\vdots 2 \end{bmatrix} \neq \frac{1}{2s-1} \begin{bmatrix} 0 \ \vdots & 0 & \vdots & 0 \\ 2s-1\vdots & 3\,(2s-1) & \vdots 2\,(2s-1) \end{bmatrix} = M_{irr}(s).$$

Example 121 *Let us consider 1x7 row matrix $M_{IIO}(s)$,*

$$M_{IIO}(s) = \left[(s-1)^2(s+2)(s+5)\right]^{-1} \bullet \begin{bmatrix} (s-1)^2(s-6)\,(17s+10) \ \vdots \\ \vdots \ -(s-1)(s-7)(17s+10) \ \vdots \\ \vdots \ -(s-1)(13s-10) \ \vdots \\ \vdots \ 10(s-1) \ \vdots \\ \vdots \ -5(s-1)(2s+5) \ \vdots \\ \vdots \ (s-1)(s^2+6s+5) \ \vdots \\ \vdots \ (s-1)(s+5) \end{bmatrix}^T . \tag{6.1}$$

It is (row) degenerate because $(s-1)$ is common factor to the denominator polynomial $(s-1)^2(s+2)(s+5)$ and to all entries of the numerator polynomial matrix that is row vector. Its (row) nondegenerate form $M_{IIOrnd}(s)$ results after the cancellation of $(s-1)$:

$$M_{IIOrnd}(s) = [(s-1)(s+2)(s+5)]^{-1} \bullet \begin{bmatrix} (s-1)(s-6)\,(17s+10) \ \vdots \\ \vdots \ -(s-7)(17s+10) \ \vdots \\ \vdots \ -(13s-10) \ \vdots \\ \vdots \ 10 \ \vdots \\ \vdots \ -5(2s+5) \ \vdots \\ \vdots \ (s^2+6s+5) \ \vdots \\ \vdots \ (s+5) \end{bmatrix}^T . \tag{6.2}$$

It is also irreducible.

Note 122 *Smith-McMillan form of a rational matrix is nondegenerate, hence irreducible, matrix. This follows from the definitions of the nondegenerate matrix, of the irreducible matrix, and from the definition of Smith-McMillan form of a rational matrix [4, pp. 298-299].*

Lemma 123 ***Basic lemma***

Let $M(.)$ be a real rational proper matrix function of s. Let $\mathbf{Z}(.)$, and $\mathbf{W}(.)$ be real rational proper vector functions of s, which are interrelated via $M(.)$,

$$\mathbf{Z}(s) = M(s)\mathbf{W}(s),\ \mathbf{Z}(s) \in \mathfrak{C}^p,\ M(s) \in \mathfrak{C}^{p\times q},\ \mathbf{W}(s) \in \mathfrak{C}^q. \tag{6.3}$$

1) Any equal pole and zero common to all elements of the same row of $M(s)$ do not influence the character of the original $\mathbf{z}(t)$ of $\mathbf{Z}(s)$ and may be cancelled.

2) Any equal pole and zero of any entry of $\mathbf{W}(s)$ do not influence the character of the corresponding entry of the original $\mathbf{z}(t)$ of $\mathbf{Z}(s)$ and may be cancelled.

3) Any equal pole and zero of any entry of $M(s)\mathbf{W}(s)$ do not influence the character of the corresponding entry of the original $\mathbf{z}(t)$ of $\mathbf{Z}(s)$ and may be cancelled.

4) The poles of the row nondegenerate form $[M(s)\mathbf{W}(s)]_{nd}$ of the matrix product $M(s)\mathbf{W}(s)$ determine the character of the original $\mathbf{z}(t)$ of $\mathbf{Z}(s)$, where $\mathbf{z}(t)$ is the inverse Laplace transform of $\mathbf{Z}(s)$,

$$\mathbf{z}(t) = \mathcal{L}^{-1}\left\{\mathbf{Z}(s)\right\} = \mathcal{L}^{-1}\left\{[M(s)\mathbf{W}(s)]_{rnd}\right\}. \tag{6.4}$$

5) If every zero of every element of every row of $M(s)$ is different from every pole of the corresponding entry of $\mathbf{W}(s)$, and every pole of every element of every row of $M(s)$ is different from every zero of the corresponding entry of $\mathbf{W}(s)$, then the row nondegenerate form $[M(s)\mathbf{W}(s)]_{rnd}$ of $M(s)\mathbf{W}(s)$ becomes the product of the row nondegenerate forms $M(s)_{rnd}$ and $\mathbf{W}(s)_{rnd}$ of $M(s)$ and $\mathbf{W}(s)$,

$$[M(s)\mathbf{W}(s)]_{rnd} = M(s)_{rnd}\mathbf{W}(s)_{rnd}. \tag{6.5}$$

Then (6.4) reduces to

$$\mathbf{z}(t) = \mathcal{L}^{-1}\left\{\mathbf{Z}(s)\right\} = \mathcal{L}^{-1}\left\{M(s)_{rnd}\mathbf{W}(s)_{rnd}\right\}. \tag{6.6}$$

Proof. Let the conditions of this Lemma 123 hold. From (6.3) and (6.4) we determine the original $\mathbf{z}(t)$. Let the ij-th element of $M(s)$ be $m_{ij}(s)$, the i-th element of $\mathbf{z}(t)$ be $z_i(t)$, the i-th element of $\mathbf{Z}(s)$ be $\varsigma_i(s)$, and the j-th element of $\mathbf{W}(s)$ be $w_j(s)$, so that

$$z_i(t) = \mathcal{L}^{-1}\left\{\varsigma_i(s)\right\} = \mathcal{L}^{-1}\left\{\sum_j m_{ij}(s)w_j(s)\right\}. \tag{6.7}$$

Since $M(.)$ is a real rational proper matrix function of s, $\mathbf{Z}(.)$ and $\mathbf{W}(.)$ are real rational proper vector functions of s, then the same holds for their entries, which can be presented in the factorized forms,

$$m_{ij}(s)=\frac{\prod_{k=1}^{\mu_{ij}}\left(s-s_{mk}^{oij}\right)}{\prod_{k=1}^{\nu_{ij}}\left(s-s_{mk}^{*ij}\right)},\ \mu_{ij}\leq\nu_{ij},\ w_j(s)=\frac{\prod_{k=1}^{\upsilon_j}\left(s-s_{wk}^{oj}\right)}{\prod_{k=1}^{\omega_j}\left(s-s_{wk}^{*j}\right)},\ \upsilon_j\leq\omega_j.$$

These equations set (6.7) into the following form

$$z_i(t)=\mathcal{L}^{-1}\left\{\varsigma_i(s)\right\}=\mathcal{L}^{-1}\left\{\sum_j\frac{\prod_{k=1}^{\mu_{ij}}\left(s-s_{mk}^{oij}\right)\prod_{k=1}^{\upsilon_j}\left(s-s_{wk}^{oj}\right)}{\prod_{k=1}^{\nu_{ij}}\left(s-s_{mk}^{*ij}\right)\prod_{k=1}^{\omega_j}\left(s-s_{wk}^{*j}\right)}\right\}.$$

We can conclude as follows. All residua of Heaviside expansion of $\varsigma_i(s) = \mathcal{L}\left\{z_i(t)\right\}$ are equal to zero in a pole that is equal to a zero of $\varsigma_i(s)$. They can be cancelled. If $s_{mk}^{oij} = s_{mk}^{*ij}$, $\forall j = 1,2,..,q$, then they should be cancelled. This proves 1). If $s_{wk}^{oj} = s_{wk}^{*j}$, $\forall j = 1,2,..,q$, then they should be also cancelled. This proves 2). The equal poles and zeros of $m_{ij}(s)w_j(s)$, $\forall j = 1,2,..,q$, do not influence $z_i(t)$. They should be cancelled, too. This proves 3). The equal poles and zeros of any entry of $M(s)\mathbf{W}(s)$ do not influence $\mathbf{z}(t)$. They should be cancelled. The result is the row nondegenerate form $[M(s)\mathbf{W}(s)]_{rnd}$ of $M(s)\mathbf{W}(s)$,

$$\mathbf{z}(t)=\mathcal{L}^{-1}\left\{\mathbf{Z}(s)\right\}=\mathcal{L}^{-1}\left\{[M(s)\mathbf{W}(s)]_{rnd}\right\},$$

i.e., (6.4). They determine the original $\mathbf{z}(t)$. The claim under 4) is correct. Let every zero of every element of every row of $M(s)$ be different from every pole of the corresponding entry of $\mathbf{W}(s)$, and let every pole of every element of every row of $M(s)$ be different from every zero of the corresponding entry of $\mathbf{W}(s)$. Then, the possible zero-pole cancellation is possible only among zeros and poles of the elements of the rows of $M(s)$, and independently of them among zeros and poles of the members of the entries of $\mathbf{W}(s)$. The cross-cancellations of the zeros/poles of the elements of the rows of $M(s)$ with poles/zeros of the members of the entries of $\mathbf{W}(s)$ is not possible. After carrying out all possible cancellations in the elements of the rows of $M(s)$ and in the components of the entries of $\mathbf{W}(s)$ we get the row nondegenerate form $[M(s)\mathbf{W}(s)]_{rnd}$ of $M(s)\mathbf{W}(s)$ as the product of the row nondegenerate forms $M(s)_{rnd}$ and $\mathbf{W}(s)_{rnd}$ of $M(s)$ and $\mathbf{W}(s)$,

$$[M(s)\mathbf{W}(s)]_{rnd} = M(s)_{rnd}\mathbf{W}(s)_{rnd},$$

which is (6.5). This and (6.4) imply (6.6) and complete the proof. ■

Example 124 *The IO system*

$$\begin{bmatrix} 1 & 1 \\ 1 & 2 \end{bmatrix} \mathbf{y}^{(2)}(t) - \begin{bmatrix} 1 & 1 \\ 1 & 2 \end{bmatrix} \mathbf{y}(t) =$$

$$= \begin{bmatrix} -1 & 2 \\ 0 & -3 \end{bmatrix} \mathbf{i}(t) + \begin{bmatrix} 1 & 1 \\ -1 & 1 \end{bmatrix} \mathbf{i}^{(1)}(t) + \begin{bmatrix} 0 & 0 \\ 1 & 0 \end{bmatrix} \mathbf{i}^{(2)}(t)$$

has the transfer function matrix

$$G_{IO}(s) = \frac{s^2-1}{(s^2-1)^2} \begin{bmatrix} -(s-1)(s-2) & s+7 \\ (s-1)^2 & -5 \end{bmatrix}.$$

Its nondegenerate form $G_{IOnd}(s)$ *reads*

$$G_{IOnd}(s) = \frac{1}{s^2-1} \begin{bmatrix} -(s-1(s-2) & s+7 \\ (s-1)^2 & -5 \end{bmatrix}.$$

It is also its row nondegenerate form $G_{IOrnd}(s)$,

$$G_{IOnd}(s) = G_{IOrnd}(s).$$

However, its column nondegenerate form $G_{IOcnd}(s)$ *is different from them,*

$$G_{IOcnd}(s) = \frac{1}{s+1} \begin{bmatrix} -(s-2) & s+7 \\ s-1 & -5 \end{bmatrix}.$$

If we wish to determine the system output response under all zero initial conditions we should use the row nondegenerate form $G_{IOrnd}(s)$ *of* $G_{IO}(s)$, *and we may not use its column nondegenerate form* $G_{IOcnd}(s)$ *because the pole* $s^* = 1$ *cannot be cancelled in the rows of* $G_{IOrnd}(s)$, *although it can be cancelled in its columns.*

Chapter 7

Definition of *F(s)*

7.1 Definition of *F(s)* in general

We present first the definition of the system *full transfer function matrix* $F(s)$ in general in order to express the essence of its meaning.

Definition 125 ***The full (complete) input-output (IO) transfer function matrix of the dynamical system in general***

The full (complete) input-output (IO) transfer function matrix of a linear time-invariant continuous-time dynamical system, *which is denoted by* $F(s)$, $F(s) \in \mathfrak{C}^{Nx(M+\varsigma)}$ *, is the complex matrix value of* ***the system full (complete) input-output (IO) matrix transfer function*** $F(.)$, $F(.) : \mathfrak{C} \rightarrow \mathfrak{C}^{Nx(M+\varsigma)}$, *which is a matrix function of the complex variable s such that it determines uniquely the (left, right) Laplace transform* $\mathbf{Y}^{(\mp)}(s)$ *of the system output* $\mathbf{y}(t)$ *as a homogenous linear function of the (left, right) Laplace transform* $\mathbf{I}^{(\mp)}(s)$ *of the system input vector* $\mathbf{i}(t)$ *for an arbitrary variation of* $\mathbf{i}(t)$, *for arbitrary initial vector values* $\mathbf{i}_{0\mp}^{\mu-1}$, *and/or* $\mathbf{r}_{0\mp}^{\alpha-1}$, *and/or* $\mathbf{x}_{0\mp}$, *and/or* $\mathbf{y}_{0\mp}^{\nu-1}$ *of the extended input vector* $\mathbf{i}^{\mu-1}(t)$, *the state vector* $\mathbf{r}^{\alpha-1}(t)$, *the state vector* $\mathbf{x}(t)$, *and the extended output vector* $\mathbf{y}^{\nu-1}(t)$ *at* $t = 0^{\mp}$, *respectively:*

$$\mathbf{Y}^{(\mp)}(s) = F(s)\left[\mathbf{I}^{(\mp)^T}(s) \vdots \left(\mathbf{i}_{0\mp}^{\mu-1}\right)^T \vdots \mathbf{x}_{0\mp}^T \vdots \left(\mathbf{y}_{0\mp}^{\nu-1}\right)^T\right]^T, \tag{7.1}$$

See Fig. 7.1.

In order to show clearly the complete analogy of the system *full transfer function matrix* $F(s)$ with the system transfer function matrix $G(s)$, and

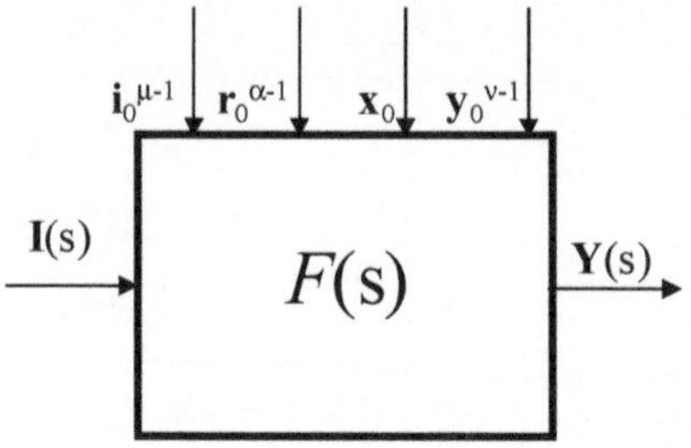

Figure 7.1: The full block of the system.

of the system *full block diagram* induced by $F(s)$ and the classical system block induced by $G(s)$, we introduce the action vector function $\mathbf{v}(.) : \mathfrak{T}_0 \longrightarrow \mathfrak{R}^{M+\varsigma}$,

$$\mathbf{v}(t) = \begin{bmatrix} \mathbf{i}(t) \\ \delta^{\mp}(t)\mathbf{C}_0 \end{bmatrix} \tag{7.2}$$

that comprises the input vector function $\mathbf{i}(t)$ and the vector $\mathbf{C}_0 \in \mathfrak{R}^{\varsigma}$ of all initial conditions: of the input ($\mathbf{i}_{0\mp}$, $\mathbf{i}_{0\mp}^{(1)}$, ..., $\mathbf{i}_{0\mp}^{(\mu-1)}$, i.e., $\mathbf{i}_{0\mp}^{\mu-1}$), of the state, i.e., of the state, ($\mathbf{r}_{0\mp}$, $\mathbf{r}_{0\mp}^{(1)}$, ..., $\mathbf{r}_{0\mp}^{(\alpha-1)}$, i.e., $\mathbf{r}_{0\mp}^{\alpha-1}$, or $\mathbf{x}_{0\mp}$) and of the output ($\mathbf{y}_{0\mp}$, $\mathbf{y}_{0\mp}^{(1)}$, ..., $\mathbf{y}_{0\mp}^{(\nu-1)}$, i.e., $\mathbf{y}_{0\mp}^{\nu-1}$) of the system, in general,

$$\mathbf{C}_0^{\mp} = \mathbf{C}_0^{\mp}\left(\mathbf{i}_{0\mp}^{\mu-1}, \mathbf{r}_{0\mp}^{\alpha-1}, \mathbf{x}_{0\mp}, \mathbf{y}_{0\mp}^{\nu-1}\right). \tag{7.3}$$

Specifically,

- For the *IO* systems:

$$\mathbf{C}_0^{\mp} = \mathbf{C}_0^{\mp}\left(\mathbf{i}_{0\mp}^{\mu-1}, \mathbf{y}_{0\mp}^{\nu-1}\right) = \begin{bmatrix} \mathbf{i}_{0\mp}^{\mu-1} \\ \mathbf{y}_{0\mp}^{\nu-1} \end{bmatrix} \in \mathfrak{R}^{\varsigma},\ \varsigma = \mu M + \nu N. \tag{7.4}$$

- For the *ISO* systems:

$$\mathbf{C}_0^{\mp} = \mathbf{C}_0^{\mp}\left(\mathbf{x}_{0\mp}\right) = \mathbf{x}_{0\mp} \in \mathfrak{R}^{\varsigma},\ \varsigma = n. \tag{7.5}$$

- For the *IIO* systems:

$$\mathbf{C}_0^{\mp} = \mathbf{C}_0^{\mp}\left(\mathbf{i}_{0\mp}^{\mu-1}, \mathbf{r}_{0\mp}^{\alpha-1}, \mathbf{y}_{0\mp}^{\nu-1}\right) = \begin{bmatrix} \mathbf{i}_{0\mp}^{\mu-1} \\ \mathbf{r}_{0\mp}^{\alpha-1} \\ \mathbf{y}_{0\mp}^{\nu-1} \end{bmatrix} \in \mathfrak{R}^{\varsigma},\ \varsigma = \mu M + \alpha\rho + \nu N. \tag{7.6}$$

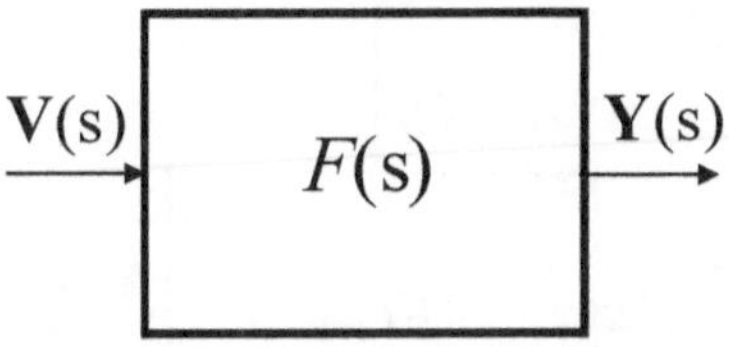

Figure 7.2: The full block of the system in the compact form.

Left/right Laplace transform of $\mathbf{v}(.)$ is $\mathbf{V}(.) : \mathfrak{C} \longrightarrow \mathfrak{C}^{M+\varsigma}$,

$$\mathbf{V}^{\mp}(s) = \begin{bmatrix} \mathbf{I}^{\mp}(s) \\ \mathbf{C}_0^{\mp} \end{bmatrix} \in \mathfrak{R}^{M+\varsigma}, \tag{7.7}$$

respectively.

We present now the equivalent definition to Definition 125.

Definition 126 ***The full (complete) input-output (IO) transfer function matrix of the dynamical system in general***

The full (complete) input-output (IO) transfer function matrix of a linear time-invariant continuous-time dynamical system, *which is denoted by $F(s)$, $F(s) \in \mathfrak{C}^{N\times(M+\varsigma)}$, is the complex matrix value of* ***the system full (complete) input-output (IO) matrix transfer function*** *$F(.)$, $F(.) : \mathfrak{C} \rightarrow \mathfrak{C}^{N\times(M+\varsigma)}$, which is a matrix function of the complex variable s such that it determines uniquely the (left, right) Laplace transform $\mathbf{Y}^{(\mp)}(s)$ of the system output $\mathbf{y}(t)$ as a homogenous linear function of the (left, right) Laplace transform $\mathbf{V}^{(\mp)}(s)$ of the overall system action vector $\mathbf{v}(t)$ for its arbitrary value and its variation,*

$$\mathbf{Y}^{(\mp)}(s) = F(s)\mathbf{V}^{(\mp)}(s), \tag{7.8}$$

See Fig. 7.2.

Comment 127 *We can present Fig. 7.1 and Fig. 7.2 as 7.3 due to (7.7).*

We will continue this presentation in Chapter 9 in order to establish *the algebra of the full block diagrams of the systems.*

7.2 Definition of $F(s)$ of the IO system

The full transfer function matrix of the IO system (3.77), (in Subsection 3.5.2),

$$A^{(\nu)}\mathbf{y}^{\nu}(t) = B^{(\mu)}\mathbf{i}^{\mu}(t), \forall\, t \in \mathfrak{T}, \tag{7.9}$$

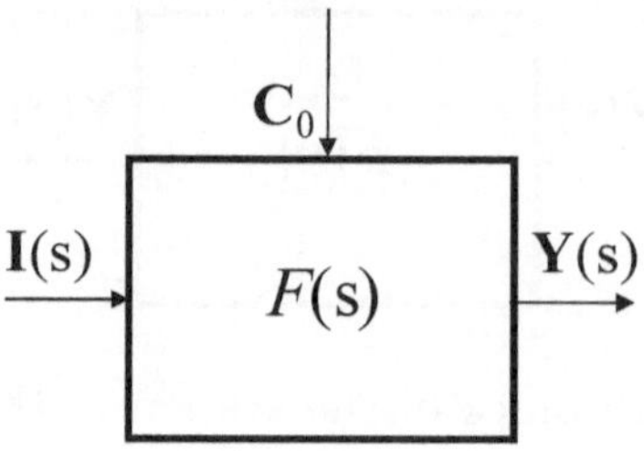

Figure 7.3: The full block of the system in the slightly extended form.

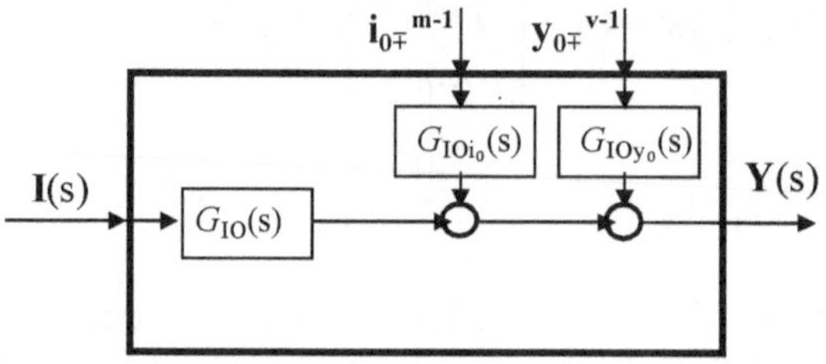

Figure 7.4: The block diagram of the *IO* system shows the system transfer function matrices relative to the input and initial conditions in which m $= \mu$ and v $= \nu$.

describes in the complex domain $\mathfrak{C}$ how the system transfers in *time* a simultaneous influence of arbitrary both input and output initial conditions $\mathbf{i}_{0\mp}$, $\mathbf{i}_{0\mp}^{(1)}$, ..., $\mathbf{i}_{0\mp}^{(\mu-1)}$, and $\mathbf{y}_{0\mp}$, $\mathbf{y}_{0\mp}^{(1)}$, ..., $\mathbf{y}_{0\mp}^{(\nu-1)}$, and of variations of the input vector function $\mathbf{i}(.)$ on the system output response $\mathbf{y}(.)$; see Fig. 7.4.

We will use the notation $Int\mathfrak{T}_0$ for the *interior* of the set $\mathfrak{T}_0$,

$$Int\mathfrak{T}_0 = \{t: \ t \in \mathfrak{T}_0, \ t > 0\}. \tag{7.10}$$

Definition 128 *a)* ***The full (complete) input-output (IO) transfer function matrix of the IO system (7.9)****, denoted by $F_{IO}(s)$, $F_{IO}(s) \in \ \mathfrak{C}^{N\mathrm{x}[(\mu+1)M+\nu N]}$, is the complex matrix value of* ***the system full (complete) IO matrix transfer function*** $F_{IO}(.)$*,* $F_{IO}(.) : \mathfrak{C} \rightarrow \mathfrak{C}^{N\mathrm{x}[(\mu+1)M+\nu N]}$*, which is a matrix function of the complex variable s such that it determines uniquely the (left, right) Laplace transform $\mathbf{Y}^{\mp}(s)$ of the system output $\mathbf{y}(t)$ as a homogenous linear function of the (left, right) Laplace transform $\mathbf{I}^{\mp}(s)$ of the system input $\mathbf{i}(t)$ for an arbitrary variation of $\mathbf{i}(t)$, for arbitrary initial vector values $\mathbf{i}_{0\mp}^{\mu-1}$ and $\mathbf{y}_{0\mp}^{\nu-1}$ of the extended input vector $\mathbf{i}^{\mu-1}(t)$ and of the extended output vector $\mathbf{y}^{\nu-1}(t)$ at $t = 0^{\mp}$,* *respectively:*

$$\mathbf{Y}^{\mp}(s) = F_{IO}(s)\left[\mathbf{I}^{(\mp)^T}(s) \vdots \left(\mathbf{i}_{0\mp}^{\mu-1}\right)^T \vdots \left(\mathbf{y}_{0\mp}^{\nu-1}\right)^T\right]^T \Longrightarrow \tag{7.11}$$

$$\mathbf{Y}^{\mp}(s) = F_{IO}(s)\left[\mathbf{I}^{(\mp)^T}(s) \vdots \mathbf{C}_{0IO}^{\mp^T}\right]^T = F_{IO}(s)\mathbf{V}_{IO}^{\mp}(s), \tag{7.12}$$

$$\mathbf{C}_{0IO}^{\mp} = \left\{ \begin{array}{c} \left[\left(\mathbf{i}_{0\mp}^{\mu-1}\right)^T \vdots \left(\mathbf{y}_{0\mp}^{\nu-1}\right)^T\right]^T, \ \mu \geq 1, \\ \mathbf{y}_{0\mp}^{\nu-1}, \ \mu = 0 \end{array} \right\}, \tag{7.13}$$

$$\mathbf{V}_{IO}^{\mp}(s) = \left[\mathbf{I}^{(\mp)^T}(s) \vdots \mathbf{C}_{0IO}^{\mp^T}\right]^T \tag{7.14}$$

b) ***The input-output (IO) transfer function matrix of the IO system (7.9)****, which is denoted by $G_{IO}(s)$, $G_{IO}(s) \in \mathfrak{C}^{N\times M}$, is the complex matrix value of* **the system IO matrix transfer function** *$G_{IO}(.)$, $G_{IO}(.) : \mathfrak{C} \to \mathfrak{C}^{N\times M}$, which is a matrix function of the complex variable s such that it determines uniquely the (left, right) Laplace transform $\mathbf{Y}^{\mp}(s)$ of the system output vector $\mathbf{y}(t)$ as a homogenous linear function of the (left, right) Laplace transform $\mathbf{I}^{\mp}(s)$ of the system input vector $\mathbf{i}(t)$ for an arbitrary variation of $\mathbf{i}(t)$, and under all zero initial conditions, that is that the initial vector values $\mathbf{i}_{0\mp}^{\mu-1} = \mathbf{0}_{\mu M}$ and $\mathbf{y}_{0\mp}^{\nu-1} = \mathbf{0}_{\nu N}$ of the extended input vector $\mathbf{i}^{\mu-1}(t)$ and of the extended output vector $\mathbf{y}^{\nu-1}(t)$ at $t = 0^{\mp}$ are equal to zero vectors, respectively:*

$$\mathbf{Y}^{\mp}(s) = G_{IO}(s)\mathbf{I}^{\mp}(s), \ \mathbf{i}_{0\mp}^{\mu-1} = \mathbf{0}_{\mu M}, \ \mathbf{y}_{0\mp}^{\nu-1} = \mathbf{0}_{\nu N}. \tag{7.15}$$

c) ***The input-output transfer function matrix (IOIC) relative to $\mathbf{i}_{0\mp}^{\mu-1}$ of the IO system (7.9)****, which is denoted by $G_{IOi_0}(s)$, $G_{IOi_0}(s) \in \mathfrak{C}^{N\times \mu M}$, is the complex matrix value of* **the system IOIC matrix transfer function** *$G_{IOi_0}(.)$* **relative to** *$\mathbf{i}_{0\mp}^{\mu-1}$, $G_{IOi_0}(.) : \mathfrak{C} \to \mathfrak{C}^{N\times \mu M}$, which is a matrix function of the complex variable s such that it determines uniquely, respectively, the (left, right) Laplace transform $\mathbf{Y}^{\mp}(s)$ of the system output $\mathbf{y}(t)$ as a homogenous linear function of an arbitrary initial vector $\mathbf{i}_{0\mp}^{\mu-1} \neq \mathbf{0}_{\mu M}$ of the extended input vector $\mathbf{i}^{\mu-1}(t)$ at $t = 0^{\mp}$ in the free regime on $Int\mathfrak{T}_0$ and for all zero output initial conditions, i.e., for $\mathbf{i}(t) = \mathbf{0}_M$, $\forall t \in Int\mathfrak{T}_0$, and $\mathbf{y}_{0\mp}^{\nu-1} \equiv \mathbf{0}_{\nu N}$:*

$$\mathbf{Y}^{\mp}(s) = G_{IOi_0}(s)\mathbf{i}_{0\mp}^{\mu-1}, \ \mathbf{i}(t) = \mathbf{0}_M, \ \forall t \in Int\mathfrak{T}_0, \ \mathbf{y}_{0\mp}^{\nu-1} \equiv \mathbf{0}_{\nu N}. \tag{7.16}$$

d) ***The input-output transfer function matrix (IOIY) relative to $\mathbf{y}_{0\mp}^{\nu-1}$ of the IO system (7.9)****, which is denoted by $G_{IOy_0}(s)$, $G_{IOy_0}(s) \in$*

$\mathfrak{C}^{N\times\nu N}$, *is the complex matrix value of* **the system IOIY matrix transfer function** $G_{IOy_0}(.)$ **relative to** $\mathbf{y}_{0\mp}^{\nu-1}$, $G_{IOy_0}(.) : \mathfrak{C} \to \mathfrak{C}^{N\times\nu N}$, *which is a matrix function of the complex variable s such that it determines uniquely, respectively, the (left, right) Laplace transform* $\mathbf{Y}^{\mp}(s)$ *of the system output* $\mathbf{y}(t)$ *as a homogenous linear function of an arbitrary initial vector* $\mathbf{y}_{0\mp}^{\nu-1}$ *of the extended output vector* $\mathbf{y}^{\nu-1}(t)$ *at* $t = 0^{\mp}$ *for the system in a free regime and under all zero input initial conditions, i.e., for* $\mathbf{i}(t) \equiv \mathbf{0}_M$ *and* $\mathbf{i}_{0\mp}^{\mu-1} = \mathbf{0}_{\mu M}$:

$$\mathbf{Y}^{\mp}(s) = G_{IOy_0}(s)\mathbf{y}_{0\mp}^{\nu-1},\ \mathbf{i}(t) \equiv \mathbf{0}_M,\ \mathbf{i}_{0\mp}^{\mu-1} = \mathbf{0}_{\mu M}. \tag{7.17}$$

e) **The input-output transfer function matrix relative to all initial conditions (IORAI) of the IO system (7.9)**, *which is denoted by* $G_{IO_0}(s)$, $G_{IO_0}(s) \in \mathfrak{C}^{N\times(\mu M+\nu N)}$, *is the complex matrix value of* **the system IORAI matrix transfer function** $G_{IO_0}(.)$ **relative to** $[\mathbf{i}_{0\mp}^{\mu-1^T}\ \ \mathbf{y}_{0\mp}^{\nu-1^T}]^T$, $G_{IO_0}(.) : \mathfrak{C} \to \mathfrak{C}^{N\times(\mu M+\nu N)}$, *which is a matrix function of the complex variable s such that it determines uniquely, respectively, the (left, right) Laplace transform* $\mathbf{Y}^{\mp}(s)$ *of the system output* $\mathbf{y}(t)$ *as a homogenous linear function of an arbitrary overall initial vector* $\mathbf{C}_{0IO} = [\mathbf{i}_{0\mp}^{\mu-1^T}\ \ \mathbf{y}_{0\mp}^{\nu-1^T}]^T \in \mathfrak{R}^{\mu M+\nu N}$ *composed of the extended input vector* $\mathbf{i}^{\mu-1}(t)$ *and of the extended output vector* $\mathbf{y}^{\nu-1}(t)$ *at* $t = 0^{\mp}$ *for the system in a free regime on* $Int\mathfrak{T}_0$, *i.e., for* $\mathbf{i}(t) = \mathbf{0}_M$, $\forall t \in Int\mathfrak{T}_0$:

$$\mathbf{Y}^{\mp}(s) = G_{IO_0}(s)\begin{bmatrix} \mathbf{i}_{0\mp}^{\mu-1} \\ \mathbf{y}_{0\mp}^{\nu-1} \end{bmatrix},\ \mathbf{i}(t) = \mathbf{0}_M,\ \forall t \in Int\mathfrak{T}_0. \tag{7.18}$$

Note 129 *The application of the superposition principle permits us to conclude immediately that* $G_{IO}(s)$, $G_{IOi_0}(s)$, *and* $G_{IOy_0}(s)$, *i.e.,* $G_{IO_0}(s)$, *are submatrices of* $F_{IO}(s)$; *see Fig. 7.5,*

$$F_{IO}(s) = \left[G_{IO}(s) \vdots \underbrace{G_{IOi_0}(s) \vdots G_{IOy_0}(s)}_{G_{IO_0}(s)} \right] = \left[G_{IO}(s) \vdots G_{IO_0}(s) \right],$$

$$G_{IO_0}(s) = \left[G_{IOi_0}(s) \vdots G_{IOy_0}(s) \right]. \tag{7.19}$$

$G_{IOi_0}(s)$ *and* $G_{IOy_0}(s)$ *compose the transfer function matrix* $G_{IO_0}(s)$ *relative to all initial conditions,*

$$G_{IO_0}(s) = \left[G_{IOi_0}(s) \vdots G_{IOy_0}(s) \right]. \tag{7.20}$$

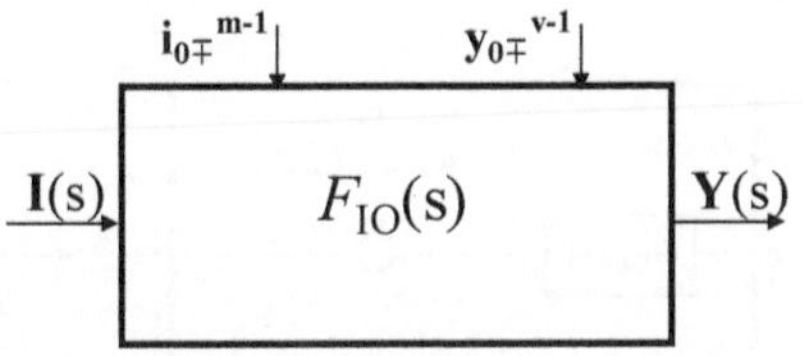

Figure 7.5: The block of the IO system shows the full transfer function matrix $F_{IO}(s)$, m $= \mu$ and v $= \nu$.

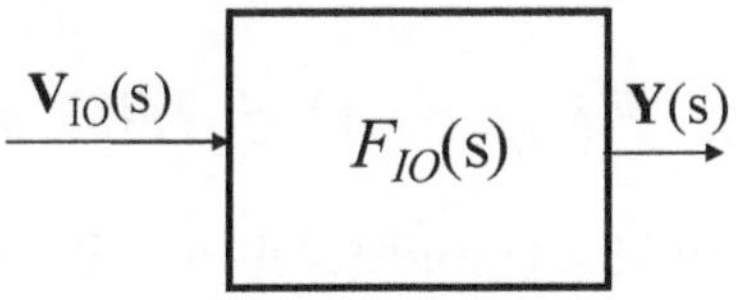

Figure 7.6: The full block of the IO system in the compact form.

We can now easily deduce (7.15) through (7.17) from (7.11) and (7.19), and vice versa.

Note 130 *We can generalize directly the classical block diagram technique if we replace the system transfer function matrix $G_{IO}(s)$ by its full transfer function matrix $F_{IO}(s)$ and if we use the vector $\mathbf{V}_{IO}^{\mp}(s)$,*

$$\mathbf{V}_{IO}^{\mp}(s) = \left[\mathbf{I}^{\mp^T}(s) \vdots \mathbf{C}_{0IO}^{\mp^T} \right]^T = \left[\mathbf{I}^{(\mp)^T}(s) \vdots \left(\mathbf{i}_{0\mp}^{\mu-1} \right)^T \vdots \left(\mathbf{y}_{0\mp}^{\nu-1} \right)^T \right]^T,$$

$$\mathbf{C}_{0IO}^{\mp} = \left[\left(\mathbf{i}_{0\mp}^{\mu-1} \right)^T \vdots \left(\mathbf{y}_{0\mp}^{\nu-1} \right)^T \right]^T,$$

instead of $\mathbf{I}^{\mp}(s)$, Fig. 7.6, or in the slightly extended form, Fig. 7.7.

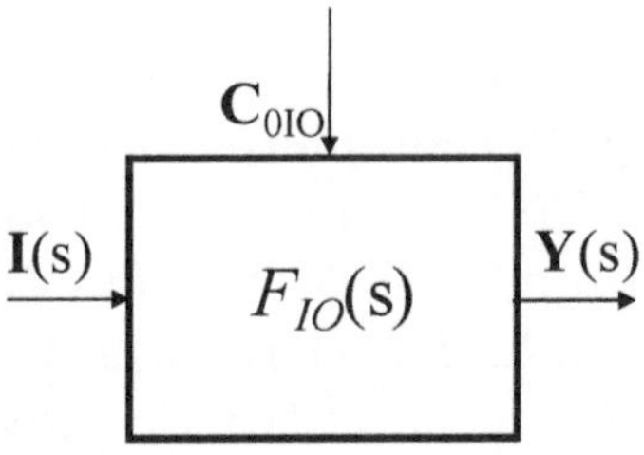

Figure 7.7: The slightly extended form of the full block of the IO system.

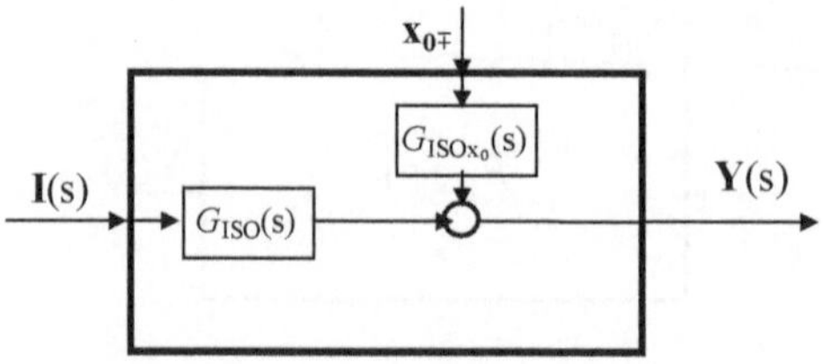

Figure 7.8: The full block diagram of the *ISO* system shows the system transfer function matrices.

7.3 Definition of *F(s)* of the *ISO* system

We repeat the *ISO* system (3.84) and (3.85) (in Subsection 3.5.3) as (7.21) and (7.22),

$$\frac{d\mathbf{x}(t)}{dt} = A\mathbf{x}(t) + B\mathbf{i}(t), \ \forall t \in \mathfrak{T}_0, \tag{7.21}$$

$$\mathbf{y}(t) = C\mathbf{x}(t) + D\mathbf{i}(t), \ \forall t \in \mathfrak{T}_0. \tag{7.22}$$

Fig. 7.8 shows the full block diagram of the *ISO* system (3.84) and (3.85). Its *full transfer function matrix* describes, in general, in the complex domain $\mathfrak{C}$ how the system transfers in *time* a simultaneous influence of an arbitrary initial state vector $\mathbf{x}_{0\mp}$, and of variations of the input vector function $\mathbf{i}(.)$ on the system output response $\mathbf{y}(.)$.

The input vector and the output vector of a fixed physical system do not depend on a form of a system mathematical description. Therefore, the system transfer function matrix $G(s)$ does not depend on the type of the system mathematical model. However, initial conditions, although arbitrary, have in general different meanings, hence forms, for the *IO* form (7.9) (Section 7.2) and for the *ISO* form (7.21) and (7.22) of the mathematical model of the same physical system. Therefore, the form of the *ISO* system full transfer function matrix depends on the form of the *ISO* system mathematical model in general.

Definition 131 *a)* ***The full (complete) input-output (IO) transfer function matrix of the ISO system (7.21) and (7.22),*** *which is denoted by* $F_{ISOIO}(s)$, $F_{ISOIO}(s) \in \mathfrak{C}^{N\times(M+n)}$, *for short* $F_{ISO}(s)$, *is the complex matrix value of* ***the system full (complete) IO matrix transfer function*** $F_{ISO}(.)$, $F_{ISO}(.) : \mathfrak{C} \to \mathfrak{C}^{N\times(M+n)}$, *which is a matrix function of the complex variable s such that it determines uniquely the (left, right) Laplace transform* $\mathbf{Y}^{\mp}(s)$ *of the system output* $\mathbf{y}(t)$ *as a homogenous linear* *function of the (left, right) Laplace transform* $\mathbf{I}^{\mp}(s)$ *of the system input* $\mathbf{i}(t)$ *for an arbitrary variation of* $\mathbf{i}(t)$*, and of arbitrary initial vector values* $\mathbf{x}_{0^{\mp}}$ *of the state vector* $\mathbf{x}(t)$ *at* $t = 0^{\mp}$*, respectively:*

$$\mathbf{Y}^{\mp}(s) = F_{ISO}(s)\left[\mathbf{I}^{\mp^T}(s) \vdots \mathbf{x}_{0^{\mp}}^T\right]^T \Longrightarrow \tag{7.23}$$

$$\mathbf{Y}^{\mp}(s) = F_{ISO}(s)\left[\mathbf{I}^{\mp^T}(s) \vdots \mathbf{C}_{0ISO}^{\mp^T}\right]^T = F_{ISO}(s)\mathbf{V}_{ISO}^{\mp}(s), \tag{7.24}$$

$$\mathbf{C}_{0ISO}^{\mp} = \mathbf{x}_{0^{\mp}},\ \mathbf{V}_{ISO}^{\mp}(s) = \left[\mathbf{I}^{\mp^T}(s) \vdots \mathbf{C}_{0ISO}^{\mp^T}\right]^T. \tag{7.25}$$

b) ***The input-output (IO) transfer function matrix of the ISO system*** *(7.21) and (7.22)***,** *which is denoted by* $G_{ISO}(s)$*,* $G_{ISO}(s) \in \mathfrak{C}^{N\times M}$*, is the complex matrix value of* ***the system IO matrix transfer function*** $G_{ISO}(.)$*,* $G_{ISO}(.) : \mathfrak{C} \rightarrow \mathfrak{C}^{N\times M}$*, which is a matrix function of the complex variable* s *such that it determines uniquely the (left, right) Laplace transform* $\mathbf{Y}^{\mp}(s)$ *of the system output* $\mathbf{y}(t)$ *as a homogenous linear function of the (left, right) Laplace transform* $\mathbf{I}^{\mp}(s)$ *of the system input* $\mathbf{i}(t)$ *for an arbitrary variation of* $\mathbf{i}(t)$*, and for zero initial state vector* $\mathbf{x}_{0^{\mp}}$ *of the state vector* $\mathbf{x}(t)$ *at* $t = 0^{\mp}$*, respectively:*

$$\mathbf{Y}^{\mp}(s) = G_{ISO}(s)\mathbf{I}^{\mp}(s),\ \mathbf{x}_{0^{\mp}} = \mathbf{0}_n. \tag{7.26}$$

c) ***The input, initial state-output (IISO) transfer function matrix relative to*** $\mathbf{x}_{0^{\mp}}$ ***of the ISO system*** *(7.21) and (7.22), which is denoted by* $G_{ISOx_0}(s)$*,* $G_{ISOx_0}(s) \in \mathfrak{C}^{N\times n}$*, is the complex matrix value of* ***the system IISO matrix transfer function*** $G_{ISOx_0}(.)$ ***relative to*** $\mathbf{x}_{0^{\mp}}$*,* $G_{ISOx_0}(.) : \mathfrak{C} \rightarrow \mathfrak{C}^{N\times n}$*, which is a matrix function of the complex variable* s *such that it determines uniquely the (left, right) Laplace transform* $\mathbf{Y}^{\mp}(s)$ *of the system output* $\mathbf{y}(t)$ *as a homogenous linear function of an arbitrary initial vector value* $\mathbf{x}_{0^{\mp}}$ *of the state vector* $\mathbf{x}(t)$ *at* $t = 0^{\mp}$ *for the system in a free regime (i.e., for* $\mathbf{i}(t) \equiv \mathbf{0}_M$*):*

$$\mathbf{Y}^{\mp}(s) = G_{ISOx_0}(s)\mathbf{x}_{0^{\mp}},\ \mathbf{i}(t) \equiv \mathbf{0}_M,\ G_{ISOx_0}(s) = G_{ISOo}(s). \tag{7.27}$$

Note 132 *Equations (7.23) - (7.25) imply (7.26) and (7.27), and vice versa, due to the superposition principle.*

Note 133 $G_{ISO}(s)$ *and* $G_{ISOo}(s)$ *compose the system full transfer function matrix* $F_{ISO}(s)$*,*

$$F_{ISO}(s) = \left[G_{ISO}(s) \vdots G_{ISOo}(s)\right],\ G_{ISOo}(s) = G_{ISOx_0}(s). \tag{7.28}$$

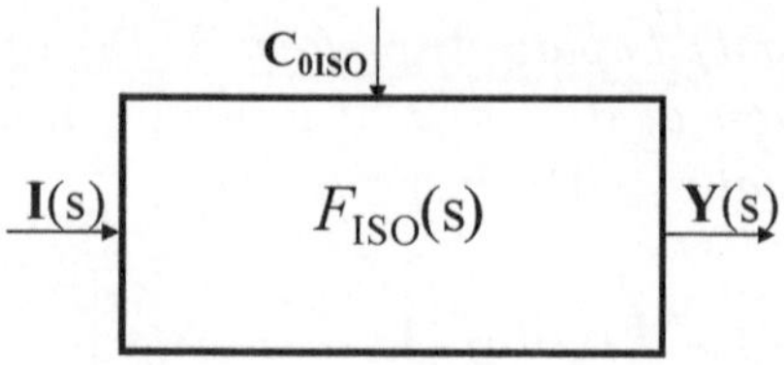

Figure 7.9: The slightly extended form of the full block of the ISO system.

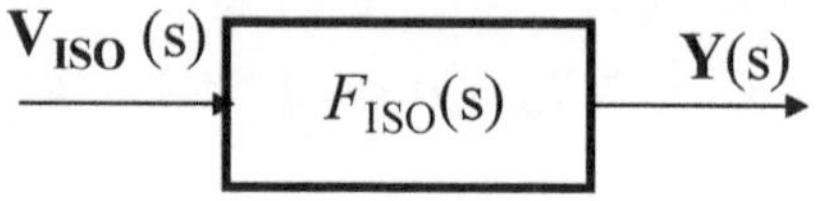

Figure 7.10: The full block of the ISO system in the compact form.

Fig. 7.9 and Fig. 7.10 represent the block of the ISO system (7.21) and (7.22).

Note 134 *When we use* $\mathbf{V}^{\mp}_{ISO}(s)$,

$$\mathbf{V}^{\mp}_{ISO}(s) = \left[\mathbf{I}^{\mp^T}(s) \vdots \mathbf{C}^{\mp^T}_{0ISO}\right]^T = \left[\mathbf{I}^{\mp^T}(s) \vdots \mathbf{x}^T_{0\mp}\right]^T,$$

instead of $\mathbf{I}^{\mp}(s)$, $F_{ISO}(s)$ *instead of* $G_{ISO}(s)$, *then we can apply the extended form of the full block of the ISO system (7.21) and (7.22), which is shown in Fig. 7.9 or the full block of the system in the compact form, Fig. 7.10.*

Note 135 *The initial state vector* $\mathbf{x}_{0\mp} = \mathbf{C}^{\mp}_{0ISO}$ *incorporates the influence of the initial input and output conditions. The latter do not appear explicitly in (7.23). Therefore, the ISO system (7.21) and (7.22) does not have (explicitly) the transfer function matrices relative to the initial input and output conditions. They are (implicitly) contained in the initial state vector* $\mathbf{x}_{0\mp}$.

Note 136 *The full IO transfer function matrix* $F_{IOISO}(s)$ *of the ISO system is the full transfer function matrix obtained from the equivalent IO system induced by the given ISO system (Appendix C.2). Its definition is given in Section 7.2, and its determination is the determination of the full transfer function matrix of the induced IO system, which is explained in Section 8.1.*

The full IO transfer function matrix $F_{IOISO}(s)$ *of the ISO system is different from the full ISO transfer function matrix* $F_{ISO}(s) \equiv F_{ISOIO}(s)$

of the same ISO system, $F_{IOISO}(s) \neq F_{ISO}(s)$*, which is the full transfer function matrix obtained from the given ISO model of the system.*

Note 137 *Besides the various system transfer function matrices related to the system output vector, we can consider the system transfer function matrices relative to the system state vector. They are defined as follows:*

Definition 138 *a)* ***The full (complete) input-state (IS) transfer function matrix of the ISO system*** *(7.21) and (7.22),* *which is denoted by* $F_{ISOIS}(s)$*,* $F_{ISOIS}(s) \in \mathfrak{C}^{nx(M+n)}$*, is the complex matrix value of* ***the system IS matrix transfer function*** $F_{ISOIS}(.)$*,* $F_{ISOIS}(.) : \mathfrak{C} \rightarrow \mathfrak{C}^{nx(M+n)}$*, which is a matrix function of the complex variable s such that it determines uniquely the (left, right) Laplace transform* $\mathbf{X}^{\mp}(s)$ *of the system state vector* $\mathbf{x}(t)$ *as a homogenous linear function of the (left, right) Laplace transform* $\mathbf{I}^{\mp}(s)$ *of the system input* $\mathbf{i}(t)$ *for an arbitrary variation of* $\mathbf{i}(t)$*, and for an arbitrary initial vector value* $\mathbf{x}_{0\mp}$ *of the state vector* $\mathbf{x}(t)$ *at* $t = 0^{\mp}$*, respectively,*

$$\mathbf{X}^{\mp}(s) = F_{ISOIS}(s)\left[\mathbf{I}^{\mp^T}(s) \vdots \mathbf{x}_{0\mp}^T\right]^T = F_{ISOIS}(s)\mathbf{V}_{ISO}^{\mp}(s). \tag{7.29}$$

b) ***The IS transfer function matrix of the ISO system*** *(7.21) and (7.22), which is denoted by* $G_{ISOIS}(s)$*,* $G_{ISOIS}(s) \in \mathfrak{C}^{nxM}$*, is the complex matrix value of* ***the system IS matrix transfer function*** $G_{ISOIS}(.)$*,* $G_{ISOIS}(.) : \mathfrak{C} \rightarrow \mathfrak{C}^{nxM}$*, which is a matrix function of the complex variable s such that it determines uniquely the (left, right) Laplace transform* $\mathbf{X}^{\mp}(s)$ *of the system state vector* $\mathbf{x}(t)$ *as a homogenous linear function of the (left, right) Laplace transform* $\mathbf{I}^{\mp}(s)$ *of the system input* $\mathbf{i}(t)$ *for an arbitrary variation of* $\mathbf{i}(t)$*, and for zero initial state vector* $\mathbf{x}_{0\mp}$ *of the state vector* $\mathbf{x}(t)$ *at* $t = 0^{\mp}$*, respectively:*

$$\mathbf{X}^{\mp}(s) = G_{ISOIS}(s)\mathbf{I}^{\mp}(s),\ \mathbf{x}_{0\mp} = \mathbf{0}_n. \tag{7.30}$$

c) ***The state-state (SS) transfer function matrix of the ISO system*** *(7.21) and (7.22), which is denoted by* $G_{ISOSS}(s)$*,* $G_{ISOSS}(s) \in \mathfrak{C}^{nxn}$*, is the complex matrix value of* ***the system SS matrix transfer function*** $G_{ISOSS}(.)$ ***relative to the initial state*** $\mathbf{x}_{0\mp}$*,* $G_{ISOSS}(.) : \mathfrak{C} \rightarrow \mathfrak{C}^{nxn}$*, which is a matrix function of the complex variable s such that it determines uniquely the (left, right) Laplace transform* $\mathbf{X}^{\mp}(s)$ *of the system state vector* $\mathbf{x}(t)$ *as a homogenous linear function of an arbitrary initial vector value* $\mathbf{x}_{0\mp}$

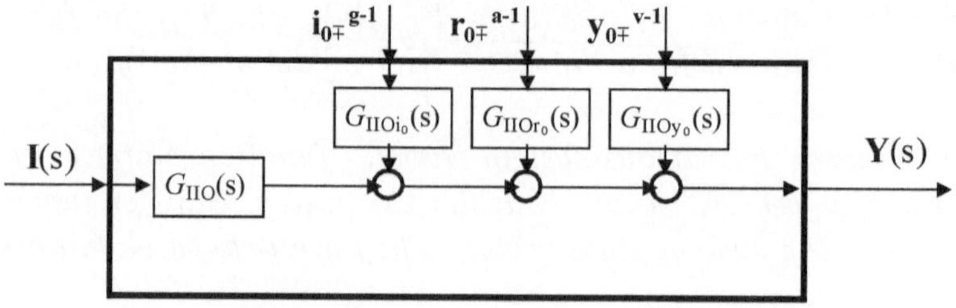

Figure 7.11: The block diagram of the *IIO* system shows the system transfer function matrices related to the input and to the extended initial vectors in which a $= \alpha$, g $= \gamma$, and v $= \nu$.

of the state vector $\mathbf{x}(t)$ *at* $t = 0^{\mp}$ *for the system in a free regime (i.e., for* $\mathbf{i}(t) \equiv \mathbf{0}_M$, $\forall t \in Int\mathfrak{T}_0$*):*

$$\mathbf{X}^{\mp}(s) = G_{ISOSS}(s)\mathbf{x}_{0^{\mp}}, \ \mathbf{i}(t) \equiv \mathbf{0}_M, \forall t \in Int\mathfrak{T}_0. \tag{7.31}$$

Note 139 *The system IS full transfer function matrix* $F_{ISOIS}(s)$ *is composed of* $G_{ISOIS}(s)$ *and* $G_{ISOSS}(s)$*,*

$$F_{ISOIS}(s) = \left[G_{ISOIS}(s) \vdots G_{ISOSS}(s) \right]. \tag{7.32}$$

7.4 Definition of *F(s)* of the *IIO* system

We defined γ in (2.40) (in Section 2.3) by

$$\gamma = \max\{\beta, \mu\}. \tag{7.33}$$

This helps us to determine *the full transfer function matrix* of the *IIO* system (3.90) (in Subsection 3.5.4) repeated as follows:

$$Q^{(\alpha)}\mathbf{r}^{\alpha}(t) = P^{(\beta)}\mathbf{i}^{\beta}(t), \ \forall t \in \mathfrak{T}_0,$$
$$E^{(\nu)}\mathbf{y}^{\nu}(t) = R^{(\alpha)}\mathbf{r}^{\alpha}(t) + T^{(\mu)}\mathbf{i}^{\mu}(t), \ \forall t \in \mathfrak{T}_0. \tag{7.34}$$

The system transmits and transfers differently the influences of the input vector $\mathbf{i}(\mathrm{t})$ and of the extended initial input vector $\mathbf{i}_{0\mp}^{\gamma-1}$, of the initial state vector $\mathbf{r}_{0\mp}^{\alpha-1}$, and of the extended initial output vector $\mathbf{y}_{0\mp}^{\nu-1}$. Therefore, the system has several different transfer function matrices related to different external actions and to the same system output; see Fig. 7.11.

Definition 140 *a)* ***The full (complete) input-output transfer (IO) function matrix of the IIO system (7.34)****, which is denoted by* $F_{IIO}(s)$,

$$F_{IIO}(s) \in \mathfrak{C}^{N\times[(\gamma+1)M+\delta\rho+\nu N]},$$

is the complex matrix value of ***the system full IO matrix transfer function*** $F_{IIO}(.)$, $F_{IIO}(.) : \mathfrak{C} \to \mathfrak{C}^{N\times[(\gamma+1)M+\alpha\rho+\nu N]}$*, which is matrix function of the complex variable* s *such that it determines uniquely the (left, right) Laplace transform* $\mathbf{Y}^{\mp}(s)$ *of the system output* $\mathbf{y}(t)$ *as a homogenous linear function of the (left, right) Laplace transform* $\mathbf{I}^{\mp}(s)$ *of the system input* $\mathbf{i}(t)$ *for an arbitrary variation of* $\mathbf{i}(t)$*, for arbitrary initial vector values* $\mathbf{i}_{0\mp}^{\gamma-1}$, $\mathbf{r}_{0\mp}^{\alpha-1}$*, and* $\mathbf{y}_{0\mp}^{\nu-1}$ *of the extended input vector* $\mathbf{i}^{\gamma-1}(t)$*, of the state vector* $\mathbf{r}^{\alpha-1}(t)$*, and of the extended output vector* $\mathbf{y}^{\nu-1}(t)$ *at* $t = 0^{\mp}$*, respectively:*

$$\mathbf{Y}^{\mp}(s) = F_{IIO}(s)\left[\mathbf{I}^{\mp^T}(s) \vdots \left(\mathbf{i}_{0\mp}^{\gamma-1}\right)^T \vdots \left(\mathbf{r}_{0\mp}^{\alpha-1}\right)^T \vdots \left(\mathbf{y}_{0\mp}^{\nu-1}\right)^T\right]^T \Longrightarrow \quad (7.35)$$

$$\mathbf{Y}^{\mp}(s) = F_{IIO}(s)\left[\mathbf{I}^{\mp^T}(s) \vdots \mathbf{C}_{0IIO}^{\mp^T}\right]^T = F_{IO}(s)\mathbf{V}_{IIO}^{\mp}(s), \quad (7.36)$$

$$\mathbf{C}_{0IIO}^{\mp} = \left[\left(\mathbf{i}_{0\mp}^{\gamma-1}\right)^T \vdots \left(\mathbf{r}_{0\mp}^{\alpha-1}\right)^T \vdots \left(\mathbf{y}_{0\mp}^{\nu-1}\right)^T\right]^T, \quad (7.37)$$

$$\mathbf{V}_{IIO}^{\mp}(s) = \left[\mathbf{I}^{\mp^T}(s) \vdots \mathbf{C}_{0IIO}^{\mp^T}\right]^T. \quad (7.38)$$

b) ***The input-output (IO) transfer function matrix of the IIO system (7.34)****, which is denoted by* $G_{IIO}(s)$, $G_{IIO}(s) \in \mathfrak{C}^{N\times M}$*, is the complex matrix value of* ***the system IO matrix transfer function*** $G_{IIO}(.)$, $G_{IIO}(.) : \mathfrak{C} \to \mathfrak{C}^{N\times M}$*, which is matrix function of the complex variable* s *such that it determines uniquely the (left, right) Laplace transform* $\mathbf{Y}^{\mp}(s)$ *of the system output* $\mathbf{y}(t)$ *as a homogenous linear function of the (left, right) Laplace transform* $\mathbf{I}^{\mp}(s)$ *of the system input* $\mathbf{i}(t)$ *for an arbitrary variation of* $\mathbf{i}(t)$*, and for all zero initial conditions, i.e., for* $\mathbf{i}_{0\mp}^{\gamma-1} = \mathbf{0}_{\gamma M}$, $\mathbf{r}_{0\mp}^{\alpha-1} = \mathbf{0}_{\alpha\rho}$*, and* $\mathbf{y}_{0\mp}^{\nu-1} = \mathbf{0}_{\nu N}$ *of the extended input vector* $\mathbf{i}^{\gamma-1}(t)$*, of the state vector* $\mathbf{r}^{\alpha-1}(t)$*, and of the extended output vector* $\mathbf{y}^{\nu-1}(t)$ *at* $t = 0^{\mp}$*, respectively:*

$$\mathbf{Y}^{\mp}(s) = G_{IIO}(s)\mathbf{I}^{\mp}(s),\ \mathbf{i}_{0\mp}^{\gamma-1} = \mathbf{0}_{\gamma M},\ \mathbf{r}_{0\mp}^{\alpha-1} = \mathbf{0}_{\alpha\rho},\ \mathbf{y}_{0\mp}^{\nu-1} = \mathbf{0}_{\nu N}. \quad (7.39)$$

c) ***The input-output (IICO) transfer function matrix relative to*** $\mathbf{i}_{0\mp}^{\gamma-1}$ ***of the IIO system (7.34)****, which is denoted by* $G_{IIOi_0}(s)$, $G_{IIOi_0}(s) \in \mathfrak{C}^{N\times(\gamma+1)M}$*, is the complex matrix value of* ***the system IICO matrix***

transfer function $G_{IIOi_0}(.)$ ***relative to*** $\mathbf{i}_{0\mp}^{\gamma-1}$, $G_{IIOi_0}(.) : \mathfrak{C} \to \mathfrak{C}^{N\times(\gamma+1)M}$, *which is matrix function of the complex variable s such that it determines uniquely the (left, right) Laplace transform* $\mathbf{Y}^{\mp}(s)$ *of the system output* $\mathbf{y}(t)$ *as a homogenous linear function of* $\mathbf{i}_{0\mp}^{\gamma-1}$ *in the free regime on* $Int\mathfrak{T}_0$ *(i.e., for* $\mathbf{i}(t) = \mathbf{0}_M$, $\forall t \in Int\mathfrak{T}_0$*), and for all other zero initial conditions, i.e., for* $\mathbf{r}_{0\mp}^{\alpha-1} = \mathbf{0}_{\alpha\rho}$, *and* $\mathbf{y}_{0\mp}^{\nu-1} = \mathbf{0}_{\nu N}$ *of the state vector* $\mathbf{r}^{\alpha-1}(t)$ *and of the extended output vector* $\mathbf{y}^{\nu-1}(t)$ *at* $t = 0^{\mp}$, *respectively:*

$$\mathbf{Y}^{\mp}(s) = G_{IIOi_0}(s)\mathbf{i}_{0\mp}^{\gamma-1},\ \mathbf{i}(t) \equiv \mathbf{0}_M,$$
$$\forall t \in Int\mathfrak{T}_0,\ \mathbf{r}_{0\mp}^{\alpha-1} = \mathbf{0}_{\alpha\rho},\ \mathbf{y}_{0\mp}^{\nu-1} = \mathbf{0}_{\nu N}. \tag{7.40}$$

d) ***The input-output (IIRO) transfer function matrix relative to*** $\mathbf{r}_{0\mp}^{\alpha-1}$ ***of the IIO system (7.34)****, which is denoted by* $G_{IIOr_0}(s)$, $G_{IIOr_0}(s) \in \mathfrak{C}^{N\times\alpha\rho}$, *is the complex matrix value of* **the system IIRO** ***matrix transfer function*** $G_{IIOr_0}(.)$ ***relative to*** $\mathbf{r}_{0\mp}^{\alpha-1}$, $G_{IIOr_0}(.) : \mathfrak{C} \to \mathfrak{C}^{N\times\alpha\rho}$, *which is a matrix function of the complex variable s such that it determines uniquely the (left, right) Laplace transform* $\mathbf{Y}^{\mp}(s)$ *of the system output* $\mathbf{y}(t)$ *as a homogenous linear function of* $\mathbf{r}_{0\mp}^{\alpha-1}$ *in the free regime (i.e., for* $\mathbf{i}(t) \equiv \mathbf{0}_M$*), and for all other zero initial conditions, i.e., for* $\mathbf{i}_{0\mp}^{\gamma-1} = \mathbf{0}_{\gamma M}$, *and* $\mathbf{y}_{0\mp}^{\nu-1} = \mathbf{0}_{\nu N}$ *of the extended input vector* $\mathbf{i}^{\gamma-1}(t)$ *and of the extended output vector* $\mathbf{y}^{\nu-1}(t)$ *at* $t = 0^{\mp}$, *respectively:*

$$\mathbf{Y}^{\mp}(s) = G_{IIOr_0}(s)\mathbf{r}_{0\mp}^{\alpha-1},\ \mathbf{i}(t) \equiv \mathbf{0}_M,\ \mathbf{i}_{0\mp}^{\gamma-1} = \mathbf{0}_{\gamma M},\ \mathbf{y}_{0\mp}^{\nu-1} = \mathbf{0}_{\nu N}. \tag{7.41}$$

e) ***The input-output (IIYO) transfer function matrix relative to*** $\mathbf{y}_{0\mp}^{\nu-1}$ ***of the IIO system (7.34)****, which is denoted by* $G_{IIOy_0}(s)$, $G_{IIOy_0}(s) \in \mathfrak{C}^{N\times\nu N}$, *is the complex matrix value of* **the system IIYO** ***matrix transfer function*** $G_{IIOy_0}(.)$ ***relative to*** $\mathbf{y}_{0\mp}^{\nu-1}$, $G_{IIOy_0}(.) : \mathfrak{C} \to \mathfrak{C}^{N\times\nu N}$, *which is a matrix function of the complex variable s such that it determines uniquely the (left, right) Laplace transform* $\mathbf{Y}^{\mp}(s)$ *of the system output* $\mathbf{y}(t)$ *as a homogenous linear function of* $\mathbf{y}_{0\mp}^{\nu-1}$ *in the free regime (i.e., for* $\mathbf{i}(t) \equiv \mathbf{0}_M$*), and for all other zero initial conditions, i.e., for* $\mathbf{i}_{0\mp}^{\gamma-1} = \mathbf{0}_{\gamma M}$, *and* $\mathbf{r}_{0\mp}^{\alpha-1} = \mathbf{0}_{\alpha\rho}$ *of the extended input vector* $\mathbf{i}^{\gamma-1}(t)$ *and of the state vector* $\mathbf{r}^{\alpha-1}(t)$ *at* $t = 0^{\mp}$, *respectively:*

$$\mathbf{Y}^{\mp}(s) = G_{IIOy_0}(s)\mathbf{y}_{0\mp}^{\nu-1},\ \mathbf{i}(t) \equiv \mathbf{0}_M,\ \mathbf{i}_{0\mp}^{\gamma-1} = \mathbf{0}_{\gamma M},\ \mathbf{r}_{0\mp}^{\alpha-1} = \mathbf{0}_{\alpha\rho}. \tag{7.42}$$

Note 141 *This definition applies also to the RS systems, the PMD systems and the GISO systems (2.39) (Section 2.3). They are special cases of the IIO systems (Appendix C.3).*

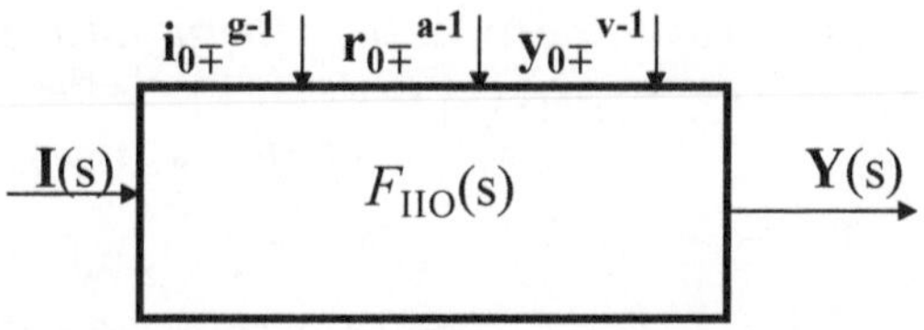

Figure 7.12: The extended full block of the *IIO* system, in which a = α, g = γ, and v = ν.

Note 142 *The transfer function matrices* $G_{IIO}(s)$, $G_{IIOi_0}(s)$, $G_{IIOr_0}(s)$, *and* $G_{IIOy_0}(s)$ *compose the system full transfer function matrix* $F_{IIO}(s)$,

$$F_{IIO}(s) = \left[G_{IIO}(s) \vdots G_{IIOi_0}(s) \vdots G_{IIOr_0}(s) \vdots G_{IIOy_0}(s) \right] \Longrightarrow \tag{7.43}$$

$$F_{IIO}(s) = \left[G_{IIO}(s) \vdots G_{IIOo}(s) \right], \tag{7.44}$$

$$G_{IIOo}(s) = \left[G_{IIOi_0}(s) \vdots G_{IIOr_0}(s) \vdots G_{IIOy_0}(s) \right], \tag{7.45}$$

in view of the system linearity, Fig. 7.12. $G_{IIOo}(s)$ *is the system transfer function matrix relative to all initial conditions. Equations (7.43) through (7.45) result from (7.39) through (7.42), and vice versa.*

Note 143 *The very useful block diagram method stays effective if the system full transfer function matrix* $F_{IIO}(s)$ *replaces the system transfer function matrix* $G_{IIO}(s)$ *and the vector* $\mathbf{V}_{IIO}^{\mp}(s)$,

$$\mathbf{V}_{IIO}^{\mp}(s) = \left[\mathbf{I}^{\mp^T}(s) \vdots \mathbf{C}_{0IIO}^{\mp^T} \right]^T =$$

$$= \left[\mathbf{I}^{\mp^T}(s) \vdots \left(\mathbf{i}_{0\mp}^{\gamma-1}\right)^T \vdots \left(\mathbf{r}_{0\mp}^{\alpha-1}\right)^T \vdots \left(\mathbf{y}_{0\mp}^{\nu-1}\right)^T \right]^T.$$

It takes the role of the Laplace transform $\mathbf{I}^{\mp}(s)$ *of the input vector* $\mathbf{i}(t)$. *Therefore,*

$$\mathbf{Y}^{\mp}(s) = \left[G_{IIO}(s) \vdots G_{IIOo}(s) \right] \left[\mathbf{I}^{\mp^T}(s) \vdots \mathbf{C}_{0IIO}^{\mp^T} \right]^T = F_{IIO}(s)\mathbf{V}_{IIO}^{\mp}(s).$$

Analogously to the *ISO* systems (Section 7.3), there exist various transfer function matrices related only to the state of the *IIO* system:

Definition 144 *a)* ***The full (complete) input-state (IS) transfer function matrix*** $F_{IIOIS}(s)$, $F_{IIOIS}(s) \in \mathfrak{C}^{\rho \times [(\beta+1)M+\alpha\rho]}$, ***of the IIO system (7.34)*** *is the complex matrix value of* ***the system full IS matrix transfer function*** $F_{IIOIS}(.)$,

$$F_{IIOIS}(.) : \mathfrak{C} \rightarrow \mathfrak{C}^{\rho \times [(\beta+1)M+\alpha\rho]},$$

which is a matrix function of the complex variable s such that it determines uniquely the (left, right) Laplace transform $\mathbf{R}^{\mp}(s)$ *of the system substate vector* $\mathbf{r}(t)$ *as a homogenous linear function of the (left, right) Laplace transform* $\mathbf{I}^{\mp}(s)$ *of the system input* $\mathbf{i}(t)$ *for an arbitrary variation of* $\mathbf{i}(t)$, *for arbitrary initial vector values* $\mathbf{i}_{0\mp}^{\beta-1}$, *and* $\mathbf{r}_{0\mp}^{\alpha-1}$ *of the extended input vector* $\mathbf{i}^{\beta-1}(t)$ *and of the state vector* $\mathbf{r}^{\alpha-1}(t)$ *at* $t = 0^{\mp}$, *respectively:*

$$\mathbf{R}^{\mp}(s) = F_{IIOIS}(s) \left[\mathbf{I}^{\mp^T}(s) \vdots \left(\mathbf{i}_{0\mp}^{\beta-1} \right)^T \vdots \left(\mathbf{r}_{0\mp}^{\alpha-1} \right)^T \right]^T. \tag{7.46}$$

b) ***The IRIS transfer function matrix of the IIO system (7.34),*** *which is denoted by* $G_{IIOIS}(s)$, $G_{IIOIS}(s) \in \mathfrak{C}^{\rho \times M}$, *is the complex matrix value of* ***the system IRIS matrix transfer function*** $G_{IIOIS}(.)$, $G_{IIOIS}(.) : \mathfrak{C} \rightarrow \mathfrak{C}^{\rho \times M}$, *which is a matrix function of the complex variable s such that it determines uniquely the (left, right) Laplace transform* $\mathbf{R}^{\mp}(s)$ *of the system substate vector* $\mathbf{r}(t)$ *as a homogenous linear function of the (left, right) Laplace transform* $\mathbf{I}^{\mp}(s)$ *of the system input* $\mathbf{i}(t)$ *for an arbitrary variation of* $\mathbf{i}(t)$, *and for all zero initial conditions, i.e., for* $\mathbf{i}_{0\mp}^{\beta-1} = \mathbf{0}_{\beta M}$ *of the extended input vector* $\mathbf{i}^{\beta-1}(t)$ *and of the state vector* $\mathbf{r}^{\alpha-1}(t)$ *at* $t = 0^{\mp}$:

$$\mathbf{R}^{\mp}(s) = G_{IIOIS}(s)\mathbf{I}^{\mp}(s),\ \mathbf{i}_{0\mp}^{\beta-1} = \mathbf{0}_{\beta M},\ \mathbf{r}_{0\mp}^{\alpha-1} = \mathbf{0}_{\alpha\rho}. \tag{7.47}$$

c) ***The IRII transfer function matrix relative to*** $\mathbf{i}_{0\mp}^{\beta-1}$ ***of the IIO system (7.34),*** *which is denoted by* $G_{IIOi_0IS}(s)$, $G_{IIOi_0IS}(s) \in \mathfrak{C}^{\rho \times (\beta+1)M}$, *is the complex matrix value of* ***the system IRII matrix transfer function*** $G_{IIOi_0IS}(.)$ ***relative to*** $\mathbf{i}_{0\mp}^{\beta-1}$, $G_{IIOi_0IS}(.) : \mathfrak{C} \rightarrow \mathfrak{C}^{\rho \times (\beta+1)M}$, *which is matrix function of the complex variable s such that it determines uniquely the (left, right) Laplace transform* $\mathbf{R}^{\mp}(s)$ *of the system substate vector* $\mathbf{r}(t)$ *as a homogenous linear function of* $\mathbf{i}_{0\mp}^{\beta-1}$ *in the free regime on* $Int\mathfrak{T}_0$ *(i.e., for* $\mathbf{i}(t) = \mathbf{0}_M$, $\forall t \in Int\mathfrak{T}_0$*), and for all other zero initial conditions, i.e., for* $\mathbf{r}_{0\mp}^{\alpha-1} = \mathbf{0}_{\alpha\rho}$, *of the state vector* $\mathbf{r}^{\alpha-1}(t)$ *at* $t = 0^{\mp}$, *respectively:*

$$\mathbf{R}^{\mp}(s) = G_{IIOi_0IS}(s)\mathbf{i}_{0\mp}^{\beta-1},\ \mathbf{r}_{0\mp}^{\alpha-1} = \mathbf{0}_{\alpha\rho},\ \mathbf{i}(t) = \mathbf{0}_M, \forall t \in Int\mathfrak{T}_0. \tag{7.48}$$

*d) **The IRIR transfer function matrix relative to $\mathbf{r}_{0\mp}^{\alpha-1}$ of the IIO system (7.34)**, which is denoted by $G_{IIOr_0IS}(s)$, $G_{IIOr_0IS}(s) \in \mathfrak{C}^{\rho x \alpha \rho}$, is the complex matrix value of **the system IRIR matrix transfer function** $G_{IIOr_0IS}(.)$ **relative to $\mathbf{r}_{0\mp}^{\alpha-1}$**, $G_{IIOr_0IS}(.) : \mathfrak{C} \rightarrow \mathfrak{C}^{\rho x \alpha \rho}$, which is a matrix function of the complex variable s such that it determines uniquely the (left, right) Laplace transform $\mathbf{R}^{\mp}(s)$ of the system substate vector $\mathbf{r}(t)$ as a homogenous linear function of $\mathbf{r}_{0\mp}^{\alpha-1}$ in the free regime (i.e., for $\mathbf{i}(t) \equiv \mathbf{0}_M$), and for all other zero initial conditions, i.e., for $\mathbf{i}_{0\mp}^{\beta-1} = \mathbf{0}_{\beta M}$, of the extended input vector $\mathbf{i}^{\beta-1}(t)$ at $t = 0^{\mp}$:*

$$\mathbf{R}^{\mp}(s) = G_{IIOr_0IS}(s)\mathbf{r}_{0\mp}^{\alpha-1},\ \mathbf{i}(t) \equiv \mathbf{0}_M,\ \mathbf{i}_{0\mp}^{\beta-1} = \mathbf{0}_{\beta M}. \tag{7.49}$$

Note 145 *The submatrices of the full IRI transfer function matrix $F_{IIOII}(s)$ are the transfer function matrices $G_{IIOII}(s)$, $G_{IIOi_0II}(s)$, and $G_{IIOr_0II}(s)$,*

$$F_{IIOIS}(s) = \left[G_{IIOIS}(s) \vdots \underbrace{G_{IIOi_0IS}(s) \vdots G_{IIOr_0IS}(s)}_{G_{IIOoIS}(s)} \right] =$$
$$= \left[G_{IIOIS}(s) \vdots G_{IIOoIS}(s) \right],$$
$$G_{IIOoIS}(s) = \left[G_{IIOi_0IS}(s) \vdots G_{IIOr_0IS}(s) \right]. \tag{7.50}$$

Note 146 $F_{IIOIS}(s) \neq F_{IIO}(s)$.

Chapter 8

Determination of $F(s)$

8.1 $F(s)$ of the IO system

The transfer function matrix $G_{IO}(s)$ of the IO system (3.76), i.e., (3.77), (in Subsection 3.5.2), which is repeated as (8.1):

$$A^{(\nu)}\mathbf{y}^{\nu}(t) = B^{(\mu)}\mathbf{i}^{\mu}(t), \ \forall t \in \mathfrak{T}_0. \tag{8.1}$$

can be set into the following *compact form* [28] and [51]:

$$G_{IO}(s) = \left(A^{(\nu)} S_N^{(\nu)}(s)\right)^{-1} \left(B^{(\mu)} S_M^{(\mu)}(s)\right), \tag{8.2}$$

by using the matrices $A^{(\nu)}$ and $B^{(\mu)}$ (2.2) (in Section 2.1),

$$A^{(\nu)} = \left[A_0 \vdots \ A_1 \vdots ... \vdots A_\nu\right] \in \mathfrak{R}^{N\text{x}(\nu+1)N},$$

$$B^{(\mu)} = \left[B_0 \vdots \ B_1 \vdots ... \vdots B_\mu\right] \in \mathfrak{R}^{N\text{x}(\mu+1)M}, \tag{8.3}$$

and the complex matrix function $S_i^{(k)}(.) : \mathfrak{C} \longrightarrow \mathfrak{C}^{\ i(k+1)\text{x}i}$, (3.41) (in Subsection 3.4.2),

$$S_i^{(k)}(s) = \left[s^0 I_i \vdots s^1 I_i \vdots s^2 I_i \vdots ... \vdots s^k I_i\right]^T \in \mathfrak{C}^{\ i(k+1)\text{x}i},$$

$$I_i = diag\{1 \ \ 1 \ \ ... \ 1\} \in \mathfrak{R}^{i\text{x}i}, \ (k,i) \in \{(\mu, M), \ \ (\nu, N)\}. \tag{8.4}$$

The compact form (8.2) of $G_{IO}(s)$ results from the Laplace transform of (8.1) and from the fact that (8.3) and (8.4) imply

$$\left(\sum_{k=0}^{k=\nu} A_k s^k\right)^{-1} \left(\sum_{k=0}^{k=\mu\leq\nu} B_k s^k\right) = \left(A^{(\nu)} S_N^{(\nu)}(s)\right)^{-1} \left(B^{(\mu)} S_M^{(\mu)}(s)\right).$$

We use also the matrix function $Z_k^{(\varsigma-1)}(.) : \mathfrak{C} \to \mathfrak{C}^{(\varsigma+1)kx\varsigma k}$, (3.42) (in Subsection 3.4.2), in order to determine the compact form of the system full transfer function matrix $F_{IO}(s)$,

$$Z_k^{(\varsigma-1)}(s) = \left\{ \begin{array}{c} \left[\begin{array}{ccccc} O_k & O_k & O_k & \dots & O_k \\ s^0 I_k & O_k & O_k & \dots & O_k \\ \dots & \dots & \dots & \dots & \dots \\ s^{\varsigma-1} I_k & s^{\varsigma-2} I_k & s^{\varsigma-3} I_k & \dots & s^0 I_k \end{array} \right], \ \varsigma \geq 1 \\ O_k, \ \varsigma < 1 \end{array} \right\},$$

$$Z_k^{(\varsigma-1)}(s) \in \mathfrak{C}^{(\varsigma+1)kx\varsigma k}, \ (\varsigma, k) \in \{(\mu, M), \ (\nu, N)\} \ . \tag{8.5}$$

We repeat Note 55 (in Subsection 3.4.2) in order to avoid any ambiguity:

Note 147 *If* $\varsigma = 0$, *then the matrix* $Z_k^{(\varsigma-1)}(s) = Z_k^{(-1)}(s)$ *should be completely omitted rather than replaced by the zero matrix. The matrix* $Z_k^{(\zeta-1)}(s)$ *is not defined for* $\zeta \leq 0$ *and should be treated as the nonexisting one.*

Theorem 148 *a) The full IO transfer function matrix* $F_{IO}(s)$ *of the IO system (8.1) has the following form:*

$$If \ \mu \geq 1, \ then \ F_{IO}(s) = F_{IOD}^{-1}(s) F_{ION}(s) =$$
$$= \left(A^{(\nu)} S_N^{(\nu)}(s)\right)^{-1} \left[B^{(\mu)} S_M^{(\mu)}(s) \ \vdots \ -B^{(\mu)} Z_M^{(\mu-1)}(s) \ \vdots \ A^{(\nu)} Z_N^{(\nu-1)}(s)\right] =$$
$$= \left[G_{IO}(s) \ \vdots \ \underbrace{G_{IOi_0}(s) \ \vdots \ G_{IOy_0}(s)}_{G_{IOo}(s)}\right], \ G_{IOo}(s) = \left[G_{IOi_0}(s) \ \vdots \ G_{IOy_0}(s)\right], \tag{8.6}$$

$$If \ \mu = 0, \ then \ F_{IO}(s) = F_{IOD}^{-1}(s) F_{ION}(s) =$$
$$= \left(A^{(\nu)} S_N^{(\nu)}(s)\right)^{-1} \left[B^{(\mu)} S_M^{(\mu)}(s) \ \vdots \ A^{(\nu)} Z_N^{(\nu-1)}(s)\right] =$$
$$= \left(A^{(\nu)} S_N^{(\nu)}(s)\right)^{-1} \left[B_0 \ \vdots \ A^{(\nu)} Z_N^{(\nu-1)}(s)\right]$$
$$= \left[G_{IO}(s) \ \vdots \ G_{IOy_0}(s)\right], \ G_{IOy_0}(s) = G_{IOo}(s). \tag{8.7}$$

so that

$$\mathbf{Y}^{\mp}(s) = F_{IO}(s)\left\{\begin{array}{c} \left[\mathbf{I}^{\mp^T}(s) \vdots \left(\mathbf{i}_{0\mp}^{\mu-1}\right)^T \vdots \left(\mathbf{y}_{0\mp}^{\nu-1}\right)^T\right]^T \; if\ \mu \geq 1, \\ \left[\mathbf{I}^{\mp^T}(s) \vdots \left(\mathbf{y}_{0\mp}^{\nu-1}\right)^T\right]^T \; if\ \mu = 0 \end{array}\right\} =$$

$$= F_{IO}(s)\mathbf{V}_{IO}(s),\ \mathbf{V}_{IO}(s) = \begin{bmatrix} \mathbf{I}(s) \\ \mathbf{C}_{0IO} \end{bmatrix}, \tag{8.8}$$

$$\mathbf{C}_{0IO} = \left\{\begin{array}{c} \begin{bmatrix} \mathbf{i}_{0\mp}^{\mu-1} \\ \mathbf{y}_{0\mp}^{\nu-1} \end{bmatrix},\ if\ \mu \geq 1, \\ \mathbf{y}_{0\mp}^{\nu-1},\ if\ \mu = 0 \end{array}\right\}. \tag{8.9}$$

b) The IO transfer function matrix $G_{IO}(s)$ of the system (3.76) is determined by

$$G_{IO}(s) = \left(A^{(\nu)}S_N^{(\nu)}(s)\right)^{-1}\left(B^{(\mu)}S_M^{(\mu)}(s)\right). \tag{8.10}$$

c) The IOIC transfer function matrix $G_{IOi_0}(s)$ of the system (3.76) is determined by

$$G_{IOi_0}(s) = \left(A^{(\nu)}S_N^{(\nu)}(s)\right)^{-1}\left\{\begin{array}{c} -B^{(\mu)}Z_M^{(\mu-1)}(s),\ if\ \mu \geq 1 \\ O,\ if\ \mu = 0, \end{array}\right\}. \tag{8.11}$$

d) The IOIY transfer function matrix $G_{IOy_0}(s)$ of the system (3.76) is determined by

$$G_{IOy_0}(s) = \left(A^{(\nu)}S_N^{(\nu)}(s)\right)^{-1}A^{(\nu)}Z_N^{(\nu-1)}(s). \tag{8.12}$$

e) The IORAI transfer function matrix $G_{IO_0}(s)$ of the system (3.76) is determined by

$$G_{IO_0}(s) = \left(A^{(\nu)}S_N^{(\nu)}(s)\right)^{-1}\left\{\begin{array}{c} \left[-B^{(\mu)}Z_M^{(\mu-1)}(s) \vdots A^{(\nu)}Z_N^{(\nu-1)}(s)\right], if\ \mu \geq 1 \\ A^{(\nu)}Z_N^{(\nu-1)}(s),\ if\ \mu = 0. \end{array}\right\} \tag{8.13}$$

The proof of this theorem is in Appendix D.

Note 149 *From $A_k \in \mathfrak{R}^{N\times N}$, $B_k \in \mathfrak{R}^{N\times M}$, $k = 0, 1, .., \nu$, $A_\nu \neq O_N$, (3.76), and (8.5) we determine the following:*

$$\deg\left[A^{(\nu)}Z_N^{(\nu-1)}(s)\right] = \nu - 1,\ \mu \geq 1 \Longrightarrow \deg\left[B^{(\mu)}Z_M^{(\mu-1)}(s)\right] = \mu - 1. \tag{8.14}$$

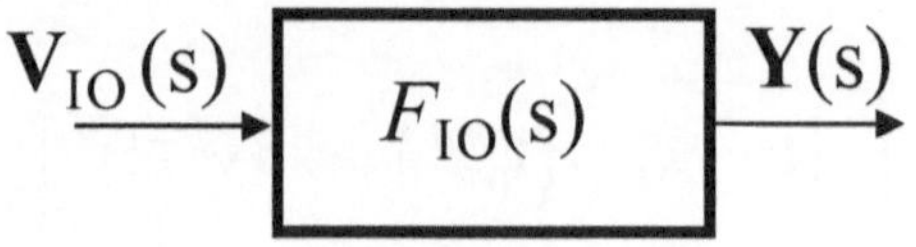

Figure 8.1: The full block of the IO system in the compact form.

Comment 150 *Equation (8.8) determines the overall action vector function $\mathbf{v}_{IO}(.)$ and its left Laplace transform $\mathbf{V}_{IO}^{-}(s)$ for the IO system,*

$$\mathbf{v}_{IO}(t) = \left\{ \begin{array}{c} \left[\mathbf{i}^T(t) \vdots \delta(t) \left(\mathbf{i}_{0^-}^{\mu-1}\right)^T \vdots \delta(t) \left(\mathbf{y}_{0^-}^{\nu-1}\right)^T\right]^T, \mu \geq 1 \\ \left[\mathbf{i}^T(t) \vdots \delta(t) \left(\mathbf{y}_{0^-}^{\nu-1}\right)^T\right]^T, \mu = 0 \end{array} \right\},$$

$$\mathbf{V}_{IO}^{-}(s) = \left\{ \begin{array}{c} \left[(\mathbf{I}^{-}(s))^T \vdots \underbrace{\left(\mathbf{i}_{0^-}^{\mu-1}\right)^T \vdots \left(\mathbf{y}_{0^-}^{\nu-1}\right)^T}_{\left(\mathbf{C}_{0IO}^{(\mp)}\right)^T} \right]^T, \mu \geq 1 \\ \left[\mathbf{I}^{\mp^T}(s) \vdots \left(\mathbf{y}_{0^-}^{\nu-1}\right)^T\right]^T = \left[\mathbf{I}^{\mp^T}(s) \vdots \mathbf{C}_{0IO}^{(\mp)^T}\right]^T, \mu = 0 \end{array} \right\} \equiv$$

$$= \left\{ \begin{array}{c} \mathbf{V}_{IO}^{-}(s; \mathbf{i}_{0^-}^{\mu-1}, \mathbf{y}_{0^-}^{\nu-1}),\ \mu \geq 1 \\ \mathbf{V}_{IO}^{-}(s; \mathbf{y}_{0^-}^{\nu-1}),\ \mu = 0 \end{array} \right\}, \tag{8.15}$$

where $\delta(t)$ is the unit Dirac impulse (Appendix B.2). Equation (8.8) takes now, for nonzero initial conditions, the classical form of $\mathbf{Y}^{-}(s)$ obtained under zero initial conditions,

$$\mathbf{Y}^{-}(s) = F_{IO}(s)\mathbf{V}_{IO}^{-}(s). \tag{8.16}$$

The system full (complete) block, valid now for arbitrary initial conditions, has the classical form established for the zero initial conditions; see Fig. 8.1. The vector functions $\mathbf{V}_{IO}^{-}(.) \equiv \mathbf{V}_{IO}^{-}(.; \mathbf{i}_{0^-}^{\mu-1}, \mathbf{y}_{0^-}^{\nu-1})$ and $\mathbf{V}_{IO}^{-}(.) = \mathbf{V}_{IO}^{-}(.; \mathbf{y}_{0^-}^{\nu-1})$ hide the nonzero initial conditions $\mathbf{i}_{0^-}^{\mu-1}$ and $\mathbf{y}_{0^-}^{\nu-1}$, so that in fact

$$\begin{aligned} \mathbf{Y}^{-}(s) &= \mathbf{Y}^{-}(s; \mathbf{i}_{0^-}^{\mu-1}, \mathbf{y}_{0^-}^{\nu-1}) = F_{IO}(s)\mathbf{V}_{IO}^{-}(s; \mathbf{i}_{0^-}^{\mu-1}, \mathbf{y}_{0^-}^{\nu-1}),\ \mu \geq 1, \\ \mathbf{Y}^{-}(s) &= \mathbf{Y}^{-}(s; \mathbf{y}_{0^-}^{\nu-1}) = F_{IO}(s)\mathbf{V}_{IO}^{-}(s; \mathbf{y}_{0^-}^{\nu-1}),\ \mu = 0. \end{aligned} \tag{8.17}$$

Comment 151 ***The generalization of the block diagram technique is the full block diagram technique***

The introduction of the generalized input vector function $\mathbf{v}_{IO}(.)$, *(8.15), permits us to use the unchanged classical block diagram technique with the full transfer function matrices* $F_{IOi}(s)$ *of the subsystems* S_i *instead of their transfer function matrices* $G_{IOi}(s)$, *in the sense of Fig. 8.1. We will establish the algebra of the corresponding full block diagrams in Chapter 9.*

Comment 152 *The system full transfer function matrix* $F_{IO}(s)$ *is the system dynamical invariant. The order, the dimension, and the parameters of the system completely determine* $F_{IO}(s)$. *It is independent of the input vector and of all initial conditions, i.e., it is independent of the generalized input vector* $\mathbf{v}_{IO}(t)$, *hence, of its Laplace transform* $\mathbf{V}_{IO}(s)$. *It has the same principal characteristics as the system transfer function matrix* $G_{IO}(s)$ *that is determined for zero initial conditions.*

Note 153 *The system full transfer function matrix* $F_{IO}(s)$ *incorporates the system transfer function matrix* $G_{IO}(s)$ *as a submatrix:*

$$F_{IO}(s) = F_{IOD}^{-1}(s)F_{ION}(s) = \underbrace{\left(A^{(\nu)}S_N^{(\nu)}(s)\right)}_{F_{IOD}(s)=G_{IOD}(s)}{}^{-1} \bullet$$

$$\bullet \underbrace{\left\{ \begin{array}{c} \left[\overbrace{B^{(\mu)}S_M^{(\mu)}(s)}^{G_{ION}(s)} \vdots -B^{(\mu)}Z_M^{(\mu-1)}(s) \vdots A^{(\nu)}Z_N^{(\nu-1)}(s) \right], \ \mu \geq 1 \\ \left[\overbrace{B^{(\mu)}S_M^{(\mu)}(s)}^{G_{ION}(s)} \vdots A^{(\nu)}Z_N^{(\nu-1)}(s) \right], \ \mu = 0 \end{array} \right\}}_{F_{ION}(s)} =$$

$$= \left\{ \begin{array}{c} \left[\begin{array}{c} G_{IO}(s) \vdots \left(A^{(\nu)}S_N^{(\nu)}(s)\right)^{-1} \left(-B^{(\mu)}Z_M^{(\mu-1)}(s)\right) \vdots \\ \vdots \left(A^{(\nu)}S_N^{(\nu)}(s)\right)^{-1} A^{(\nu)}Z_N^{(\nu-1)}(s) \end{array} \right], \\ \mu \geq 1 \\ \left[G_{IO}(s) \vdots \left(A^{(\nu)}S_N^{(\nu)}(s)\right)^{-1} A^{(\nu)}Z_N^{(\nu-1)}(s) \right], \ \mu = 0 \end{array} \right\}. \quad (8.18)$$

$F_{IOD}(s)$ and $F_{ION}(s)$ are the denominator and the numerator polynomial matrices of the system full transfer function matrix $F_{IO}(s)$,

$$F_{IOD}(s) = A^{(\nu)} S_N^{(\nu)}(s),$$

$$F_{ION}(s) = \left[B^{(\mu)} S_M^{(\mu)}(s) \vdots -B^{(\mu)} Z_M^{(\mu-1)}(s) \vdots A^{(\nu)} Z_N^{(\nu-1)}(s) \right], \ \mu \geq 1$$

$$F_{ION}(s) = \left[B^{(\mu)} S_M^{(\mu)}(s) \vdots A^{(\nu)} Z_N^{(\nu-1)}(s) \right], \ \mu = 0,$$

and

$$G_{IO}(s) = G_{IOD}^{-1}(s) G_{ION}(s), \ G_{IOD}(s) = F_{IOD}(s).$$

Example 154 *The given IO system is described by*

$$y^{(2)}(t) + y^{(1)}(t) - 2y(t) = 2i^{(2)}(t) - 14i^{(1)}(t) + 12i(t).$$

Its description and characteristics in the complex domain read

$$(s^2 + s - 2)Y^{\mp}(s) - (s+1)y_{0\mp} - y_{0\mp}^{(1)} =$$
$$= (2s^2 - 14s + 12)I^{\mp}(s) - (2s - 14)i_{0\mp} - 2i_{0\mp}^{(1)} \Longrightarrow$$

$$Y^{\mp}(s) = (s^2 + s - 2)^{-1} \bullet$$
$$\bullet \left[2s^2 - 14s + 12 \vdots -(2s-14) \vdots -2 \vdots s+1 \vdots 1 \right] \bullet$$
$$\bullet \left[I^{\mp}(s) \vdots \underbrace{\underbrace{i_{0\mp} \vdots i_{0\mp}^{(1)}}_{\mathbf{i}_{0\mp}^{1T}} \vdots \underbrace{y_{0\mp} \vdots y_{0\mp}^{(1)}}_{\mathbf{y}_{0\mp}^{1T}}}_{\mathbf{C}_{oIO}} \right]^T =$$

$$= F_{IO}(s)\mathbf{V}^{\mp}(s), \ \mathbf{V}^{\mp}(s) = \left[I^{\mp}(s) \vdots \mathbf{C}_{oIO}^T \right]^T \Longrightarrow$$

$$\mathbf{C}_{oIO} = \begin{bmatrix} \mathbf{i}_{0\mp}^1 \\ \mathbf{y}_{0\mp}^1 \end{bmatrix}, \ F_{IO}(s) = (s^2 + s - 2)^{-1} \bullet$$
$$\bullet \left[2s^2 - 14s + 12 \vdots -(2s-14) \vdots -2 \vdots s+1 \vdots 1 \right],$$

$$G_{IO}(s) = \frac{2s^2 - 14s + 12}{s^2 + s - 2} = \frac{2(s-1)(s-6)}{(s-1)(s+2)} \Longrightarrow G_{IOnd}(s) = \frac{s-6}{s+2}.$$

The system transfer function matrix $G_{IO}(s)$ has the same zero $s^0 = 1$ and pole $s^* = 1$. They can be cancelled. However, the cancellation is not possible in the system full transfer function matrix $F_{IO}(s)$.

Example 155 *C. A. Desoer and M. Vidyasagar wrote in [19, p. 58] that the following two ratios of polynomials,*

$$\frac{s+1}{s(s+2)} \text{ and } \frac{(s+1)(s-1)}{s(s+2)(s-1)} \tag{8.19}$$

correspond to the same rational function $f(.) : \mathfrak{T} \longrightarrow \mathfrak{R}$. They have two common zeros $s_1^0 = -1$ and $s_2^0 = \infty$, and two common poles $s_1^* = 0$ and $s_2^* = -2$. The second ratio has an additional positive real pole $s_3^* = 1$. However, if they represent the system transfer functions $G_1(.)$ and $G_2(.)$,

$$G_1(s) = \frac{s+1}{s(s+2)} \text{ and } G_2(s) = \frac{(s+1)(s-1)}{s(s+2)(s-1)}, \tag{8.20}$$

then they do not correspond to the same system. $G_1(s)$ is nondegenerate, while $G_2(s)$ is degenerate. The former is the nondegenerate form of the latter. $G_1(.)$ is the transfer function of the second-order $SISO$ system described by

$$y^{(2)}(t) + 2y^{(1)}(t) = i(t) + i^{(1)}(t), \tag{8.21}$$

while $G_2(.)$ is the transfer function of the third-order $SISO$ system determined by

$$y^{(3)}(t) + y^{(2)}(t) - 2y^{(1)}(t) = -i(t) + i^{(2)}(t). \tag{8.22}$$

The full transfer function matrix $F_1(s)$ of the former reads

$$F_1(s) = (s^2+s)^{-1}\left[s+1 \ \vdots\ -1 \ \vdots\ s+2 \ \vdots\ 1\right], \tag{8.23}$$

while the full transfer function matrix $F_2(s)$ of the latter is found as

$$F_2(s) = (s^3 + s^2 - 2s)^{-1} \bullet$$
$$\bullet \left[s^2-1 \ \vdots\ -s \ \vdots\ -1 \ \vdots\ s^2+s-2 \ \vdots\ s+1 \ \vdots\ 1\right]. \tag{8.24}$$

Both $F_1(s)$ and $F_2(s)$ are row nondegenerate and nondegenerate.

Example 156 *A MIMO IO system is described by*

$$\mathbf{y}^{(1)}(t)+\left[\begin{array}{c:c} 2 & 0 \\ 0 & 1 \end{array}\right]\mathbf{y}(t)=\left[\begin{array}{c:c} 2 & 0 \\ 0 & 0 \end{array}\right]\mathbf{i}(t)+\mathbf{i}^{(1)}(t),\ \mathbf{y}=\left[\begin{array}{c} y_1 \\ y_2 \end{array}\right],\ \mathbf{i}=\left[\begin{array}{c} i_1 \\ i_2 \end{array}\right].$$

Its transfer function matrix $G_{IO}(s)$,

$$G_{IO}(s)=\left[\begin{array}{cc} 1 & 0 \\ 0 & \frac{s}{s+1} \end{array}\right],\ full\ rank\ G_{IO}(s)=2,$$

is rank defective for $s=0$,

$$rank\ G_{IO}(0)=1.$$

However, the system full transfer function matrix $F_{IO}(s)$,

$$F_{IO}(s)=\left[\begin{array}{cccccc} 1 & 0 & -\frac{1}{s+2} & 0 & \frac{1}{s+2} & 0 \\ 0 & \frac{s}{s+1} & 0 & -\frac{1}{s+1} & 0 & \frac{1}{s+1} \end{array}\right],$$

has the full rank over the field of complex numbers,

$$rankF_{IO}(s)\equiv full\ rankF_{IO}(s)=2.$$

In this example,

$$G_{IO}(s)=\left[\begin{array}{cc} 1 & 0 \\ 0 & \frac{s}{s+1} \end{array}\right],\ G_{IOi_0}(s)=\left[\begin{array}{cc} -\frac{1}{s+2} & 0 \\ 0 & -\frac{1}{s+1} \end{array}\right],$$

$$G_{IOy_0}(s)=\left[\begin{array}{cc} \frac{1}{s+2} & 0 \\ 0 & \frac{1}{s+1} \end{array}\right],\ G_{IO_0}(s)=\left[\begin{array}{cccc} -\frac{1}{s+2} & 0 & \frac{1}{s+2} & 0 \\ 0 & -\frac{1}{s+1} & 0 & \frac{1}{s+1} \end{array}\right].$$

Example 157 *Appendix E contains an example of the determination of* $F_{IO}(s)$ *of a second-order MIMO system.*

Example 158 *The IO system of Example 31 (in Section 2.1)*

$$\left[\begin{array}{c:c:c} 3 & 0 & 0 \\ 0 & 0 & 0 \\ 0 & 0 & 0 \end{array}\right]\mathbf{y}^{(2)}(t)+\left[\begin{array}{c:c:c} 0 & 0 & 0 \\ 0 & 1 & 0 \\ 0 & 0 & 0 \end{array}\right]\mathbf{y}^{(1)}(t)+\left[\begin{array}{c:c:c} 0 & 0 & 0 \\ 0 & 0 & 0 \\ 0 & 0 & 1 \end{array}\right]\mathbf{y}(t)=$$

$$=\left[\begin{array}{c:c} 2 & 0 \\ 0 & 1 \\ 1 & 0 \end{array}\right]\mathbf{i}(t)+\left[\begin{array}{c:c} 1 & 0 \\ 0 & 1 \\ 1 & 1 \end{array}\right]\mathbf{i}^{(2)}(t)$$

yields

$$\nu = 2,\ \mu = 2,\ N = 3,\ M = 2,\ \det A_\nu = \det A_2 = \det \begin{bmatrix} 3 \vdots 0 \vdots 0 \\ 0 \vdots 0 \vdots 0 \\ 0 \vdots 0 \vdots 0 \end{bmatrix} = 0,$$

$$A^{(2)} = \begin{bmatrix} 0 \vdots 0 \vdots 0 & 0 \vdots 0 \vdots 0 & 3 \vdots 0 \vdots 0 \\ 0 \vdots 0 \vdots 0 & 0 \vdots 1 \vdots 0 & 0 \vdots 0 \vdots 0 \\ 0 \vdots 0 \vdots 1 & 0 \vdots 0 \vdots 0 & 0 \vdots 0 \vdots 0 \end{bmatrix},$$

$$S_3^{(2)}(s) = \begin{bmatrix} 1 & 0 & 0 \\ 0 & 1 & 0 \\ 0 & 0 & 1 \\ s & 0 & 0 \\ 0 & s & 0 \\ 0 & 0 & s \\ s^2 & 0 & 0 \\ 0 & s^2 & 0 \\ 0 & 0 & s^2 \end{bmatrix},$$

$$A^{(2)} S_3^{(2)}(s) = \begin{bmatrix} 3s^2 \vdots 0 \vdots 0 \\ 0 \vdots s \vdots 0 \\ 0 \vdots 0 \vdots 1 \end{bmatrix},\ \deg \left[A^{(2)} S_3^{(2)}(s) \right] = 2,$$

$$adj \left(A^{(2)} S_3^{(2)}(s) \right) = \begin{bmatrix} s \vdots 0 \vdots 0 \\ 0 \vdots 3s^2 \vdots 0 \\ 0 \vdots 0 \vdots 3s^3 \end{bmatrix},\ \deg \left[adj \left(A^{(2)} S_3^{(2)}(s) \right) \right] = 3,$$

$$\det \left[A^{(2)} S_3^{(2)}(s) \right] = 3s^3,\ \deg \left\{ \det \left[A^{(2)} S_3^{(2)}(s) \right] \right\} = 3,$$

$$B^{(2)} = \begin{bmatrix} 2 \vdots 0 & 0 \vdots 0 & 1 \vdots 0 \\ 0 \vdots 1 & 0 \vdots 0 & 0 \vdots 1 \\ 1 \vdots 0 & 0 \vdots 0 & 1 \vdots 1 \end{bmatrix},\ S_2^{(2)}(s) = \begin{bmatrix} 1 & 0 \\ 0 & 1 \\ s & 0 \\ 0 & s \\ s^2 & 0 \\ 0 & s^2 \end{bmatrix},$$

$$B^{(2)}S_2^{(2)}(s) = \begin{bmatrix} 2+s^2 \vdots & 0 \\ 0 & \vdots\, 1+s^2 \\ 1+s^2 \vdots & s^2 \end{bmatrix}, \ \deg\left[B^{(2)}S_2^{(2)}(s)\right] = 2,$$

$$B^{(2)}Z_2^{(2-1)}(s) = \begin{bmatrix} 2 \vdots 0 & 0 \vdots 0 & 1 \vdots 0 \\ 0 \vdots 1 & 0 \vdots 0 & 0 \vdots 1 \\ 1 \vdots 0 & 0 \vdots 0 & 1 \vdots 1 \end{bmatrix} \begin{bmatrix} 0 \vdots 0 \vdots 0 \vdots 0 \\ 0 \vdots 0 \vdots 0 \vdots 0 \\ 1 \vdots 0 \vdots 0 \vdots 0 \\ 0 \vdots 1 \vdots 0 \vdots 0 \\ s \vdots 0 \vdots 1 \vdots 0 \\ 0 \vdots s \vdots 0 \vdots 1 \end{bmatrix} =$$

$$= \begin{bmatrix} s \vdots 0 & 1 \vdots 0 \\ 0 \vdots s & 0 \vdots 1 \\ s \vdots s & 1 \vdots 1 \end{bmatrix}, \ \deg B^{(2)}Z_2^{(2-1)}(s) = deg(s) = 1,$$

$$A^{(2)}Z_3^{(2-1)}(s) = \begin{bmatrix} 0 \vdots 0 \vdots 0 & 0 \vdots 0 \vdots 0 & 3 \vdots 0 \vdots 0 \\ 0 \vdots 0 \vdots 0 & 0 \vdots 1 \vdots 0 & 0 \vdots 0 \vdots 0 \\ 0 \vdots 0 \vdots 1 & 0 \vdots 0 \vdots 0 & 0 \vdots 0 \vdots 0 \end{bmatrix} \begin{bmatrix} 0 \vdots 0 \vdots 0 \vdots 0 \vdots 0 \vdots 0 \\ 0 \vdots 0 \vdots 0 \vdots 0 \vdots 0 \vdots 0 \\ 0 \vdots 0 \vdots 0 \vdots 0 \vdots 0 \vdots 0 \\ 1 \vdots 0 \vdots 0 \vdots 0 \vdots 0 \vdots 0 \\ 0 \vdots 1 \vdots 0 \vdots 0 \vdots 0 \vdots 0 \\ 0 \vdots 0 \vdots 1 \vdots 0 \vdots 0 \vdots 0 \\ s \vdots 0 \vdots 0 \vdots 1 \vdots 0 \vdots 0 \\ 0 \vdots s \vdots 0 \vdots 0 \vdots 1 \vdots 0 \\ 0 \vdots 0 \vdots s \vdots 0 \vdots 0 \vdots 1 \end{bmatrix} =$$

$$= \begin{bmatrix} 3s \vdots 0 \vdots 0 & 3 \vdots 0 \vdots 0 \\ 0 \vdots 1 \vdots 0 & 0 \vdots 0 \vdots 0 \\ 0 \vdots 0 \vdots 0 & 0 \vdots 0 \vdots 0 \end{bmatrix}, \ \deg\left[A^{(2)}Z_3^{(2-1)}(s)\right] = \deg(3s) = 1,$$

and

$$F_{IO}(s) = \frac{\begin{bmatrix} s & \vdots & 0 & \vdots & 0 \\ 0 & \vdots & 3s^2 & \vdots & 0 \\ 0 & \vdots & 0 & \vdots & 3s^3 \end{bmatrix}}{3s^3} \bullet$$

$$\bullet \begin{bmatrix} 2+s^2 & 0 & -s & 0 & -1 & 0 & 3s & 0 & 0 & 3 & 0 & 0 \\ 0 & 1+s^2 & 0 & -s & 0 & -1 & 0 & 1 & 0 & 0 & 0 & 0 \\ 1+s^2 & s^2 & -s^2 & -s & -1 & -1 & 0 & 0 & 0 & 0 & 0 & 0 \end{bmatrix}.$$

The full transfer function matrix $F_{IO}(s)$ is improper because the degree of its numerator matrix (which is equal to 5) exceeds the degree of its denominator polynomial (which is equal to 3).

Example 159 *The IO system of Example 32 (in Section 2.1)*

$$\begin{bmatrix} 3 & \vdots & 0 & \vdots & 1 \\ 2 & \vdots & 0 & \vdots & 0 \\ 0 & \vdots & 1 & \vdots & 1 \end{bmatrix} \mathbf{y}^{(2)}(t) + \begin{bmatrix} 0 & \vdots & 0 & \vdots & 0 \\ 0 & \vdots & 1 & \vdots & 0 \\ 0 & \vdots & 0 & \vdots & 0 \end{bmatrix} \mathbf{y}^{(1)}(t) + \begin{bmatrix} 0 & \vdots & 0 & \vdots & 0 \\ 0 & \vdots & 0 & \vdots & 0 \\ 0 & \vdots & 0 & \vdots & 1 \end{bmatrix} \mathbf{y}(t) =$$

$$= \begin{bmatrix} 2 & \vdots & 0 \\ 0 & \vdots & 1 \\ 1 & \vdots & 0 \end{bmatrix} \mathbf{i}(t) + \begin{bmatrix} 1 & \vdots & 0 \\ 0 & \vdots & 1 \\ 1 & \vdots & 1 \end{bmatrix} \mathbf{i}^{(2)}(t)$$

induces

$$\nu = 2,\ \mu = 2,\ N = 3,\ M = 2,\ \det A_\nu = \det A_2 = \det \begin{bmatrix} 3 & \vdots & 0 & \vdots & 1 \\ 2 & \vdots & 0 & \vdots & 0 \\ 0 & \vdots & 1 & \vdots & 1 \end{bmatrix} = 2 \neq 0,$$

$$A^{(2)} S_3^{(2)}(s) = \begin{bmatrix} 3s^2 & 0 & s^2 \\ 2s^2 & s & 0 \\ 0 & s^2 & s^2+1 \end{bmatrix},\ \deg \left[A^{(2)} S_3^{(2)}(s) \right] = 2,$$

$$adj \left(A^{(2)} S_3^{(2)}(s) \right) = \begin{bmatrix} s+s^3 & s^4 & -s^3 \\ -2s^4-2s^2 & 3s^4+3s^2 & 2s^4 \\ 2s^4 & -3s^4 & 3s^3 \end{bmatrix},$$

$$\deg\left[adj\left(A^{(2)}S_3^{(2)}(s)\right)\right] = 4,$$

$$\det\left[A^{(2)}S_3^{(2)}(s)\right] = 2s^6 + 3s^5 + 3s^3, \ \deg\left\{\det\left[A^{(2)}S_3^{(2)}(s)\right]\right\} = 6,$$

$$B^{(2)}S_2^{(2)}(s) = \begin{bmatrix} 2+s^2 & 0 \\ 0 & 1+s^2 \\ 1+s^2 & s^2 \end{bmatrix}, \ \deg\left[B^{(2)}S_2^{(2)}(s)\right] = 2,$$

$$B^{(2)}Z_2^{(2-1)}(s) = \begin{bmatrix} 2 \vdots 0 & 0 \vdots 0 & 1 \vdots 0 \\ 0 \vdots 1 & 0 \vdots 0 & 0 \vdots 1 \\ 1 \vdots 0 & 0 \vdots 0 & 1 \vdots 1 \end{bmatrix} \begin{bmatrix} 0 \vdots 0 \vdots 0 \vdots 0 \\ 0 \vdots 0 \vdots 0 \vdots 0 \\ 1 \vdots 0 \vdots 0 \vdots 0 \\ 0 \vdots 1 \vdots 0 \vdots 0 \\ s \vdots 0 \vdots 1 \vdots 0 \\ 0 \vdots s \vdots 0 \vdots 1 \end{bmatrix} =$$

$$= \begin{bmatrix} s \vdots 0 & 1 \vdots 0 \\ 0 \vdots s & 0 \vdots 1 \\ s \vdots s & 1 \vdots 1 \end{bmatrix}, \ \deg B^{(2)}Z_2^{(2-1)}(s) = deg(s) = 1,$$

$$A^{(2)}Z_3^{(2-1)}(s) =$$

$$= \begin{bmatrix} 0 & 0 & 0 & 0 & 0 & 0 & 3 & 0 & 1 \\ 0 & 0 & 0 & 0 & 1 & 0 & 2 & 0 & 0 \\ 0 & 0 & 1 & 0 & 0 & 0 & 0 & 1 & 1 \end{bmatrix} \begin{bmatrix} 0 \vdots 0 \vdots 0 \vdots 0 \vdots 0 \vdots 0 \\ 0 \vdots 0 \vdots 0 \vdots 0 \vdots 0 \vdots 0 \\ 0 \vdots 0 \vdots 0 \vdots 0 \vdots 0 \vdots 0 \\ 1 \vdots 0 \vdots 0 \vdots 0 \vdots 0 \vdots 0 \\ 0 \vdots 1 \vdots 0 \vdots 0 \vdots 0 \vdots 0 \\ 0 \vdots 0 \vdots 1 \vdots 0 \vdots 0 \vdots 0 \\ s \vdots 0 \vdots 0 \vdots 1 \vdots 0 \vdots 0 \\ 0 \vdots s \vdots 0 \vdots 0 \vdots 1 \vdots 0 \\ 0 \vdots 0 \vdots s \vdots 0 \vdots 0 \vdots 1 \end{bmatrix} =$$

$$= \begin{bmatrix} 3s & 0 & s & 3 & 0 & 1 \\ 2s & 1 & 0 & 2 & 0 & 0 \\ 0 & s & s & 0 & 1 & 1 \end{bmatrix}, \ \deg\left[A^{(2)}Z_3^{(2-1)}(s)\right] = \deg(3s) = 1,$$

and

$$F_{IO}(s) = \frac{\begin{bmatrix} s+s^3 & s^4 & -s^3 \\ -2s^2-2s^4 & 3s^2+3s^4 & 2s^4 \\ 2s^4 & -3s^4 & 3s^3 \end{bmatrix}}{2s^6+3s^5+3s^3} \bullet$$

$$\bullet \begin{bmatrix} 2+s^2 & 0 & -s & 0 & -1 & 0 & 3s & 0 & s & 3 & 0 & 1 \\ 0 & 1+s^2 & 0 & -s & 0 & -1 & 2s & 1 & 0 & 2 & 0 & 0 \\ 1+s^2 & s^2 & -s & -s & -1 & -1 & 0 & s & s & 0 & 1 & 1 \end{bmatrix}.$$

The degree of the numerator matrix polynomial is equal to six, which is also the degree of the denominator polynomial. The full transfer function matrix $F_{IO}(s)$ is proper in this case.

8.2 $F(s)$ of the *ISO* system

The *ISO* system (3.84) and (3.85) (in Subsection 3.5.3) is repeated as (8.25) and (8.26),

$$\frac{d\mathbf{x}(t)}{dt} = A\mathbf{x}(t) + B\mathbf{i}(t), \ \forall t \in \mathfrak{T}_0, \tag{8.25}$$

$$\mathbf{y}(t) = C\mathbf{x}(t) + D\mathbf{i}(t), \ \forall t \in \mathfrak{T}_0, \tag{8.26}$$

We allow at first an arbitrary rank of the matrix C in (8.26).

The *IO* transfer function $G_{ISO}(s)$ has the well-known form,

$$G_{ISO}(s) = C(sI_n - A)^{-1}B + D. \tag{8.27}$$

Theorem 160 *a) The full IO transfer function matrix $F_{ISO}(s)$ of the ISO system (8.25) and (8.26) has the following form in general:*

$$F_{ISO}(s) = \left[C(sI_n - A)^{-1}B + D \vdots C(sI_n - A)^{-1} \right] = \\ = F_{ISOD}^{-1}(s)F_{ISON}(s), \tag{8.28}$$

so that

$$\mathbf{Y}^{\mp}(s) = F_{ISO}(s) \left[\mathbf{I}^{\mp^T}(s) \vdots \mathbf{x}_{0\mp}^T \right]^T = F_{ISO}(s)\mathbf{V}_{ISO}^{\mp}(s;\mathbf{x}_{0\mp}), \tag{8.29}$$

$$\mathbf{V}_{ISO}^{\mp}(s;\mathbf{x}_{0\mp}) = \begin{bmatrix} \mathbf{I}^{\mp}(s) \\ \mathbf{C}_{0ISO}^{\mp} \end{bmatrix}, \ \mathbf{C}_{0ISO}^{\mp} = \mathbf{x}_{0\mp}. \tag{8.30}$$

b) *The IISO transfer function matrix* $G_{ISOx_0}(s)$ *relative to* $\mathbf{x}_0$ *of the ISO system (8.25) and (8.26) has the following form in general:*

$$G_{ISOx_0}(s) = C(sI_n - A)^{-1}. \tag{8.31}$$

c) *The full (complete) IS transfer function matrix* $F_{ISOIS}(s)$ *of the ISO system (8.25) and (8.26) reads*

$$F_{ISOIS}(s) = \left[(sI_n - A)^{-1}B \,\vdots\, (sI_n - A)^{-1}\right] = \\ = F_{ISOISD}^{-1}(s)F_{ISOISN}(s). \tag{8.32}$$

d) *The IS transfer function matrix* $G_{ISOIS}(s)$ *of the ISO system (8.25) and (8.26) obeys*

$$G_{ISOIS}(s) = (sI_n - A)^{-1}B\ . \tag{8.33}$$

e) *The SS transfer function matrix* $G_{ISOSS}(s)$ *relative to* $x_{0\mp}$ *of the ISO system (8.25) and (8.26) fulfills*

$$G_{ISOSS}(s) = (sI_n - A)^{-1}. \tag{8.34}$$

Although the proof of Theorem 160 is almost evident, we present it immediately in order to illustrate the procedure of the determination of the *ISO* system full transfer function matrix $F_{ISO}(s)$ and to show the origin of the difference between it and the system transfer function matrix $G_{ISO}(s)$.

Proof. a) The left and right Laplace transforms of (8.25) and (8.26) yield:

$$\mathbf{X}^{\mp}(s) = (sI_n - A)^{-1}\left(\mathbf{x}_{0\mp} + B\mathbf{I}^{(\mp)}(s)\right) \Longrightarrow$$

$$\mathbf{X}^{\mp}(s) = \left[(sI_n - A)^{-1}B \,\vdots\, (sI_n - A)^{-1}\right] \underbrace{\begin{bmatrix} \mathbf{I}^{\mp}(s) \\ \mathbf{x}_{0\mp} \end{bmatrix}}_{\mathbf{V}_{ISO}^{\mp}(s;\mathbf{x}_{0\mp})}$$

$$\mathbf{Y}^{\mp}(s) = C\left[(sI_n - A)^{-1}\left(\mathbf{x}_{0\mp} + B\mathbf{I}^{(\mp)}(s)\right)\right] + D\mathbf{I}^{\mp}(s) =$$

$$= \underbrace{\left[\overbrace{C(sI_n - A)^{-1}B + D}^{G_{ISO}(s)} \vdots \overbrace{C(sI_n - A)^{-1}}^{G_{ISOx_0}(s)}\right]}_{F_{ISO}(s)} \underbrace{\begin{bmatrix} \mathbf{I}^{\mp}(s) \\ \mathbf{x}_{0\mp} \end{bmatrix}}_{\mathbf{V}^{\mp}_{ISO}(s;\mathbf{x}_{0\mp})} =$$

$$= [det(sI_n - A)]^{-1}\left[Cadj(sI_n - A)B + Ddet(sI_n - A) \vdots Cadj(sI_n - A)\right] \bullet$$

$$\bullet \underbrace{\begin{bmatrix} \mathbf{I}^{\mp}(s) \\ \mathbf{C}^{\mp}_{0ISO} \end{bmatrix}}_{\mathbf{V}^{\mp}_{ISO}(s;\mathbf{C}^{\mp}_{0ISO})} = F^{-1}_{ISOD}(s)F_{ISON}(s)\mathbf{V}^{\mp}_{ISO}(s;\mathbf{x}_{0\mp}),\ \mathbf{C}^{\mp}_{0ISO} = \mathbf{x}_{0\mp},$$

$$F_{ISOD}(s) = det(sI_n - A),$$

$$F_{ISON}(s) = \left[Cadj(sI_n - A)B + Ddet(sI_n - A) \vdots Cadj(sI_n - A)\right],$$

$$F_{ISO}(s) = \left[C(sI_n - A)^{-1}B + D \vdots C(sI_n - A)^{-1}\right] =$$

$$= [det(sI_n - A)]^{-1}\left[Cadj(sI_n - A)B + Ddet(sI_n - A) \vdots Cadj(sI_n - A)\right]. \tag{8.35}$$

These results and (7.23), (Definition 131, in Section 7.3), prove the statement under a) of the theorem, i.e., Equations (8.28) and (8.29).

b) (8.31) results directly from a) and from the definition of $G_{ISOx_0}(s)$ (7.27), c) of Definition 131, in Section 7.3).

c) The first two equations in (8.35) and a) of Definition 138 (in Section 7.3) imply (8.32).

d) Equation (8.33) results from the first equation in (8.35) and b) of Definition 138 (in Section 7.3).

e) The combination of the first equation in (8.35) and c) of Definition 138 (in Section 7.3) gives (8.34). ■

Comment 161 *The input vector function* $\mathbf{i}(.)$ *and the initial state vector* $\mathbf{x}_{0\mp}$ *are all the actions on the system. This inspires us to introduce the action vector function* $\mathbf{v}_{ISO}(.)$ *for the* ISO *system by*

$$\mathbf{v}_{ISO}(t;\mathbf{x}_0) = \begin{bmatrix} \mathbf{i}(t) \\ \delta(t)\mathbf{x}_{0-} \end{bmatrix} \in \mathfrak{R}^{M+n},\ \mathbf{V}^{-}_{ISO}(s;\mathbf{x}_0) = \begin{bmatrix} \mathbf{I}^{-}(s) \\ \mathbf{x}_{0-} \end{bmatrix} \in \mathfrak{C}^{M+n}. \tag{8.36}$$

This permitted to set (8.29) into the compact form, i.e.,

$$\mathbf{Y}^{-}(s) = \mathbf{Y}^{-}(s;\mathbf{x}_0) = F_{ISO}(s)\mathbf{V}^{-}_{ISO}(s;\mathbf{x}_0). \tag{8.37}$$

This is the well-known form of the classical relationship between $\mathbf{Y}^{-}(s)$ *and* $\mathbf{I}^{-}(s)$ *for the zero initial state vector,* $\mathbf{x}_{0^-} = \mathbf{0}_N$, *which is expressed via the system transfer function matrix* $G_{ISO}(s)$,

$$\mathbf{Y}^{-}(s) = \mathbf{Y}^{-}(s;\mathbf{0}_N) = G_{ISO}(s)\mathbf{I}^{-}(s),\ \mathbf{x}_{0^-} = \mathbf{0}_N. \tag{8.38}$$

We emphasize once more that (8.38) is valid only for the zero initial state vector, $\mathbf{x}_{0^-} = \mathbf{0}_N$, *while (8.29), i.e., (8.37), holds for arbitrary initial conditions.*

Comment 162 *The system full transfer function matrix* $F_{ISO}(s)$ *incorporates the system transfer function matrix* $G_{ISO}(s)$:

$$F_{ISO}(s) = \left[G_{ISO}(s) \vdots G_{ISOx_0}(s) \right].$$

Note 163 ***The full block diagram technique is the generalization of the classical block diagram technique***

The action vector $\mathbf{v}_{ISO}(t;\mathbf{x}_0)$ *permits us to apply directly the block diagram technique if* $\mathbf{i}(t)$ *is replaced by* $\mathbf{v}_{ISO}(t;\mathbf{x}_0)$, *i.e.,* $\mathbf{I}(s)$ *is replaced by* $\mathbf{V}^{-}_{ISO}(s;\mathbf{x}_0)$, *and* $F_{ISO}(s)$ *replaces* $G_{ISO}(s)$; *see Fig. 7.10.*

Note 164 *The system transfer function matrix* $G_{ISO}(s)$ *is invariant relative to a mathematical description of a fixed physical system. This does not apply to the system full transfer function matrix* $F_{ISO}(s)$ *in general, which depends on the choice of the state variables, i.e., on the choice of the state vector* $\mathbf{x}$. *This is clear from Equation (8.29). It shows the influence of the initial state vector* $\mathbf{x}_0$, *rather than the initial output vector* $\mathbf{y}_0$, *on the output response. Therefore, the transfer function matrix* $G_{ISOx_0}(s)$ *relative to* $\mathbf{x}_0$ *shows how* $\mathbf{x}_0$ *influences the output vector* $\mathbf{y}$. *The form of* $G_{ISOx_0}(s)$ *depends on the choice of* $\mathbf{x}$. *Let* $F_{IOISO}(s)$ *and* $G_{IOISO}(s)$ *be the full transfer function matrix and the transfer function matrix obtained from the IO mathematical model of the given ISO system. Then:*

$$F_{IOISO}(s) \neq F_{ISO}(s) \text{ and } G_{IOISO}(s) = G_{ISO}(s) \neq G_{ISOx_0}(s) \text{ in general.}$$

For a special class of the systems we can express $\mathbf{x}_0$ in terms of $\mathbf{i}_0$ and $\mathbf{y}_0$. Recall $n \geq N$ so that the full rank of C equals N. Hence, the full rank of $C^TC \in \mathfrak{R}^{nxn}$ is also N that implies that $detC^TC \neq 0$ is possible if and only if $N = n$. Consequently, we can solve uniquely (8.26) for $\mathbf{x}$ if and only if both $N = n$ and $detC \neq 0$.

Theorem 165 *If $N = n$ and the matrix C is nonsingular, then the following statements hold:*

a) The full IO transfer function matrix $F_{ISOsp}(s)$ of the ISO system (8.25) and (8.26) has the following form in this special case (the subscript "sp"):

$$F_{ISOsp}(s) =$$

$$= \left[C(sI_n - A)^{-1}B + D \vdots - C(sI_n - A)^{-1}C^{-1}D \vdots C(sI_n - A)^{-1}C^{-1} \right], \tag{8.39}$$

so that

$$\mathbf{Y}^{\mp}(s) = F_{ISOsp}(s) \begin{bmatrix} \mathbf{I}^{\mp}(s) \\ \mathbf{i}_{0\mp} \\ \mathbf{y}_{0\mp} \end{bmatrix}. \tag{8.40}$$

b) The transfer function matrix $G_{ISOi_0sp}(s)$ relative to $\mathbf{i}_0$ of the system (8.25) and (8.26) has the following form:

$$G_{ISOi_0sp}(s) = -C(sI_n - A)^{-1}C^{-1}D = -(sI_n - CAC^{-1})^{-1}D. \tag{8.41}$$

c) The transfer function matrix $G_{ISOy_0}(s)$ relative to $\mathbf{y}_0$ of the system (8.25) and (8.26) has the following special form:

$$G_{ISOy_0sp}(s) = C(sI_n - A)^{-1}C^{-1} = (sI_n - CAC^{-1})^{-1}. \tag{8.42}$$

Proof. Let $N = n$ and let C be nonsingular. This permits us to resolve (8.26) for $\mathbf{x}(t)$ at the initial moment $t_0 = 0^{\mp}$, i.e., to resolve $\mathbf{y}_{0\mp} = C\mathbf{x}_{0\mp} + D\mathbf{i}_{0\mp}$ for $\mathbf{x}_{0\mp}$,

$$\mathbf{x}_{0\mp} = -C^{-1}D\mathbf{i}_{0\mp} + C^{-1}\mathbf{y}_{0\mp}.$$

This equation, (8.29) and (8.35) imply (8.39) through (8.42) ■

Comment 166 *We determined the action vector $\mathbf{v}_{ISO}(t) = \mathbf{v}_{ISO}(t; \mathbf{x}_0)$ in general and its left/right Laplace transform $\mathbf{V}^{\mp}_{ISO}(s; \mathbf{x}_0)$ in (8.36) for the ISO system so that (8.29) becomes in the special case for* $\det C \neq 0$,

$$\mathbf{Y}^{\mp}(s) = F_{ISOsp}(s)\mathbf{V}^{\mp}_{ISOsp}(s; \mathbf{i}_0; \mathbf{y}_0). \tag{8.43}$$

From (8.40) follows that the action vector $\mathbf{v}_{ISOsp}(t) = \mathbf{v}_{ISOsp}(t; \mathbf{i}_0; \mathbf{y}_0)$ and its left/right Laplace transform $\mathbf{V}^{\mp}_{ISOsp}(s; \mathbf{i}_0; \mathbf{y}_0)$ are determined for the

ISO system in the special case by

$$\mathbf{v}_{ISOsp}(t;\mathbf{i}_0;\mathbf{y}_0)=\left[\mathbf{i}^T(t)\ \vdots\ \delta(t)\mathbf{i}_{0^-}^T\ \vdots\ \delta(t)\mathbf{y}_{0^-}^T\right]^T,$$

$$\mathbf{V}_{ISOsp}^{-}(s;\mathbf{i}_0;\mathbf{y}_0)=\left[\left(\mathbf{I}^-(s)\right)^T\ \vdots\ \mathbf{i}_{0^-}^T\ \vdots\ \mathbf{y}_{0^-}^T\right]^T. \tag{8.44}$$

This permitted us to set (8.40) into the classical form (8.38),

$$\mathbf{Y}^{\mp}(s)=F_{ISOsp}(s)\mathbf{V}_{ISOsp}^{\mp}(s;\mathbf{i}_0;\mathbf{y}_0). \tag{8.45}$$

Example 167 *Given $G_{ISO}(s)=\left(s^2-1\right)^{-1}(s-1)=(s+1)^{-1}$. Four different (state space, i.e., ISO) realizations (A,B,C,D) of $G_{ISO}(s)$ are determined in [4, p. 395]. We show at first how to determine the full transfer function matrix and other transfer function matrices for each.*

1) $\left\{A_1=\begin{bmatrix}0&1\\1&0\end{bmatrix},\ B_1=\begin{bmatrix}0\\1\end{bmatrix},\ C_1=\left[-1\ \vdots\ 1\right],\ D_1=0\right\}$
$\Longrightarrow$

$$Y_1^{\mp}(s)=\left[-1\vdots 1\right]\begin{bmatrix}s&-1\\-1&s\end{bmatrix}^{-1}\left\{\begin{bmatrix}0\\1\end{bmatrix}I^{\mp}(s)+\mathbf{x}_{0\mp}\right\}=$$

$$=\left(s^2-1\right)^{-1}\left[s-1\ \vdots\ 1-s\ \vdots\ s-1\right]\begin{bmatrix}I^{\mp}(s)\\x_{10\mp}\\x_{20\mp}\end{bmatrix},$$

$$Y_1^{\mp}(s)=F_{ISO1}(s)\mathbf{V}_{ISO1}^{\mp}(s;\mathbf{x}_0),\ \mathbf{V}_{ISO1}^{\mp}(s;\mathbf{x}_0)=\begin{bmatrix}I^{\mp}(s)\\\mathbf{x}_{0\mp}\end{bmatrix}\Longrightarrow$$

$$F_{ISO1}(s)=\left(s^2-1\right)^{-1}\left[s-1\ \vdots\ 1-s\ \vdots\ s-1\right]=$$

$$=\left[C_1(sI_2-A_1)^{-1}B_1+D_1\ \vdots\ C_1(sI_2-A_1)^{-1}\right],$$

$$\Longrightarrow$$

$$F_{ISO1nd}(s)=\left[\frac{1}{s+1}\ \vdots\ -\frac{1}{s+1}\ \vdots\ \frac{1}{s+1}\right],$$

$$G_{ISO1}(s)=\left[C_1(sI_2-A_1)^{-1}B_1+D_1\ \right]=\frac{s-1}{s^2-1}\Longrightarrow G_{ISO1nd}(s)=\frac{1}{s+1},$$

$$G_{ISOx_01}(s)=C_1(sI_2-A_1)^{-1}=\frac{\left[1-s\ \vdots\ -1+s\right]}{s^2-1},$$

$$F_{ISOIS1}(s) = \left[(sI_2 - A_1)^{-1}B_1 \vdots (sI_2 - A_1)^{-1}\right] = \frac{\begin{bmatrix} 1 & s & 1 \\ s & 1 & s \end{bmatrix}}{s^2 - 1},$$

$$G_{ISOIS1}(s) = (sI_2 - A_1)^{-1}B_1 = \frac{\begin{bmatrix} 1 \\ s \end{bmatrix}}{s^2 - 1},$$

$$G_{ISOSS1}(s) = (sI_2 - A_1)^{-1} = \frac{\begin{bmatrix} s & 1 \\ 1 & s \end{bmatrix}}{s^2 - 1}.$$

We cannot show the influence of the initial output value $y_{0\mp}$ *on the system response because we cannot express the state variables* x_1 *and* x_2 *in terms of the output* y_1 *due to* $y_1 = -x_1 + x_2$, *i.e., due to* $rankC_1 = 1 = M_1 < 2 = n_1$.

Notice that the given ISO system description, i.e.,

$$\begin{bmatrix} \frac{dx_1}{dt} \\ \frac{dx_2}{dt} \end{bmatrix} = \begin{bmatrix} 0 & 1 \\ 1 & 0 \end{bmatrix}\begin{bmatrix} x_1 \\ x_2 \end{bmatrix} + \begin{bmatrix} 0 \\ 1 \end{bmatrix} i,\ y_1 = \begin{bmatrix} -1 \vdots 1 \end{bmatrix}\begin{bmatrix} x_1 \\ x_2 \end{bmatrix}$$

allows

$$\begin{aligned} x_1^{(1)} &= x_2,\ x_2^{(1)} = x_1 + i, \\ y_1 &= -x_1 + x_2 \Longrightarrow \\ y_1^{(1)} &= -x_1^{(1)} + x_2^{(1)} = -x_2 + x_1 + i = \\ &= -y_1 + i \Longrightarrow \\ & y_1^{(1)} + y_1 = i. \end{aligned}$$

This yields

$$\begin{aligned} sY_1^{\mp}(s) - y_{10\mp} + Y_1^{\mp}(s) &= I^{\mp}(s) \Longrightarrow Y_1^{\mp}(s) = \frac{1}{s+1}\begin{bmatrix} 1 \vdots 1 \end{bmatrix}\begin{bmatrix} I^{\mp}(s) \\ y_{10\mp} \end{bmatrix} = \\ &= F_{IOISO1}(s)\mathbf{V}^{\mp}_{IOISO1}(s; y_{10\mp}) \Longrightarrow \end{aligned}$$

$$F_{IOISO1}(s) = \left[\frac{1}{s+1} \vdots \frac{1}{s+1}\right] \neq F_{ISO1}(s), G_{IOISO1}(s) = \frac{1}{s+1} = G_{ISO1nd}(s),$$

$$\mathbf{V}^{\mp}_{IOISO1}(s; y_{10\mp}) = \begin{bmatrix} I^{\mp}(s) \\ y_{10\mp} \end{bmatrix} \neq \begin{bmatrix} I^{\mp}(s) \\ \mathbf{x}_{0\mp} \end{bmatrix} = \mathbf{V}^{\mp}_{ISO1}(s; \mathbf{x}_0).$$

$G_{IOISO1}(s)$ is the nondegenerate form $G_{ISO1nd}(s)$ of $G_{ISO1}(s)$. $F_{IOISO1}(s)$ is different from $F_{ISO1}(s)$ and from the nondegenerate form $F_{ISO1nd}(s)$ of $F_{ISO1}(s)$, that is,

$$F_{ISO1nd}(s) = \left[\frac{1}{s+1} \vdots -\frac{1}{s+1} \vdots \frac{1}{s+1}\right].$$

Notice that

$$Y_1^{\mp}(s) = F_{ISO1}(s)\mathbf{V}_{ISO1}^{\mp}(s;\mathbf{x}_0) = \left[\frac{s-1}{s^2-1} \vdots -\frac{s-1}{s^2-1} \vdots \frac{s-1}{s^2-1}\right]\left[\begin{array}{c} I^{\mp}(s) \\ \mathbf{x}_{0\mp} \end{array}\right] =$$

$$= \left[\frac{1}{s+1} \vdots \frac{1}{s+1} \vdots \frac{1}{s+1}\right]\left[\begin{array}{c} I^{\mp}(s) \\ -x_{10\mp} \\ x_{20\mp} \end{array}\right],$$

and

$$y_1 = -x_1 + x_2 \Longrightarrow y_{10} = -x_{10} + x_{20}$$

imply

$$Y_1^{\mp}(s) = \left[\frac{1}{s+1} \vdots \frac{1}{s+1}\right]\left[\begin{array}{c} I^{\mp}(s) \\ y_{10\mp} \end{array}\right] = F_{IOISO1}(s)\mathbf{V}_{IOISO1}^{\mp}(s;y_{10\mp}).$$

This shows the equivalence between $F_{ISO1}(s)$ and $F_{IOISO1}(s)$ in this example,

$$F_{ISO1}(s)\mathbf{V}_{ISO1}^{\mp}(s;\mathbf{x}_0) = F_{IOISO1}(s)\mathbf{V}_{IOISO1}^{\mp}(s;y_{10\mp}) = Y_1^{\mp}(s).$$

2) $\left\{A_2 = \left[\begin{array}{cc} 0 & 1 \\ 1 & 0 \end{array}\right],\ B_2 = \left[\begin{array}{c} -1 \\ 1 \end{array}\right],\ C_2 = \left[0 \vdots 1\right],\ D_2 = 0\right\}$
$\Longrightarrow$

$$Y_2^{\mp}(s) = \left[0 \vdots 1\right]\left[\begin{array}{cc} s & -1 \\ -1 & s \end{array}\right]^{-1}\left\{\left[\begin{array}{c} -1 \\ 1 \end{array}\right] I^{\mp}(s) + \mathbf{x}_{0\mp}\right\} =$$

$$= \underbrace{\left(s^2-1\right)^{-1}\left[s-1 \vdots 1 \vdots s\right]}_{F_{ISO2}(s)}\left[\begin{array}{c} I^{\mp}(s) \\ x_{10\mp} \\ x_{20\mp} \end{array}\right],$$

$$Y_2^{\mp}(s) = F_{ISO2}(s)\mathbf{V}_{ISO2}^{\mp}(s;\mathbf{x}_0),\ \mathbf{V}_{ISO2}^{\mp}(s;\mathbf{x}_0) = \left[\begin{array}{c} I^{\mp}(s) \\ \mathbf{x}_{0\mp} \end{array}\right] \Longrightarrow$$

$$F_{ISO2}(s) = \left(s^2-1\right)^{-1}\left[s-1 \vdots 1 \vdots s\right] = F_{ISO2nd}(s),$$

$$G_{ISO2}(s) = \frac{s-1}{s^2-1},\ G_{ISO2nd}(s) = \frac{1}{s+1},$$

$$G_{ISOx_02}(s) = \frac{\begin{bmatrix} 1 \vdots s \end{bmatrix}}{s^2-1},\ F_{ISOIS2}(s) = \frac{\begin{bmatrix} 1-s & s & 1 \\ -1+s & 1 & s \end{bmatrix}}{s^2-1},$$

$$G_{ISOIS2}(s) = \frac{\begin{bmatrix} 1-s \\ -1+s \end{bmatrix}}{s^2-1},\ G_{ISOSS2}(s) = \frac{\begin{bmatrix} s & 1 \\ 1 & s \end{bmatrix}}{s^2-1}.$$

We find the IO system model as follows:

$$\frac{dx_1}{dt} = x_2 - i,\ \frac{dx_2}{dt} = x_1 + i,\ y_2 = x_2 \Longrightarrow$$
$$\frac{dy_2}{dt} = x_1 + i \Longrightarrow \frac{d^2y_2}{dt^2} = x_2 - i + i^{(1)} = y_2 - i + i^{(1)} \Longrightarrow$$
$$y_2^{(2)} - y_2 = i^{(1)} - i.$$

This IO system model implies

$$s^2Y_2^{\mp}(s) - sy_{20\mp} - y_{20\mp}^{(1)} - Y_2^{\mp}(s) = sI^{\mp}(s) - i_{0\mp} - I^{\mp}(s) \Longrightarrow$$
$$Y_2^{\mp}(s) = \frac{1}{s^2-1}\begin{bmatrix} s-1 \vdots -1 \vdots s \vdots 1 \end{bmatrix}\begin{bmatrix} I^{\mp}(s) \\ i_{0\mp} \\ y_{20\mp} \\ y_{20\mp}^{(1)} \end{bmatrix} =$$
$$= F_{IOISO2}(s)\mathbf{V}_{IOISO2}^{\mp}(s; i_0; \mathbf{y}_{20\mp}^{1}) \Longrightarrow$$

$$F_{IOISO2}(s) = \begin{bmatrix} \frac{s-1}{s^2-1} \vdots -\frac{1}{s^2-1} \vdots \frac{s}{s^2-1} \vdots \frac{1}{s^2-1} \end{bmatrix} \neq F_{ISO2}(s),$$

$$\mathbf{V}_{IOISO2}^{\mp}(s; i_0; \mathbf{y}_{20\mp}^{1}) = \begin{bmatrix} I^{\mp}(s) \\ i_{0\mp} \\ y_{20\mp} \\ y_{20\mp}^{(1)} \end{bmatrix} \neq \begin{bmatrix} I^{\mp}(s) \\ \mathbf{x}_{0\mp} \end{bmatrix} = \mathbf{V}_{ISO2}^{\mp}(s; \mathbf{x}_0),$$

$$G_{IOISO2}(s) = \frac{s-1}{s^2-1} = G_{ISO2}(s).$$

Besides,

$$Y_2^{\mp}(s) = \frac{1}{s^2-1}\left[s-1 \vdots -1 \vdots s \vdots 1\right]\begin{bmatrix} I^{\mp}(s) \\ i_{0\mp} \\ y_{20\mp} \\ y_{20\mp}^{(1)} \end{bmatrix} =$$

$$= F_{IOISO2}(s)\mathbf{V}_{IOISO2}^{\mp}(s; i_0; \mathbf{y}_{20\mp}^1) =$$

$$= \frac{1}{s^2-1}\left[(s-1)\,I^{\mp}(s) \; - i_{0\mp} \; + \; sx_{20} \; + \; x_{10} + i_{0\mp}\right] =$$

$$= \frac{1}{s^2-1}\left[(s-1)\,I^{\mp}(s) \; + \; x_{10} \; + \; sx_{20} \;\right] =$$

$$= \left[\frac{s-1}{s^2-1} \vdots \frac{1}{s} \vdots \frac{s}{s^2-1}\right]\begin{bmatrix} I^{\mp}(s) \\ x_{10} \\ x_{20} \end{bmatrix} = F_{ISO2}(s)\mathbf{V}_{ISO2}^{\mp}(s; \mathbf{x}_0).$$

This shows the equivalence between $F_{ISO2}(s)$ and $F_{IOISO2}(s)$ in this case.

3) $\left\{A_3 = \begin{bmatrix} 1 & 0 \\ 0 & -1 \end{bmatrix}, \; B_3 = \begin{bmatrix} 0 \\ 1 \end{bmatrix}, \; C_3 = \left[0 \vdots 1\right], \; D_3 = 0\right\}$
$\Longrightarrow$

$$Y_3^{\mp}(s) = \left[0 \vdots 1\right]\begin{bmatrix} s-1 & 0 \\ 0 & s+1 \end{bmatrix}^{-1}\left\{\begin{bmatrix} 0 \\ 1 \end{bmatrix} I^{\mp}(s) + \mathbf{x}_{0\mp}\right\} =$$

$$= \left(s^2-1\right)^{-1}\left[s-1 \vdots 0 \vdots s-1\right]\begin{bmatrix} I^{\mp}(s) \\ x_{10\mp} \\ x_{20\mp} \end{bmatrix}_{x_{20\mp}=y_{30\mp}}$$

$$= \underbrace{\left(s^2-1\right)^{-1}\left[s-1 \vdots 0 \vdots s-1\right]}_{F_{ISO3}(s)}\begin{bmatrix} I^{\mp}(s) \\ x_{10\mp} \\ y_{30\mp} \end{bmatrix},$$

$$F_{ISO3}(s) = \left(s^2-1\right)^{-1}\left[s-1 \vdots 0 \vdots s-1\right] \Longrightarrow$$

$$F_{ISO3nd}(s) = \left[\frac{1}{s+1} \vdots 0 \vdots \frac{1}{s+1}\right].$$

$$Y_3^{\mp}(s) = F_{ISO3}(s)\mathbf{V}_{ISO3}^{\mp}(s;\mathbf{x}_0),\ \mathbf{V}_{ISO3}^{\mp}(s;\mathbf{x}_0) = \begin{bmatrix} I^{\mp}(s) \\ \mathbf{x}_{0\mp} \end{bmatrix} \Longrightarrow$$

$$F_{ISO3}(s) = \left(s^2-1\right)^{-1}\left[s-1 \vdots 0 \vdots s-1\right] \Longrightarrow$$

$$G_{ISO3}(s) = \frac{s-1}{s^2-1},\ G_{ISOx_03}(s) = \frac{\left[0 \vdots s-1\right]}{s^2-1},$$

$$F_{ISOIS3}(s) = \frac{\begin{bmatrix} 0 & s+1 & 0 \\ -1+s & 0 & s-1 \end{bmatrix}}{s^2-1} \Longrightarrow$$

$$G_{ISOIS3}(s) = \frac{\begin{bmatrix} 0 \\ -1+s \end{bmatrix}}{s^2-1},\ G_{ISOSS3}(s) = \frac{\begin{bmatrix} s+1 & 0 \\ 0 & s-1 \end{bmatrix}}{s^2-1}.$$

The nondegenerate form $F_{ISO3nd}(s)$ of $F_{ISO3}(s)$ reads

$$F_{ISO3nd}(s) = \frac{1}{s+1}\left[1 \vdots 0 \vdots 1\right].$$

We determine now the IO model of the system,

$$\frac{dx_1}{dt} = x_1,\ \frac{dx_2}{dt} = -x_2 + i,\ y_3 = x_2 \Longrightarrow$$
$$y_3^{(1)} + y_3 = i \Longrightarrow$$
$$Y_3^{\mp}(s) = \frac{1}{s+1}\left[1 \vdots 1\right]\begin{bmatrix} I^{\mp}(s) \\ y_{30\mp} \end{bmatrix} = F_{IOISO3}(s)\mathbf{V}_{IOISO3}^{\mp}(s;y_{30\mp}) \Longrightarrow$$
$$F_{IOISO3}(s) = \left[\frac{1}{s+1} \vdots \frac{1}{s+1}\right],\ \mathbf{V}_{IOISO3}^{\mp}(s;y_{30\mp}) = \begin{bmatrix} I^{\mp}(s) \\ y_{30\mp} \end{bmatrix}.$$

Notice that we can write

$$Y_3^{\mp}(s) = \underbrace{\frac{1}{s+1}\left[1 \vdots 1\right]}_{F_{ISO3ind}(s)}\ \underbrace{\begin{bmatrix} I^{\mp}(s) \\ y_{30\mp} \end{bmatrix}}_{\mathbf{V}_{IOISO3}^{\mp}(s;y_{30\mp})} = \underbrace{\frac{1}{s+1}\left[1 \vdots 0 \vdots 1\right]}_{F_{ISO3ind}(s)}\begin{bmatrix} I^{\mp}(s) \\ x_{10\mp} \\ y_{30\mp} \end{bmatrix} =$$
$$= F_{ISO3ind}(s)\begin{bmatrix} I^{\mp}(s) \\ x_{10\mp} \\ x_{20\mp} \end{bmatrix}.$$

We may conclude the equivalence between $F_{IOISO3}(s)$ and the nondegenerate form $F_{ISO3nd}(s)$ of $F_{ISO3}(s)$ in this example.

4) $\{A_4 = [-1],\ B_4 = [1],\ C_4 = [1],\ D_4 = 0\} \Longrightarrow$

$$Y_4^{\mp}(s) = [1]\,[s+1]^{-1}\left\{[1]\,I^{\mp}(s) + \mathbf{x}_{0\mp}\right\} =$$
$$= (s+1)^{-1}\begin{bmatrix} 1 \vdots 1 \end{bmatrix}\begin{bmatrix} I^{\mp}(s) \\ x_{10\mp} \end{bmatrix}_{x_{10\mp}=y_{40\mp}} =$$
$$= (s+1)^{-1}\begin{bmatrix} 1 \vdots 1 \end{bmatrix}\begin{bmatrix} I^{\mp}(s) \\ y_{40\mp} \end{bmatrix} =$$
$$= F_{ISO4}(s)\mathbf{V}_{ISO4}^{\mp}(s;\mathbf{x}_0),\ \mathbf{V}_{ISO4}^{\mp}(s;\mathbf{x}_0) = \begin{bmatrix} I^{\mp}(s) \\ x_{10\mp} \end{bmatrix} \Longrightarrow$$

$$F_{ISO4}(s) = (s+1)^{-1}\begin{bmatrix} 1 \vdots 1 \end{bmatrix} = F_{IO4}(s),$$
$$G_{ISO4}(s) = \frac{1}{s+1},$$

$$G_{ISOx_04}(s) = \frac{1}{s+1},\ F_{ISOIS4}(s) = \frac{\begin{bmatrix} 1 \vdots 1 \end{bmatrix}}{s+1},$$
$$G_{ISOIS4}(s) = \frac{1}{s+1},\ G_{ISOSS4}(s) = \frac{1}{s+1}.$$

When we replace $\mathbf{x}_{0\mp} = (x_{0\mp}) = (x_{10\mp})$ *by* $y_{40\mp}$ *due to* $x_{10\mp} = y_{40\mp}$ *then, formally, there is not an explicit influence of the initial state variable on the system output response.*

The state space model under (iv) corresponds to the following first-order IO differential equation and the full transfer function matrix $F_{IO4}(s)$*:*

$$y_4^{(1)}(t) + y_4(t) = i(t) \Longrightarrow$$
$$(s+1)\,Y_4^{\mp}(s) - y_{40\mp} = I^{\mp}(s) \Longrightarrow$$
$$Y_4^{\mp}(s) = \underbrace{(s+1)^{-1}\begin{bmatrix} 1 \vdots 1 \end{bmatrix}}_{F_{IOISO4}(s)}\begin{bmatrix} I^{\mp}(s) \\ y_{40\mp} \end{bmatrix},\ F_{IOISO4}(s) = \frac{1}{s+1}\begin{bmatrix} 1 \vdots 1 \end{bmatrix},$$

$$Y_4^{\mp}(s) = F_{IOISO4}(s)\mathbf{V}_{IOISO4}^{\mp}(s;y_{40}),\ \mathbf{V}_{IOISO4}^{\mp}(s;y_{40}) = \begin{bmatrix} I^{\mp}(s) \\ y_{40\mp} \end{bmatrix} \Longrightarrow$$

$$F_{IOISO4}(s) = (s+1)^{-1}\begin{bmatrix} 1 \vdots 1 \end{bmatrix} = F_{ISO4}(s),$$

$$G_{IOISO4}(s) = \frac{1}{s+1} = G_{ISO4}(s).$$

Example 168 *Kalman considered a time-varying LC network in [80, Example 1, pp. 163-165], which is neither completely controllable nor observable. Without losing these properties, we accept that all network parameters are constant, i.e., $C(t) \equiv C$ and $L(t) \equiv L$, so that the system description reads*

$$\frac{dx_1}{dt} = -\frac{1}{L}x_1 + u \Longrightarrow X_1(s) = (s+\frac{1}{L})^{-1}\left[x_{10} + U(s)\right],$$
$$\frac{dx_2}{dt} = -\frac{1}{L}x_2 \Longrightarrow X_2(s) = (s+\frac{1}{L})^{-1}x_{20},$$

$$y = \frac{2}{L}x_2 + u \Longrightarrow Y(s) = \frac{2}{L}(s+\frac{1}{L})^{-1}x_{20} + U(s) \Longrightarrow$$
$$Y(s) = \begin{bmatrix} 1 \vdots \frac{2}{L}(s+\frac{1}{L})^{-1} \end{bmatrix}\begin{bmatrix} U(s) \\ x_{20} \end{bmatrix} \Longrightarrow$$

$$G_{ISO}(s) \equiv 1,\ F_{ISO}(s) = \begin{bmatrix} 1 \vdots \frac{2}{L}\left(s+\frac{1}{L}\right)^{-1} \end{bmatrix}.$$

The transfer function $G_{ISO}(s)$ leads to the conclusion that the system is static. However, the full transfer function matrix $F_{ISO}(s)$ shows that the system is dynamic. If we write $F_{ISO}(s)$ in the form

$$F_{ISO}(s) = (s+\frac{1}{L})^{-1}\begin{bmatrix} s+\frac{1}{L} \vdots \frac{2}{L} \end{bmatrix}$$

then

$$G_{ISO}(s) = \frac{s+\frac{1}{L}}{s+\frac{1}{L}} \equiv 1 = G_{ISOnd}(s)$$

shows also that the system is dynamic, and that it is not completely controllable and observable. $F_{ISO}(s)$ is not either degenerate or reducible, while $G_{ISO}(s)$ is both degenerate and reducible.

Example 169 *Kalman presented also the following ISO system in [80, Example 8, pp. 188, 189]:*

$$\frac{d\mathbf{x}}{dt} = \begin{bmatrix} 0 & 1 & 0 \\ 5 & 0 & 2 \\ -2 & 0 & -2 \end{bmatrix} \mathbf{x} + \begin{bmatrix} 0 \\ 0 \\ 0.5 \end{bmatrix} i,$$

$$y = \begin{bmatrix} -2 \vdots 1 \vdots 0 \end{bmatrix} \mathbf{x}.$$

We apply the last Equation (8.35)

$$F_{ISO}(s) = [det(sI_n - A)]^{-1} \bullet$$
$$\bullet \left[Cadj(sI_n - A)B + Ddet(sI_n - A) \vdots Cadj(sI_n - A) \right] =$$

$$= \left(s^3 + 2s^2 - 5s - 6\right)^{-1} \bullet$$
$$\bullet \left[s - 2 \vdots -2s^2 + s + 6 \vdots s^2 - 4 \vdots 2s - 4 \right] =$$
$$= [(s+1)(s-2)(s+3)]^{-1} \bullet$$
$$\bullet \left[s - 2 \vdots (s-2)(-2s-3) \vdots (s-2)(s+2) \vdots 2(s-2) \right] =$$
$$= \frac{s-2}{s-2} [(s+1)(s+3)]^{-1} \left[1 \vdots -2s - 3 \vdots s + 2 \vdots 2 \right] \Longrightarrow$$

$$G_{ISO}(s) = [det(sI_n - A)]^{-1} [Cadj(sI_n - A)B + Ddet(sI_n - A)] =$$
$$= \left(s^3 + 2s^2 - 5s - 6\right)^{-1} (s-2) = \frac{s-2}{s-2} [(s+1)(s+3)]^{-1}.$$

Since $G_{ISO}(s)$ is reducible, i.e., since it degenerates to

$$G_{ISOird}(s) = \frac{1}{(s+1)(s+3)} = G_{ISOnd}(s),$$

then it follows that the system is not completely controllable and observable. In this example, $F_{ISO}(s)$ is also both reducible and degenerate. After cancelling the same zero and pole at $s = 2$, we determine the nondegenerate form $F_{ISOnd}(s)$ of $F_{ISO}(s)$,

$$F_{ISOnd}(s) = \frac{\left[1 \vdots -2s - 3 \vdots s + 2 \vdots 2 \right]}{(s+1)(s+3)}.$$

It is also the irreducible form $F_{ISOird}(s)$ of $F_{ISO}(s)$,

$$F_{ISOird}(s) = F_{ISOnd}(s).$$

8.3 $F(s)$ of the *IIO* system

The *IIO* system (3.89), i.e., (3.90), (in Subsection 3.5.4) repeated as

$$\begin{aligned} Q^{(\alpha)}\mathbf{r}^{\alpha}(t) &= P^{(\beta)}\mathbf{i}^{\beta}(t),\ \forall t \in \mathfrak{T}_0, \\ E^{(\nu)}\mathbf{y}^{\nu}(t) &= R^{(\alpha)}\mathbf{r}^{\alpha}(t) + T^{(\mu)}\mathbf{i}^{\mu}(t),\ \forall t \in \mathfrak{T}_0, \end{aligned} \tag{8.46}$$

has two full transfer function matrices:

- The full *IS* transfer function matrix $F_{IIOIS}(s)$ that relates $\mathbf{R}^{\mp}(s)$ to $\mathbf{I}^{\mp}(s)$, $\mathbf{i}_{0\mp}^{\gamma-1}$ and $\mathbf{r}_{0\mp}^{\alpha-1}$, (Definition 144, in Section 7.4),

$$\mathbf{R}^{\mp}(s) = F_{IIOIS}(s)\left[\mathbf{I}^{\mp T}(s) \vdots \left(\mathbf{i}_{0\mp}^{\beta-1}\right)^T \vdots \left(\mathbf{r}_{0\mp}^{\alpha-1}\right)^T\right]^T \tag{8.47}$$

$$\mathbf{R}^{\mp}(s) = F_{IIOIS}(s)\mathbf{V}_{IIOIS}^{\mp}(s), \tag{8.48}$$

$$\mathbf{V}_{IIOIS}^{\mp}(s;\mathbf{i}_{0\mp}^{\beta-1};\mathbf{r}_{0\mp}^{\alpha-1}) = \left[\mathbf{I}^{\mp T}(s) \vdots \underbrace{\left(\mathbf{i}_{0\mp}^{\beta-1}\right)^T \vdots \left(\mathbf{r}_{0\mp}^{\alpha-1}\right)^T}_{\mathbf{C}_{oIIOIS}^{\mp T}}\right]^T, \tag{8.49}$$

$$\mathbf{C}_{oIIOIS}^{\mp} = \begin{bmatrix} \mathbf{i}_{0\mp}^{\beta-1} \\ \mathbf{r}_{0\mp}^{\alpha-1} \end{bmatrix},\ \mathbf{V}_{IIOIS}^{\mp}(s;\mathbf{C}_{oIIOIS}^{\mp}) = \begin{bmatrix} \mathbf{I}^{\mp}(s) \\ \mathbf{C}_{oIIOIS}^{\mp} \end{bmatrix}, \tag{8.50}$$

  where $\mathbf{V}_{IIOIS}^{-}(s)$ is the left Laplace transform of the vector function $\mathbf{v}_{IIOIS}(.)$ of all the actions on the system,

$$\mathbf{v}_{IIOIS}(t;\mathbf{i}_{0\mp}^{\beta-1};\mathbf{r}_{0\mp}^{\alpha-1}) = \left[\mathbf{i}^T(t) \vdots \delta(t)\left(\mathbf{i}_{0-}^{\beta-1}\right)^T \vdots \delta(t)\left(\mathbf{r}_{0-}^{\alpha-1}\right)^T\right]^T. \tag{8.51}$$

- The full *IO* transfer function matrix $F_{IIO}(s)$ that relates $\mathbf{Y}^{\mp}(s)$ to

$\mathbf{I}^{\mp}(s)$, $\mathbf{i}_{0\mp}^{\gamma-1}$, $\mathbf{r}_{0\mp}^{\alpha-1}$, and $\mathbf{y}_0^{\nu-1}$, (Definition 140, in Section 7.4),

$$\mathbf{Y}^{\mp}(s) = F_{IIO}(s)\left[\mathbf{I}^{\mp T}(s) \vdots \left(\mathbf{i}_{0\mp}^{\gamma-1}\right)^T \vdots \left(\mathbf{r}_{0\mp}^{\alpha-1}\right)^T \vdots \left(\mathbf{y}_{0\mp}^{\nu-1}\right)^T\right]^T,$$

$$\mathbf{Y}^{\mp}(s) = F_{IIO}(s)\mathbf{V}_{IIO}^{\mp}(s), \tag{8.52}$$

$$\mathbf{V}_{IIO}^{\mp}(s;\mathbf{i}_{0\mp}^{\gamma-1};\mathbf{r}_{0\mp}^{\alpha-1};\mathbf{y}_0^{\nu-1}) = \left[\mathbf{I}^{\mp T}(s) \vdots \underbrace{\left(\mathbf{i}_{0\mp}^{\gamma-1}\right)^T \vdots \left(\mathbf{r}_{0\mp}^{\alpha-1}\right)^T \vdots \left(\mathbf{y}_{0\mp}^{\nu-1}\right)^T}_{\mathbf{C}_{oIIO}^{\mp T}}\right]^T, \tag{8.53}$$

$$\mathbf{C}_{oIIO}^{\mp} = \begin{bmatrix} \mathbf{i}_{0\mp}^{\gamma-1} \\ \mathbf{r}_{0\mp}^{\alpha-1} \\ \mathbf{y}_{0\mp}^{\nu-1} \end{bmatrix} = \begin{bmatrix} \mathbf{C}_{oIIOIS}^{\mp} \\ \mathbf{y}_{0\mp}^{\nu-1} \end{bmatrix},$$

$$\mathbf{V}_{IIO}^{\mp}(s;\mathbf{C}_{oIIO}^{\mp}) = \begin{bmatrix} \mathbf{I}^{\mp}(s) \\ \mathbf{C}_{oIIO}^{\mp} \end{bmatrix} = \begin{bmatrix} \mathbf{V}_{IIOIS}^{\mp}(s;\mathbf{C}_{oIIOIS}^{\mp}) \\ \mathbf{y}_{0\mp}^{\nu-1} \end{bmatrix}, \tag{8.54}$$

where $\mathbf{V}_{IIO}^{-}(s)$ is the left Laplace transform of the vector function $\mathbf{v}_{IIO}(.)$ of all the actions on the system,

$$\mathbf{v}_{IIO}(t;\mathbf{i}_{0\mp}^{\gamma-1};\mathbf{r}_{0\mp}^{\alpha-1};\mathbf{y}_0^{\nu-1}) =$$
$$= \left[\mathbf{i}^T(t) \vdots \delta(t)\left(\mathbf{i}_{0\mp}^{\gamma-1}\right)^T \vdots \delta(t)\left(\mathbf{r}_{0\mp}^{\alpha-1}\right)^T \vdots \delta(t)\left(\mathbf{y}_{0\mp}^{\nu-1}\right)^T\right]^T. \tag{8.55}$$

Theorem 170 *The full IS transfer function matrix $F_{IIOIS}(s)$ of the system (3.89), i.e., of (8.46), is determined by (8.57),*

$$F_{IIOIS}(s) =$$
$$= \left(Q^{(\alpha)}S_{\rho}^{(\alpha)}(s)\right)^{-1}\left[P^{(\beta)}S_M^{(\beta)}(s) \vdots -P^{(\beta)}Z_M^{(\beta-1)}(s) \vdots Q^{(\alpha)}Z_{\rho}^{(\alpha-1)}(s)\right] \tag{8.56}$$

$$F_{IIOIS}(s) = \left[G_{IIOIS}(s) \vdots G_{IIOoIS}(s)\right] = F_{IIOISD}^{-1}(s)F_{IIOISN}(s),$$

$$G_{IIOoIS}(s) = \left[G_{IIOi_0IS}(s) \vdots G_{IIOr_0IS}(s)\right], \tag{8.57}$$

together with (8.58) through (8.60):

- *IRIS transfer function matrix* $G_{IIOIS}(s)$,

$$G_{IIOIS}(s) = \left(Q^{(\alpha)}S_{\rho}^{(\alpha)}(s)\right)^{-1} P^{(\beta)}S_{M}^{(\beta)}(s) = \\ = G_{IIOISD}^{-1}(s)G_{IIOISN}(s), \tag{8.58}$$

- *IRII transfer function matrix* $G_{IIOi_0IS}(s)$,

$$G_{IIOi_0IS}(s) = -\left(Q^{(\alpha)}S_{\rho}^{(\alpha)}(s)\right)^{-1} P^{(\beta)}Z_{M}^{(\beta-1)}(s) = \\ = G_{IIOi_0ISD}^{-1}(s)G_{IIOi_0ISN}(s), \tag{8.59}$$

- *IRIR transfer function matrix* $G_{IIOr_0IS}(s)$,

$$G_{IIOr_0IS}(s) = \left(Q^{(\alpha)}S_{\rho}^{(\alpha)}(s)\right)^{-1} Q^{(\alpha)}Z_{\rho}^{(\alpha-1)}(s) = \\ = G_{IIOr_0ISD}^{-1}(s)G_{IIOr_0ISN}(s), \tag{8.60}$$

so that the denominator matrix polynomial $F_{IIOISD}(s)$ *and the numerator matrix polynomial* $F_{IIOISN}(s)$ *of* $F_{IIOIS}(s)$,

$$F_{IIOIS}(s) = F_{IIODIS}^{-1}(s)F_{IIOISN}(s),$$

read:

$$F_{IIOISD}(s) = Q^{(\alpha)}S_{\rho}^{(\alpha)}(s), \\ F_{IIOISN}(s) = \\ = \left[P^{(\beta)}S_{M}^{(\beta)}(s) \vdots -P^{(\beta)}Z_{M}^{(\beta-1)}(s) \vdots Q^{(\alpha)}Z_{\rho}^{(\alpha-1)}(s)\right]. \tag{8.61}$$

The proof is in the first part of Appendix F.

Theorem 171 *The full IO transfer function matrix* $F_{IIO}(s)$ *of the system (3.89), i.e., (8.46), is determined by (8.62),*

$$F_{IIO}(s) = \left[G_{IIO}(s) \vdots G_{IIOo}(s)\right] = F_{IIOD}^{-1}(s)F_{IION}(s), \\ G_{IIOo}(s) = \left[G_{IIOi_0}(s) \vdots G_{IIOr_0}(s) \vdots G_{IIOy_0}(s)\right], \tag{8.62}$$

together with (8.63) through (8.67), which determine the IIO system:

- *IO transfer function matrix* $G_{IIO}(s)$,

$$G_{IIO}(s) = \left(E^{(\nu)}S_N^{(\nu)}(s)\right)^{-1} \bullet$$
$$\bullet \left[R^{(\alpha)}S_\rho^{(\alpha)}(s)\left(Q^{(\alpha)}S_\rho^{(\alpha)}(s)\right)^{-1}P^{(\beta)}S_M^{(\beta)}(s) + T^{(\mu)}S_M^{(\mu)}(s)\right] =$$
$$= G_{IIOD}^{-1}(s)G_{IION}(s), \tag{8.63}$$

- *IICO transfer function matrix* $G_{IIOi_0}(s)$,

$$G_{IIOi_0}(s) = -\left(E^{(\nu)}S_N^{(\nu)}(s)\right)^{-1} \bullet$$
$$\bullet \left(\begin{array}{c} \left[R^{(\alpha)}S_\rho^{(\alpha)}(s)\left(Q^{(\alpha)}S_\rho^{(\alpha)}(s)\right)^{-1}P^{(\beta)}Z_M^{(\beta-1)}(s) \vdots O_{N(\gamma-\beta)}\right] + \\ + \left[T^{(\mu)}Z_M^{(\mu-1)}(s) \vdots O_{N(\gamma-\mu)}\right] \end{array} \right) =$$
$$= G_{IIOi_0D}^{-1}(s)G_{IIOi_0N}(s), \tag{8.64}$$

- *IIRO transfer function matrix* $G_{IIOr_0}(s)$,

$$G_{IIOr_0}(s) = \left(E^{(\nu)}S_N^{(\nu)}(s)\right)^{-1} \bullet$$
$$\bullet \left(\begin{array}{c} R^{(\alpha)}S_\rho^{(\alpha)}(s)\left(Q^{(\alpha)}S_\rho^{(\alpha)}(s)\right)^{-1}Q^{(\alpha)}Z_\rho^{(\alpha-1)}(s) - \\ -R^{(\alpha)}Z_\rho^{(\alpha-1)}(s) \end{array} \right) =$$
$$= G_{IIOr_0D}^{-1}(s)G_{IIOr_0N}(s), \tag{8.65}$$

- *IIYO transfer function matrix* $G_{IIOy_0}(s)$,

$$G_{IIOy_0}(s) = \left(E^{(\nu)}S_N^{(\nu)}(s)\right)^{-1}E^{(\nu)}Z_N^{(\nu-1)}(s) =$$
$$= G_{IIOy_0D}^{-1}(s)G_{IIOy_0N}(s), \tag{8.66}$$

so that the denominator matrix polynomial $F_{IIOD}(s)$ *and the numerator matrix polynomial* $F_{IION}(s)$ *of* $F_{IIO}(s), F_{IIO}(s) = F_{IIOD}^{-1}(s)F_{IION}(s)$, *read:*

$$F_{IIOD}(s) = \left(E^{(\nu)}S_N^{(\nu)}(s)\right),$$

$$F_{IION}(s) =$$

$$\begin{bmatrix} \left[R^{(\alpha)}S_\rho^{(\alpha)}(s)\left(Q^{(\alpha)}S_\rho^{(\alpha)}(s)\right)^{-1}P^{(\beta)}S_M^{(\beta)}(s) + T^{(\mu)}S_M^{(\mu)}(s)\right]^T \\ \begin{bmatrix} -\left[R^{(\alpha)}S_\rho^{(\alpha)}(s)\left(Q^{(\alpha)}S_\rho^{(\alpha)}(s)\right)^{-1}P^{(\beta)}Z_M^{(\beta-1)}(s)\vdots O_{N\times(\gamma-\beta)}\right] - \\ -\left[T^{(\mu)}Z_M^{(\mu-1)}(s) \vdots O_{N\times(\gamma-\mu)}\right] \end{bmatrix}^T \\ \begin{bmatrix} \left[R^{(\alpha)}S_\rho^{(\alpha)}(s)\left(Q^{(\alpha)}S_\rho^{(\alpha)}(s)\right)^{-1}Q^{(\alpha)}Z_\rho^{(\alpha-1)}(s)\right] - \\ -R^{(\alpha)}Z_\rho^{(\alpha-1)}(s) \end{bmatrix}^T \\ \left(E^{(\nu)}Z_N^{(\nu-1)}(s)\right)^T \end{bmatrix}^T. \tag{8.67}$$

We present proof in the second part of Appendix F.

Note 172 *This theorem holds also for the RS (2.37), and the PMD systems (2.36), (in Section 2.3), because they are the special cases (Appendix C.3) of the IIO systems (2.35), (in Section 2.3), i.e., (3.90) (in Subsection 3.5.4),*

$$Q^{(\alpha)}\mathbf{r}^\alpha(t) = P^{(\beta)}\mathbf{i}^\beta(t),$$
$$E^{(\nu)}\mathbf{y}^\nu(t) = R^{(\alpha)}\mathbf{r}^\alpha(t) + T^{(\mu)}\mathbf{i}^\mu(t). \tag{8.68}$$

Note 173 ***IIO block diagram technique generalization***

These equations show that the general relationship between the Laplace transform of the output vector function $\mathbf{y}(.)$ *and the Laplace transform* $\mathbf{V}_{IIO}^{\mp}(s)$ *of the action vector function* $\mathbf{v}(.)$ *of the system under* ***nonzero initial conditions*** *has the same form as the relationship between the Laplace transform* $\mathbf{Y}(s)$ *of the output vector function* $\mathbf{y}(.)$ *and the Laplace transform* $\mathbf{I}(s)$ *of the input vector function* $\mathbf{i}(.)$ *under* ***all zero initial conditions****. In the former case, the full transfer function matrix* $F_{IIO}(s)$ *takes the role of the transfer function matrix* $G_{IIO}(s)$ *in the latter case; see Fig. 8.2. The use of* $F_{IIO}(s)$ *and* $\mathbf{V}_{IIO}^{\mp}(s) = \mathbf{V}_{IIO}^{\mp}(s; \mathbf{i}_{0\mp}^{\gamma-1}; \mathbf{r}_{0\mp}^{\alpha-1}; \mathbf{y}_0^{\nu-1})$ *permits the generalization of the block diagram technique to nonzero initial conditions. The former case holds for arbitrary initial conditions. The latter case is valid only under all zero initial conditions.*

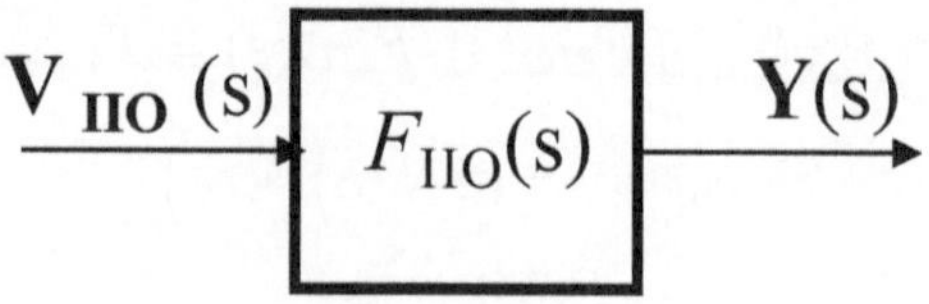

Figure 8.2: The generalized block of the *IIO* system.

Example 174 *Let the IIO system be described by*

$$r^{(2)}(t)+4r^{(1)}(t)-5r(t)=3i^{(2)}(t)-21i^{(1)}(t)+18i(t),$$
$$y^{(2)}(t)+y^{(1)}(t)-2y(t)=$$
$$=5r^{(2)}(t)-5r^{(1)}(t)+2i^{(2)}(t)-14i^{(1)}(t)+12i(t). \quad (8.69)$$

Its description and characteristics in the complex domain follow (Appendix G):

$$F_{IIO}(s)=\left[(s-1)^2(s+2)(s+5)\right]^{-1}\bullet\begin{bmatrix} (s-1)^2(s-6)(17s+10)\ \vdots \\ \vdots\ -(s-1)(s-7)(17s+10)\ \vdots \\ \vdots\ -(s-1)(13s-10)\ \vdots \\ \vdots\ 10(s-1)\ \vdots \\ \vdots\ -5(s-1)(2s+5)\ \vdots \\ \vdots\ (s-1)(s^2+6s+5)\ \vdots \\ \vdots\ (s-1)(s+5) \end{bmatrix}^T. \quad (8.70)$$

We can cancel $(s-1)$ because it is common factor to the denominator polynomial and to all entries of the numerator polynomial matrix,

$$F_{IIOnd}(s)=\left[(s-1)(s+2)(s+5)\right]^{-1}\bullet\begin{bmatrix} (s-1)(s-6)(17s+10)\ \vdots \\ \vdots\ -(s-7)(17s+10)\ \vdots \\ \vdots\ -(13s-10)\ \vdots \\ \vdots\ 10\ \vdots \\ \vdots\ -5(2s+5)\ \vdots \\ \vdots\ (s^2+6s+5)\ \vdots \\ \vdots\ (s+5) \end{bmatrix}^T. \quad (8.71)$$

This is the nondegenerate form $F_{IIOnd}(s)$ of $F_{IIO}(s)$. The system transfer function $G_{IIO}(.)$ follows now from (8.70),

$$G_{IIO}(s)=\left[(s-1)^2(s+2)(s+5)\right]^{-1}(s-1)^2(s-6)(17s+10). \quad (8.72)$$

We can cancel now $(s-1)^2$ in $G_{IIO}(s)$, which is not possible either in $F_{IIO}(s)$ or in its nondegenerate form $F_{IIOnd}(s)$,

$$G_{IIO}(s)=\left[(s+2)(s+5)\right]^{-1}(s-6)(17s+10). \quad (8.73)$$

The left Laplace transform $\mathbf{V}^{-}_{IIO}(s)$ of the action vector function $\mathbf{v}_{IIO}(.)$ and the action vector function itself result from (G.7) (Appendix G),

$$\mathbf{V}^{-}_{IIO}(s;\mathbf{i}^1_{0-};\mathbf{r}^1_{0-};\mathbf{y}^1_{0-})=\left[I^{-}(s)\;\vdots\;\underbrace{i_{0-}\;\vdots\;i^{(1)}_{0-}}_{\mathbf{i}^{1T}_{0-}}\;\vdots\;\underbrace{r_{0-}\;\vdots\;r^{(1)}_{0-}}_{\mathbf{r}^{1T}_{0-}}\;\vdots\;\underbrace{y_{0-}\;\vdots\;y^{(1)}_{0-}}_{\mathbf{y}^{1T}_{0-}}\right]^T,$$

$$\mathbf{v}_{IIO}(t;\mathbf{i}^1_{0\mp};\mathbf{r}^1_{0\mp};\mathbf{y}^1_{0})=\begin{bmatrix} i(t)\\ \delta(t)\mathbf{i}^1_{0\mp}\\ \delta(t)\mathbf{r}^1_{0\mp}\\ \delta(t)\mathbf{y}^1_{0\mp}\end{bmatrix}. \quad (8.74)$$

Equation (G.8) (in Appendix G) relates $\mathbf{Y}^{-}(s)$ to $\mathbf{V}^{-}_{IIO}(s;\mathbf{i}^1_{0\mp};\mathbf{r}^1_{0\mp};\mathbf{y}^1_{0})$ via $F_{IIO}(s)$,

$$\mathbf{Y}^{-}(s)=F_{IIO}(s)\mathbf{V}^{-}_{IIO}(s;\mathbf{i}^1_{0\mp};\mathbf{r}^1_{0\mp};\mathbf{y}^1_{0}). \quad (8.75)$$

Note 175 *The determination of the IIO system full IS transfer function matrix and its submatrices is left to the interested reader for exercise. Hint: follow the determination of the full IS transfer function matrix and its submatrices of the IO system (Section 8.2).*

8.4 Conclusion: Common general form of *F(s)*

The preceding results, (7.11), (8.8), (8.16), (8.28), (8.29), (8.37), and (8.52)-(8.54), show that for all the considered systems, the following relationship holds between $\mathbf{Y}^{-}(s)$ and $\mathbf{V}^{-}(s)$,

$$\mathbf{Y}^{-}(s)=F(s)\mathbf{V}^{-}(s),\; F(s)=F_D^{-1}(s)F_N(s),$$
$$F(s)\in\mathfrak{C}^{N\times W},\,\mathbf{V}^{-}(s)\in\mathfrak{C}^{W},\;\mathbf{V}^{-}(s)=\begin{bmatrix}\mathbf{I}^{-}(s)\\ \mathbf{C}^{-}_{0}\end{bmatrix},\; W=(L+1)U, \quad (8.76)$$

where $\mathbf{C}_0^-$ is the vector of all initial conditions, $F_D(s)$ and $F_N(s)$ are, respectively, the denominator polynomial matrix and the numerator polynomial matrix of $F(s)$. Their general forms read

$$
\begin{gathered}
F_D(s) = A_D^{(J)} S_N^{(J)}(s) = \sum_{k=0}^{k=J} A_{Dk} s^k \in \mathfrak{C}^{N\times N}, \\
A_{Dk} \in \mathfrak{R}^{N\times N},\ A_D^{(J)} \in \mathfrak{R}^{N\times (J+1)N}, \\
F_N(s) = B_N^{(L)} S_W^{(L)}(s) = \sum_{k=0}^{k=L} B_{Nk} s^k \in \mathfrak{C}^{N\times W}, \\
B_{Nk} \in \mathfrak{R}^{N\times U},\ B_N^{(L)} \in \mathfrak{R}^{N\times (L+1)U},\ (L+1)\,U = W.
\end{gathered}
\tag{8.77}
$$

This means that

$$
\Delta(s) = detF_D(s) = det\left[A_D^{(J)} S_N^{(J)}(s)\right] = det\left[\sum_{k=0}^{k=J} A_{Dk} s^k\right] \tag{8.78}
$$

is the characteristic polynomial of the system in general.

Conclusion 176 ***The block diagram technique generalization***

The use of the system full transfer function matrix $F(s)$ and Laplace transform $\mathbf{V}^-(s)$ of the generalized action vector $\mathbf{v}(t)$ (which incorporates the input vector and the vectors of all initial conditions) enables the generalization of the classic block diagram technique to incorporate all initial conditions. We should simply replace the system transfer function matrix $G(s)$ by the system full transfer function matrix $F(s)$ and Laplace transform of the input vector $\mathbf{i}(t)$ by Laplace transform $\mathbf{V}^-(s)$ of the action vector $\mathbf{v}(t)$.

Chapter 9

Full block diagram algebra

9.1 Introduction

We use (8.76), (in Section 8.4) in the following form

$$\mathbf{Y}(s) = F(s)\mathbf{V}(s), \tag{9.1}$$

where in general

$$F(s) = \left[G(s) \ \vdots \ G_o(s) \right], \tag{9.2}$$

and

$$\mathbf{V}(s) = \begin{bmatrix} \mathbf{I}(s) \\ \mathbf{C}_0 \end{bmatrix}. \tag{9.3}$$

The matrix $G(s)$ is the system transfer function matrix with respect to the input vector,

$$G(s) = \left\{ \begin{array}{c} G_{IO}(s) \ for \ the \ IO \ system, \\ G_{ISO}(s) \ for \ the \ ISO \ system, \\ G_{IIO}(s) \ for \ the \ IIO \ system \end{array} \right\}.$$

The matrix $G_o(s)$ is the system transfer function matrix with respect to all initial conditions at $t = 0$ acting on the system,

- $G_o(s)$ is $G_{IO_0}(s)$, (7.20), (in Section 7.2),

$$G_o(s) = G_{IO_0}(s) = \left[G_{IOi_0}(s) \vdots G_{IOy_0}(s) \right]. \tag{9.4}$$

- $G_o(s)$ is $G_{ISO_0}(s)$, (7.27), (in Section 7.3),

$$G_o(s) = G_{ISO_0}(s) = G_{ISOx_0}(s), \tag{9.5}$$

- $G_o(s)$ is $G_{IIO_0}(s)$, (7.45), (in Section 7.4),

$$G_o(s) = G_{IIO_0}(s) = \left[G_{IIOi_0}(s) \vdots G_{IIOr_0}(s) \vdots G_{IIOy_0}(s) \right]. \tag{9.6}$$

The vector $\mathbf{C}_0$ composed of all initial conditions at $t = 0$ acting on the system has the special forms,

- for the *IO* system (3.77), (in Subsection 3.5.2) $\mathbf{C}_0$ is $\mathbf{C}_{0IO}$, (7.13), (Section 7.2),

$$\mathbf{C}_0 = \mathbf{C}_{0IO} = \left\{ \begin{array}{c} \begin{bmatrix} \mathbf{i}_{0\mp}^{\mu-1} \\ \mathbf{y}_{0\mp}^{\nu-1} \end{bmatrix}, \ if\ \mu \geq 1, \\ \mathbf{y}_{0\mp}^{\nu-1}, \ if\ \mu = 0 \end{array} \right\}, \tag{9.7}$$

- for the *ISO* system (3.84) and (3.85), (in Subsection 3.5.3) $\mathbf{C}_0$ is $\mathbf{C}_{0ISO}$, (7.25), (Section 7.3),

$$\mathbf{C}_0 = \mathbf{C}_{0ISO} = \mathbf{x}_{0(\mp)} \tag{9.8}$$

- for the *IIO* system (3.90) (in Subsection 3.5.4) $\mathbf{C}_0$ is $\mathbf{C}_{0IIO}$, (7.37), (in Section 7.4),

$$\mathbf{C}_0 = \mathbf{C}_{0IIO} = \begin{bmatrix} \mathbf{i}_{0(\mp)}^{\gamma-1} \\ \mathbf{r}_{0(\mp)}^{\alpha-1} \\ \mathbf{y}_{0(\mp)}^{\nu-1} \end{bmatrix}. \tag{9.9}$$

Claim 177 *Equations (9.1) through (9.9) enable us to present* ***the full (complete) block diagram of the system;*** *see Fig. 9.1, and* ***the full (complete) block of the system;*** *see Fig. 9.2. They are expressed in terms of the system full transfer function matrix* $F(s) = \left[G(s) \ \vdots \ G_o(s) \right]$ *and the vectors* $\mathbf{I}(s)$ *and* $\mathbf{C}_0$*, or in terms of the system full transfer function matrix* $F(s)$ *and the vector* $\mathbf{V}(s)$*, respectively. They are valid for arbitrary initial conditions, hence for both zero and nonzero initial conditions. They generalize the classical block diagram and the block of the system expressed in terms of* $G(s)$ *and* $\mathbf{I}(s)$ *, which are valid exclusively for all zero initial conditions, i.e., for*

$$\mathbf{C}_o = \mathbf{0},$$

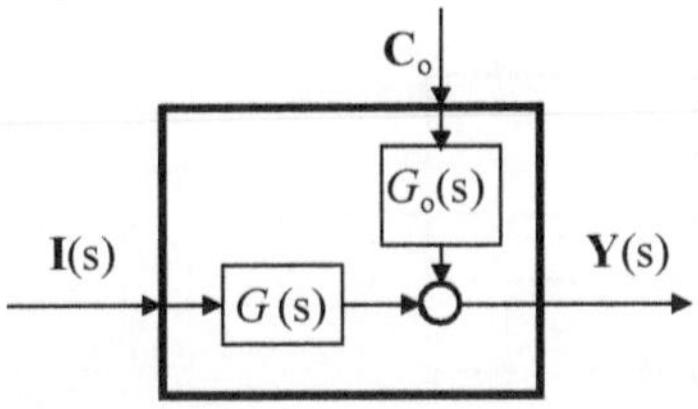

$$\mathbf{Y}(s) = G(s)\mathbf{I}(s) + G_0(s)\mathbf{C}_0 = [G(s)\ G_0(s)]\begin{bmatrix}\mathbf{I}(s)\\ \mathbf{C}_0\end{bmatrix}.$$

Figure 9.1: The full (complete) block diagram of the system.

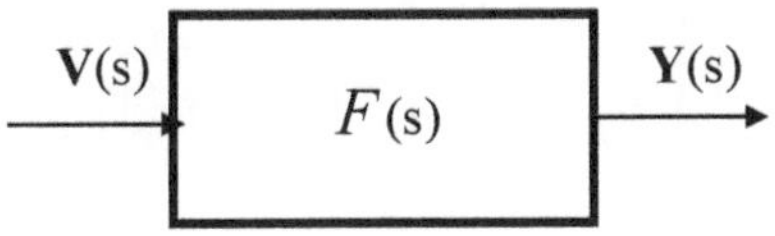

$$F(s) = [G(s)\ \ G_0(s)],\ \mathbf{V}(s) = [\mathbf{I}^T(s)\ \ C^T{}_0]^T,\ \mathbf{Y}(s) = F(s)\mathbf{V}(s)$$

Figure 9.2: The full (complete) block of the system.

hence only for

$$\mathbf{V}(s) = \begin{bmatrix}\mathbf{I}(s)\\ \mathbf{0}\end{bmatrix}.$$

We will establish general rules of **the algebra of the full block diagrams of the systems** in the next sections. The preceding consideration is the basis of the new technique-**the technique of the full block diagrams of the systems.**

9.2 Parallel connection

Fig. 9.3 presents the full block diagram of the parallel connection of k systems. We will discover in what follows the rule how to simplify the full block diagram of the parallel connection of k systems by determining the full transfer function matrix $F(s)$ of the whole system, the vector $\mathbf{C}_0$ of all initial conditions and the Laplace transform $\mathbf{V}(s)$ of the overall action vector $\mathbf{v}(t)$.

Theorem 178 *The full transfer function matrix $F(s)$ of the parallel connection of k systems, see Fig. 9.3, the resulting IO transfer function matrix*

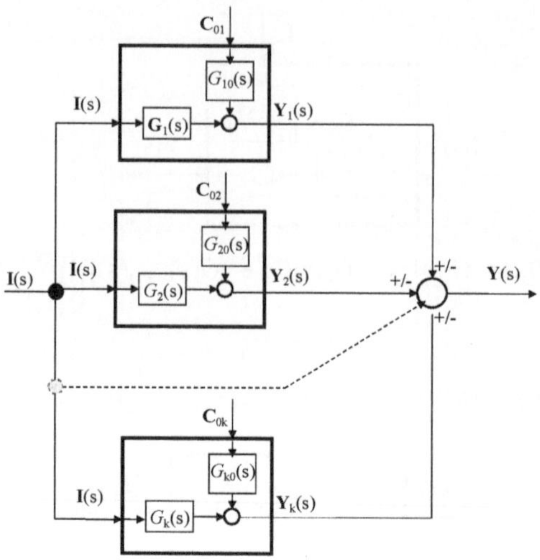

Figure 9.3: The full block diagram of the parallel connection.

$G(s)$ relative to the input vector $\mathbf{i}$*, the resulting transfer function matrix $G_o(s)$ relative to all initial conditions, the vector* $\mathbf{C}_0$ *of all initial conditions, and the vector* $\mathbf{V}(s)$ *representing the Laplace transform of the overall action vector* $\mathbf{v}(t)$*, obey the following; see Fig. 9.4:*

$$F(s) = \left[G(s) \;\vdots\; G_o(s) \right], \tag{9.10}$$

$$G(s) = \sum_{i=1}^{i=k} \left(\pm G_i(s) \right), \tag{9.11}$$

$$G_o(s) = \left[\pm G_{1o}(s) \vdots \pm G_{2o}(s) \vdots \ldots \pm G_{ko}(s) \right], \tag{9.12}$$

$$\mathbf{C}_0 = \begin{bmatrix} \mathbf{C}_{01} \\ \mathbf{C}_{02} \\ \ldots \\ \mathbf{C}_{0k} \end{bmatrix}, \ \mathbf{V}(s) = \begin{bmatrix} \mathbf{I}(s) \\ \mathbf{C}_0 \end{bmatrix}. \tag{9.13}$$

so that

$$\mathbf{Y}(s) = \left[G(s) \;\vdots\; G_o(s) \right] \begin{bmatrix} \mathbf{I}(s) \\ \mathbf{C}_0 \end{bmatrix} = F(s)\mathbf{V}(s). \tag{9.14}$$

Proof. We refer to the full block diagram of the parallel connection of

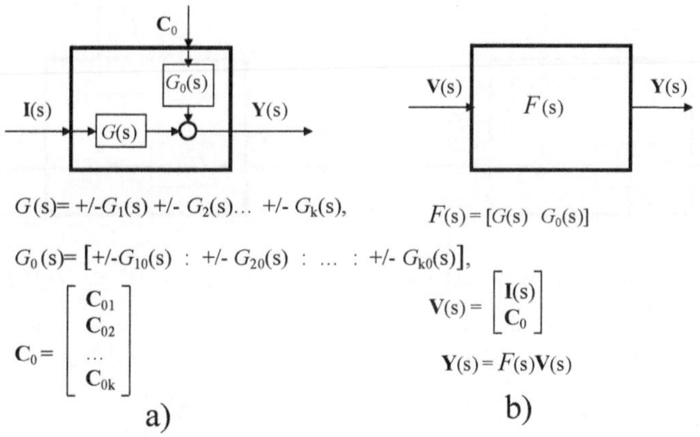

Figure 9.4: The equivalent full block diagram on the left under a) and the full block on the right under b) of the parallel connection of k systems.

k systems, Fig. 9.3 and to (9.1) through (9.9), (in Section 9.1),

$$\mathbf{Y}_i(s) = \begin{bmatrix} G_i(s) \;\vdots\; G_{io}(s) \end{bmatrix} \begin{bmatrix} \mathbf{I}(s) \\ \mathbf{C}_{0i} \end{bmatrix} = F_i(s)\mathbf{V}_i(s), \;\; \mathbf{V}_i(s) = \begin{bmatrix} \mathbf{I}(s) \\ \mathbf{C}_{0i} \end{bmatrix},$$

$$\mathbf{Y}(s) = \sum_{i=1}^{i=k} (\pm \mathbf{Y}_i(s)).$$

$$\Longrightarrow$$

$$\mathbf{Y}(s) = \sum_{i=1}^{i=k} \begin{bmatrix} \pm G_i(s) \;\vdots\; \pm G_{io}(s) \end{bmatrix} \begin{bmatrix} \mathbf{I}(s) \\ \mathbf{C}_{0i} \end{bmatrix} = \sum_{i=1}^{i=k} (\pm F_i(s))\, \mathbf{V}_i(s) \Longrightarrow$$

$$\mathbf{Y}(s) = \underbrace{\left(\sum_{i=1}^{i=k} \pm G_i(s) \right)}_{G(s)} \mathbf{I}(s) + \underbrace{\sum_{i=1}^{i=k} (\pm G_{io}(s))\, \mathbf{C}_{0i}}_{G_o(s)\mathbf{C}_0}.$$

From these equations, we find the following formulae:

$$G(s) = \sum_{i=1}^{i=k} \pm G_i(s),$$

$$G_o(s) = \begin{bmatrix} \pm G_{1o}(s) \vdots \pm G_{2o}(s) \vdots \ldots \pm G_{ko}(s) \end{bmatrix},$$

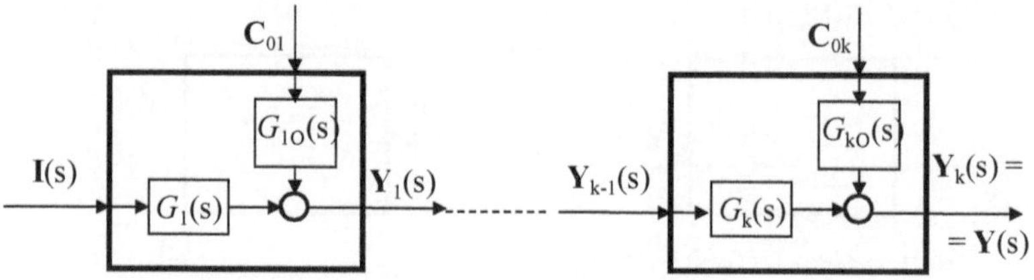

Figure 9.5: The full block diagram of the connection of k systems in series.

$$\mathbf{C}_0 = \begin{bmatrix} \mathbf{C}_{01} \\ \mathbf{C}_{02} \\ \cdots \\ \mathbf{C}_{0k} \end{bmatrix}, \ \mathbf{V}(s) = \begin{bmatrix} \mathbf{I}(s) \\ \mathbf{C}_0 \end{bmatrix},$$

which prove (9.10)-(9.14). ■

9.3 Connection in series

Fig. 9.5 shows the full block diagram of the connection of k systems in series. It will permit us to simplify the block diagram of the whole connection of the systems to be replaced by the equivalent full block represented by the resulting full transfer function matrix $F(s)$ of the whole connection and by the Laplace transform $\mathbf{V}(s)$ of the overall action vector $\mathbf{v}(t)$, which replaces the Laplace transform $\mathbf{I}(s)$ of the input vector $\mathbf{i}(t)$. Besides, we will determine the vector $\mathbf{C}_0$ of all initial conditions together with the corresponding transfer function matrices $G(s)$ and $G_o(s)$ relative to $\mathbf{I}(s)$ and to $\mathbf{C}_0$, respectively.

Theorem 179 *The full transfer function matrix $F(s)$ of the connection of k systems in series, see Fig. 9.5, the resulting IO transfer function matrix $G(s)$ relative to the input vector $\mathbf{i}$, the resulting transfer function matrix $G_o(s)$ relative to all initial conditions, the vector $\mathbf{C}_0$ of all initial conditions, and the vector $\mathbf{V}(s)$ representing the Laplace transform of the overall action vector $\mathbf{v}(t)$, obey the following; see Fig. 9.6 and Fig. 9.7:*

$$F(s) = \left[G(s) \ \vdots \ G_o(s) \right], \tag{9.15}$$

$$G(s) = \prod_{i=k}^{i=1} G_i(s), \tag{9.16}$$

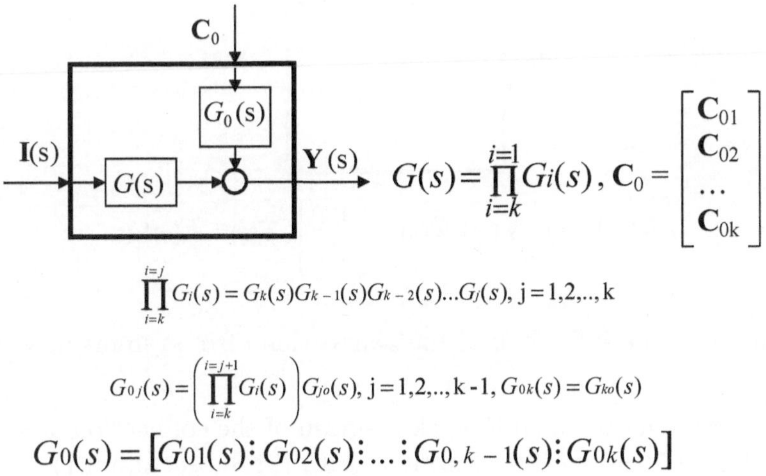

Figure 9.6: The equivalent block diagram of the connection of k systems in series.

$$G_o(s) = \left[\begin{array}{c} \underbrace{\left(\prod_{i=k}^{i=2} G_i(s)\right) G_{1o}(s)}_{G_{01}(s)} \vdots \underbrace{\left(\prod_{i=k}^{i=3} G_i(s)\right) G_{2o}(s)}_{G_{02}(s)} \vdots \\ \vdots \underbrace{G_k(s)G_{k-1,o}(s)}_{G_{0,k-1}(s)} \vdots \underbrace{G_{ko}(s)}_{G_{0k}(s)} \end{array} \right] =$$
$$= \left[G_{01}(s) \vdots G_{02}(s) \vdots ... \vdots G_{0,k-1}(s) \vdots G_{0k}(s) \right], \tag{9.17}$$

$$G_{0j}(s) = \prod_{i=k}^{i=j+1} G_i(s)G_{jo}(s), \ j = 1, 2, .., k-1, \ G_{0k}(s) = G_{ko}(s), \tag{9.18}$$

$$\mathbf{C}_0 = \begin{bmatrix} \mathbf{C}_{01} \\ \mathbf{C}_{02} \\ ... \\ \mathbf{C}_{0k} \end{bmatrix}, \ \mathbf{V}(s) = \begin{bmatrix} \mathbf{I}(s) \\ \mathbf{C}_0 \end{bmatrix}. \tag{9.19}$$

so that

$$\mathbf{Y}(s) = \left[G(s) \ \vdots \ G_o(s) \right] \begin{bmatrix} \mathbf{I}(s) \\ \mathbf{C}_0 \end{bmatrix} = F(s)\mathbf{V}(s). \tag{9.20}$$

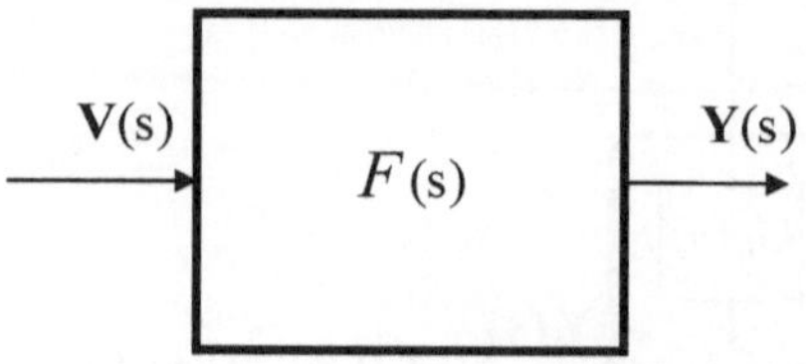

$$F(s) = [G(s) \;\; G_0(s)],\ \mathbf{V}(s) = \begin{bmatrix} \mathbf{I}(s) \\ \mathbf{C}_0 \end{bmatrix},\ \mathbf{Y}(s) = F(s)\mathbf{V}(s)$$

Figure 9.7: The full block of the connection of k systems in series.

Proof. We refer to the full block diagram of the connection of k systems in series, Fig. 9.5 and to (9.1) through (9.9), (in Subsection 9.1),

$$\mathbf{Y}_1(s) = \begin{bmatrix} G_1(s) & \vdots & G_{1o}(s) \end{bmatrix} \begin{bmatrix} \mathbf{I}(s) \\ \mathbf{C}_{01} \end{bmatrix} = F_1(s)\mathbf{V}_1(s),$$
$$\mathbf{V}_1(s) = \begin{bmatrix} \mathbf{I}(s) \\ \mathbf{C}_{01} \end{bmatrix},\ F_1(s) = \begin{bmatrix} G_1(s) & \vdots & G_{1o}(s) \end{bmatrix}$$

$$\mathbf{Y}_i(s) = \begin{bmatrix} G_i(s) & \vdots & G_{io}(s) \end{bmatrix} \begin{bmatrix} \mathbf{Y}_{i-1}(s) \\ \mathbf{C}_{0i} \end{bmatrix} = F_i(s)\mathbf{V}_i(s),$$
$$\mathbf{V}_i(s) = \begin{bmatrix} \mathbf{Y}_{i-1}(s) \\ \mathbf{C}_{0i} \end{bmatrix},\ F_i(s) = \begin{bmatrix} G_i(s) & \vdots & G_{io}(s) \end{bmatrix},$$
$$i = 2, 3, ..., k-1,$$

$$\mathbf{Y}(s) = \mathbf{Y}_k(s) = \begin{bmatrix} G_k(s) \vdots G_{ko}(s) \end{bmatrix} \begin{bmatrix} \mathbf{Y}_{k-1}(s) \\ \mathbf{C}_{0k} \end{bmatrix} = F_k(s)\mathbf{V}_k(s),$$
$$F_k(s) = \begin{bmatrix} G_k(s) \vdots G_{ko}(s) \end{bmatrix},\ \mathbf{V}_k(s) = \begin{bmatrix} \mathbf{Y}_{k-1}(s) \\ \mathbf{C}_{0k} \end{bmatrix} \Longrightarrow$$

$$\mathbf{Y}(s) = G_k(s)\mathbf{Y}_{k-1}(s) + G_{ko}(s)\mathbf{C}_{0k} \Longrightarrow$$
$$\mathbf{Y}(s) = G_k(s)\left[G_{k-1}(s)\mathbf{Y}_{k-2}(s) + G_{k-1,o}(s)\mathbf{C}_{0,k-1}\right] + G_{ko}(s)\mathbf{C}_{0k}$$
$$\Longrightarrow$$

$$\mathbf{Y}(s) = G_k(s)G_{k-1}(s)\mathbf{Y}_{k-2}(s) + G_k(s)G_{k-1,o}(s)\mathbf{C}_{0,k-1} + G_{ko}(s)\mathbf{C}_{0k}$$
$$\Longrightarrow$$

$$\mathbf{Y}(s) = G_k(s)G_{k-1}(s)\left[G_{k-2}(s)\mathbf{Y}_{k-3}(s) + G_{k-2,o}(s)\mathbf{C}_{0,k-2}\right] + \\ +G_k(s)G_{k-1,o}(s)\mathbf{C}_{0,k-1} + G_{ko}(s)\mathbf{C}_{0k} \Longrightarrow$$

$$\mathbf{Y}(s) = G_k(s)G_{k-1}(s)G_{k-2}(s)\mathbf{Y}_{k-3}(s) + \\ +G_k(s)G_{k-1}(s)G_{k-2,o}(s)\mathbf{C}_{0,k-2} + \\ +G_k(s)G_{k-1,o}(s)\mathbf{C}_{0,k-1} + G_{ko}(s)\mathbf{C}_{0k}.$$

By continuing this calculation, we arrive at

$$\mathbf{Y}(s) = G_k(s)G_{k-1}(s)G_{k-2}(s)...G_2(s)\mathbf{Y}_1(s) + \\ +G_k(s)G_{k-1}(s)G_{k-2}(s)...G_3(s)G_{20}(s)\mathbf{C}_{0,2} + \\ .... \\ +G_k(s)G_{k-1}(s)G_{k-2,o}(s)\mathbf{C}_{0,k-2} + \\ +G_k(s)G_{k-1,o}(s)\mathbf{C}_{0,k-1} + G_{ko}(s)\mathbf{C}_{0k}.$$

This, and

$$\mathbf{Y}_1(s) = G_1(s)\mathbf{I}(s) + G_{1o}(s)\mathbf{C}_{01}$$

furnish

$$\mathbf{Y}(s) = G_k(s)G_{k-1}(s)G_{k-2}(s)...G_2(s)G_1(s)\mathbf{I}(s) + \\ +G_k(s)G_{k-1}(s)G_{k-2}(s)...G_2(s)G_{1o}(s)\mathbf{C}_{01} + \\ +G_k(s)G_{k-1}(s)G_{k-2}(s)...G_3(s)G_{20}(s)\mathbf{C}_{0,2} + \\ .... \\ +G_k(s)G_{k-1}(s)G_{k-2,o}(s)\mathbf{C}_{0,k-2} + \\ +G_k(s)G_{k-1,o}(s)\mathbf{C}_{0,k-1} + G_{ko}(s)\mathbf{C}_{0k}.$$

We can put this in the matrix-vector form,

$$\mathbf{Y}(s) = \underbrace{\left[G_k(s)G_{k-1}(s)G_{k-2}(s)...G_2(s)G_1(s)\right]}_{G(s)}\mathbf{I}(s) +$$

$$+\underbrace{\begin{bmatrix} \left(G_k(s)G_{k-1}(s)G_{k-2}(s)...G_2(s)G_{1o}(s)\right)^T \\ \left(G_k(s)G_{k-1}(s)G_{k-2}(s)...G_3(s)G_{20}(s)\right)^T \\ .... \\ \left(G_k(s)G_{k-1}(s)G_{k-2,o}(s)\right)^T \\ \left(G_k(s)G_{k-1,o}(s)\right)^T \\ G_{ko}^T(s) \end{bmatrix}^T}_{G_o^T(s)} \underbrace{\begin{bmatrix} \mathbf{C}_{01} \\ \mathbf{C}_{02} \\ ... \\ \mathbf{C}_{0,k-2} \\ \mathbf{C}_{0,k-1} \\ \mathbf{C}_{0k} \end{bmatrix}}_{\mathbf{C}_0} \Longrightarrow$$

$$\mathbf{Y}(s) = \left[G(s) \ \vdots \ G_o(s) \right] \left[\begin{array}{c} \mathbf{I}(s) \\ \mathbf{C}_0 \end{array} \right] = F(s)\mathbf{V}(s),$$

which yield

$$G(s) = \prod_{i=k}^{i=1} G_i(s),$$

$$G_o(s) = \left[\underbrace{\left(\prod_{i=k}^{i=2} G_i(s) \right) G_{1o}(s)}_{G_{0,1}(s)} \vdots \underbrace{\left(\prod_{i=k}^{i=3} G_i(s) \right) G_{2o}(s)}_{G_{0,2}(s)} \vdots \ldots \vdots \underbrace{G_k(s) G_{k-1,o}(s)}_{G_{0,k-1}(s)} \vdots \underbrace{G_{ko}(s)}_{G_{0,k}(s)} \right],$$

$$F(s) = \left[G(s) \ \vdots \ G_o(s) \right],$$

$$\mathbf{C}_0 = \left[\begin{array}{c} \mathbf{C}_{01} \\ \mathbf{C}_{02} \\ \cdots \\ \mathbf{C}_{0,k-2} \\ \mathbf{C}_{0,k-1} \\ \mathbf{C}_{0k} \end{array} \right], \ \mathbf{V}(s) = \left[\begin{array}{c} \mathbf{I}(s) \\ \mathbf{C}_0 \end{array} \right].$$

These equations prove (9.16)-(9.20). ∎

9.4 Feedback connection

seeFig. 9.8 represents the full block diagram of the feedback connection of *two* systems. The overall system has two input vectors $\mathbf{i}_1(t)$ and $\mathbf{i}_2(t)$, and two output vectors $\mathbf{y}_1(t)$ and $\mathbf{y}_2(t)$. Their Laplace transforms are $\mathbf{I}_1(s)$, $\mathbf{I}_2(s)$, $\mathbf{Y}_1(s)$ and $\mathbf{Y}_2(s)$, respectively. We wish to find the full transfer function matrix $F(s)$ of the whole connection, its main transfer function submatrices $G_{R1}(s)$, $G_{R1o}(s)$, $G_{R2}(s)$, and $G_{R2o}(s)$, i.e., $G(s)$, and $G_o(s)$, as well as the corresponding vectors $\mathbf{V}_1(s)$, $\mathbf{V}_2(s)$, $\mathbf{C}_{01}$ and $\mathbf{C}_{02}$.

Theorem 180 *The full transfer function matrix $F(s)$ of the feedback connection of two systems, see Fig. 9.8, the resulting IO transfer function matrix $G(s)$ relative to the overall input vector $\mathbf{i}$, the resulting transfer function matrix $G_o(s)$ relative to all initial conditions, the vector $\mathbf{C}_0$ of all initial*

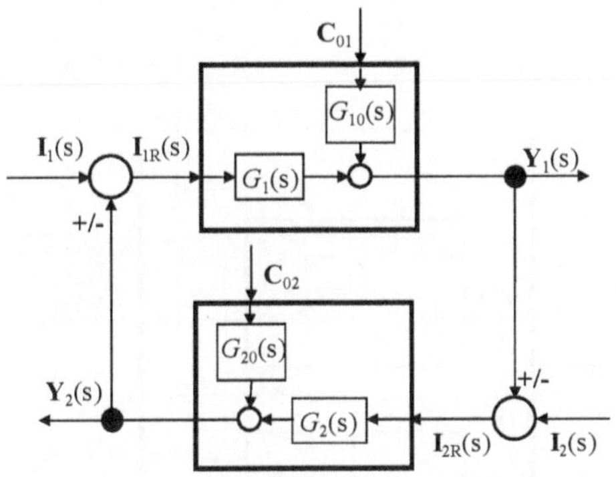

Figure 9.8: The full block diagram of the feedback connection of two systems.

conditions, and the vector $\mathbf{V}(s)$ *representing the Laplace transform of the overall action vector* $\mathbf{v}(t)$*, obey the following; see Fig. 9.9 and Fig. 9.10:*

$$F(s) = \begin{bmatrix} F_1(s) \\ F_2(s) \end{bmatrix} = \begin{bmatrix} G_{R1}(s) & G_{R1o}(s) \\ G_{R2}(s) & G_{R2o}(s) \end{bmatrix} = \begin{bmatrix} G(s) & G_o(s) \end{bmatrix}, \tag{9.21}$$

$$G_{R1}(s) = (I_{N_1} \mp G_1(s)G_2(s))^{-1} \left[G_1(s) \vdots \pm G_1(s)G_2(s) \right], \tag{9.22}$$

$$G_{R1o}(s) = (I_{N_1} \mp G_1(s)G_2(s))^{-1} \left[G_{1o}(s) \vdots \pm G_1(s)G_{2o}(s) \right], \tag{9.23}$$

$$G_{R2}(s) = (I_{N_2} \mp G_2(s)G_1(s))^{-1} \left[\pm G_2(s)G_1(s) \vdots G_2(s) \right], \tag{9.24}$$

$$G_{R2o}(s) = (I_{N_2} \mp G_2(s)G_1(s))^{-1} \left[\pm G_2(s)G_{1o}(s) \vdots G_{2o}(s) \right], \tag{9.25}$$

$$G(s) = \begin{bmatrix} G_{R1}(s) \\ G_{R2}(s) \end{bmatrix}, \; G_o(s) = \begin{bmatrix} G_{R1o}(s) \\ G_{R2o}(s) \end{bmatrix}, \tag{9.26}$$

$$\mathbf{C}_0 = \begin{bmatrix} \mathbf{C}_{01} \\ \mathbf{C}_{02} \end{bmatrix}, \; \mathbf{V}(s) = \begin{bmatrix} \mathbf{I}(s) \\ \mathbf{C}_0 \end{bmatrix}. \tag{9.27}$$

so that

$$\mathbf{Y}(s) = \left[G(s) \;\vdots\; G_o(s) \right] \begin{bmatrix} \mathbf{I}(s) \\ \mathbf{C}_0 \end{bmatrix} = F(s)\mathbf{V}(s). \tag{9.28}$$

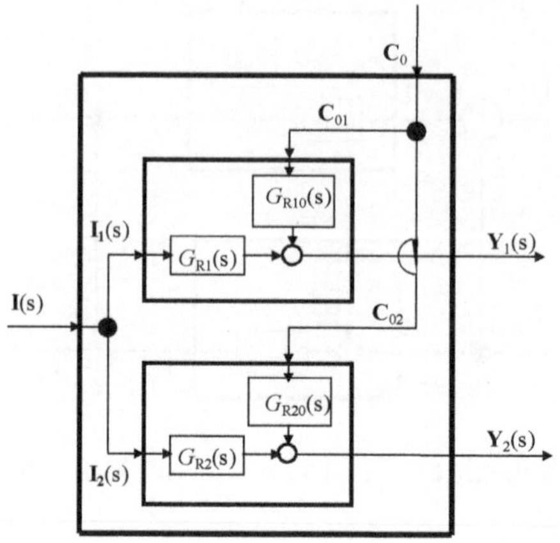

Figure 9.9: The equivalent full block diagram of the feedback connection of *two* systems.

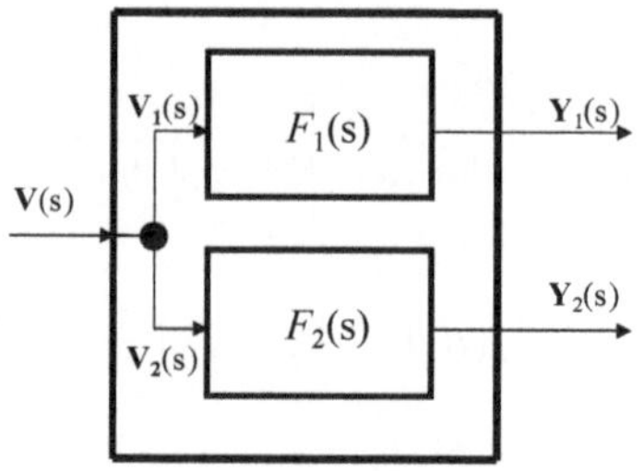

Figure 9.10: The equivalent full block diagram of the feedback connection of *two* systems.

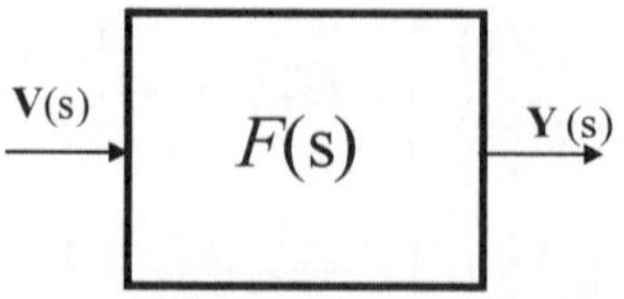

$$F(s) = \begin{bmatrix} F_1(s) \\ F_2(s) \end{bmatrix}, \mathbf{Y}(s) = F(s)\mathbf{V}(s)$$

Figure 9.11: The full block of the feedback connection of two systems.

Proof. We refer to the full block diagram of the feedback connection of two systems, Fig. 9.8 and to (9.1) through (9.9), (in Section 9.1),

$$\mathbf{Y}_i(s) = \begin{bmatrix} G_i(s) & \vdots & G_{io}(s) \end{bmatrix} \begin{bmatrix} \mathbf{I}_{iR}(s) \\ \mathbf{C}_{0i} \end{bmatrix} = F_i(s)\mathbf{V}_i(s),\ i = 1, 2,$$

$$\mathbf{V}_i(s) = \begin{bmatrix} \mathbf{I}_{iR}(s) \\ \mathbf{C}_{0i} \end{bmatrix},\ F_i(s) = \begin{bmatrix} G_i(s) & \vdots & G_{io}(s) \end{bmatrix},\ i = 1, 2,$$

$$\mathbf{Y}_1(s) = \begin{bmatrix} G_1(s) & \vdots & G_{1o}(s) \end{bmatrix} \begin{bmatrix} \underbrace{\mathbf{I}_1(s) \pm \mathbf{Y}_2(s)}_{\mathbf{I}_{1R}(s)} \\ \mathbf{C}_{01} \end{bmatrix},$$

$$\mathbf{Y}_2(s) = \begin{bmatrix} G_2(s) & \vdots & G_{2o}(s) \end{bmatrix} \begin{bmatrix} \underbrace{\mathbf{I}_2(s) \pm \mathbf{Y}_1(s)}_{\mathbf{I}_{2R}(s)} \\ \mathbf{C}_{02} \end{bmatrix} \Longrightarrow$$

$$\begin{aligned} \mathbf{Y}_1(s) &= G_1(s)\left[\mathbf{I}_1(s) \pm \mathbf{Y}_2(s)\right] + G_{1o}(s)\mathbf{C}_{01} = \\ &= G_1(s)\mathbf{I}_1(s) \pm G_1(s)\mathbf{Y}_2(s) + G_{1o}(s)\mathbf{C}_{01}, \end{aligned} \tag{9.29}$$

$$\begin{aligned} \mathbf{Y}_2(s) &= G_2(s)\left[\mathbf{I}_2(s) \pm \mathbf{Y}_1(s)\right] + G_{2o}(s)\mathbf{C}_{02} = \\ &= G_2(s)\mathbf{I}_2(s) \pm G_2(s)\mathbf{Y}_1(s) + G_{2o}(s)\mathbf{C}_{02} \end{aligned} \tag{9.30}$$

$$\Longrightarrow$$

$$\begin{gathered} \mathbf{Y}_1(s) = G_1(s)\mathbf{I}_1(s) \pm \\ \pm G_1(s)\left[G_2(s)\mathbf{I}_2(s) \pm G_2(s)\mathbf{Y}_1(s) + G_{2o}(s)\mathbf{C}_{02}\right] + G_{1o}(s)\mathbf{C}_{01} \Longrightarrow \end{gathered}$$

$$\begin{gathered} \mathbf{Y}_1(s) = G_1(s)\mathbf{I}_1(s) \pm G_1(s)G_2(s)\mathbf{I}_2(s) \pm G_1(s)G_2(s)\mathbf{Y}_1(s) \pm \\ \pm G_1(s)G_{2o}(s)\mathbf{C}_{02} + G_{1o}(s)\mathbf{C}_{01} \Longrightarrow \end{gathered}$$

$$\begin{gathered} \left(I_{N_1} \mp G_1(s)G_2(s)\right)\mathbf{Y}_1(s) = \\ = G_1(s)\mathbf{I}_1(s) \pm G_1(s)G_2(s)\mathbf{I}_2(s) \pm G_1(s)G_{2o}(s)\mathbf{C}_{02} + G_{1o}(s)\mathbf{C}_{01} \Longrightarrow \end{gathered}$$

$$\mathbf{Y}_1(s) = \left(I_{N_1} \mp G_1(s)G_2(s)\right)^{-1} \begin{pmatrix} G_1(s)\mathbf{I}_1(s) \pm G_1(s)G_2(s)\mathbf{I}_2(s) \pm \\ \pm G_1(s)G_{2o}(s)\mathbf{C}_{02} + G_{1o}(s)\mathbf{C}_{01} \end{pmatrix} \Longrightarrow$$

$$\mathbf{Y}_1(s) = \underbrace{(I_{N_1} \mp G_1(s)G_2(s))^{-1}\left[G_1(s) \vdots \pm G_1(s)G_2(s)\right]}_{G_{R1}(s)}\underbrace{\left[\begin{array}{c}\mathbf{I}_1(s)\\ \mathbf{I}_2(s)\end{array}\right]}_{\mathbf{I}(s)}+$$

$$+\underbrace{(I_{N_1} \mp G_1(s)G_2(s))^{-1}\left[G_{1o}(s) \vdots \pm G_1(s)G_{2o}(s)\right]}_{G_{R1o}(s)}\underbrace{\left[\begin{array}{c}\mathbf{C}_{01}\\ \mathbf{C}_{02}\end{array}\right]}_{\mathbf{C}_0} \Longrightarrow$$

$$G_{R1}(s) = (I_{N_1} \mp G_1(s)G_2(s))^{-1}\left[G_1(s) \vdots \pm G_1(s)G_2(s)\right],$$

$$G_{R1o}(s) = (I_{N_1} \mp G_1(s)G_2(s))^{-1}\left[G_{1o}(s) \vdots \pm G_1(s)G_{2o}(s)\right],$$

$$\mathbf{I}(s) = \left[\begin{array}{c}\mathbf{I}_1(s)\\ \mathbf{I}_2(s)\end{array}\right], \ \mathbf{C}_0 = \left[\begin{array}{c}\mathbf{C}_{01}\\ \mathbf{C}_{02}\end{array}\right],$$

and

$$\mathbf{Y}_1(s) = \underbrace{\left[G_{R1}(s) \vdots G_{R1o}(s)\right]}_{F_1(s)}\underbrace{\left[\begin{array}{c}\mathbf{I}(s)\\ \mathbf{C}_0\end{array}\right]}_{\mathbf{V}(s)} \Longrightarrow$$

$$F_1(s) = \left[G_{R1}(s) \vdots G_{R1o}(s)\right], \ \mathbf{V}(s) = \left[\begin{array}{c}\mathbf{I}(s)\\ \mathbf{C}_0\end{array}\right],$$

so that

$$\mathbf{Y}_1(s) = F_1(s)\mathbf{V}(s).$$

Analogously, (9.29) and (9.30) yield

$$\mathbf{Y}_2(s) = G_2(s)\mathbf{I}_2(s) \pm G_2(s)\left(G_1(s)\mathbf{I}_1(s) \pm G_1(s)\mathbf{Y}_2(s) + G_{1o}(s)\mathbf{C}_{01}\right) +$$
$$+G_{2o}(s)\mathbf{C}_{02}$$
$$\Longrightarrow$$

$$\mathbf{Y}_2(s) = G_2(s)\mathbf{I}_2(s) \pm G_2(s)G_1(s)\mathbf{I}_1(s) \pm G_2(s)G_1(s)\mathbf{Y}_2(s) \pm$$
$$\pm G_2(s)G_{1o}(s)\mathbf{C}_{01} + G_{2o}(s)\mathbf{C}_{02} \Longrightarrow$$

$$\left(I_{N_2} \mp G_2(s)G_1(s)\right)\mathbf{Y}_2(s) =$$
$$= \pm G_2(s)G_1(s)\mathbf{I}_1(s) + G_2(s)\mathbf{I}_2(s)) \pm G_2(s)G_{1o}(s)\mathbf{C}_{01} + G_{2o}(s)\mathbf{C}_{02} \Longrightarrow$$

$$\mathbf{Y}_2(s) = (I_{N_2} \mp G_2(s)G_1(s))^{-1} \begin{pmatrix} \pm G_2(s)G_1(s)\mathbf{I}_1(s) + G_2(s)\mathbf{I}_2(s))\pm \\ \pm G_2(s)G_{1o}(s)\mathbf{C}_{01} + G_{2o}(s)\mathbf{C}_{02} \end{pmatrix}$$

$$\Longrightarrow$$

$$\mathbf{Y}_2(s) = \underbrace{(I_{N_2} \mp G_2(s)G_1(s))^{-1}\left[\pm G_2(s)G_1(s) \vdots G_2(s)\right]}_{G_{R2}(s)}\underbrace{\begin{bmatrix} \mathbf{I}_1(s) \\ \mathbf{I}_2(s) \end{bmatrix}}_{\mathbf{I}(s)}+$$

$$\underbrace{+(I_{N_2} \mp G_2(s)G_1(s))^{-1}\left[\pm G_2(s)G_{1o}(s) \vdots G_{2o}(s)\right]}_{G_{R2o}(s)}\underbrace{\begin{bmatrix} \mathbf{C}_{01} \\ \mathbf{C}_{02} \end{bmatrix}}_{\mathbf{C}_0} \Longrightarrow$$

$$G_{R2}(s) = (I_{N_2} \mp G_2(s)G_1(s))^{-1}\left[\pm G_2(s)G_1(s) \vdots G_2(s)\right],$$

$$G_{R2o}(s) = (I_{N_2} \mp G_2(s)G_1(s))^{-1}\left[\pm G_2(s)G_{1o}(s) \vdots G_{2o}(s)\right],$$

and

$$\mathbf{Y}_2(s) = \underbrace{\left[G_{R2}(s) \vdots G_{R2o}(s)\right]}_{F_2(s)}\underbrace{\begin{bmatrix} \mathbf{I}(s) \\ \mathbf{C}_0 \end{bmatrix}}_{\mathbf{V}(s)} \Longrightarrow$$

$$\mathbf{Y}_2(s) = F_2(s)\mathbf{V}(s).$$

By continuing this calculation, we arrive at

$$\mathbf{Y}(s) = \begin{bmatrix} \mathbf{Y}_1(s) \\ \mathbf{Y}_2(s) \end{bmatrix} = \underbrace{\begin{bmatrix} F_1(s) \\ F_2(s) \end{bmatrix}}_{F(s)}\mathbf{V}(s) = F(s)\mathbf{V}(s) \Longrightarrow$$

$$F(s) = \begin{bmatrix} F_1(s) \\ F_2(s) \end{bmatrix} = \begin{bmatrix} G_{R1}(s) & G_{R1o}(s) \\ G_{R2}(s) & G_{R2o}(s) \end{bmatrix}.$$

These equations prove (9.21)-(9.28). ■

Chapter 10

Physical meaning of $F(s)$

10.1 The IO system

The full transfer function matrix $F(s)$ has an important physical meaning, additional to that expressed in its definition. Let $\mathbf{1} = [1 \;\; 1 \; ... \; 1]^T$ be of the appropriate dimension. For the definition of Dirac unit impulse $\delta(.)$, see Appendix B.2.

Definition 181 *A matrix function $\Psi_{IO}(.) : \mathfrak{T} \longrightarrow \mathfrak{R}^{N\times[(\mu+1)M+\nu N]}$ is **the full fundamental matrix function** of the IO system (3.77), (in Subsection 3.5.2), if and only if it obeys both (i) and (ii) for an arbitrary input vector function $\mathbf{i}(.)$, and for arbitrary initial conditions $\mathbf{i}_{0^-}^{\mu-1}$ and $\mathbf{y}_{0^-}^{\nu-1}$,*

(i)

$$\mathbf{y}(t;\mathbf{y}_{0^-}^{\nu-1};\mathbf{i}) = \int_{0^-}^{t} \left\{ \Psi_{IO}(\tau) \begin{bmatrix} \mathbf{i}(t-\tau) \\ \delta(t-\tau)\mathbf{i}_{0^-}^{\mu-1} \\ \delta(t-\tau)\mathbf{y}_{0^-}^{\nu-1} \end{bmatrix} \right\} d\tau =$$

$$= \int_{0^-}^{t} \left[\Gamma_{IO}(\tau)\mathbf{i}(t-\tau)d\tau\right] + \Gamma_{IOi_0}(t)\mathbf{i}_{0^-}^{\mu-1} + \Gamma_{IOy_0}(t)\mathbf{y}_{0^-}^{\nu-1}, \tag{10.1}$$

$$\Psi_{IO}(t) = \left[\Gamma_{IO}(t) \vdots \Gamma_{IOi_0}(t) \vdots \Gamma_{IOy_0}(t)\right],$$

$$\Gamma_{IO}(t) \in \mathfrak{R}^{N\times M},\; \Gamma_{IOi_0}(t) \in \mathfrak{R}^{N\times\mu M}, \Gamma_{IOy_0}(t) \in \mathfrak{R}^{N\times\nu N}, \tag{10.2}$$

(ii)

$$\Gamma_{IOi_0}(0^-) = \left[\Gamma_{IOi_01} \vdots O_{N,(\mu-1)M}\right] \text{ where}$$

$$\Gamma_{IOi_01}(0^-)\mathbf{i}_{0^-} = -\int_{0^-}^{0^-} \left[\Gamma_{IO}(\tau)\mathbf{i}(t-\tau)d\tau\right],$$

$$\Gamma_{IOy_0}(0^-) \equiv \left[I_N \vdots O_{N,(\nu-1)N}\right]. \tag{10.3}$$

Note 182 *The second Equation (10.1) under (i) of Definition 181 results from its first equation and from the properties of $\delta(.)$ (in Subsection B.2):*

$$\mathbf{y}(t;\mathbf{y}_{0^-}^{\nu-1};\mathbf{i}) = \left(\int_{0^-}^{t} \Gamma_{IO}(\tau)\mathbf{i}(t-\tau)d\tau\right) + \Gamma_{IOi_0}(t)\mathbf{i}_{0^-}^{\mu-1} + \Gamma_{IOy_0}(t)\mathbf{y}_{0^-}^{\nu-1},\ t \in \mathfrak{T}_0. \tag{10.4}$$

Theorem 183 *(i) The full fundamental matrix function $\Gamma_{IO}(.)$ of the IO system (3.77) is the inverse of the left Laplace transform of the system full transfer function matrix $F_{IO}(s)$,*

$$\Psi_{IO}(t) = \mathcal{L}^{-1}\left\{F_{IO}(s)\right\}. \tag{10.5}$$

(ii) The full transfer function matrix $F_{IO}(s)$ of the IO system (3.77) is the left Laplace transform of the system fundamental matrix $\Psi_{IO}(t)$,

$$F_{IO}(s) = \mathcal{L}^{-}\left\{\Psi_{IO}(t)\right\}. \tag{10.6}$$

(iii) The submatrices $\Gamma_{IO}(t)$, $\Gamma_{IOi_0}(t)$ and $\Gamma_{IOy_0}(t)$ are the inverse Laplace transforms of $G_{IO}(s)$, $G_{IOi_0}(s)$ and $G_{IOy_0}(s)$, respectively,

$$\Gamma_{IO}(t) = \mathcal{L}^{-1}\left\{G_{IO}(s)\right\} = \mathcal{L}^{-1}\left\{\Phi_{IO}(s)\left[B^{\mu}S_M^{(\mu)}(s)\right]\right\},$$

$$\Gamma_{IOi_0}(t) = \mathcal{L}^{-1}\left\{G_{IOi_0}(s)\right\} = -\mathcal{L}^{-1}\left\{\Phi_{IO}(s)\left[B^{\mu}Z_M^{(\mu-1)}(s)\right]\right\},$$

$$\Gamma_{IOy_0}(t) = \mathcal{L}^{-1}\left\{G_{IOy_0}(s)\right\} = \mathcal{L}^{-1}\left\{\Phi_{IO}(s)\left[A^{\nu}Z_N^{(\nu-1)}(s)\right]\right\}, \tag{10.7}$$

*where $\Phi_{IO}(s)$ is the left Laplace transform of the IO **system fundamental matrix** $\Phi_{IO}(t)$,*

$$\Phi_{IO}(s) = \mathcal{L}^{-}\left\{\Phi_{IO}(t)\right\},\ \Phi_{IO}(t) = \mathcal{L}^{-1}\left\{\Phi_{IO}(s)\right\}, \tag{10.8}$$

and

$$\Phi_{IO}(s) = \left(A^{(\nu)}S_N^{(\nu)}(s)\right)^{-1},\ \Phi_{IO}(t) = \mathcal{L}^{-1}\left\{\left(A^{(\nu)}S_N^{(\nu)}(s)\right)^{-1}\right\}. \tag{10.9}$$

(iv) The IO system full fundamental matrix $\Gamma_{IO}(t)$ *and its fundamental matrix* $\Phi_{IO}(t)$ *are linked as follows:*

$$\Gamma_{IO}(t) = \mathcal{L}^{-1}\left\{\Phi_{IO}(s)\left[B^{\mu}S_M^{(\mu)}(s) \vdots - B^{\mu}Z_M^{(\mu-1)}(s) \vdots A^{\nu}Z_N^{(\nu-1)}(s)\right]\right\},$$

$$\Gamma_{IO}(s) = \Phi_{IO}(s)\left[B^{\mu}S_M^{(\mu)}(s) \vdots - B^{\mu}Z_M^{(\mu-1)}(s) \vdots A^{\nu}Z_N^{(\nu-1)}(s)\right]. \quad (10.10)$$

Appendix I contains the proof.

This theorem expresses a physical meaning of the full transfer function matrix of the *IO* system (3.77).

Example 184 *Let us refer to Example 154 (in Section 8.1). The given second-order SISO IO system (i.e., the equivalent IO form of the ISO system) is described by*

$$y^{(2)}(t) + y^{(1)}(t) - 2y(t) = 2i^{(2)}(t) - 14i^{(1)}(t) + 12i(t).$$

Its full transfer function matrix was found to read

$$F_{IO}(s) =$$

$$= \left[\underbrace{\frac{2s^2 - 14s + 12}{s^2 + s - 2}}_{G_{IO}(s)} \vdots \underbrace{\frac{-(2s-14)}{s^2+s-2} \vdots \frac{-2}{s^2+s-2}}_{G_{IOio}(s)} \vdots \underbrace{\frac{s+1}{s^2+s-2} \vdots \frac{1}{s^2+s-2}}_{G_{IOyo}(s)}\right]$$

i.e.,

$$F_{IO}(s) = \left[G_{IO}(s) \vdots G_{IOio}(s) \vdots G_{IOyo}(s)\right] =$$

$$= \left[\begin{array}{c} \underbrace{\frac{2(s-1)(s-6)}{(s-1)(s+2)}}_{G_{IO}(s)} \vdots \\ \underbrace{\left[\frac{-2(s-7)}{(s-1)(s+2)} \vdots \frac{-2}{(s-1)(s+2)}\right]}_{G_{IOio}(s)} \vdots \\ \underbrace{\left[\frac{s+1}{(s-1)(s+2)} \vdots \frac{1}{(s-1)(s+2)}\right]}_{G_{IOyo}(s)} \end{array}\right]^T.$$

Notice that $F_{IO}(.)$ is only proper, not strictly proper, real rational matrix function. Its inverse Laplace transform $\mathcal{L}^{-1}\{F_{IO}(s)\}$,

$$\mathcal{L}^{-1}\{F_{IO}(s)\} = \mathcal{L}^{-1}\left\{\left[G_{IO}(s) \vdots G_{IOio}(s) \vdots G_{IOyo}(s)\right]\right\} =$$

$$= \begin{bmatrix} \underbrace{\mathcal{L}^{-1}\left\{\dfrac{2(s-1)(s-6)}{(s-1)(s+2)}\right\}\vdots}_{\mathcal{L}^{-1}\{G_{IO}(s)\}} \\ \begin{bmatrix} \underbrace{\mathcal{L}^{-1}\left\{\dfrac{-2(s-7)}{(s-1)(s+2)}\right\} \vdots \mathcal{L}^{-1}\left\{\dfrac{-2}{(s-1)(s+2)}\right\}}_{\mathcal{L}^{-1}\{G_{IOio}(s)\}} \end{bmatrix} \vdots \\ \begin{bmatrix} \underbrace{\mathcal{L}^{-1}\left\{\dfrac{s+1}{(s-1)(s+2)}\right\} \vdots \mathcal{L}^{-1}\left\{\dfrac{1}{(s-1)(s+2)}\right\}}_{\mathcal{L}^{-1}\{G_{IOyo}(s)\}} \end{bmatrix} \end{bmatrix}^T .$$

determines the full fundamental matrix function $\Psi_{IO}(.)$ of the system in view of (10.4):

$$\Psi_{IO}(t) = \left[\Gamma_{IO}(t) \vdots \Gamma_{IOi_0}(t) \vdots \Gamma_{IOy_0}(t)\right] =$$

$$= \begin{bmatrix} \underbrace{2\left[\delta(t) - 8e^{-2t}\right]\vdots}_{\Gamma_{IOi}(t)} \\ \begin{bmatrix} \underbrace{\left(4e^{t} - 6e^{-2t}\right) \vdots -\dfrac{2}{3}\left(e^{t} - e^{-2t}\right)}_{\Gamma_{IOi_0}(t)} \end{bmatrix}^T \\ \begin{bmatrix} \underbrace{\dfrac{1}{3}\left(2e^{t} + e^{-2t}\right) \vdots \dfrac{1}{3}\left(e^{t} - e^{-2t}\right)}_{\Gamma_{IOy_0}(t)} \end{bmatrix} \end{bmatrix}^T .$$

Notice that

$$\int_{0^-}^{t=0^-} \Gamma_{IO}(\tau)i(t-\tau)d\tau = \int_{0^-}^{t=0^-} 2\left[\delta(\tau) - 8e^{-2\tau}\right] i(t-\tau)d\tau =$$

$$= 2\int_{0^-}^{t=0^-} \delta(t)i(t-\tau)dt = 2i_{0^-},$$

$$\Gamma_{IOi_0}(0^-)\mathbf{i}^1(0^-) =$$

$$= \left[4e^t - 6e^{-2t} \vdots -\frac{2}{3}\left(e^t - e^{-2t}\right)\right]_{t=0^-} \mathbf{i}^1(0^-) =$$

$$= \left[-2 \vdots 0\right] \mathbf{i}^1(0^-) = -2i_{0^-},$$

$$\Gamma_{IOy_0}(0^-)\mathbf{y}^1(0^-) = \left[\frac{1}{3}\left(2e^t + e^{-2t}\right) \vdots \frac{1}{3}\left(e^t - e^{-2t}\right)\right]_{t=0^-} \mathbf{y}^1(0^-) =$$

$$= \left[\frac{1}{3}(3) \vdots \frac{1}{3}(0)\right] \mathbf{y}^1(0^-) = \left[1 \vdots 0\right] \mathbf{y}^1(0^-) = y_{0^-} \Longrightarrow$$

$$y(0^-;\mathbf{y}^1_{0^-};i) = \int_{0^-}^{t=0^-} \Gamma_{IO}(\tau)i(t-\tau)d\tau + \Gamma_{IOi_0}(0^-)\mathbf{i}^1(0^-) +$$

$$+\Gamma_{IOy_0}(0^-)\mathbf{y}^1(0^-) = 2i_{0^-} - 2i_{0^-} + y_{0^-} = y_{0^-}.$$

$\Gamma_{IO}(t) = 2\left[\delta(t) - 8e^{-2t}\right]$ *contains Dirac unit impulse* $\delta(t)$ *because it is the inverse Laplace transform of the proper ratio*

$$\frac{2(s-1)(s-6)}{(s-1)(s+2)} = 2\frac{s-6}{s+2} = 2\frac{s+2-8}{s+2} = 2\left(1 - \frac{8}{s+2}\right),$$

which implies that $F_{IO}(.)$ *is only proper, not strictly proper, real rational matrix function. However,* $\Gamma_{IOi_0}(t)$ *and* $\Gamma_{IOy_0}(t)$ *do not contain Dirac impulse. They are matrix exponential functions. Their Laplace transforms are strictly proper rational functions* $G_{IOio}(s)$ *and* $G_{IOyo}(s)$*, respectively.*

10.2 The ISO system

Definition 185 *A matrix function* $\Psi_{ISOIS}(.) : \mathfrak{T} \longrightarrow \mathfrak{R}^{n\times(M+n)}$ *is* ***the IS full fundamental matrix function*** *of the ISO system (3.84) and (3.85), (in Subsection 3.5.3), if and only ifit obeys both (i) and (ii) for an arbitrary input vector function* $\mathbf{i}(.)$*, and for an arbitrary initial state vector* $\mathbf{x}_{0^-}$*:*

(i)

$$\mathbf{x}(t;\mathbf{x}_{0^-};\mathbf{i}) = \int_{0^-}^{t} \left\{ \Psi_{ISOIS}(\tau) \left[\begin{array}{c} \mathbf{i}(t-\tau) \\ \delta(t-\tau)\mathbf{x}_{0^-} \end{array} \right] \right\} d\tau =$$

$$= \int_{0^-}^{t} \Gamma_{ISOIS}(\tau)\mathbf{i}(t-\tau)d\tau + \Gamma_{ISOSS}(t)\mathbf{x}_{0^-},$$

$$\Gamma_{ISOIS}(t) \in \mathfrak{R}^{nxM},\ \Gamma_{ISOSS}(t) \in \mathfrak{R}^{nxn}, \tag{10.11}$$

(ii)

$$\Gamma_{ISOIS}(0^-) = O_{n,M}, \Gamma_{ISOSS}(0^-) = I_n. \tag{10.12}$$

It is well known, [4], [13], [19], [77], [85], [86], [93], [106], [107] that

$$\Gamma_{ISOIS}(t) = e^{At}B,\ \Gamma_{ISOSS}(t) = e^{At} = \Phi_{ISOSS}(t), \tag{10.13}$$

where $\Phi_{ISOSS}(t) = e^{At}$ is **the** ISO system **state fundamental matrix**, for short its **fundamental matrix**. It can be computed from **the resolvent matrix** $(sI - A)^{-1}$, e.g., [91], of the ISO system (3.84) and (3.85),

$$\Gamma_{ISOSS}(t) = e^{At} = \Phi_{ISOSS}(t) = \mathcal{L}^{-1}\left\{(sI - A)^{-1}\right\}, \tag{10.14}$$

and vice versa,

$$\Phi_{ISOSS}(s) = (sI - A)^{-1} = \mathcal{L}^{-}\left\{\Gamma_{ISOSS}(t)\right\} = \mathcal{L}^{-}\left\{e^{At}\right\} = \mathcal{L}^{-}\left\{\Phi_{ISOSS}(t)\right\}. \tag{10.15}$$

The transfer function matrix $\Phi_{ISOSS}(s) = \Gamma_{ISOSS}(s)$, which is simultaneously the resolvent matrix of the matrix A, is **the** ISO **system state fundamental transfer function matrix**, for short its **fundamental transfer function matrix.**

Theorem 186 *(i) The IS full fundamental matrix function $\Psi_{ISOIS}(.)$ of the ISO system (3.84) and (3.85) is the inverse of the left Laplace transform of the system full IS transfer function matrix $F_{ISOIS}(s)$,*

$$\Psi_{ISOIS}(t) = \mathcal{L}^{-1}\left\{F_{ISOIS}(s)\right\}. \tag{10.16}$$

(ii) The full transfer function matrix $F_{ISOIS}(s)$ of the ISO system (3.84) and (3.85) is the left Laplace transform of the system full fundamental matrix $\Psi_{ISOIS}(t)$,

$$F_{ISOIS}(s) = \mathcal{L}^{-}\left\{\Psi_{ISOIS}(t)\right\}. \tag{10.17}$$

The proof is in Appendix J.

Definition 187 *A matrix function* $\Psi_{ISO}(.) : \mathfrak{T} \longrightarrow \mathfrak{R}^{N\times(M+n)}$ *is* ***the IO full fundamental matrix function*** *of the* ISO *system (3.84) and (3.85) if and only if it obeys both (i) and (ii) for an arbitrary input vector function* $\mathbf{i}(.)$*, and for an arbitrary initial state vector* $\mathbf{x}_{0^-}$*,*

(i)

$$\mathbf{y}(t;\mathbf{x}_{0^-};\mathbf{i}) = \int_{0^-}^{t} \left\{ \Psi_{ISO}(\tau) \left[\begin{array}{c} \mathbf{i}(t-\tau) \\ \delta(t-\tau)\mathbf{x}_{0^-} \end{array} \right] \right\} d\tau =$$

$$= \int_{0^-}^{t} \left\{ \left[\Gamma_{ISO}(\tau)\mathbf{i}(t-\tau) \right] d\tau \right\} + \Gamma_{ISOx_o}(t)\mathbf{x}_{0^-}, \tag{10.18}$$

(ii)

$$\Psi_{ISO}(t) = \left[\Gamma_{ISO}(t) \vdots \Gamma_{ISOx_o}(t) \right], \ \Gamma_{ISO}(t) \in \mathfrak{R}^{N\times M}, \ \Gamma_{ISOx_o}(t) \in \mathfrak{R}^{N\times n}. \tag{10.19}$$

Note 188 *The equations of Definition 187 and the properties of* $\delta(.)$ *(Subsection B.2) permit*

$$\mathbf{y}(t;\mathbf{x}_{0^-};\mathbf{i}) = \left(\int_{0^-}^{t} \Gamma_{ISO}(\tau)\mathbf{i}(t-\tau)d\tau \right) + \Gamma_{ISOx_o}(t)\mathbf{x}_{0^-}, \ t \in \mathfrak{T}_0. \tag{10.20}$$

Note 189 *From the linear systems theory, [4], [13], [19], [77], [85], [86], [93], [106], [107], and from (10.20) follow:*

$$\Gamma_{ISO}(t) = Ce^{At}B + \delta(t)D, \ \Gamma_{ISOx_o}(t) = Ce^{At}. \tag{10.21}$$

Theorem 190 *(i) The IO full fundamental matrix function* $\Psi_{ISO}(.)$ *of the* ISO *system (3.84) and (3.85) is the inverse of the left Laplace transform of the system full transfer function matrix* $F_{ISO}(s)$*,*

$$\Psi_{ISO}(t) = \mathcal{L}^{-1}\left\{ F_{ISO}(s) \right\}. \tag{10.22}$$

(ii) The full transfer function matrix $F_{ISO}(s)$ *of the* ISO *system (3.84) and (3.85) is the left Laplace transform of the system full fundamental matrix* $\Psi_{ISO}(t)$*,*

$$F_{ISO}(s) = \mathcal{L}^{-}\left\{ \Psi_{ISO}(t) \right\}. \tag{10.23}$$

(iii) The transfer function matrices $\Gamma_{ISO}(s)$ *and* $\Gamma_{ISOx_o}(s)$ *of the* ISO *system (3.84) and (3.85) are the left Laplace transforms of the system transfer function matrices* $\Gamma_{ISO}(t)$ *and* $\Gamma_{ISOx_o}(t)$*, respectively,*

$$\Gamma_{ISO}(s) = \mathcal{L}^{-}\left\{ \Gamma_{ISO}(t) \right\} = C(sI-A)^{-1}B + D = C\Phi_{ISOSS}(s)B + D,$$
$$\Gamma_{ISOx_o}(s) = \mathcal{L}^{-}\left\{ \Gamma_{ISOx_o}(t) \right\} = C(sI-A)^{-1} = C\Phi_{ISOSS}(s), \tag{10.24}$$

and vice versa,

$$\Gamma_{ISO}(t) = \mathcal{L}^{-1}\left\{G_{ISO}(s)\right\} = Ce^{At}B + \delta(t)D = C\Phi_{ISOSS}(t)B + \delta(t)D,$$
$$\Gamma_{ISOx_o}(t) = \mathcal{L}^{-1}\left\{\Gamma_{ISOx_o}(s)\right\} = Ce^{At} = C\Phi_{ISOSS}(t). \quad (10.25)$$

The proof is in Appendix K.

Theorem 190 shows a physical meaning of the IO full transfer function matrix of the ISO system (3.84) and (3.85).

Note 191 *Equations (3.85), (10.20) and (10.21), written for* $t = 0$, *establish the relationship between* $\mathbf{y}_0 = \mathbf{y}(0)$ *and* $\mathbf{i}_0 = \mathbf{i}(0)$ *linked with* $\mathbf{x}_0 = \mathbf{x}(0)$,

$$\mathbf{y}_{0^-} = \int_{0^-}^{0^-} \Gamma_{ISO}(\tau)\mathbf{i}(t-\tau)d\tau + \Gamma_{ISOx_0}(0^-)\mathbf{x}_{0^-} = C\mathbf{x}_{0^-} + D\mathbf{i}_{0^-} \Longrightarrow$$
$$\int_{0^-}^{0^-} \Gamma_{ISO}(t)dt = \int_{0^-}^{0^-} \left(Ce^{At}B + \delta(t)D\right)dt = D,\ \Gamma_{ISOx_0}(0^-) = C. \quad (10.26)$$

Note 192 *In a special case,* $N = n$ *and* $det\Gamma_{ISOx_0}(0^-) = detC \neq 0$. *This permits us to resolve (10.26) for* $\mathbf{x}_{0^-}$ *in terms of* $\mathbf{y}_{0^-}$,

$$\mathbf{x}_{0^-} = C^{-1}\left(\mathbf{y}_{0^-} - D\mathbf{i}_{0^-}\right). \quad (10.27)$$

This equation transforms (8.28) and (8.29), (in Section 8.2), into

$$\mathbf{Y}^-(s) = F_{ISOsp}(s)\left[\left(\mathbf{I}^-(s)\right)^T \vdots\ \mathbf{i}_{0^-}^T \vdots\ \mathbf{y}_{0^-}^T\right]^T,$$
$$F_{ISOsp}(s) =$$
$$= \left[C(sI_n - A)^{-1}B + D \vdots (sI_n - CAC^{-1})^{-1}D \vdots (sI_n - CAC^{-1})^{-1}\right], \quad (10.28)$$

and

$$\mathbf{Y}^-(s) = F_{ISOsp}(s)\mathbf{V}_{ISOsp}^-(s),\ \mathbf{V}_{ISOsp}(s) = \left[\left(\mathbf{I}^-(s)\right)^T \vdots\ \mathbf{i}_{0^-}^T \vdots\ \mathbf{y}_{0^-}^T\right]^T. \quad (10.29)$$

Equation (10.28) agrees with Equation (8.39) (Theorem 165, in Subsection 8.2).

10.3 The IIO system

Definition 193 *A matrix function* $\Psi_{IIOIS}(.)\colon \mathfrak{T} \longrightarrow \mathfrak{R}^{\rho \times [(\beta+1)M+\alpha\rho]}$,

$$\Psi_{IIOIS}(t) = [\Gamma_{IIOIS}(t) \vdots \Gamma_{IIOi_oIS}(t) \vdots \Gamma_{IIOr_oIS}(t)],$$

is ***the full*** IS ***fundamental matrix function of the*** IIO ***system (3.90)*** *(Subsection 3.5.4) if and only if it obeys both (i) and (ii) for an arbitrary input vector function* $\mathbf{i}(.)$*, and for arbitrary initial conditions* $\mathbf{i}_{0^-}^{\mu-1}$ *and* $\mathbf{r}_{0^-}^{\alpha-1}$*:*

(i)

$$\mathbf{r}(t;\mathbf{r}_{0^-}^{\alpha-1};\mathbf{i}) = \int_{0^-}^{t} \left\{ \Psi_{IIOIS}(\tau) \left[\begin{array}{c} \mathbf{i}(t-\tau) \\ \delta(t-\tau)\mathbf{i}_{0^-}^{\beta-1} \\ \delta(t-\tau)\mathbf{r}_0^{\alpha-1} \end{array} \right] \right\} d\tau =$$

$$= \int_{0^-}^{t} [\Gamma_{IIOIS}(\tau)\mathbf{i}(t-\tau)d\tau] + \Gamma_{IIOi_oIS}(t)\mathbf{i}_{0^-}^{\beta-1} + \Gamma_{IIOr_oIS}(t)\mathbf{r}_0^{\alpha-1}, \quad (10.30)$$

(ii)

$$\Gamma_{IIOi_oIS}(0^-) = \left[\Gamma_{IIOi_0IS1} \vdots O_{N,(\mu-1)M}\right] \text{ where}$$

$$\Gamma_{IIOi_0IS1}\mathbf{i}_{0^-} = -\int_{0^-}^{0^-} [\Gamma_{IIOIS}(\tau)\mathbf{i}(t-\tau)d\tau],$$

$$\Gamma_{IIOr_oIS}(0^-) = \left[I_\rho \vdots O_{\rho,(\alpha-1)\rho}\right]. \quad (10.31)$$

Theorem 194 *(i) The full* IS *fundamental matrix function* $\Psi_{IIOIS}(.)$ *of the* IIO *system (3.90) is the inverse of the left Laplace transform of the system full* IS *transfer function matrix* $F_{IIOIS}(s)$*,*

$$\Psi_{IIOIS}(t) = \mathcal{L}^{-1}\{F_{IIOIS}(s)\}. \quad (10.32)$$

(ii) The full IS *transfer function matrix* $F_{IIOIS}(s)$ *of the* IIO *system (3.90) is the left Laplace transform of the system full* IS *fundamental matrix* $\Psi_{IIOIS}(t)$*,*

$$F_{IIOIS}(s) = \mathcal{L}^{-}\{\Psi_{IIOIS}(t)\}. \quad (10.33)$$

(iii) The full IS *transfer function matrix* $F_{IIOIS}(s)$ *of the system (3.89),*

i.e., of (8.46), is determined by (8.56), (8.57),

$$F_{IIOIS}(s) =$$
$$= (\Phi_{IIOIS}(s))^{-1}\left[P^{(\beta)}S_M^{(\beta)}(s) \vdots -P^{(\beta)}Z_M^{(\beta-1)}(s) \vdots Q^{(\alpha)}Z_\rho^{(\alpha-1)}(s)\right] \quad (10.34)$$
$$F_{IIOIS}(s) = \left[G_{IIOIS}(s) \vdots G_{IIOoIS}(s)\right] = F_{IIOISD}^{-1}(s)F_{IIOISN}(s),$$
$$G_{IIOoIS}(s) = \left[G_{IIOi_0IS}(s) \vdots G_{IIOr_0IS}(s)\right], \quad (10.35)$$

together with (10.36) through (10.39):

- *IRIS transfer function matrix* $G_{IIOIS}(s)$,

$$G_{IIOIS}(s) = (\Phi_{IIOIS}(s))^{-1}P^{(\beta)}S_M^{(\beta)}(s) =$$
$$= G_{IIOISD}^{-1}(s)G_{IIOISN}(s), \quad (10.36)$$

- *IRII transfer function matrix* $G_{IIOi_0IS}(s)$,

$$G_{IIOi_0IS}(s) = -(\Phi_{IIOIS}(s))^{-1}P^{(\beta)}Z_M^{(\beta-1)}(s) =$$
$$= G_{IIOi_0ISD}^{-1}(s)G_{IIOi_0ISN}(s), \quad (10.37)$$

- *IRIR transfer function matrix* $G_{IIOr_0IS}(s)$,

$$G_{IIOr_0IS}(s) = -(\Phi_{IIOIS}(s))^{-1}Q^{(\alpha)}Z_\rho^{(\alpha-1)}(s) =$$
$$= G_{IIOr_0ISD}^{-1}(s)G_{IIOr_0ISN}(s), \quad (10.38)$$

so that the denominator matrix polynomial $F_{IIOISD}(s)$ *and the numerator matrix polynomial* $F_{IIOISN}(s)$ *of* $F_{IIOIS}(s)$,

$$F_{IIOIS}(s) = F_{IIODIS}^{-1}(s)F_{IIOISN}(s),$$

read:

$$F_{IIOISD}(s) = \Phi_{IIOIS}(s),$$
$$F_{IIOISN}(s) =$$
$$= \left[P^{(\beta)}S_M^{(\beta)}(s) \vdots -P^{(\beta)}Z_M^{(\beta-1)}(s) \vdots Q^{(\alpha)}Z_\rho^{(\alpha-1)}(s)\right]. \quad (10.39)$$

where $\Phi_{IIOIS}(s)$ *is the left Laplace transform of the IIO* ***system fundamental IS matrix*** $\Phi_{IIOIS}(t)$,

$$\Phi_{IIOIS}(s) = \mathcal{L}^{-}\{\Phi_{IIOIS}(t)\},\ \Phi_{IIOIS}(t) = \mathcal{L}^{-1}\{\Phi_{IiOIS}(s)\}, \quad (10.40)$$

and

$$\Phi_{IIOIS}(s) = \left(Q^{(\alpha)}S_{\rho}^{(\alpha)}(s)\right)^{-1}, \ \Phi_{IIOIS}(t) = \mathcal{L}^{-1}\left\{\left(Q^{(\alpha)}S_{\rho}^{(\alpha)}(s)\right)^{-1}\right\}. \tag{10.41}$$

The proof of this theorem is analogous to the proofs of Theorem 183 (Section 10.1) and Theorem 186 (Section 10.2).

Definition 195 *A matrix function* $\Psi_{IIO}(.) : \mathfrak{T} \longrightarrow \mathfrak{R}^{N\text{x}[(\gamma+1)M+\alpha\rho+\nu N]}$ *is* ***the IO full fundamental matrix function of the IIO system (3.90)*** *if and only if it obeys both (i) and (ii) for an arbitrary input vector function* $\mathbf{i}(.)$, *and for arbitrary initial conditions* $\mathbf{i}_{0^-}^{\gamma-1}$, $\mathbf{r}_{0^-}^{\alpha-1}$, *and* $\mathbf{y}_{0^-}^{\nu-1}$,

(i)

$$\mathbf{y}(t;\mathbf{r}_{0^-}^{\alpha-1};\mathbf{y}_0^{\nu-1};\mathbf{i}) =$$

$$= \int_{0^-}^{t}\left\{\Psi_{IIO}(\tau)\begin{bmatrix}\mathbf{i}(t-\tau)\\ \delta(t-\tau)\mathbf{i}_{0^-}^{\gamma-1}\\ \delta(t-\tau)\mathbf{r}_{0^-}^{\alpha-1}\\ \left\{\begin{matrix}\delta(t-\tau)\mathbf{y}_{0^-}^{\nu-1}, \ \nu\geq 1,\\ \mathbf{0}_{\nu N} \ , \ \nu=0\end{matrix}\right\}\end{bmatrix}\right\}d\tau,$$

$$\Psi_{IIO}(t) = \left[\Gamma_{IIO}(t) \vdots \Gamma_{IIOi_0}(t) \vdots \Gamma_{IIOr_0}(t) \vdots \Gamma_{IIOy_0}(t)\right],$$

$$\Gamma_{IIO}(t) \in \mathfrak{R}^{N\text{x}M}, \Gamma_{IIOi_0}(t) \in \mathfrak{R}^{N\text{x}\gamma M}, \Gamma_{IIOr_0}(t) \in \mathfrak{R}^{N\text{x}\alpha\rho},$$

$$\Gamma_{IIOy_0}(t) \in \mathfrak{R}^{N\text{x}\nu N}, \tag{10.42}$$

(ii)

$$\Gamma_{IIOi_0}(0^-) = \left[\Gamma_{IIOi_01} \vdots O_{N\text{x}(\gamma-1)M}\right],$$

where

$$\Gamma_{IIOi_01}(0^-)\mathbf{i}_{0^-} = -\int_{0^-}^{0^-}\left[\Gamma_{IIO}(\tau)\mathbf{i}(t-\tau)d\tau\right],$$

$$\Gamma_{IIOr_0}(0^-) = O_{N\text{x}\alpha\rho},$$

$$\Gamma_{IIOy_0}(0^-) = \left\{\begin{matrix}\left[I_N \vdots O_{N\text{x}(\nu-1)N}\right], \ \nu \geq 1,\\ O_N \ , \ \nu = 0\end{matrix}\right\}. \tag{10.43}$$

Note 196 *The equations under (i) of this definition and the properties of $\delta(.)$ (Subsection B.2) permit*

$$\mathbf{y}(t;\mathbf{r}_{0^-}^{\alpha-1};\mathbf{y}_0^{\nu-1};\mathbf{i}) = \left(\int_{0^-}^{t}\Gamma_{IIO}(\tau)\mathbf{i}(t-\tau)d\tau\right) + \Gamma_{IIOi_0}(t)\mathbf{i}_{0^-}^{\gamma-1} +$$

$$+\Gamma_{IIOr_0}(t)\mathbf{r}_{0^-}^{\alpha-1} + \left\{ \begin{array}{c} \Gamma_{IIOy_0}(t)\mathbf{y}_{0^-}^{\nu-1},\ \nu \geq 1, \\ \mathbf{0}_N\ ,\ \nu = 0 \end{array} \right\},\ \forall t \in \mathfrak{T}_0. \tag{10.44}$$

Theorem 197 *(i) The IO full fundamental matrix function $\Psi_{IIO}(.)$ of the IIO system (3.90) is the inverse of the left Laplace transform of the system IO full transfer function matrix $F_{IIO}(s)$,*

$$\Psi_{IIO}(t) = \mathcal{L}^{-1}\left\{F_{IIO}(s)\right\}. \tag{10.45}$$

(ii) The full transfer function matrix $F_{IIO}(s)$ of the IIO system (3.90) is the left Laplace transform of the system IO full fundamental matrix $\Psi_{IIO}(t)$,

$$F_{IIO}(s) = \mathcal{L}^{-}\left\{\Psi_{IIO}(t)\right\}. \tag{10.46}$$

(iii) The IIO system IO transfer matrix functions

$$\Gamma_{IIO}(s) = \mathcal{L}^{-}\left\{\Gamma_{IIO}(t)\right\},\ \Gamma_{IIOi_0}(s) = \mathcal{L}^{-}\left\{\Gamma_{IIOi_0}(t)\right\}$$

and $\Gamma_{IIOy_0}(s) = \mathcal{L}^{-}\left\{\Gamma_{IIOy_0}(t)\right\}$ are linked with the system transfer functions as follows:

$$\begin{aligned} \Gamma_{IIO}(s) &= G_{IIO}(s),\ \Gamma_{IIOi_0}(s) = G_{IIOi_0}(s), \\ \Gamma_{IIOr_0}(s) &= G_{IIOr_0}s),\ \Gamma_{IIOy_0}(s) = G_{IIOy_0}(s), \end{aligned} \tag{10.47}$$

and

$$\Gamma_{IIO}(t) = \mathcal{L}^{-1}\left\{G_{IIO}(s)\right\} =$$

$$= \mathcal{L}^{-1}\left\{ \begin{array}{c} \Phi_{IIO}(s)\bullet \\ \bullet\left[R^{(\alpha)}S_\rho^{(\alpha)}(s)\left(Q^{(\alpha)}S_\rho^{(\alpha)}(s)\right)^{-1}P^{(\beta)}S_M^{(\beta)}(s) + T^{(\mu)}S_M^{(\mu)}(s)\right] \end{array} \right\} =$$

$$= \mathcal{L}^{-1}\left\{\Phi_{IIO}(s)\left[R^{(\alpha)}S_\rho^{(\alpha)}(s)\left(Q^{(\alpha)}S_\rho^{(\alpha)}(s)\right)^{-1}P^{(\beta)}S_M^{(\beta)}(s) + T^{(\mu)}S_M^{(\mu)}(s)\right]\right\}, \tag{10.48}$$

$$\Gamma_{IIOi_0}(t) = \mathcal{L}^{-1}\left\{G_{IIOi_0}(s)\right\} =$$
$$= \mathcal{L}^{-1}\left\{\begin{array}{c} -\Phi_{IIO}(s)\bullet \\ \bullet\left(\begin{array}{c}\left[\begin{array}{c} R^{(\alpha)}S_{\rho}^{(\alpha)}(s)\left(Q^{(\alpha)}S_{\rho}^{(\alpha)}(s)\right)^{-1}P^{(\beta)}Z_{M}^{(\beta-1)}(s)\vdots \\ \vdots O_{N(\gamma-\beta)}\end{array}\right]+ \\ +\left[T^{(\mu)}Z_{M}^{(\mu-1)}(s)\vdots O_{N(\gamma-\mu)}\right],\end{array}\right)\end{array}\right\} \tag{10.49}$$

$$\Gamma_{IIOr_0}(t) = \mathcal{L}^{-1}\left\{G_{IIOr_0}s)\right\} =$$
$$= \mathcal{L}^{-1}\left\{\begin{array}{c} -\Phi_{IIO}(s)\bullet \\ \bullet\left(\begin{array}{c}\left[R^{(\alpha)}S_{\rho}^{(\alpha)}(s)\left(Q^{(\alpha)}S_{\rho}^{(\alpha)}(s)\right)^{-1}Q^{(\alpha)}Z_{\rho}^{(\alpha-1)}(s)\right]- \\ -R^{(\alpha)}Z_{\rho}^{(\alpha-1)}(s)\end{array}\right)\end{array}\right\} \tag{10.50}$$

$$\Gamma_{IIOy_0}(t) = \mathcal{L}^{-1}\left\{\Phi_{IIO}(s)E^{(\nu)}Z_{N}^{(\nu-1)}(s)\right\}, \tag{10.51}$$

*where $\Phi_{IIO}(s)$ is the left Laplace transform of the IIO **system fundamental matrix** $\Phi_{IO}(t)$,*

$$\Phi_{IIO}(s) = \mathcal{L}^{-}\left\{\Phi_{IIO}(t)\right\},\ \Phi_{IIO}(t) = \mathcal{L}^{-1}\left\{\Phi_{IIO}(s)\right\}, \tag{10.52}$$

and

$$\Phi_{IIO}(s) = \left(E^{(\nu)}S_{N}^{(\nu)}(s)\right)^{-1},\ \Phi_{IIO}(t) = \mathcal{L}^{-1}\left\{\left(E^{(\nu)}S_{N}^{(\nu)}(s)\right)^{-1}\right\}. \tag{10.53}$$

This theorem is proved in the same way as Theorem 183 (Section 10.1) by noting that the system transfer function matrices $G_{IIO}(s)$, $G_{IIOi_0}(s)$, $G_{IIOr_0}s)$ and $G_{IIOy_0}(s)$ are determined by (8.63)-(8.66).

Equations (10.47) imply

$$\Gamma_{IIO}(t) = \mathcal{L}^{-1}\left\{G_{IIO}(s)\right\},\ \Gamma_{IIOi_0}(t) = \mathcal{L}^{-1}\left\{G_{IIOi_0}(s)\right\},$$
$$\Gamma_{IIOr_0}(t) = \mathcal{L}^{-1}\left\{G_{IIOr_0}s)\right\},\ \Gamma_{IIOy_0}(t) = \mathcal{L}^{-1}\left\{G_{IIOy_0}(s)\right\}. \tag{10.54}$$

Theorem 197 explains a physical meaning of the IO full transfer function matrix of the IIO system (3.90) and how we can effectively determine it.

Chapter 11

System matrix and equivalence

11.1 System matrix of the IO system

The *IO* system (3.77) (in Subsection 3.5.2) can be set into the following form:

$$A^{(\nu)}\mathbf{r}^{\nu}(t) = B^{(\mu)}\mathbf{i}^{\mu}(t),\ \mathbf{y}(t) = \mathbf{r}(t),\ \forall t \in \mathfrak{T}_0. \tag{11.1}$$

We know that the Laplace transforms of these equations read:

$$\begin{gathered} A^{(\nu)}S_N^{(\nu)}(s)\mathbf{R}^{\mp}(s) = \\ = B^{(\mu)}S_M^{(\mu)}(s)\mathbf{I}^{\mp}(s) - B^{(\mu)}Z_M^{(\mu-1)}(s)\mathbf{i}_{0\mp}^{\mu-1} + A^{(\nu)}Z_N^{(\nu-1)}(s)\mathbf{r}_{0\mp}^{\nu-1},\ \mathbf{r}_{0\mp}^{\nu-1} = \mathbf{y}_{0\mp}^{\nu-1}, \\ \mathbf{Y}^{\mp}(s) = \mathbf{R}^{\mp}(s). \end{gathered} \tag{11.2}$$

Let us first accept *zero initial conditions*, under which Equations (11.2) become

$$\begin{gathered} A^{(\nu)}S_N^{(\nu)}(s)\mathbf{R}^{\mp}(s) - B^{(\mu)}S_M^{(\mu)}(s)\mathbf{I}^{\mp}(s) = \mathbf{0}_N, \\ -\mathbf{R}^{\mp}(s) = -\mathbf{Y}^{\mp}(s), \end{gathered}$$

or equivalently,

$$\underbrace{\begin{bmatrix} A^{(\nu)}S_N^{(\nu)}(s) & \vdots & B^{(\mu)}S_M^{(\mu)}(s) \\ -I_N & \vdots & O_{N\times M} \end{bmatrix}}_{P_{IO}(s;\mathbf{0}_{(\mu-1)M};\mathbf{0}_{(\nu-1)N})} \begin{bmatrix} \mathbf{R}^{\mp}(s) \\ -\mathbf{I}^{\mp}(s) \end{bmatrix} = \begin{bmatrix} \mathbf{0}_N \\ -\mathbf{Y}^{\mp}(s) \end{bmatrix}. \tag{11.3}$$

The matrix function $P_{IO}(.;\mathbf{0}_{(\mu-1)M};\mathbf{0}_{(\nu-1)N}):\mathfrak{C}\longrightarrow\mathfrak{C}^{2N\times(M+N)}$ is ***the system matrix function under all zero initial conditions of the IO system (11.1)***, for short: ***the system matrix function of the IO system (11.1)***. Its matrix value $P_{IO}(s;\mathbf{0}_{(\mu-1)M};\mathbf{0}_{(\nu-1)N})$,

$$P_{IO}(s;\mathbf{0}_{(\mu-1)M};\mathbf{0}_{(\nu-1)N})\in\mathfrak{C}^{2N\times(M+N)},$$

is ***the system matrix under all zero initial conditions of the IO system (11.1)***, for short: ***the system matrix of the IO system (11.1)***,

$$P_{IO}(s;\mathbf{0}_{(\mu-1)M};\mathbf{0}_{(\nu-1)N})=\begin{bmatrix} A^{(\nu)}S_N^{(\nu)}(s) & \vdots & B^{(\mu)}S_M^{(\mu)}(s) \\ -I_N & \vdots & O_{N\times M}\end{bmatrix}. \tag{11.4}$$

It is related to the IO system transfer function matrix $G_{IO}(s)$,

$$G_{IO}(s)=\underbrace{\left[A^{(\nu)}S_N^{(\nu)}(s)\right]}_{G_{IOD}(s)}{}^{-1}\underbrace{B^{(\mu)}S_M^{(\mu)}(s)}_{G_{ION}(s)}=G_{IOD}^{-1}(s)G_{ION}(s)\Longrightarrow$$

$$P_{IO}(s;\mathbf{0}_{(\mu-1)M};\mathbf{0}_{(\nu-1)N})=\begin{bmatrix} G_{IOD}(s) & \vdots & G_{ION}(s) \\ -I_N & \vdots & O_{N\times M}\end{bmatrix}. \tag{11.5}$$

Let us now allow *arbitrary initial conditions* for which we refer to Equations (11.2) that we can write as

$$\underbrace{\begin{bmatrix} A^{(\nu)}S_N^{(\nu)}(s) & B^{(\mu)}S_M^{(\mu)}(s) & B^{(\mu)}Z_M^{(\mu-1)}(s) & -A^{(\nu)}Z_N^{(\nu-1)}(s) \\ -I_N & O_{N\times M} & O_{N\times\mu M} & O_{N\times\nu N}\end{bmatrix}}_{P_{IO}(s;\mathbf{i}_{0\mp}^{\mu-1};\mathbf{y}_{0\mp}^{\nu-1})}\bullet$$

$$\bullet\begin{bmatrix}\mathbf{R}^{\mp}(s)\\ -\mathbf{I}^{\mp}(s)\\ \mathbf{i}_{0\mp}^{\mu-1}\\ \mathbf{y}_{0\mp}^{\nu-1}\end{bmatrix}=\begin{bmatrix}\mathbf{0}_N\\ -\mathbf{Y}^{\mp}(s)\end{bmatrix}. \tag{11.6}$$

In view of this, we define ***the full system matrix function***

$$P_{IO}(.;\mathbf{i}_{0\mp}^{\mu-1};\mathbf{y}_{0\mp}^{\nu-1}):\mathfrak{C}\longrightarrow\mathfrak{C}^{2Nx[(\mu+1)M+(\nu+1)N]}$$

under arbitrary initial conditions of the IO system (11.1), for short: ***the full system matrix function $P_{IO}(.)$ of the IO system (11.1),*** by

$$P_{IO}(s)\equiv P_{IO}(s;\mathbf{i}_{0\mp}^{\mu-1};\mathbf{y}_{0\mp}^{\nu-1})=$$

$$=\begin{bmatrix} A^{(\nu)}S_N^{(\nu)}(s) & B^{(\mu)}S_M^{(\mu)}(s) & B^{(\mu)}Z_M^{(\mu-1)}(s) & -A^{(\nu)}Z_N^{(\nu-1)}(s) \\ -I_N & O_{N\times M} & O_{N\times\mu M} & O_{N\times\nu N}\end{bmatrix}\Longleftrightarrow$$

$$\Longleftrightarrow P_{IO}(s)\begin{bmatrix} \mathbf{R}^{\mp}(s) \\ -\mathbf{V}_{IO}^{\mp}(s) \end{bmatrix} = \begin{bmatrix} \mathbf{0}_N \\ -\mathbf{Y}^{\mp}(s) \end{bmatrix} \text{ and}$$

$$\mathbf{V}_{IO}^{\mp}(s) = \begin{bmatrix} \mathbf{I}^{\mp}(s) \\ \mathbf{i}_{0\mp}^{\mu-1} \\ \mathbf{y}_{0\mp}^{\nu-1} \end{bmatrix} \in \mathfrak{C}^{2Nx(\mu M+\nu N)}. \tag{11.7}$$

$P_{IO}(s;\mathbf{i}_{0\mp}^{\mu-1};\mathbf{y}_{0\mp}^{\nu-1})$,

$$P_{IO}(s;\mathbf{i}_{0\mp}^{\mu-1};\mathbf{y}_{0\mp}^{\nu-1}) \in \mathfrak{C}^{2N\times[(\mu+1)M+(\nu+1)N]},$$

for short $P_{IO}(s)$,

$$P_{IO}(s) \equiv P_{IO}(s;\mathbf{i}_{0\mp}^{\mu-1};\mathbf{y}_{0\mp}^{\nu-1}),$$

is the matrix value of $P_{IO}(.;\mathbf{i}_{0\mp}^{\mu-1};\mathbf{y}_{0\mp}^{\nu-1})$. It is called ***the full system matrix of the IO system (11.1) under arbitrary initial conditions***, for short: ***the full system matrix $P_{IO}(s)$ of the IO system (11.1).***

Note 198 *The full system matrix function $P_{IO}(.)$ and the full system matrix $P_{IO}(s)$ should be distinguished from the system full matrix transfer function $F_{IO}(.)$ and the system full transfer function matrix $F_{IO}(s)$, respectively. They are interrelated as follows due to (8.18), (in Section 8.1), and (11.7):*

$$F_{IO}(s) = F_{IOD}^{-1}(s)F_{ION}(s) =$$

$$= \underbrace{\left(A^{(\nu)}S_N^{(\nu)}(s)\right)^{-1}}_{F_{IOD}(s)} \underbrace{\left[B^{(\mu)}S_M^{(\mu)}(s) \,\vdots\, B^{(\mu)}Z_M^{(\mu-1)}(s) \,\vdots\, A^{(\nu)}Z_N^{(\nu-1)}(s)\right]}_{F_{ION}(s)} \Longrightarrow \tag{11.8}$$

$$P_{IO}(s) = \begin{bmatrix} F_{IOD}(s) & \vdots & F_{ION}(s) \\ -I_N & \vdots & O_{N\times[(\mu+1)M+\nu N]} \end{bmatrix}. \tag{11.9}$$

Note 199 *The system matrix $P_{IO}(s;\mathbf{0}_{(\mu-1)M};\mathbf{0}_{(\nu-1)N})$ (11.4) and the full system matrix $P_{IO}(s)$ (11.7) have the same structure, but their submatrices are different, as well as their dimensions.*

The polynomial submatrices of $P_{IO}(s)$ have the following forms:

$$T_{IO}(s) = A^{(\nu)}S_N^{(\nu)}(s) = F_{IOD}(s) \in \mathfrak{C}^{N\times N},$$

$$U_{IO}(s) = \left[B^{(\mu)} S_M^{(\mu)}(s) \vdots - B^{(\mu)} Z_M^{(\mu-1)}(s) \vdots A^{(\nu)} Z_N^{(\nu-1)}(s) \right] =$$
$$= F_{ION}(s) \in \mathfrak{C}^{N \times [(\mu+1)M + \nu N]},$$

$$V_{IO}(s) = I_N, \; W_{IO}(s) = O_{N \times [(\mu+1)M+\nu N]}. \tag{11.10}$$

They will be called *the full Rosenbrock submatrices of* $P_{IO}(s)$ by following Rosenbrock [110, p. 43] (for more details see Section 11.2 in the sequel). They transform $P_{IO}(s)$, (11.7), of the IO system (11.1), into the following *Rosenbrock form* $P_{IOR}(s)$:

$$P_{IOR}(s) = \begin{bmatrix} T_{IO}(s) & \vdots \; U_{IO}(s) \\ -V_{IO}(s) & \vdots \; W_{IO}(s) \end{bmatrix} \Longrightarrow$$
$$P_{IOR}(s) = \begin{bmatrix} F_{IOD}(s) & \vdots & F_{ION}(s) \\ -I_N & \vdots & O_{N \times [(\mu+1)M+\nu N]} \end{bmatrix}, \tag{11.11}$$

where $F_{IOD}(s)$ and $F_{ION}(s)$ are the denominator and the numerator polynomial matrices of the system full transfer function matrix $F_{IO}(s)$, Note 153 (Section 8.1). This and (11.8) imply

$$F_{IO}(s) = F_{IOD}^{-1}(s) F_{ION}(s) = V_{IO}(s) T_{IO}^{-1}(s) U_{IO}(s) + W_{IO}(s). \tag{11.12}$$

The second Equation (11.11) is in fact Equation (11.9). Hence,

$$P_{IOR}(s) = P_{IO}(s). \tag{11.13}$$

By following Rosenbrock (the definition of strictly system equivalent matrices and Theorem 3.1 in [110, p. 52]) we state:

Definition 200 ***Strictly equivalent full system matrices***
Let $2N \times [(\mu+1)M + (\nu+1)N]$ *full system matrix* $P_{IO}(s)$ *be given in its Rosenbrock form* $P_{IOR}(s)$ *(11.11). Let* $M(s)$ *and* $N(s)$ *be* $N \times N$ *unimodular polynomial matrices. Let* $X(s)$ *and* $Y(s)$ *be also polynomial matrices, respectively* $N \times N$ *and* $N \times [(\mu+1)M + \nu N]$*. If* $P_{IO}(s)$ *and*

$$P_1(s) = \begin{bmatrix} T_1(s) & \vdots \; U_1(s) \\ -V_1(s) & \vdots \; W_1(s) \end{bmatrix} \tag{11.14}$$

are related by the transformation

$$\begin{bmatrix} M(s) & O_N \\ X(s) & I_N \end{bmatrix} \begin{bmatrix} T_{IO}(s) & U_{IO}(s) \\ -V_{IO}(s) & W_{IO}(s) \end{bmatrix} \bullet$$

$$\bullet \begin{bmatrix} N(s) & Y(s) \\ O_{[(\mu+1)M+\nu N]\times N} & I_{[(\mu+1)M+\nu N]} \end{bmatrix} = \begin{bmatrix} T_1(s) & U_1(s) \\ -V_1(s) & W_1(s) \end{bmatrix} \tag{11.15}$$

then $P_{IO}(s)$ and $P_1(s)$ are ***strictly system equivalent.***

Theorem 201 *Two full system matrices which are strictly system equivalent correspond to the same system full transfer function matrix $F_{IO}(s)$.*

Proof. We refer to Definition 200. Let $2N \times [(\mu+1)M + (\nu+1)N]$ full system matrix $P_{IO}(s)$ be given in its Rosenbrock form $P_{IOR}(s)$ (11.11). Equation (11.15) can be written as

$$\begin{bmatrix} M(s) & O_N \\ X(s) & I_N \end{bmatrix} \begin{bmatrix} T_{IO}(s) & U_{IO}(s) \\ -V_{IO}(s) & W_{IO}(s) \end{bmatrix} \bullet$$

$$\bullet \begin{bmatrix} N(s) & Y(s) \\ O_{[(\mu+1)M+\nu N]\times N} & I_{[(\mu+1)M+\nu N]} \end{bmatrix} = \begin{bmatrix} T_1(s) & U_1(s) \\ -V_1(s) & W_1(s) \end{bmatrix} =$$

$$= \begin{bmatrix} M(s)T_{IO}(s) & M(s)U_{IO}(s) \\ X(s)T_{IO}(s) - V_{IO}(s) & X(s)U_{IO}(s) + W_{IO}(s) \end{bmatrix} \bullet$$

$$\bullet \begin{bmatrix} N(s) & Y(s) \\ O_{[(\mu+1)M+\nu N]\times N} & I_{[(\mu+1)M+\nu N]} \end{bmatrix} =$$

$$= \begin{bmatrix} MT_{IO}N & MT_{IO}Y + MU_{IO} \\ (XT_{IO} - V_{IO})\,N & (XT_{IO} - V_{IO})\,Y + \ XU_{IO} + W_{IO} \end{bmatrix}.$$

This result, Equations (11.10) and (11.12) that request

$$F_1(s) = V_1(s)T_1^{-1}(s)U_1(s) + W_1(s),$$

and (11.15) yield the following in view of the nonsingularity of the unimodular matrices:

$$F_1(s) = -\,(XT_{IO} - V_{IO})\,N\,(MT_{IO}N)^{-1}\,(MT_{IO}Y + MU_{IO}) +$$
$$+\,(XT_{IO} - V_{IO})\,Y + \ XU_{IO} + W_{IO} =$$

$$= -\,(XT_{IO} - V_{IO})\,T_{IO}^{-1}M^{-1}M\,(T_{IO}Y + U_{IO}) +$$
$$+\,(XT_{IO} - V_{IO})\,Y + \ XU_{IO} + W_{IO} =$$

$$= -(XT_{IO} - V_{IO})T_{IO}^{-1}(T_{IO}Y + U_{IO}) + \\ + XT_{IO}Y - V_{IO}Y + XU_{IO} + W_{IO} =$$

$$= -XT_{IO}Y - XU_{IO} + V_{IO}Y + V_{IO}T_{IO}^{-1}U_{IO} + \\ + XT_{IO}Y - V_{IO}Y + XU_{IO} + W_{IO} =$$

$$= V_{IO}T_{IO}^{-1}U_{IO} + W_{IO} = F_{IOD}^{-1}(s)F_{ION}(s) = F_{IO}(s).$$

This ends the proof. ■

This theorem generalizes Rosenbrock's Theorem 3.1 [110, p. 52] in the framework of the IO systems.

Example 202 *For the given IO system in Example 154 we determined*

$$F_{IO}(s) = (s^2 + s - 2)^{-1} \bullet \\ \bullet \left[2s^2 - 14s + 12 \vdots -(2s - 14) \vdots -2 \vdots s + 1 \vdots 1 \right].$$

The full system matrix $P_{IO}(s)$ has the following form induced by $F_{IO}(s)$ in view of (11.8) and (11.9):

$$P_{IO}(s) = \\ = \begin{bmatrix} s^2 + s - 2 & 2s^2 - 14s + 12 & -(2s - 14) & -2 & s + 1 & 1 \\ -1 & 0 & 0 & 0 & 0 & 0 \end{bmatrix}.$$

Note 203 *Desoer and Vidyasagar wrote in [19, p. 58] that the following two ratios of polynomials,*

$$\frac{s + 1}{s(s + 2)} \text{ and } \frac{(s + 1)(s - 1)}{s(s + 2)(s - 1)} \tag{11.16}$$

correspond to the same rational function $f(.)$. They have two common zeros $s_1^0 = -1$ and $s_2^0 = \infty$, and two common poles $s_1^* = 0$ and $s_2^* = -2$. The second ratio has an additional positive real pole $s_3^* = 1$. However, if they represent the system transfer functions $G_1(.)$ and $G_2(.)$,

$$G_1(s) = \frac{s + 1}{s(s + 2)} \text{ and } G_2(s) = \frac{(s + 1)(s - 1)}{s(s + 2)(s - 1)}, \tag{11.17}$$

then they do not correspond to the same system. $G_1(.)$ is the transfer function of the second-order SISO system described by

$$y^{(2)}(t) + 2y^{(1)}(t) = i(t) + i^{(1)}(t), \tag{11.18}$$

while $G_2(.)$ is the transfer function of the third-order $SISO$ system determined by

$$y^{(3)}(t) + y^{(2)}(t) - 2y^{(1)}(t) = -i(t) + i^{(2)}(t). \tag{11.19}$$

The full transfer function matrix $F_1(s)$ of the former reads

$$F_1(s) = (s^2+s)^{-1} \left[s+1 \vdots -1 \vdots s+2 \vdots 1 \right], \tag{11.20}$$

while the full transfer function matrix $F_2(s)$ of the latter is found as

$$F_2(s) = (s^3+s^2-2s)^{-1} \bullet$$

$$\bullet \left[s^2-1 \vdots -s \vdots -1 \vdots s^2+s-2 \vdots s+1 \vdots 1 \right] \neq F_1(s). \tag{11.21}$$

They imply the following full system functions, respectively:

$$P_{IO1}(s) = \begin{bmatrix} s^2+s & s+1 & -1 & s+2 & 1 \\ -1 & 0 & 0 & 0 & 0 \end{bmatrix},$$

$$P_{IO2}(s) =$$

$$= \begin{bmatrix} s^3+s^2-2s & s^2-1 & -s & -1 & s^2+s-2 & s+1 & 1 \\ -1 & 0 & 0 & 0 & 0 & 0 & 0 \end{bmatrix} \neq$$

$$\neq P_{IO1}(s).$$

Example 204 *For the $MIMO$ IO system described by*

$$\mathbf{y}^{(1)}(t) + \begin{bmatrix} 2 \vdots 0 \\ 0 \vdots 0 \end{bmatrix} \mathbf{y}(t) = \begin{bmatrix} 0 \vdots 0 \\ 0 \vdots 2 \end{bmatrix} \mathbf{i}(t) + \mathbf{i}^{(1)}(t),$$

$$\mathbf{y} = \begin{bmatrix} y_1 \\ y_2 \end{bmatrix}, \mathbf{i} = \begin{bmatrix} i_1 \\ i_2 \end{bmatrix}.$$

we find the transfer function matrix $G_{IO}(s)$,

$$G_{IO}(s) = \begin{bmatrix} \frac{s}{s+2} & 0 \\ 0 & \frac{s+2}{s} \end{bmatrix}, \ fullrank G_{IO}(s) = 2,$$

and the full transfer function matrix $F_{IO}(s)$,

$$F_{IO}(s) = \begin{bmatrix} \frac{s}{s+2} & 0 & -\frac{1}{s+2} & 0 & \frac{1}{s+2} & 0 \\ 0 & \frac{s+2}{s} & 0 & -\frac{1}{s} & 0 & \frac{1}{s} \end{bmatrix} =$$

$$= \begin{bmatrix} \frac{1}{s+2} & 0 \\ 0 & \frac{1}{s} \end{bmatrix} \begin{bmatrix} s & 0 & -1 & 0 & 1 & 0 \\ 0 & s+2 & 0 & -1 & 0 & 1 \end{bmatrix} =$$

$$= \underbrace{\begin{bmatrix} s+2 & 0 \\ 0 & s \end{bmatrix}^{-1}}_{F_{IOD}(s)} \underbrace{\begin{bmatrix} s \vdots & 0 & \vdots -1 \vdots & 0 & \vdots \; 1 & \vdots \, 0 \\ 0 \vdots & s+2 \vdots & 0 & \vdots -1 \vdots & 0 & \vdots \, 1 \end{bmatrix}}_{F_{ION}(s)}.$$

We determine now the full system matrix $P_{IO}(s)$,

$$P_{IO}(s) = \begin{bmatrix} s+2 & 0 & s & 0 & -1 & 0 & 1 & 0 \\ 0 & s & 0 & s+2 & 0 & -1 & 0 & 1 \\ 1 & 0 & 0 & 0 & 0 & 0 & 0 & 0 \\ 0 & 1 & 0 & 0 & 0 & 0 & 0 & 0 \end{bmatrix}.$$

11.2 System matrix of the ISO System

Rosenbrock defined in [110, p. 43] *the system matrix function* $P_R(.)$ *under all zero initial conditions* $\mathbf{x}_0 = \mathbf{0}_n$ for Rosenbrock systems RS (2.37) (in Section 2.3), which we will denote also by $P_R(.;\mathbf{0}_n)$,

$$P_R(.) = P_R(.;\mathbf{0}_n).$$

This has been largely accepted in the literature.

The ISO systems form a subclass of Rosenbrock systems RS (2.37) so that we can apply Rosenbrock definition of the system matrix function $P_{ISOR}(.;\mathbf{0}_n)$ of (2.37) to the ISO system (3.84) and (3.85) (in Subsection 3.5.3) under all zero initial conditions,

$$P_{ISOR}(s;\mathbf{0}_n) = \begin{bmatrix} sI_n - A & B \\ -C & D \end{bmatrix} =$$

$$= \begin{bmatrix} T_{ISOR}(s;\mathbf{0}_n) & U_{ISOR}(s;\mathbf{0}_n) \\ -V_{ISOR}(s;\mathbf{0}_n) & W_{ISOR}(s;\mathbf{0}_n) \end{bmatrix} \Longleftrightarrow$$

$$\Longleftrightarrow P_{ISOR}(s;\mathbf{0}_n) \begin{bmatrix} \mathbf{X}^{\mp}(s) \\ -\mathbf{I}^{\mp}(s) \end{bmatrix} = \begin{bmatrix} \mathbf{0}_n \\ -\mathbf{Y}^{\mp}(s) \end{bmatrix}. \tag{11.22}$$

$T_{ISOR}(s;\mathbf{0}_n)$, $U_{ISOR}(s;\mathbf{0}_n)$, $V_{ISOR}(s;\mathbf{0}_n)$, and $W_{ISOR}(s;\mathbf{0}_n)$ are *Rosenbrock polynomial submatrices of* $P_{ISOR}(s;\mathbf{0}_n)$, [110, p. 52], of the ISO system (3.84) and (3.85),

$$T_{ISOR}(s;\mathbf{0}_n) = sI_n - A, \; U_{ISOR}(s;\mathbf{0}_n) = B,$$
$$V_{ISOR}(s;\mathbf{0}_n) = C, \; W_{ISOR}(s;\mathbf{0}_n) = D. \tag{11.23}$$

Skelton [114, p. 134] permitted nonzero initial state vector, $\mathbf{x}_0 \neq \mathbf{0}_n$, for the *ISO* system (3.84) and (3.85). In that case, Rosenbrock system matrix $P_{ISOR}(s;\mathbf{0}_n)$ (11.22) should be replaced by the following ***full system matrix*** $P_{ISOR}(s;\mathbf{x}_0)$, $P_{ISOR}(s;\mathbf{x}_0) \in \mathfrak{C}^{(N+n)\times(M+2n)}$, ***in the general Rosenbrock form***

$$P_{ISOR}(s;\mathbf{x}_0) = \begin{bmatrix} T_{ISOR}(s;\mathbf{x}_0) & U_{ISOR}(s;\mathbf{x}_0) \\ -V_{ISOR}(s;\mathbf{x}_0) & W_{ISOR}(s;\mathbf{x}_0) \end{bmatrix}, \tag{11.24}$$

$$P_{ISOR}(s;\mathbf{x}_0) = \begin{bmatrix} sI_n - A & B & I_n \\ -C & D & O_{N\times n} \end{bmatrix} =$$

$$= \left[P_{ISOR}(s;\mathbf{0}) \vdots \begin{matrix} I_n \\ O_{N\times n} \end{matrix} \right] =$$

$$= \begin{bmatrix} T_{ISOR}(s;\mathbf{x}_0) & U_{ISOR}(s;\mathbf{x}_0) \\ -V_{ISOR}(s;\mathbf{x}_0) & W_{ISOR}(s;\mathbf{x}_0) \end{bmatrix} \Longleftrightarrow$$

$$\Longleftrightarrow P_{ISOR}(s;\mathbf{x}_0) \begin{bmatrix} \mathbf{X}^{\mp}(s) \\ -\mathbf{V}^{\mp}_{ISOR}(s;\mathbf{x}_{0\mp}) \end{bmatrix} = \begin{bmatrix} \mathbf{0}_n \\ -\mathbf{Y}^{\mp}(s) \end{bmatrix},$$

$$where\ \mathbf{V}_{ISOR}(s;\mathbf{x}_{0\mp}) = \begin{bmatrix} \mathbf{I}^{\mp}(s) \\ \mathbf{x}_{0\mp} \end{bmatrix} \in \mathfrak{C}^{M+n}. \tag{11.25}$$

Rosenbrock submatrices changed from their forms in (11.23) to the following *full Rosenbrock submatrices* of $P_{ISOR}(s;\mathbf{x}_0)$,

$$T_{ISOR}(s;\mathbf{x}_0) = (sI_n - A) \in \mathfrak{C}^{n\times n},$$
$$U_{ISOR}(s;\mathbf{x}_0) = \left[B \vdots I_n \right] \in \mathfrak{C}^{n\times(M+n)},$$
$$V_{ISOR}(s;\mathbf{x}_0) = C \in \mathfrak{C}^{N\times n},$$
$$W_{ISOR}(s;\mathbf{x}_0) = \left[D \vdots O_{N\times n} \right] \in \mathfrak{C}^{N\times(M+n)}. \tag{11.26}$$

These equations and (8.28) (in Section 8.2) imply:

$$F_{ISO}(s) = V_{ISOR}(s;\mathbf{x}_0) T^{-1}_{ISOR}(s;\mathbf{x}_0) U_{ISOR}(s;\mathbf{x}_0) + W_{ISOR}(s;\mathbf{x}_0) =$$
$$= F^{-1}_{ISOD}(s) F_{ISON}(s). \tag{11.27}$$

We will broaden in the sequel Rosenbrock's definition of strictly system equivalent matrices and Theorem 3.1 in [110, p. 52].

Example 205 *Given* $G_{ISO}(s) = \left(s^2-1\right)^{-1}(s-1) = (s+1)^{-1}$. *Four different (state space, i.e., ISO) realizations* (A,B,C,D) *of* $G_{ISO}(s)$ *are determined in [4, p. 395], Example 167 (in Section 8.2). We show at first how to determine the full system matrix for each in the Rosenbrock form (11.25).*

1) $A_1 = \begin{bmatrix} 0 & 1 \\ 1 & 0 \end{bmatrix}$, $B_1 = \begin{bmatrix} 0 \\ 1 \end{bmatrix}$, $C_1 = \begin{bmatrix} -1 \vdots 1 \end{bmatrix}$, $D_1 = 0$, $\mathbf{x} = \begin{bmatrix} x_1 \\ x_2 \end{bmatrix}$, $\mathbf{y} = [y]$

$\Longrightarrow$

$$P_{ISOR1}(s;\mathbf{x}_0) = \begin{bmatrix} T_{ISOR1}(s;\mathbf{x}_0) & U_{ISOR1}(s;\mathbf{x}_0) \\ -V_{ISOR1}(s;\mathbf{x}_0) & W_{ISOR1}(s;\mathbf{x}_0) \end{bmatrix} =$$

$$= \begin{bmatrix} s & -1 & 0 & 1 & 0 \\ -1 & s & 1 & 0 & 1 \\ 1 & -1 & 0 & 0 & 0 \end{bmatrix}, \ \mathbf{V}_{ISOR1}(s;\mathbf{x}_{0\mp}) = \begin{bmatrix} I^{\mp}(s) \\ \mathbf{x}_{0\mp} \end{bmatrix}.$$

2) $A_2 = \begin{bmatrix} 0 & 1 \\ 1 & 0 \end{bmatrix}$, $B_2 = \begin{bmatrix} -1 \\ 1 \end{bmatrix}$, $C_2 = \begin{bmatrix} 0 \vdots 1 \end{bmatrix}$, $D_2 = 0$, $\mathbf{x} = \begin{bmatrix} x_1 \\ x_2 \end{bmatrix}$, $\mathbf{y} = [y]$

$\Longrightarrow$

$$P_{ISOR2}(s;\mathbf{x}_0) = \begin{bmatrix} T_{ISOR2}(s;\mathbf{x}_0) & U_{ISOR2}(s;\mathbf{x}_0) \\ -V_{ISOR2}(s;\mathbf{x}_0) & W_{ISOR2}(s;\mathbf{x}_0) \end{bmatrix} =$$

$$= \begin{bmatrix} s & -1 & -1 & 1 & 0 \\ -1 & s & 1 & 0 & 1 \\ 0 & -1 & 0 & 0 & 0 \end{bmatrix}, \ \mathbf{V}_{ISOR2}(s;\mathbf{x}_{0\mp}) = \begin{bmatrix} I^{\mp}(s) \\ \mathbf{x}_{0\mp} \end{bmatrix}.$$

3) $A_3 = \begin{bmatrix} 1 & 0 \\ 0 & -1 \end{bmatrix}$, $B_3 = \begin{bmatrix} 0 \\ 1 \end{bmatrix}$, $C_3 = \begin{bmatrix} 0 \vdots 1 \end{bmatrix}$, $D_3 = 0$, $\mathbf{x} = \begin{bmatrix} x_1 \\ x_2 \end{bmatrix}$, $\mathbf{y} = [y]$

$\Longrightarrow$

$$P_{ISOR3}(s;\mathbf{x}_0) = \begin{bmatrix} T_{ISOR3}(s;\mathbf{x}_0) & U_{ISOR3}(s;\mathbf{x}_0) \\ -V_{ISOR3}(s;\mathbf{x}_0) & W_{ISOR3}(s;\mathbf{x}_0) \end{bmatrix} =$$

$$= \begin{bmatrix} s-1 & 0 & 0 & 1 & 0 \\ 0 & s+1 & 1 & 0 & 1 \\ 0 & -1 & 0 & 0 & 0 \end{bmatrix}, \ \mathbf{V}_{ISOR3}(s;\mathbf{x}_{0\mp}) = \begin{bmatrix} I^{\mp}(s) \\ \mathbf{x}_{0\mp} \end{bmatrix}.$$

4) $A_4 = [-1]$, $B_4 = [1]$, $C_4 = [1]$, $D_4 = 0$, $\mathbf{x} = [x]$, $\mathbf{y} = [y] \Longrightarrow$

$$P_{ISOR4}(s;\mathbf{x}_0) = \begin{bmatrix} T_{ISOR4}(s;\mathbf{x}_0) & U_{ISOR4}(s;\mathbf{x}_0) \\ -V_{ISOR4}(s;\mathbf{x}_0) & W_{ISOR4}(s;\mathbf{x}_0) \end{bmatrix} =$$

$$= \begin{bmatrix} s+1 \vdots 1 \vdots 1 \\ -1 \vdots 0 \vdots 0 \end{bmatrix}, \ \mathbf{V}_{ISOR4}(s;\mathbf{x}_{0\mp}) = \begin{bmatrix} I^{\mp}(s) \\ x_{0\mp} \end{bmatrix}.$$

Definition 206 ***Strictly equivalent full system matrices***

Let $(N+n)\mathrm{x}(M+2n)$ full system matrix $P_{ISO}(s)$ be given in its Rosenbrock form $P_{ISOR}(s)$ (11.25). Let $M(s)$ and $N(s)$ be $n\mathrm{x}n$ unimodular polynomial matrices. Let $X(s)$ and $Y(s)$ be also polynomial matrices, respectively $N\mathrm{x}n$ and $n\mathrm{x}(M+n)$. If $P_{ISOR}(s)$ and

$$P_1(s) = \begin{bmatrix} T_1(s) & U_1(s) \\ -V_1(s) & W_1(s) \end{bmatrix} \in \mathfrak{C}^{(n+N)\mathrm{x}(M+n)} \tag{11.28}$$

are related by the transformation

$$\begin{bmatrix} M(s) & O_{n\mathrm{x}N} \\ X(s) & I_N \end{bmatrix} \begin{bmatrix} T_{ISOR}(s) & U_{ISOR}(s) \\ -V_{ISOR}(s) & W_{ISOR}(s) \end{bmatrix} \bullet$$
$$\bullet \begin{bmatrix} N(s) & Y(s) \\ O_{(M+n)\mathrm{x}n} & I_{M+n} \end{bmatrix} = \begin{bmatrix} T_1(s) & U_1(s) \\ -V_1(s) & W_1(s) \end{bmatrix} \tag{11.29}$$

then $P_{ISOR}(s)$ and $P_1(s)$ are ***strictly system equivalent.***

Theorem 207 *Two full system matrices which are strictly system equivalent correspond to the same system full transfer function matrix $F_{ISO}(s)$.*

Proof. We refer to Definition 206. Let $(n+N)\times(M+n)$ system matrix $P_{ISO}(s)$ be given in its Rosenbrock form $P_{ISOR}(s)$ (11.25) and (11.26). Equation (11.29) can be written as

$$\begin{bmatrix} M(s) & O_{n\mathrm{x}N} \\ X(s) & I_N \end{bmatrix} \begin{bmatrix} T_{ISOR}(s) & U_{ISOR}(s) \\ -V_{ISOR}(s) & W_{ISOR}(s) \end{bmatrix} \bullet$$
$$\bullet \begin{bmatrix} N(s) & Y(s) \\ O_{(M+n)\mathrm{x}n} & I_{M+n} \end{bmatrix} = \begin{bmatrix} T_1(s) & U_1(s) \\ -V_1(s) & W_1(s) \end{bmatrix} =$$

$$= \begin{bmatrix} M(s)T_{ISOR}(s) & M(s)U_{ISOR}(s) \\ X(s)T_{ISOR}(s) - V_{ISOR}(s) & X(s)U_{ISOR}(s) + W_{ISOR}(s) \end{bmatrix} \bullet$$
$$\bullet \begin{bmatrix} N(s) & Y(s) \\ O_M & I_M \end{bmatrix} =$$

$$= \begin{bmatrix} MT_{ISOR}N & \vdots & MT_{ISOR}Y + MU_{ISOR} \\ (XT_{ISOR} - V_{ISOR})N & \vdots & (XT_{ISOR} - V_{ISOR})Y + \ XU_{ISOR} + W_{ISOR} \end{bmatrix}.$$

This result, Equations (11.12):

$$F_1(s) = V_1(s)T_1^{-1}(s)U_1(s) + W_1(s),$$

and (11.29) yield the following in view of the nonsingularity of the unimodular matrices:

$$\begin{gathered}F_1(s) = -\left(XT_{ISOR} - V_{ISOR}\right)N\left(MT_{ISOR}N\right)^{-1}\left(MT_{ISOR}Y + MU_{ISOR}\right) + \\ +\left(XT_{ISOR} - V_{ISOR}\right)Y + \ XU_{ISOR} + W_{ISOR} = \\ = -XT_{ISOR}Y - XU_{ISOR} + V_{ISOR}Y + V_{ISOR}T_{ISOR}^{-1}U_{ISOR} + \\ +XT_{ISOR}Y - V_{ISOR}Y + XU_{ISOR} + W_{ISOR} = \\ = V_{ISOR}T_{ISOR}^{-1}U_{ISOR} + W_{ISOR} = F_{ISOD}^{-1}(s)F_{ISON}(s) = F_{ISO}(s)\end{gathered}$$

due to (11.27). ■

This theorem generalizes Rosenbrock's Theorem 3.1 [110, p. 52] in the framework of the *ISO* systems.

Let $\mathbf{r}(t) \equiv \mathbf{y}(t)$. Hence, $\mathbf{R}^{\mp}(s) \equiv \mathbf{Y}^{\mp}(s)$. We can combine this with (8.28), (8.29), and (8.36) (in Section 8.2) into the *IO* form:

$$\mathbf{R}^{\mp}(s) - \left[C(sI_n - A)^{-1}B + D \vdots C(sI_n - A)^{-1}\right]\mathbf{V}_{ISOIO}^{\mp}(s) = \mathbf{0}_N \Longrightarrow$$

$$\Delta(s)\mathbf{R}^{\mp}(s) +$$

$$+\left[Cadj\left(sI_n - A\right)B + D\Delta(s) \vdots Cadj\left(sI_n - A\right)\right]\left(-\mathbf{V}_{ISOIO}^{\mp}(s)\right) = \mathbf{0}_N,$$

$$\mathbf{Y}^{\mp}(s) = \mathbf{R}^{\mp}(s),\ where\ \mathbf{V}_{ISOIO}^{\mp}(s;\mathbf{x}_{0\mp}) = \begin{bmatrix}\mathbf{I}^{\mp}(s) \\ \mathbf{x}_{0\mp}\end{bmatrix} \in \mathfrak{C}^{M+n}. \quad (11.30)$$

These equations imply the following *IO full system matrix* $P_{ISOIO}(s)$ *of the ISO system:*

$$\overbrace{\begin{bmatrix}\Delta(s)I_N \vdots Cadj(sI_n - A)B + D\Delta(s) \vdots Cadj(sI_n - A) \\ -I_N \quad \vdots \quad O_{N\times M} \quad \vdots \quad O_{N\times n}\end{bmatrix}}^{P_{ISOIO}(s)}\begin{bmatrix}\mathbf{R}^{\mp}(s) \\ -\mathbf{V}_{ISO}^{\mp}(s)\end{bmatrix} =$$

$$= \begin{bmatrix}\mathbf{0}_N \\ -\mathbf{Y}^{\mp}(s)\end{bmatrix}. \quad (11.31)$$

If we define *the IO full polynomial submatrices* $T_{ISOIO}(s;\mathbf{x}_0)$, $U_{ISOIO}(s;\mathbf{x}_0)$, $V_{ISOIO}(s;\mathbf{x}_0)$, and $W_{ISOIO}(s;\mathbf{x}_0)$ of $P_{ISOIO}(s;\mathbf{x}_0)$ due to (8.28) (in Section

8.2) by

$$T_{ISOIO}(s;\mathbf{x}_0) = \Delta(s)I_N = F_{ISOD}(s),$$

$$U_{ISOIO}(s;\mathbf{x}_0) = \left[Cadj(sI_n - A)B + D\Delta(s) \vdots Cadj(sI_n - A)\right] = F_{ISON}(s),$$

$$V_{ISOIO}(s;\mathbf{x}_0) = I_N,\ W_{ISOIO}(s;\mathbf{x}_0) = O_{N\text{x}(M+n)}, \tag{11.32}$$

then (11.31) furnishes

$$P_{ISOIO}(s;\mathbf{x}_0) = \begin{bmatrix} T_{ISOIO}(s;\mathbf{x}_0) \vdots U_{ISOIO}(s;\mathbf{x}_0) \\ -V_{ISOIO}(s;\mathbf{x}_0) \vdots W_{ISOIO}(s;\mathbf{x}_0) \end{bmatrix}. \tag{11.33}$$

Comment 208 *Equations (11.26), (11.32), and (11.33) prove that the full system matrices $P_{ISOR}(s;\mathbf{x}_0)$ in (11.25) and $P_{ISOIO}(s;\mathbf{x}_0)$ in (11.31), i.e., in (11.33), have the same form and structure as the general Rosenbrock system matrix (11.24) [110, p. 52].*

Definition 209 ***Strictly equivalent IO full system matrices: Rosenbrock form***

Let $(N+n)\text{x}(M+2n)$ full system matrix $P_{ISOIO}(s)$ be given in the Rosenbrock form (11.25). Let $M(s)$ and $N(s)$ be $n\text{x}n$ unimodular polynomial matrices. Let $X(s)$ and $Y(s)$ be also polynomial matrices, respectively $N\text{x}n$ and $n\text{x}(M+n)$. If $P_{ISOIO}(s)$ and $P_1(s)$ (11.28) are related by the transformation

$$\begin{bmatrix} M(s) & O_{n\text{x}N} \\ X(s) & I_N \end{bmatrix} \begin{bmatrix} T_{ISOIO}(s) & U_{ISOIO}(s) \\ -V_{ISOIO}(s) & W_{ISOIO}(s) \end{bmatrix} \bullet$$

$$\bullet \begin{bmatrix} N(s) & Y(s) \\ O_{(M+n)\text{x}n} & I_{M+n} \end{bmatrix} = \begin{bmatrix} T_1(s) & U_1(s) \\ -V_1(s) & W_1(s) \end{bmatrix} \tag{11.34}$$

*then $P_{ISOIO}(s)$ and $P_1(s)$ are **strictly system equivalent**.*

Theorem 210 *Two strictly system equivalent IO full system matrices correspond to the same system full IO transfer function matrix $F_{ISOIO}(s)$.*

The proof of this theorem is the same as that of Theorem 207.

Example 211 *Given $G_{ISO}(s) = \left(s^2-1\right)^{-1}(s-1) = (s+1)^{-1}$. Four different (state space, i.e., ISO) realizations (A,B,C,D) of $G_{ISO}(s)$ are determined in [4, p. 395], Example 167 (in Section 8.2). We use the full transfer* *function matrix* $F_{ISO}(s)$ *determined for each in Example 167 in order to deduce from it the full system matrix* $P_{ISOIO}(s;\mathbf{x}_0)$.

1) $A_1 = \begin{bmatrix} 0 & 1 \\ 1 & 0 \end{bmatrix}$, $B_1 = \begin{bmatrix} 0 \\ 1 \end{bmatrix}$, $C_1 = \begin{bmatrix} -1 \vdots 1 \end{bmatrix}$, $D_1 = 0$, $\mathbf{x} = \begin{bmatrix} x_1 \\ x_2 \end{bmatrix}$, $\mathbf{y} = [y]$
$\Longrightarrow$

$$F_{ISO1}(s) = \left(s^2 - 1\right)^{-1} \begin{bmatrix} s-1 \vdots 1-s \vdots s-1 \end{bmatrix} \Longrightarrow$$

$$P_{ISOIO1}(s;\mathbf{x}_0) = \begin{bmatrix} T_{ISOIO1}(s;\mathbf{x}_0) & U_{ISOIO1}(s;\mathbf{x}_0) \\ -V_{ISOIO1}(s;\mathbf{x}_0) & W_{ISOIO1}(s;\mathbf{x}_0) \end{bmatrix} =$$

$$= \begin{bmatrix} s^2-1 \vdots s-1 \vdots 1-s \vdots s-1 \\ -1 \vdots 0 \vdots 0 \vdots 0 \end{bmatrix} \neq P_{ISOR1}(s;\mathbf{x}_0),$$

$$\mathbf{V}_{ISOIO1}(s;\mathbf{x}_{0\mp}) = \begin{bmatrix} I^{\mp}(s) \\ \mathbf{x}_{0\mp} \end{bmatrix}.$$

2) $A_2 = \begin{bmatrix} 0 & 1 \\ 1 & 0 \end{bmatrix}$, $B_2 = \begin{bmatrix} -1 \\ 1 \end{bmatrix}$, $C_2 = \begin{bmatrix} 0 \vdots 1 \end{bmatrix}$, $D_2 = 0$, $\mathbf{x} = \begin{bmatrix} x_1 \\ x_2 \end{bmatrix}$, $\mathbf{y} = [y]$
$\Longrightarrow$

$$F_{ISO2}(s) = \left(s^2 - 1\right)^{-1} \begin{bmatrix} s-1 \vdots 1 \vdots s \end{bmatrix} \Longrightarrow$$

$$P_{ISOIO2}(s;\mathbf{x}_0) = \begin{bmatrix} T_{ISOIO2}(s;\mathbf{x}_0) & U_{ISOIO2}(s;\mathbf{x}_0) \\ -V_{ISOIO2}(s;\mathbf{x}_0) & W_{ISOIO2}(s;\mathbf{x}_0) \end{bmatrix} =$$

$$= \begin{bmatrix} s^2-1 \vdots s-1 \vdots 1 \vdots s \\ -1 \vdots 0 \vdots 0 \vdots 0 \end{bmatrix} \neq P_{ISOR2}(s;\mathbf{x}_0),$$

$$\mathbf{V}_{ISOIO2}(s;\mathbf{x}_{0\mp}) = \begin{bmatrix} I^{\mp}(s) \\ \mathbf{x}_{0\mp} \end{bmatrix}.$$

3) $A_3 = \begin{bmatrix} 1 & 0 \\ 0 & -1 \end{bmatrix}$, $B_3 = \begin{bmatrix} 0 \\ 1 \end{bmatrix}$, $C_3 = \begin{bmatrix} 0 \vdots 1 \end{bmatrix}$, $D_3 = 0$, $\mathbf{x} = \begin{bmatrix} x_1 \\ x_2 \end{bmatrix}$, $\mathbf{y} = [y]$
$\Longrightarrow$

$$F_{ISO3}(s) = \left(s^2 - 1\right)^{-1} \begin{bmatrix} s-1 \vdots 0 \vdots s-1 \end{bmatrix} \Longrightarrow$$

$$P_{ISOIO3}(s;\mathbf{x}_0) = \begin{bmatrix} T_{ISOIO3}(s;\mathbf{x}_0) & U_{ISOIO3}(s;\mathbf{x}_0) \\ -V_{ISOIO3}(s;\mathbf{x}_0) & W_{ISOIO3}(s;\mathbf{x}_0) \end{bmatrix} =$$

$$= \begin{bmatrix} s^2-1 \vdots s-1 \vdots 0 \vdots s-1 \\ -1 \vdots \quad 0 \quad \vdots 0 \vdots \quad 0 \end{bmatrix} \neq P_{ISOR3}(s;\mathbf{x}_0),$$

$$\mathbf{V}_{ISOIO3}(s;\mathbf{x}_{0\mp}) = \begin{bmatrix} I^{\mp}(s) \\ \mathbf{x}_{0\mp} \end{bmatrix}.$$

4) $A_4 = [-1]$, $B_4 = [1]$, $C_4 = [1]$, $D_4 = 0$, $\mathbf{x} = [x]$, $\mathbf{y} = [y] \Longrightarrow$

$$F_{ISO4}(s) = (s+1)^{-1}\begin{bmatrix} 1 \vdots 1 \end{bmatrix} \Longrightarrow$$

$$P_{ISOIO4}(s;\mathbf{x}_0) = \begin{bmatrix} T_{ISOIO4}(s;\mathbf{x}_0) & U_{ISOIO4}(s;\mathbf{x}_0) \\ -V_{ISOIO4}(s;\mathbf{x}_0) & W_{ISOIO4}(s;\mathbf{x}_0) \end{bmatrix} =$$

$$= \begin{bmatrix} s+1 \vdots 1 \vdots 1 \\ -1 \vdots 0 \vdots 0 \end{bmatrix} = P_{ISOR4}(s;\mathbf{x}_0),$$

$$\mathbf{V}_{ISOIO4}(s;\mathbf{x}_{0\mp}) = \begin{bmatrix} I^{\mp}(s) \\ x_{0\mp} \end{bmatrix}.$$

Note 212 *The full system matrix functions* $P_{ISOIO}(s;\mathbf{x}_0)$ *and* $P_{ISOR}(s;\mathbf{x}_0)$ *are different in the cases 1) through 3) of Examples 205 and 211 because the subsidiary vectors* $\mathbf{r} = \mathbf{y} \in R^N$ *and* $\mathbf{x} \in R^n$ *are different in general due to* $\mathbf{x} \neq \mathbf{y}$.

Note 213 *The cases 1) through 3) of Examples 205 and 211 show that*

$$P_{ISOR}(s;\mathbf{x}_0) \neq P_{ISOIO}(s;\mathbf{x}_0) \text{ in general.}$$

However, the case 4) of the same Examples discovers that

$$P_{ISOR}(s;\mathbf{x}_0) = P_{ISOIO}(s;\mathbf{x}_0)$$

is possible in special cases.

11.3 System matrix of the IIO system

Let $\eta = \alpha$ in (3.90) (in Section 3.5.4). We apply the Laplace transform to the *IIO* system (3.90),

$$Q^{(\alpha)}\mathbf{r}^{\alpha}(t) = P^{(\beta)}\mathbf{i}^{\beta}(t),\ E^{(\nu)}\mathbf{y}^{\nu}(t) = R^{(\alpha)}\mathbf{r}^{\alpha}(t) + T^{(\mu)}\mathbf{i}^{\mu}(t),\ \forall t \in \mathfrak{T}_0,$$

and obtain

$$Q^{(\alpha)}S_{\rho}^{(\alpha)}(s)\mathbf{R}^{\mp}(s) = P^{(\beta)}S_{M}^{(\beta)}(s)\mathbf{I}^{\mp}(s)-$$
$$-P^{(\beta)}Z_{M}^{(\beta-1)}(s)\mathbf{i}_{0\mp}^{\beta-1} + Q^{(\alpha)}Z_{\rho}^{(\alpha-1)}(s)\mathbf{r}_{0\mp}^{\alpha-1},$$

$$E^{(\nu)}S_{N}^{(\nu)}(s)\mathbf{Y}^{\mp}(s) = R^{(\alpha)}S_{\rho}^{(\alpha)}(s)\mathbf{R}^{\mp}(s) + T^{(\mu)}S_{M}^{(\mu)}(s)\mathbf{I}^{\mp}(s)-$$
$$-R^{(\alpha)}Z_{\rho}^{(\alpha-1)}(s)\mathbf{r}_{0\mp}^{\alpha-1} - T^{(\mu)}Z_{M}^{(\mu-1)}(s)\mathbf{i}_{0\mp}^{\mu-1} + E^{(\nu)}Z_{N}^{(\nu-1)}(s)\mathbf{y}_{0\mp}^{\nu-1},$$

or equivalently due to (7.33) (in Subsection 7.4),

$$Q^{(\alpha)}S_{\rho}^{(\alpha)}(s)\mathbf{R}^{\mp}(s) - P^{(\beta)}S_{M}^{(\beta)}(s)\mathbf{I}^{\mp}(s)+$$
$$+\left[P^{(\beta)}Z_{M}^{(\beta-1)}(s) \vdots O_{\rho\times[(\gamma-\beta)M]}\right]\mathbf{i}_{0\mp}^{\gamma-1} - Q^{(\alpha)}Z_{\rho}^{(\alpha-1)}(s)\ \mathbf{r}_{0\mp}^{\alpha-1} = \mathbf{0}_{\rho},$$

$$-R^{(\alpha)}S_{\rho}^{(\alpha)}(s)\mathbf{R}^{\mp}(s) - T^{(\mu)}S_{M}^{(\mu)}(s)\mathbf{I}^{\mp}(s)+$$
$$+\left[T^{(\mu)}Z_{M}^{(\mu-1)}(s) \vdots O_{N\times[(\gamma-\mu)M]}\right]\mathbf{i}_{0\mp}^{\gamma-1} + R^{(\alpha)}Z_{\rho}^{(\alpha-1)}(s)\ \mathbf{r}_{0\mp}^{\alpha-1}-$$
$$-E^{(\nu)}Z_{N}^{(\nu-1)}(s)\mathbf{y}_{0\mp}^{\nu-1} = -E^{(\nu)}S_{N}^{(\nu)}(s)\mathbf{Y}^{\mp}(s).$$

This permits us to define ***the full system matrix function*** $P_{IIO}(.)$ ***of the IIO system*** by

$$P_{IIO}(s) \in \mathfrak{C}^{(N+\rho)\times[(\gamma+1)M+(\alpha+1)\rho+\nu N]},$$

$$P_{IIO}(s) =$$

$$\begin{bmatrix} \left(Q^{(\alpha)}S_{\rho}^{(\alpha)}(s)\right)^T & -\left(R^{(\alpha)}S_{\rho}^{(\alpha)}(s)\right)^T \\ \left(P^{(\beta)}S_{M}^{(\beta)}(s)\right)^T & \left(T^{(\mu)}S_{M}^{(\mu)}(s)\right)^T \\ \left[-P^{(\beta)}Z_{M}^{(\beta-1)}(s) \vdots O_A\right]^T & \left[-T^{(\mu)}Z_{M}^{(\mu-1)}(s) \vdots O_C\right]^T \\ \left(Q^{(\alpha)}Z_{\rho}^{(\alpha-1)}(s)\right)^T & -\left(R^{(\alpha)}Z_{\rho}^{(\alpha-1)}(s)\right)^T \\ (O_B)^T & \left(E^{(\nu)}Z_{N}^{(\nu-1)}(s)\right)^T \end{bmatrix}^T$$

$$O_A = O_{\rho\times[(\gamma-\beta)M]},\ O_B = O_{\rho\times\nu N},\ O_C = O_{N\times[(\gamma-\mu)M]}$$

$$\Longleftrightarrow$$

$$P_{IIO}(s)\begin{bmatrix} \mathbf{R}^{\mp}(s) \\ -\mathbf{V}_{IIO}^{\mp}(s;\mathbf{i}_{0\mp}^{\gamma-1};\mathbf{r}_{0\mp}^{\alpha-1};\mathbf{y}_{0\mp}^{\nu-1}) \end{bmatrix} = \begin{bmatrix} \mathbf{0}_{\rho} \\ -E^{(\nu)}S_{N}^{(\nu)}(s)\mathbf{Y}^{\mp}(s) \end{bmatrix},$$

$$where\ \mathbf{V}_{IIO}^{\mp}(s;\mathbf{i}_{0\mp}^{\gamma-1};\mathbf{r}_{0\mp}^{\alpha-1};\mathbf{y}_{0\mp}^{\nu-1}) = \begin{bmatrix} \mathbf{I}^{\mp}(s) \\ \mathbf{i}_{0\mp}^{\gamma-1} \\ \mathbf{r}_{0\mp}^{\alpha-1} \\ \mathbf{y}_{0\mp}^{\nu-1} \end{bmatrix}. \tag{11.35}$$

Rosenbrock like submatrices are determined for the *IIO* system (3.90) from (11.35) by

$$T_{IIOR}(s) = Q^{(\alpha)}S_{\rho}^{(\alpha)}(s),\ T_{IIOR}(s) \in \mathfrak{C}^{\rho\times\rho},$$

$$U_{IIOR}(s) =$$

$$= \left[P^{(\beta)}S_{M}^{(\beta)}(s) \vdots \left[-P^{(\beta)}Z_{M}^{(\beta-1)}(s) \vdots O_{\rho\times[(\gamma-\beta)M]}\right] \vdots Q^{(\alpha)}Z_{\rho}^{(\alpha-1)}(s) \vdots O_{\rho\times\nu N}\right],$$

$$U_{IIOR}(s) \in \mathfrak{C}^{\rho\times[\alpha\rho+(\gamma+1)M+\nu N]},$$

$$V_{IIOR}(s) = R^{(\alpha)}S_{\rho}^{(\alpha)}(s),\ V_{IIOR}(s) \in \mathfrak{C}^{N\times\rho},$$

$$W_{IIOR}(s) =$$
$$= \left[T^{(\mu)} S_M^{(\mu)}(s) \vdots \left[-T^{(\mu)} Z_M^{(\mu-1)}(s) \vdots O_C\right] \vdots - R^{(\alpha)} Z_\rho^{(\alpha-1)}(s) \vdots E^{(\nu)} Z_N^{(\nu-1)}(s)\right],$$
$$O_C = O_{N\times[(\gamma-\mu)M]},\ W_{IIOR}(s) \in \mathfrak{C}^{N\times[\alpha\rho+(\gamma+1)M+\nu N]}.$$

They transform $P_{IIO}(s)$ (11.35) into Rosenbrock form $P_{IIOR}(s)$,

$$P_{IIOR}(s) = \begin{bmatrix} T_{IIOR}(s) \vdots & U_{IIOR}(s) \\ -V_{IIOR}(s) \vdots & W_{IIOR}(s) \end{bmatrix}. \tag{11.36}$$

Definition 214 *Let the full system matrix $P_{IIO}(s)$,*

$$P_{IIO}(s) \in \mathfrak{R}^{(N+\rho)\times[(\alpha+1)\rho+(\gamma+1)M+\nu N]},$$

be given in its Rosenbrock form $P_{IIOR}(s)$ (11.36). Let $M(s)$ and $N(s)$ be $\rho\times\rho$ unimodular polynomial matrices. Let $X(s)$ and $Y(s)$ be also polynomial matrices, $X(s) \in \mathfrak{R}^{N\times\rho}$ and $Y(s) \in \mathfrak{R}^{\rho\times[\alpha\rho+(\gamma+1)M+\nu N]}$. If $P_{IIO}(s)$ and

$$P_1(s) = \begin{bmatrix} T_1(s) & \vdots\ U_1(s) \\ -V_1(s) & \vdots\ W_1(s) \end{bmatrix} \tag{11.37}$$

are related by the transformation

$$\begin{bmatrix} M(s) & \vdots\ O_{\rho\times N} \\ X(s) & \vdots\ I_N \end{bmatrix} \begin{bmatrix} T_{IIOR}(s) & \vdots\ U_{IIOR}(s) \\ -V_{IIOR}(s) & \vdots\ W_{IIOR}(s) \end{bmatrix} \bullet$$
$$\bullet \begin{bmatrix} N(s) & \vdots & Y(s) \\ O_{[\alpha\rho+(\gamma+1)M+\nu N]\times\rho} & \vdots & I_{[\alpha\rho+(\gamma+1)M+\nu N]} \end{bmatrix} =$$
$$= \begin{bmatrix} T_1(s) & \vdots\ U_1(s) \\ -V_1(s) & \vdots\ W_1(s) \end{bmatrix} \tag{11.38}$$

*then $P_{IIO}(s)$ and $P_1(s)$ are **strictly system equivalent**.*

Theorem 215 *Two full system matrices which are strictly system equivalent correspond to the same system full transfer function matrix $F_{IIO}(s)$.*

The proof is the same as the proof of Theorem 201 (in Section 11.1).

Antsaklis and Michel [4, pp. 553, 554] determined the system matrix function $P_{PMD}(.)$ of the PMD system under all zero initial conditions. It results in Rosenbrock form for nonzero initial conditions from $P_{IIOR}(.)$ as a special case,

$$P_{PMDR}(s) = \begin{bmatrix} T_{PMDR}(s) & \vdots & U_{PMDR}(s) \\ -V_{PMDR}(s) & \vdots & W_{PMDR}(s) \end{bmatrix}, \tag{11.39}$$

$$T_{PMDR}(s) = Q^{(\alpha)} S_{\rho}^{(\alpha)}(s),$$

$$U_{PMDR}(s) = \begin{bmatrix} \left(P^{(\beta)} S_M^{(\beta)}(s)\right)^T \vdots \\ \vdots - \left[P^{(\beta)} Z_M^{(\beta-1)}(s) \vdots O_{\rho \times [(\gamma-\beta)M]}\right]^T \vdots \\ \vdots \left[Q^{(\alpha)} Z_{\rho}^{(\alpha-1)}(s)\right]^T \end{bmatrix}^T,$$

$$V_{PMDR}(s) = R^{(\alpha)} S_{\rho}^{(\alpha)}(s),$$

$$W_{PMDR}(s) = \begin{bmatrix} \left(T^{(\mu)} S_M^{(\mu)}(s)\right)^T \vdots \\ \vdots - \left[T^{(\mu)} Z_M^{(\mu-1)}(s) \vdots O_{N \times [(\gamma-\mu)M]}\right]^T \vdots \\ \vdots - \left[R^{(\alpha)} Z_{\rho}^{(\alpha-1)}(s)\right]^T \end{bmatrix}^T, \tag{11.40}$$

$$\mathbf{V}_{PMDR}^{\mp}(s; \mathbf{i}_{0\mp}^{\gamma-1}; \mathbf{r}_{0\mp}^{\alpha-1}) = \begin{bmatrix} \mathbf{I}^{\mp}(s) \\ \mathbf{i}_{0\mp}^{\gamma-1} \\ \mathbf{r}_{0\mp}^{\alpha-1} \end{bmatrix}. \tag{11.41}$$

Note 216 *Theorem 215 is directly applicable to* $P_{PMDR}(s)$ *(11.39).*

Chapter 12

Realizations of $F(s)$

12.1 Dynamical and least dimension of a system

In view of the essential differences between the system transfer function matrix $G(s)$ and the system full transfer function matrix $F(s)$, and in order to avoid any confusion or misinterpretation, we state the following two definitions.

Definition 217 *The* ***system characteristic polynomial*** *is the characteristic polynomial of the system full transfer function matrix* $F(s)$.

Definition 218 *The* ***system minimal polynomial*** *is the minimal polynomial of the system full transfer function matrix* $F(s)$.

Note 219 *The system minimal polynomial can be different from the minimal polynomial of the system transfer function matrix* $G(s)$.

We will define *the dynamical dimension* and *the least dimension* of the system by following H. H. Rosenbrock, [110, pp. 30, 47, 48], and by noting that Rosenbrock used the term *order* in the sense of *dimension*. However, we accepted to distinguish *the dimension of the system* from *the order of the system* that we use in the classical mathematical sense of the order of a differential equation that describes a physical dynamical system [Definition 29 and Definition 30 (in Section 2.1) for the *IO* system (2.1); Definition 36 and Definition 37 for the *ISO* system (2.17) and (2.18) (in Section 2.2); Definition 40 and Definition 41 (in Section 2.3) for the *IIO* system (2.22)]. Besides, we will define the (minimal) dynamical dimension of a system realization in the same sense as the (minimal) system dimension, respectively:

Definition 220 *(a)* ***The dynamical dimension of the system (realization)*** *denoted by* **ddim** *is, respectively, the number of initial conditions that determine uniquely the output response of the system (realization) to an arbitrary input vector function* $\mathbf{i}(.)$ *and to arbitrary initial conditions; or equivalently,* ***the dynamical dimension of the system (realization)*** *is the degree of the characteristic polynomial of the system (realization).*

(b) ***The least (the minimal) dynamical dimension of the system (realization)*** *denoted by* **mddim** *is the number of independent initial conditions that must be known (i.e., the minimal number of the initial conditions that should be known) in order to determine uniquely the output response of the system (realization) to an arbitrary input vector function* $\mathbf{i}(.)$ *and to arbitrary initial conditions; or equivalently,* ***the least (the minimal) dynamical dimension of the system (realization)*** *is the degree of the minimal polynomial of the system (realization).*

This is the general definition. It is valid for the IO systems, for the ISO systems, and for the IIO systems.

Note 221 *The dimension of the basic vector, which together with its derivatives, describes the complete system dynamics, (which is the dimension* N *of the output vector* $\mathbf{y}$ *for the* IO *systems, the dimension* n *of the state vector* $\mathbf{x}$ *for the* ISO *systems, the dimension* $\rho+N$ *of the vectors* $\mathbf{r}$ *and* $\mathbf{y}$ *for the* IIO *systems) and the system order, (which is* ν *for the* IO *systems, one for the* ISO *systems, and* $\alpha+\nu$ *for the* IIO *systems), determine jointly the dynamical dimension of the system (realization):*

$$\text{the dynamical dimension } ddim_{IO} \text{ of the } IO \text{ system obeys}$$
$$\nu \leq ddim_{IO} \leq \nu N,$$
$$\text{the dynamical dimension } ddim_{ISO} \text{ of the } ISO \text{ system is its dimension,}$$
$$ddim_{ISO} = n,$$
$$\text{the dynamical dimension } ddim_{IIO} \text{ of the } IIO \text{ system satisfies}$$
$$\rho + N \leq ddim_{IIO} \leq \alpha\rho + \nu N.$$

Example 222 *The dynamical dimension of the* IO *three-dimensional* *second-order system (Example 31, in Section 2.1)*

$$\begin{bmatrix} 3 \vdots 0 \vdots 0 \\ 0 \vdots 0 \vdots 0 \\ 0 \vdots 0 \vdots 0 \end{bmatrix} \mathbf{Y}^{(2)}(t) + \begin{bmatrix} 0 \vdots 0 \vdots 0 \\ 0 \vdots 1 \vdots 0 \\ 0 \vdots 0 \vdots 0 \end{bmatrix} \mathbf{Y}^{(1)}(t) + \begin{bmatrix} 0 \vdots 0 \vdots 0 \\ 0 \vdots 0 \vdots 0 \\ 0 \vdots 0 \vdots 1 \end{bmatrix} \mathbf{Y}(t) =$$

$$= \begin{bmatrix} 2 \vdots 0 \\ 0 \vdots 1 \\ 1 \vdots 0 \end{bmatrix} \mathbf{I}(t) + \begin{bmatrix} 1 \vdots 0 \\ 0 \vdots 1 \\ 1 \vdots 1 \end{bmatrix} \mathbf{I}^{(2)}(t)$$

is determined by the degree of the system characteristic polynomial,

$$deg \left[det \left(\sum_{k=0}^{k=\nu=2} A_k s^k \right) \right] = deg \begin{vmatrix} 3s^2 & 0 & 0 \\ 0 & s & 0 \\ 0 & 0 & 1 \end{vmatrix} = deg \left(3s^3 \right) = 3 = \eta > \nu = 2.$$

It equals 3, $ddim_{IO} = 3$*. It is bigger than the system order (2), and equal to the system dimension (*$dim_{IO} = 3$*), hence less than their product* $2x3 = 6$.

Example 223 *The second-order three dimensional IO system (Example 32, in Section 2.1)*

$$\begin{bmatrix} 3 \vdots 0 \vdots 1 \\ 2 \vdots 0 \vdots 0 \\ 0 \vdots 1 \vdots 1 \end{bmatrix} \mathbf{Y}^{(2)}(t) + \begin{bmatrix} 0 \vdots 0 \vdots 0 \\ 0 \vdots 1 \vdots 0 \\ 0 \vdots 0 \vdots 0 \end{bmatrix} \mathbf{Y}^{(1)}(t) + \begin{bmatrix} 0 \vdots 0 \vdots 0 \\ 0 \vdots 0 \vdots 0 \\ 0 \vdots 0 \vdots 1 \end{bmatrix} \mathbf{Y}(t) =$$

$$= \begin{bmatrix} 2 \vdots 0 \\ 0 \vdots 1 \\ 1 \vdots 0 \end{bmatrix} \mathbf{I}(t) + \begin{bmatrix} 1 \vdots 0 \\ 0 \vdots 1 \\ 1 \vdots 1 \end{bmatrix} \mathbf{I}^{(2)}(t)$$

induces

$$deg \left[det \left(\sum_{k=0}^{k=\nu=2} A_k s^k \right) \right] = deg \begin{vmatrix} 3s^2 & 0 & s^2 \\ 2s^2 & s & 0 \\ 0 & s^2 & s^2+1 \end{vmatrix} =$$
$$= deg \left(2s^6 + 3s^5 + 3s^3 \right) = 6 = \eta = 2x3 = \nu N.$$

Its dynamical dimension equals 6, $ddim_{IO} = 6$*. In this case, the product of the system order (2) and of the system dimension (*$dim_{IO} = 3$*) equals the degree of the system characteristic polynomial; i.e., equals the system dynamical dimension,* $\nu N = 2x3 = 6 = \eta = ddim_{IO}$.

The Definition 220 implies

Proposition 224 *A system realization is* ***the minimal system realization*** *(or equivalently:* ***the irreducible system realization****) if and only if its characteristic polynomial is its minimal polynomial.*

We will treat in more details this topic separately for the *IO* systems, the *ISO* systems, and the *IIO* systems.

12.2 On realization and minimal realization

12.2.1 Minimal realization of the transfer function matrix

The system transfer function matrix $G(s)$ is valid and useful only under nonzero input and under all zero initial conditions. Its realization is meaningful only under all zero initial conditions. C.-T. Chen noted correctly that *the dynamic equation realization gives only the same zero-state response of the system [13, p. 155]*. He proved the following theorem in the framework of the *ISO* systems (see also [4, Theorem 3.3, p. 391]):

Theorem 225 *[13, Theorem 4-10, p. 157] A transfer function matrix $G(s)$ is realizable by a finite-dimensional linear time-invariant dynamical equation if and only if $G(s)$ is a proper rational matrix (the degree of the numerator of every its entry $g_{ij}(s)$ does not exceed the degree of its denominator polynomial $\Delta(s)$, i.e., $lim[|g_{ij}(s)| : |s| \longrightarrow \infty] < \infty$, $\forall i, j$).*

In the framework of the *ISO* systems we have used in the system and control theories the following definitions and the related criteria:

Definition 226 *A matrix quadruple (A,B,C,D) is* ***a realization of a given ISO system transfer function matrix*** *$G_{ISO}(s)$ if and only if it is a realization of an ISO dynamical system, the transfer function matrix of which is $G_{ISO}(s)$, i.e., if and only if*

$$C\left(sI - A\right)^{-1} B + D = G_{ISO}(s).$$

Definition 227 *[4, Definition 3.2, p. 394], [13, p. 232]* ***The minimal (least-dimension, irreducible) ISO realization of a transfer function matrix*** *$G(s)$ is its ISO realization with the least possible dynamical dimension.*

Theorem 228 *[4, Theorem 3.9, p. 395], [13, p. 233] For an n-dimensional ISO realization of $G(s)$ to be minimal, it is necessary and sufficient that it is both controllable and observable.*

Comment 229 *The dynamical dimension of the ISO system equals to the dimension of the system state vector $\mathbf{x}$ because the order of the ISO system is minimal by the definition-the first order.*

However, the order of the IO systems and the IIO systems can be greater than one. Therefore, the minimum of the realization concerns both the dimension of the vector describing the system dynamics (that is the output vector $\mathbf{y}$ for the IO systems) and the system order in general.

The actual linear control theory and the linear system theory deal only with the *ISO* minimal realization of the transfer function matrix $G(s)$ [4, Theorem 3.3, p. 391], [13] and [80]. We will treat it also in the framework of the *IO* systems and the *IIO* systems, all for arbitrary initial conditions.

12.2.2 Realization and minimal realization of the full transfer function matrix and the system

A dynamical system is subjected to the actions of both initial conditions and the input vector in reality. The use of the full transfer function matrix $F(s)$ appears adequate to the reality rather than the application of the transfer function matrix $G(s)$.

We should distinguish the (minimal) realization of the system transfer function matrix $G(s)$ from the (minimal) realization of the system itself and of its full transfer function matrix $F(s)$.

Definition 34 (in Section 2.2) and Definition 226 justify the following general definition:

Definition 230 ***A realization of the system full transfer function matrix*** *$F(s)$ is a realization of a linear dynamical system if and only if its full transfer function matrix is the given $F(s)$.*

This is the general definition. It holds for the *IO* systems, the *ISO* systems, and the *IIO* systems.

Definition 231 *A realization of the system full transfer function matrix $F(s)$ is also a realization of the system itself, and vice versa.*

A realization of $G(s)$ need not be a realization of the corresponding $F(s)$. Theorem 225 is not valid for $F(s)$. We will investigate this for different classes of the systems in the sequel. We refer to Definition 220 (in Section 12.1) in what follows.

Definition 232 *If and only if there is a realization of a linear dynamical system such that its full transfer function matrix is the nondegenerate form $F_{nd}(s)$ of a given $F(s)$, then such realization is* ***the minimal (the least dynamical dimensional, the least order, the irreducible) realization of the system full transfer function matrix*** $F(s)$.

Definition 233 *The minimal (the least dynamical dimensional, the least–order, the irreducible) realization of the system full transfer function matrix $F(s)$ is also the minimal realization of the system itself, and vice versa.*

The preceding definitions together with Definition 220, in Section 12.1 lead to the following:

Comment 234 *The minimal system realization ensures the minimal dynamical dimension of the system.*

Theorem 228 is not applicable to $F(s)$ even in the framework of the ISO systems.

All submatrices of $F(s)$ should be compatible in the sense that every submatrix of $F(s)$ is well related to the corresponding input action vector (either to the Laplace transform of the input vector or to the corresponding initial conditions vector). This condition is trivially satisfied for $G(s)$ because it is its only one (single) submatrix related to the (single) input action (i.e., related to the Laplace transform of the input vector) since all the initial conditions are then considered equal to zero vectors.

We will study the (minimal) realization of $F(s)$ for every class of the systems considered herein.

12.3 Realizations of $F(s)$ of IO systems

We should distinguish the minimal realization of the system transfer function matrix $G(s)$, which is the well-known notion, from the minimal realization of the system full transfer function matrix $F(s)$, which we defined in general (Definition 220, in Section 12.1 and Definition 232, in Section 12.2), and which we will specify in more details for the IO system (3.76), i.e., (3.77), (in Subsection 3.5.2), in the sequel.

We refer to Definition 220 (in Section 12.1), Definition 230 and Definition 232 (in Section 12.2).

Definition 235 *If and only if there is a quadruple* $\left(\nu,\ \mu,\ A^{(\nu)},\ B^{(\mu)}\right)$ *that determines an IO realization of an IO dynamical linear system such that its full transfer function matrix is the nondegenerate form* $F_{IOnd}(s)$ *of the given* $F_{IO}(s)$ *then such IO realization is* ***the minimal (irreducible) IO realization of the system full transfer function matrix*** $F_{IO}(s)$.

Comment 236 *The minimal IO system realization ensures the minimal dynamical dimension of the IO system.*

The conditions for the realizability of $F_{IO}(s)$ are different from the realizability conditions of $G_{ISO}(s)$ (Theorem 228 in Section 12.2). Example 158 (in Section 8.1) illustrates that the full system transfer function matrix $F_{IO}(s)$ can be improper, which violates the necessary and sufficient condition for the realizability of the transfer function matrix $G(s)$ (Theorem 225, in Subsection 12.2.1).

Theorem 237 *In order for a quadruple* $\left(\nu,\ \mu,\ A^{(\nu)},\ B^{(\mu)}\right)$ *to determine the (minimal) input-output (IO) system realization (3.76), it is necessary and sufficient that*

$$\left(A^{(\nu)}S_N^{(\nu)}(s)\right)^{-1}\left\{\begin{array}{c}\left[B^{(\mu)}S_M^{(\mu)}(s)\vdots-B^{(\mu)}Z_M^{(\mu-1)}(s)\vdots\ A^{(\nu)}Z_N^{(\nu-1)}(s)\right]\\ iff\ \mu\geq1\\ \left[B^{(\mu)}S_M^{(\mu)}(s)\ \vdots\ A^{(\nu)}Z_N^{(\nu-1)}(s)\right],\ iff\ \mu=0\end{array}\right\}$$

is equal to (the nondegenerate form of) the given full system transfer function matrix $F_{IO}(s)$*, respectively,*

$$\left(A^{(\nu)}S_N^{(\nu)}(s)\right)^{-1}\left\{\begin{array}{c}\left[B^{(\mu)}S_M^{(\mu)}(s)\vdots-B^{(\mu)}Z_M^{(\mu-1)}(s)\vdots A^{(\nu)}Z_N^{(\nu-1)}(s)\right]\\ iff\ \mu\geq1\\ \left[B^{(\mu)}S_M^{(\mu)}(s)\ \vdots\ A^{(\nu)}Z_N^{(\nu-1)}(s)\right],\ iff\ \mu=0\end{array}\right\}=$$
$$=F_{IO}(s),$$

$$\left(\begin{array}{c}\left(A^{(\nu)}S_N^{(\nu)}(s)\right)^{-1}\left\{\begin{array}{c}\left[B^{(\mu)}S_M^{(\mu)}(s)\vdots-B^{(\mu)}Z_M^{(\mu-1)}(s)\vdots A^{(\nu)}Z_N^{(\nu-1)}(s)\right]\\ iff\ \mu\geq1\\ \left[B^{(\mu)}S_M^{(\mu)}(s)\ \vdots\ A^{(\nu)}Z_N^{(\nu-1)}(s)\right],\ iff\ \mu=0\end{array}\right\}\\ =F_{IOnd}(s)\end{array}\right).$$

Proof. *Necessity.* Let a quadruple $\left(\nu,\ \mu,\ A^{(\nu)},\ B^{(\mu)}\right)$ determine the (minimal) input-output system realization (3.76). Then, the following holds due to Definition 230 (in Section 12.2), (Definition 113, Chapter 6, Definition 217 through Definition 220, Proposition 224, in Section 12.1, Definition 232 in Section 12.2, and Definition 235), respectively, and (8.6) and (8.7) (Theorem 148, in Section 8.1):

$$\left(A^{(\nu)}S_N^{(\nu)}(s)\right)^{-1}\left\{\begin{array}{c}\left[B^{(\mu)}S_M^{(\mu)}(s)\vdots -B^{(\mu)}Z_M^{(\mu-1)}(s)\vdots A^{(\nu)}Z_N^{(\nu-1)}(s)\right],\\ iff\ \mu\geq 1\\ \left[B^{(\mu)}S_M^{(\mu)}(s)\ \vdots\ A^{(\nu)}Z_N^{(\nu-1)}(s)\right],\ iff\ \mu=0\end{array}\right\}=$$

$$=\left[\det\left(A^{(\nu)}S_N^{(\nu)}(s)\right)\right]^{-1}\bullet\left\{adj\left(A^{(\nu)}S_N^{(\nu)}(s)\right)\right\}\bullet$$

$$\bullet\left\{\begin{array}{c}\left[B^{(\mu)}S_M^{(\mu)}(s)\ \vdots\ -B^{(\mu)}Z_M^{(\mu-1)}(s)\ \vdots\ A^{(\nu)}Z_N^{(\nu-1)}(s)\right],\ \mu\geq 1\\ \left[B^{(\mu)}S_M^{(\mu)}(s)\ \vdots\ A^{(\nu)}Z_N^{(\nu-1)}(s)\right],\ \mu=0\end{array}\right\}=$$

$$=F_{IO}(s)\ \ (=F_{IOnd}(s)),$$

which proves necessity of the condition.

Sufficiency. Let the conditions hold. The quadruple $\left(\nu,\ \mu,\ A^{(\nu)},\ B^{(\mu)}\right)$ determines (the minimal) input-output realization of $F_{IO}(s)$, which is also (the minimal) *IO* realization of the *IO* system (3.76) due to Definition 230 (in Section 12.2), Definition 232 (in Section 12.2), and Definition 235, respectively. ■

Note 238 *The (full) IO transfer function matrix of the ISO system is, respectively, the (full) transfer function matrix obtained from the IO system induced by the given ISO system (Note 136, in Section 7.3). It carries the subscript IO. However, the subscript ISO denotes the (full) transfer function matrix obtained from the ISO system.*

Example 239 *We will consider Antsaklis-Michel ISO systems of Example 167 (in Section 8.2, [4, Example 3.8]) in the framework of the IO systems. The given first three input-state-output (ISO) realizations (A, B, C, D) yield the degenerate transfer functions*

$$G_{ISO1}(s)=G_{ISO2}(s)=G_{ISO3}(s)=G_{ISO1-3}(s)=\frac{s-1}{s^2-1}$$

that are reducible to their nondegenerate forms

$$G_{ISO1nd}(s) = G_{ISO2nd}(s) = G_{ISO3nd}(s) = G_{ISO1-3nd}(s) = \frac{1}{s+1}.$$

The degenerate transfer functions imply the following IO system:

$$y^{(2)}(t) - y(t) = -i(t) + i^{(1)}(t), \tag{12.1}$$

and its IO realization (Definition 27, in Subsection 2.1)

$$\left(\nu, \mu, A^{(\nu)}, B^{(\mu)}\right) = \left(2 \vdots 1 \vdots \begin{bmatrix} -1 \vdots 0 \vdots 1 \end{bmatrix} \vdots \begin{bmatrix} -1 \vdots 1 \end{bmatrix}\right). \tag{12.2}$$

We can apply (8.6) (in Section 8.1) to $F_{IOISO1-3}(s)$ by using the IO realization (12.2) of $G_{ISO1-3}(s)$ in order to determine the full IO transfer function matrix

$$F_{IOISO1-3}(s) =$$

$$= \left(A^{(\nu)} S_N^{(\nu)}(s)\right)^{-1} \left[B^{(\mu)} S_M^{(\mu)}(s) \vdots - B^{(\mu)} Z_M^{(\mu-1)}(s) \vdots A^{(\nu)} Z_N^{(\nu-1)}(s)\right] =$$

$$= \left(A^{(2)} S_1^{(2)}(s)\right)^{-1} \left[B^{(1)} S_1^{(1)}(s) \vdots - B^{(1)} Z_1^{(1-1)}(s) \vdots A^{(2)} Z_1^{(2-1)}(s)\right],$$

$$A^{(2)} S_1^{(2)}(s) = \begin{bmatrix} -1 \vdots 0 \vdots 1 \end{bmatrix} \begin{bmatrix} 1 \\ s \\ s^2 \end{bmatrix} = \left[s^2 - 1\right],$$

$$B^{(1)} S_1^{(1)}(s) = \begin{bmatrix} -1 \vdots 1 \end{bmatrix} \begin{bmatrix} 1 \\ s \end{bmatrix} = [s - 1],$$

$$B^{(1)} Z_1^{(1-1)}(s) = \begin{bmatrix} -1 \vdots 1 \end{bmatrix} \begin{bmatrix} 0 \\ 1 \end{bmatrix} = [1],$$

$$A^{(2)} Z_1^{(2-1)}(s) = \begin{bmatrix} -1 \vdots 0 \vdots 1 \end{bmatrix} \begin{bmatrix} 0 & 0 \\ 1 & 0 \\ s & 1 \end{bmatrix} = \begin{bmatrix} s \vdots 1 \end{bmatrix} \Longrightarrow$$

$$F_{IOISO1-3}(s) = \frac{\begin{bmatrix} s-1 \vdots -1 \vdots s \vdots 1 \end{bmatrix}}{s^2 - 1}. \tag{12.3}$$

The so obtained full IO transfer function matrix $F_{IOISO1-3}(s)$ (12.3), which resulted from the realization (12.2) of $G_{ISO1-3}(s)$, is nondegenerate,

$$F_{IOISO1-3}(s) = \frac{\begin{bmatrix} s-1 \vdots -1 \vdots s \vdots 1 \end{bmatrix}}{s^2-1} = F_{IOISO1-3nd}(s). \tag{12.4}$$

The IO system (12.1) induces directly $F_{IO1-3}(s)$ as follows:

$$(s^2-1)Y(s) = (-1+s)I(s) - i_0 + sy_0 + y_0^{(1)} \Longrightarrow$$

$$Y(s) = (s^2-1)^{-1}\begin{bmatrix} s-1 \vdots -1 \vdots s \vdots 1 \end{bmatrix}\begin{bmatrix} I(s) \\ i_0 \\ y_0 \\ y_0^{(1)} \end{bmatrix} \Longrightarrow$$

$$F_{IO1-3}(s) = (s^2-1)^{-1}\begin{bmatrix} s-1 \vdots -1 \vdots s \vdots 1 \end{bmatrix} = F_{IOISO1-3}(s). \tag{12.5}$$

This verifies (12.3) and confirms the validity of (12.4) for $F_{IO1-3}(s)$,

$$F_{IO1-3}(s) = (s^2-1)^{-1}\begin{bmatrix} s-1 \vdots -1 \vdots s \vdots 1 \end{bmatrix} = F_{IO1-3nd}(s).$$

However, the nondegenerate form $G_{ISO1-3nd}(s)$ of $G_{ISO1-3}(s)$ yields its minimal IO system in the form of the reduced differential equation

$$y^{(1)}(t) + y(t) = i(t), \tag{12.6}$$

and its IO realization, which we will call the reduced IO realization (the subscript "rd"), reads

$$(\nu_{1-3rd}, \mu_{1-3rd}, A_{1-3rd}^{(\nu_{1-3rd})}, B_{1-3rd}^{(\mu_{\min})}) = \left(1 \vdots 0 \vdots \begin{bmatrix} 1 \vdots 1 \end{bmatrix} \vdots [1]\right). \tag{12.7}$$

The reduced IO system (12.6), i.e., its IO realization (12.7), determines the corresponding $F_{IOi-iiird}(s)$,

$$Y(s) = (s+1)^{-1}\begin{bmatrix} 1 \vdots 1 \end{bmatrix}\begin{bmatrix} I(s) \\ y_0 \end{bmatrix} \Longrightarrow$$

$$F_{IO1-3rd}(s) = (s+1)^{-1}\begin{bmatrix} 1 \vdots 1 \end{bmatrix}, \tag{12.8}$$

which is nondegenerate, hence the irreducible, and different in this case from the nondegenerate form $F_{IO1-3nd}(s)$ *of* $F_{IO1-3}(s)$,

$$F_{IO1-3rd}(s) = (s+1)^{-1}\left[1 \vdots 1\right] \neq$$

$$\neq (s^2-1)^{-1}\left[s-1 \vdots -1 \vdots s \vdots 1\right] = F_{IO1-3nd}(s) = F_{IO1-3}(s).$$

$F_{IO1-3rd}(s)$ *is related only to* y_0, *but* $F_{IO1-3nd}(s)$ *corresponds to* i_0, y_0, *and* $y_0^{(1)}$. *This is the consequence of different orders of the IO models (12.6) and (12.1) to which they are related, respectively.*

However, the fourth ISO realization (A, B, C, D) *implies directly the nondegenerate transfer function*

$$G_{ISO4}(s) = G_{ISO4nd}(s) = \frac{1}{s+1} = G_{ISOnd}(s) = G_{ISO1-3nd}(s).$$

It induces the IO system (12.6) and the full IO transfer function matrix,

$$F_{IOISO4}(s) = (s+1)^{-1}\left[1 \vdots 1\right] = F_{IO4nd}(s) = F_{IO1-3rd}(s). \tag{12.9}$$

Example 240 *Let*

$$\mathbf{i} \in \mathfrak{R}^2 \Longrightarrow M=2,\ \mathbf{y} \in \mathfrak{R}^3 \Longrightarrow N=3,$$

$$H(s) = \begin{bmatrix} \frac{s^2+1}{3s^2+3s^4+2s^5} & \frac{s}{3s^2+2s^3+3} & -\frac{1}{3s^2+2s^3+3} \\ \frac{-2s^2-2}{3s+3s^3+2s^4} & \frac{3s^2+3}{3s+3s^3+2s^4} & 2\frac{s}{3s^2+2s^3+3} \\ 2\frac{s}{3s^2+2s^3+3} & -3\frac{s}{3s^2+2s^3+3} & \frac{3}{3s^2+2s^3+3} \end{bmatrix} \bullet$$

$$\bullet \begin{bmatrix} 2+s^2 \vdots & 0 & -s \vdots & 0 & -1 \vdots & 0 & 3s \vdots 0 \vdots 3+s & s \vdots 3 \vdots 1 \\ 0 & \vdots 1+s^2 & 0 \vdots & -s & 0 \vdots & -1 & 2s \vdots 1 \vdots \ 0 & 2 \vdots 0 \vdots 0 \\ 1+s^2 \vdots & s^2 & -s \vdots & -s & -1 \vdots & \text{-1} & 0 \vdots s \vdots \ s & 0 \vdots 1 \vdots 1 \end{bmatrix}.$$

We wish to test whether $H(s)$ *can be the full transfer function matrix* $F_{IO}(s)$ *of an IO system. We set it at first in the form*

$$H(s) = \frac{\begin{bmatrix} s^2+1 & s^3 & -s^2 \\ -2s^3-2s & 3s^3+3s & 2s^3 \\ 2s^3 & -3s^3 & 3s^2 \end{bmatrix}}{2s^5+3s^4+3s^2} \bullet$$

$$\bullet \begin{bmatrix} 2+s^2 & 0 & -s & 0 & -1 & 0 & 3s & 0 & 3+s & s & 3 & 1 \\ 0 & 1+s^2 & 0 & -s & 0 & -1 & 2s & 1 & 0 & 2 & 0 & 0 \\ 1+s^2 & s^2 & -s & -s & -1 & -1 & 0 & s & s & 0 & 1 & 1 \end{bmatrix},$$

and now in the form

$$H(s) = \begin{bmatrix} 3s^2 & 0 & s^2 \\ 2s^2 & s & 0 \\ 0 & s^2 & s^2+1 \end{bmatrix}^{-1} \bullet$$

$$\bullet \begin{bmatrix} 2+s^2 & 0 & -s & 0 & -1 & 0 & 3s & 0 & 3+s & s & 3 & 1 \\ 0 & 1+s^2 & 0 & -s & 0 & -1 & 2s & 1 & 0 & 2 & 0 & 0 \\ 1+s^2 & s^2 & -s & -s & -1 & -1 & 0 & s & s & 0 & 1 & 1 \end{bmatrix},$$

from which we deduce:

$$\sum_{k=0}^{k=\nu} A_k s^k = A^{(\nu)} S_N^{(\nu)}(s) = A^{(\nu)} S_3^{(\nu)}(s) = \begin{bmatrix} 3s^2 & 0 & s^2 \\ 2s^2 & s & 0 \\ 0 & s^2 & s^2+1 \end{bmatrix} \Longrightarrow$$

$$\nu = 2,\ A_0 = \begin{bmatrix} 0 \vdots 0 \vdots 0 \\ 0 \vdots 0 \vdots 0 \\ 0 \vdots 0 \vdots 1 \end{bmatrix},\ A_1 = \begin{bmatrix} 0 \vdots 0 \vdots 0 \\ 0 \vdots 1 \vdots 0 \\ 0 \vdots 0 \vdots 0 \end{bmatrix},\ A_2 = \begin{bmatrix} 3 \vdots 0 \vdots 1 \\ 2 \vdots 0 \vdots 0 \\ 0 \vdots 1 \vdots 1 \end{bmatrix}$$

$$\sum_{k=0}^{k=\mu} B_k s^k = B^{(\mu)} S_M^{(\mu)}(s) = B^{(\mu)} S_2^{(\mu)}(s) = \begin{bmatrix} 2+s^2 & 0 \\ 0 & 1+s^2 \\ 1+s^2 & s^2 \end{bmatrix} \Longrightarrow$$

$$\mu = 2,\ B_0 = \begin{bmatrix} 2 \vdots 0 \\ 0 \vdots 1 \\ 1 \vdots 0 \end{bmatrix},\ B_1 = \begin{bmatrix} 0 \vdots 0 \\ 0 \vdots 0 \\ 0 \vdots 0 \end{bmatrix},\ B_2 = \begin{bmatrix} 1 \vdots 0 \\ 0 \vdots 1 \\ 1 \vdots 1 \end{bmatrix}.$$

These data yield

$$A^{(2)} S_3^{(2)}(s) = \begin{bmatrix} 3s^2 & 0 & s^2 \\ 2s^2 & s & 0 \\ 0 & s^2 & s^2+1 \end{bmatrix},\ B^{(2)} S_2^{(2)}(s) = \begin{bmatrix} 2+s^2 & 0 \\ 0 & 1+s^2 \\ 1+s^2 & s^2 \end{bmatrix},$$

$$
A^{(2)}Z_3^{(2-1)}(s) = \begin{bmatrix} 0 \vdots 0 \vdots 0 & 0 \vdots 0 \vdots 0 & 3 \vdots 0 \vdots 1 \\ 0 \vdots 0 \vdots 0 & 0 \vdots 1 \vdots 0 & 2 \vdots 0 \vdots 0 \\ 0 \vdots 0 \vdots 1 & 0 \vdots 0 \vdots 0 & 0 \vdots 1 \vdots 1 \end{bmatrix} \begin{bmatrix} 0 \vdots 0 \vdots 0 \vdots 0 \vdots 0 \vdots 0 \\ 0 \vdots 0 \vdots 0 \vdots 0 \vdots 0 \vdots 0 \\ 0 \vdots 0 \vdots 0 \vdots 0 \vdots 0 \vdots 0 \\ 1 \vdots 0 \vdots 0 \vdots 0 \vdots 0 \vdots 0 \\ 0 \vdots 1 \vdots 0 \vdots 0 \vdots 0 \vdots 0 \\ 0 \vdots 0 \vdots 1 \vdots 0 \vdots 0 \vdots 0 \\ s \vdots 0 \vdots 0 \vdots 1 \vdots 0 \vdots 0 \\ 0 \vdots s \vdots 0 \vdots 0 \vdots 1 \vdots 0 \\ 0 \vdots 0 \vdots s \vdots 0 \vdots 0 \vdots 1 \end{bmatrix} =
$$

$$
= \begin{bmatrix} 3s \vdots 0 \vdots s & 3 \vdots 0 \vdots 1 \\ 2s \vdots 1 \vdots 0 & 2 \vdots 0 \vdots 0 \\ 0 \vdots s \vdots s & 0 \vdots 1 \vdots 1 \end{bmatrix},
$$

$$
-B^{(2)}Z_2^{(2-1)}(s) = -\begin{bmatrix} 2 \vdots 0 & 0 \vdots 0 & 1 \vdots 0 \\ 0 \vdots 1 & 0 \vdots 0 & 0 \vdots 1 \\ 1 \vdots 0 & 0 \vdots 0 & 1 \vdots 1 \end{bmatrix} \begin{bmatrix} 0 \vdots 0 \vdots 0 \vdots 0 \\ 0 \vdots 0 \vdots 0 \vdots 0 \\ 1 \vdots 0 \vdots 0 \vdots 0 \\ 0 \vdots 1 \vdots 0 \vdots 0 \\ s \vdots 0 \vdots 1 \vdots 0 \\ 0 \vdots s \vdots 0 \vdots 1 \end{bmatrix} =
$$

$$
= -\begin{bmatrix} s & 0 & 1 & 0 \\ 0 & s & 0 & 1 \\ s & s & 1 & 1 \end{bmatrix}.
$$

The obtained data determine the full transfer function matrix $F_{IO}(s)$ of an IO system

$$
F_{IO}(s) = \left[A^{(2)}S_3^{(2)}(s)\right]^{-1}\left[B^{(2)}S_2^{(2)}(s) \vdots -B^{(2)}Z_2^{(2-1)}(s) \vdots A^{(2)}Z_3^{(2-1)}(s)\right] =
$$

$$= \begin{bmatrix} 3s^2 & 0 & s^2 \\ 2s^2 & s & 0 \\ 0 & s^2 & s^2+1 \end{bmatrix}^{-1} \bullet$$

$$\bullet \begin{bmatrix} 2+s^2 & 0 & -s & 0 & -1 & 0 & 3s & 0 & s & 3 & 0 & 1 \\ 0 & 1+s^2 & 0 & -s & 0 & -1 & 2s & 1 & 0 & 2 & 0 & 0 \\ 1+s^2 & s^2 & -s & -s & -1 & -1 & 0 & s & s & 0 & 1 & 1 \end{bmatrix} \neq$$

$$\neq \begin{bmatrix} 3s^2 & 0 & s^2 \\ 2s^2 & s & 0 \\ 0 & s^2 & s^2+1 \end{bmatrix}^{-1} \bullet$$

$$\bullet \begin{bmatrix} 2+s^2 & 0 & -s & 0 & -1 & 0 & 3s & 0 & 3+s & s & 3 & 1 \\ 0 & 1+s^2 & 0 & -s & 0 & -1 & 2s & 1 & 0 & 2 & 0 & 0 \\ 1+s^2 & s^2 & -s & -s & -1 & -1 & 0 & s & s & 0 & 1 & 1 \end{bmatrix} = H(s).$$

The given $H(s)$ cannot be the full transfer function matrix $F_{IO}(s)$ of any IO system because it does not obey Theorem 237. Notice that

$$\left[A^{(2)}S_3^{(2)}(s)\right]^{-1} B^{(2)}S_2^{(2)}(s) =$$

$$\begin{bmatrix} 3s^2 & \vdots & 0 & \vdots & s^2 \\ 2s^2 & \vdots & s & \vdots & 0 \\ 0 & \vdots & s^2 & \vdots & s^2+1 \end{bmatrix}^{-1} \begin{bmatrix} 2+s^2 & \vdots & 0 \\ 0 & \vdots & 1+s^2 \\ 1+s^2 & \vdots & s^2 \end{bmatrix} =$$

$$= \frac{\begin{bmatrix} s^2+1 & \vdots & s^3 & \vdots & -s^2 \\ -2s^3-2s & \vdots & 3s^3+3s & \vdots & 2s^3 \\ 2s^3 & \vdots & -3s^3 & \vdots & 3s^2 \end{bmatrix} \begin{bmatrix} 2+s^2 & \vdots & 0 \\ 0 & \vdots & 1+s^2 \\ 1+s^2 & \vdots & s^2 \end{bmatrix}}{2s^5+3s^4+3s^2}$$

is proper so that it can be the transfer function matrix $G_{IO}(s)$ of an IO system in spite $H(s)$ cannot be the full transfer function matrix $F_{IO}(s)$ of any IO system. This warns us that the IO realizability of $G_{IO}(s)$ does not guarantee the IO realizability of $H(s)$.

Conclusion 241 *If an $H(s)$ matrix is given, and a realizable transfer function matrix $G_{IO}(s)$ of an IO system is its first submatrix, then it does not guarantee that $H(s)$ can be the full transfer function matrix $F_{IO}(s)$ of any IO system.*

However, if an $H(s)$ matrix can be the full transfer function matrix $F_{IO}(s)$ of an IO system, then its first submatrix is the transfer function matrix $G_{IO}(s)$ of the same IO system.

Conclusion 242 *The full transfer function matrix $F_{IO}(s)$ of an IO system can be improper, while its transfer function matrix $G_{IO}(s)$ cannot (Example 158, in Section 8.1).*

Conclusion 243 *The preceding analysis shows the incompleteness of the IO system studies via its transfer function matrix $G_{IO}(s)$. For the complete analysis or synthesis of the IO system in the complex domain, we should use its full transfer function matrix $F_{IO}(s)$.*

12.4 Realizations of $F(s)$ of ISO systems

We should distinguish the minimal realization of the system transfer function matrix $G(s)$, which is a well-defined notion, from the minimal realization of the system, which we defined in general (Definition 220, in Section 12.1), and which we will specify in more details for the ISO system (3.84) and (3.85) (in Subsection 3.5.3) in the sequel.

We refer to Definition 220 (in Section 12.1), Definition 230 and Definition 232 (in Section 12.2).

Definition 244 *If and only if there is a quintuple $(n,\ A, B, C, D)$ that determines an ISO realization of an ISO dynamical linear system such that its full transfer function matrix is the nondegenerate form $F_{ISOnd}(s)$ of the given $F_{ISO}(s)$, then such ISO realization is* ***the minimal (irreducible) ISO realization of the full transfer function matrix*** *$F_{ISO}(s)$.*

Note 245 *The dynamical dimension of the ISO system and its realization reduce to their dimensions since their orders are fixed to one by the definition.*

The conditions for the realizability of $F_{ISO}(s)$ are different from the realizability conditions of $F_{IO}(s)$ (Theorem 237 in in Section 12.3).

Theorem 246 *In order for a quintuple $(n,\ A, B, C, D)$ to determine an ISO system realization (3.84) and (3.85) it is necessary and sufficient that*

$$\left[C(sI_n - A)^{-1}B + D \vdots C(sI_n - A)^{-1}\right] \tag{12.10}$$

is proper and that it is the given system full ISO transfer function matrix $F_{ISO}(s)$,

$$\left[C(sI_n - A)^{-1}B + D \vdots C(sI_n - A)^{-1}\right] = F_{ISO}(s). \tag{12.11}$$

Proof. *Necessity.* Let a quintuple $(n,\ A, B, C, D)$ determine an ISO system realization (3.84) and (3.85). Then it is an ISO realization of the system full ISO transfer function matrix $F_{ISO}(s)$, i.e., let (12.11) hold, in view of Definition 230 and Comment 233 (in Section 12.2). The first submatrix in (12.10) is the system transfer function $G_{ISO}(s)$. It is proper (Theorem 225, in Section 12.2). The first submatrix is proper. Let us set the matrix (12.10) into the following form:

$$\left[\frac{Cadj(sI_n - A)B + \det(sI_n - A)D \vdots Cadj(sI_n - A)}{\det(sI_n - A)}\right].$$

Since

$$deg\left[adj(sI_n - A)\right] < \deg\left[\det(sI_n - A)\right]$$

then

$$deg\left[Cadj(sI_n - A)B\right] < \deg\left[\det(sI_n - A)\right],$$
$$deg\left[Cadj(sI_n - A)\right] < \deg\left[\det(sI_n - A)\right]$$

and

$$deg\left[\det(sI_n - A)D\right] = \deg\left[\det(sI_n - A)\right].$$

Hence, the matrix (12.10),

$$\left[C(sI_n - A)^{-1}B + D \vdots C(sI_n - A)^{-1}\right],$$

is proper, which verifies that $C(sI_n - A)^{-1}B + D$ obeys Theorem 225 (in Section 12.2).

Sufficiency. Let the conditions hold. Hence, $(n,\ A, B, C, D)$ is the ISO system realization (3.84) and (3.85) in view of Definition 34 (in Section 2.2), Definition 230, Comment 233 (in Section 12.2) and Definition 244. ■

This theorem does not guarantee that the quintuple $(n,\ A, B, C, D)$, which determines the matrix (12.10), is the minimal realization of the ISO system (3.84) and (3.85).

Theorem 247 *In order for a quintuple $(n,\ A,B,C,D)$ to determine a minimal ISO system realization (3.84) and (3.85) it is necessary and sufficient that*

$$\left[C(sI_n-A)^{-1}B+D \vdots C(sI_n-A)^{-1}\right] \tag{12.12}$$

is proper and that it is the nondegenerate form $F_{ISOnd}(s)$ of the given system full ISO transfer function matrix $F_{ISO}(s)$,

$$\left[C(sI_n-A)^{-1}B+D \vdots C(sI_n-A)^{-1}\right] = F_{ISOnd}(s). \tag{12.13}$$

Proof. *Necessity.* Let a quintuple $(n,\ A,B,C,D)$ determine a minimal ISO system realization (3.84) and (3.85). Hence, the matrix 12.12 is proper due to Theorem 246 and it is the nondegenerate form $F_{ISOnd}(s)$ of the system full ISO transfer function matrix $F_{ISO}(s)$ due to Definition 232 (in Section 12.2) and Definition 244.

Sufficiency. Let the conditions hold. The conditions of Theorem 246 are satisfied. The matrix 12.12 is minimal ISO system realization (3.84) and (3.85) due to Definition 232 and Definition 244. ■

Example 248 *We analyze further Antsaklis-Michel systems presented in Example 167 (in Section 8.2), [4, Example 3.8], and Example 239 (in Section 12.3).*

Are the minimal ISO realizations of the transfer functions $G_{ISO1-3}(s)$ and of $G_{ISO4}(s)$ also the ISO minimal realizations of the full transfer functions $F_{ISO1-4}(s)$?

Are the full ISO transfer functions $F_{ISO1}(s)$ through $F_{ISO4}(s)$ also the full IO transfer functions $F_{ISOIO1}(s)$ through $F_{ISOIO4}(s)$?

Are the nondegenerate forms $F_{ISO1nd}(s)$ through $F_{ISO4nd}(s)$ of $F_{ISO1}(s)$ through $F_{ISO4}(s)$ realizable?

Let us consider each case separately.

1) $\quad A_1=\begin{bmatrix}0 & 1\\ 1 & 0\end{bmatrix},\ B_1=\begin{bmatrix}0\\ 1\end{bmatrix},\ C_1=\left[-1 \vdots 1\right],\ D_1=0 \Longrightarrow$

$$F_{ISO1}(s)=\left(s^2-1\right)^{-1}\left[s-1 \vdots 1-s \vdots s-1\right],\ G_{ISO1}(s)=\frac{s-1}{s^2-1}. \tag{12.14}$$

They follow from the ISO system realization. They are degenerate. Their nondegenerate forms read

$$F_{ISO1nd}(s)=(s+1)^{-1}\left[1 \vdots -1 \vdots 1\right],\ G_{ISO1}(s)=\frac{1}{s+1}. \tag{12.15}$$

The transfer function $G_{ISO1}(s)$ induces, via (12.1), Example 239 (in Section 12.3),

$$\nu_1 = 2,\ \mu_1 = 1,\ N_1 = 1,\ M_1 = 1,\ A_1^{(2)} = \left[-1 \vdots 0 \vdots 1\right],\ B_1^{(1)} = \left[-1 \vdots 1\right]. \quad (12.16)$$

The degenerate $F_{ISO1}(s)$ (12.14) should obey the following IO condition due to (8.6) (in Section 8.1) and due to Theorem 237 in order to be also $F_{ISOIO1}(s)$ that follows from the IO realization (12.16) of the given ISO system, Example 239,

$$F_{ISO1}(s) = \left(s^2-1\right)^{-1}\left[s-1 \vdots 1-s \vdots s-1\right] =$$

$$= \left(A_i^{(\nu)} S_N^{(\nu)}(s)\right)^{-1}\left[B_i^{(\mu)} S_M^{(\mu)}(s) \vdots -B_i^{(\mu)} Z_M^{(\mu-1)}(s) \vdots A_i^{(\nu)} Z_N^{(\nu-1)}(s)\right] =$$

$$= F_{ISOIO1}(s),$$

where

$$A_1^{(2)} S_1^{(2)}(s) = \left[-1 \vdots 0 \vdots 1\right]\begin{bmatrix} 1 \\ s \\ s^2 \end{bmatrix} = \left[-1+s^2\right] = \left[s^2-1\right],$$

$$B_1^{(1)} S_1^{(1)}(s) = \left[-1 \vdots 1\right]\begin{bmatrix} 1 \\ s \end{bmatrix} = \left[-1+s\right] = \left[s-1\right],$$

$$-B_1^{(1)} Z_1^{(1-1)}(s) = -\left[-1 \vdots 1\right]\begin{bmatrix} 0 \\ 1 \end{bmatrix} = \left[-1\right] \neq \left[1-s\right],$$

$$A_1^{(2)} Z_1^{(2-1)}(s) = \left[-1 \vdots 0 \vdots 1\right]\begin{bmatrix} 0 \vdots 0 \\ 1 \vdots 0 \\ s \vdots 1 \end{bmatrix} = \left[s \vdots 1\right] \neq \left[s-1\right]$$

$$\Longrightarrow$$

$$F_{ISOIO1}(s) = \left(s^2-1\right)^{-1}\left[s-1 \vdots -1 \vdots s \vdots 1\right].$$

Since $-B_1^{(1)} Z_1^{(1-1)}(s) = [-1]$ is different from the second term $[1-s]$ of the given $F_{ISO1}(s)$, and $A_1^{(2)} Z_1^{(2-1)}(s) = \left[s \vdots 1\right]$ is different from the third term $[s-1]$ of the given $F_{ISO1}(s)$, it follows that the degenerate $F_{ISO1}(s)$ is IO *unrealizable, i.e., there is not an IO realization of* $F_{ISO1}(s)$. *There is not an IO linear differential equation, which implies the given degenerate form of* $F_{ISO1}(s)$,

$$F_{ISO1}(s) = \left(s^2-1\right)^{-1}\left[s-1 \vdots 1-s \vdots s-1\right] \neq$$
$$\neq \left(s^2-1\right)^{-1}\left[s-1 \vdots -1 \vdots s \vdots 1\right] = F_{ISOIO1}(s) = F_{IOISO1nd}(s).$$

We should remind ourselves that $F_{ISO1}(s)$ *was obtained from the ISO system instead of the corresponding IO system.*

We can deduce the same result by comparing $F_{ISO1}(s)$ *with* $F_{IOISO1-3}(s)$ *(12.3), Example 239 (in Section 12.3), which is therein obtained directly from (12.1),*

$$F_{ISO1}(s) = \left(s^2-1\right)^{-1}\left[s-1 \vdots 1-s \vdots s-1\right] \neq$$
$$\neq \left(s^2-1\right)^{-1}\left[s-1 \vdots -1 \vdots s \vdots 1\right] = F_{ISOIO1-3}(s) = F_{ISOIO1-3nd}(s).$$

This verifies IO nonrealizability of $F_{ISO1}(s)$.

The nondegenerate, hence the irreducible, form $G_{ISO1nd}(s)$ *of* $G_{ISO1}(s)$

$$G_{ISO1nd}(s) = \frac{1}{s+1}$$

implies

$$\nu_{1\min} = 1,\ \mu_{1\min} = 0,\ N_{1\min} = 1,\ M_{1\min} = 1,$$
$$A^{(1)}_{1\min} = \left[1 \vdots 1\right],\ B^{(0)}_{1\min} = [1\ ].$$

These data and (8.6) (in Section 8.1) yield

$$A^{(1)}_{1\min}S^{(1)}_1(s) = \left[1 \vdots 1\right]\begin{bmatrix} 1 \\ s \end{bmatrix} = [1+s] = [s+1],$$
$$B^{(0)}_{1\min}S^{(0)}_1(s) = [1]\,[1] = [1] = [1],$$
$$-B^{(0)}_{1\min}Z^{(0-1)}_1(s) \text{ does not exist},$$
$$A^{(1)}_{1\min}Z^{(1-1)}_1(s) = \left[1 \vdots 1\right]\begin{bmatrix} 0 \\ 1 \end{bmatrix} = [1] = \ [1],$$

and,

$$F_{ISOIO1rd}(s) = \left[A^{(1)}_{1\min}S^{(1)}_1(s)\right]^{-1}\left[B^{(0)}_{1\min}S^{(0)}_1(s) \vdots A^{(1)}_{1\min}Z^{(1-1)}_1(s)\right] =$$
$$= (s+1)^{-1}\left[1 \vdots 1\right].$$

Since $-B^{(0)}_{1\min}Z^{(0-1)}_1(s)$ *does not exist, then it is different from the second submatrix* $[-1]$ *of the nondegenerate form* $F_{ISO1nd}(s)$ *of* $F_{ISO1}(s)$. *We conclude that the nondegenerate form* $F_{ISO1nd}(s)$ *(12.15) of* $F_{ISO1}(s)$ *is not IO realizable,*

$$F_{ISO1nd}(s) = (s+1)^{-1}\left[1 \vdots -1 \vdots 1\right] \neq (s+1)^{-1}\left[1 \vdots 1\right] = F_{ISOIO1rd}(s),$$

in spite the nondegenerate form $G_{ISO1nd}(s)$ *of* $G_{ISO1}(s)$ *is IO realizable,*

$$G_{ISO1nd}(s) = \frac{1}{s+1} = G_{IO1nd}(s).$$

However, $G_{ISO1nd}(s)$ *yields the same quadruple* $(\nu_{1\min}, \mu_{1\min}, A^{(1)}_{1\min}, B^{(0)}_{1\min})$ *that determines*

$$F_{ISOIO1rd}(s) = (s+1)^{-1}\left[1 \vdots 1\right].$$

It is equal to the reduced full IO transfer function matrix $F_{IO1-3d}(s)$ *(12.9), Example 239 (in Section 12.3), which is obtained directly from the IO system (12.6), Example 239 (in Section 12.3),*

$$F_{ISOIO1rd}(s) = (s+1)^{-1}\left[1 \vdots 1\right] = F_{IO1-3rd}(s),$$

but it is different from the nondegenerate full ISO transfer function matrix $F_{ISO1nd}(s)$,

$$F_{ISO1nd}(s) = (s+1)^{-1}\left[1 \vdots -1 \vdots 1\right] \neq (s+1)^{-1}\left[1 \vdots 1\right] = F_{ISOIO1rd}(s).$$

This shows that the IO realization either of $F_{ISO}(s)$ *or of* $F_{ISOnd}(s)$ *need not exist in spite an IO realization of* $G_{ISO}(s)$ *exists, and in spite the IO model determined by* $G_{ISO}(s)$ *implies the well-defined* $F_{ISOIO}(s)$. *This is clear if we have in mind the different meaning of the initial conditions relative to which hold* $F_{ISO}(s)$ *and* $F_{ISOIO}(s)$. *However,* $F_{ISO}(s)$ *can induce a* *reduced IO realization that gives the same $F_{ISOIOrd}(s)$ as $F_{IOrd}(s)$ obtained from the corresponding IO system.*

2) $$A_2 = \begin{bmatrix} 0 & 1 \\ 1 & 0 \end{bmatrix}, \; B_2 = \begin{bmatrix} -1 \\ 1 \end{bmatrix}, \; C_2 = \begin{bmatrix} 0 \vdots 1 \end{bmatrix}, \; D_2 = 0 \Longrightarrow$$

$$F_{ISO2}(s) = \left(s^2 - 1\right)^{-1} \begin{bmatrix} s - 1 \vdots 1 \vdots s \end{bmatrix} = F_{ISO2nd}(s),$$

$$G_{ISO2}(s) = \frac{s-1}{s^2-1} = G_{ISO1}(s) \Longrightarrow G_{ISO2nd}(s) = \frac{1}{s+1}.$$

The transfer function matrix $G_{ISO2}(s)$ is degenerate, but the full transfer function matrix $F_{ISO2}(s)$ is nondegenerate, $F_{ISO2}(s) = F_{ISO2nd}(s)$.

The degenerate transfer function matrix $G_{ISO2}(s)$ implies the IO system (12.1), Example 239 (in Section 12.3), which holds only under all zero initial conditions.

The nondegenerate form $G_{ISO2nd}(s)$ of $G_{ISO2}(s)$ reads

$$G_{ISO2nd}(s) = \frac{1}{s+1} = G_{IO2nd}(s).$$

It yields the IO system (12.6), (12.7), Example 239 (in Section 12.3), which is valid also only under all zero initial conditions.

The transfer function matrix $G_{ISO2}(s)$, which is degenerate, furnishes

$$\nu_2 = 2, \; \mu_2 = 1, \; N_2 = 1, \; M_2 = 1,$$
$$A_2^{(2)} = \begin{bmatrix} -1 \vdots 0 \vdots 1 \end{bmatrix}, \; B_2^{(1)} = \begin{bmatrix} -1 \vdots 1 \end{bmatrix},$$

which together with (8.6), (in Section 8.1), imposes the following IO conditions:

$$A_2^{(2)} S_1^{(2)}(s) = \begin{bmatrix} -1 \vdots 0 \vdots 1 \end{bmatrix} \begin{bmatrix} 1 \\ s \\ s^2 \end{bmatrix} = \left[-1 + s^2\right] = \left[s^2 - 1\right],$$

$$B_2^{(1)} S_1^{(1)}(s) = \begin{bmatrix} -1 \vdots 1 \end{bmatrix} \begin{bmatrix} 1 \\ s \end{bmatrix} = [-1 + s] = [s - 1],$$

$$-B_2^{(1)} Z_1^{(1-1)}(s) = -\begin{bmatrix} -1 \vdots 1 \end{bmatrix} \begin{bmatrix} 0 \\ 1 \end{bmatrix} = [-1] \neq [1],$$

$$A_2^{(2)} Z_1^{(1-1)}(s) = \begin{bmatrix} -1 \vdots 0 \vdots 1 \end{bmatrix} \begin{bmatrix} 0 \vdots 0 \\ 1 \vdots 0 \\ s \vdots 1 \end{bmatrix} = \begin{bmatrix} s \vdots 1 \end{bmatrix} \neq [s].$$

The result is

$$F_{ISOIO2}(s) = (s^2-1)^{-1}\left[s-1 \vdots -1 \vdots s \vdots 1\right] = F_{IOISO2nd}(s),$$

which is nondegenerate and different from $F_{ISO2}(s)$:

$$F_{ISO2}(s) = \left(s^2-1\right)^{-1}\left[s-1 \vdots 1 \vdots s\right] \neq$$
$$\neq (s^2-1)^{-1}\left[s-1 \vdots -1 \vdots s \vdots 1\right] = F_{ISOIO2}(s).$$

The full transfer function matrix $F_{ISO2}(s)$ *obtained from the* ISO *model is not* IO *realizable in spite* $G_{ISO2}(s)$ *is* IO *realizable.*

The minimal IO *realization of* $G_{ISO1}(s) = G_{ISO2}(s)$ *was found in 1) to be*

$$A^{(1)}_{1\min}S^{(1)}_{1}(s) = A^{(1)}_{2\min}S^{(1)}_{1}(s) = \left[1 \vdots 1\right]\left[\begin{array}{c} 1 \\ s \end{array}\right] = [1+s] = [s+1],$$
$$B^{(0)}_{1\min}S^{(0)}_{1}(s) = B^{(0)}_{2\min}S^{(0)}_{1}(s) = [1]\,[1] = [1] = [1],$$
$$-B^{(0)}_{1\min}Z^{(0-1)}_{1}(s) = -B^{(0)}_{2\min}Z^{(0-1)}_{1}(s) \text{ does not exist},$$
$$A^{(1)}_{1\min}Z^{(1-1)}_{1}(s) = A^{(1)}_{2\min}Z^{(1-1)}_{1}(s) = \left[1 \vdots 1\right]\left[\begin{array}{c} 0 \\ 1 \end{array}\right] = [1] = [1].$$

They imply

$$F_{IO2rd}(s) = \left[A^{(1)}_{2\min}S^{(1)}_{1}(s)\right]^{-1}\left[B^{(0)}_{2\min}S^{(0)}_{1}(s) \vdots A^{(1)}_{2\min}Z^{(1-1)}_{1}(s)\right] =$$
$$= (s+1)^{-1}\left[1 \vdots 1\right],$$

which is different from

$$F_{ISOIO2}(s) = (s^2-1)^{-1}\left[s-1 \vdots -1 \vdots s \vdots 1\right] = F_{IOISO2nd}(s).$$

The minimal IO *realization of* $G_{ISO2}(s)$ *is not the minimal realization of* $F_{ISO2}(s)$ *and* $F_{ISO2}(s)$ *is* IO *unrealizable in view of Theorem 237 (in Section 12.3).*

3) $A_3 = \left[\begin{array}{cc} 1 & 0 \\ 0 & -1 \end{array}\right],\ B_3 = \left[\begin{array}{c} 0 \\ 1 \end{array}\right],\ C_3 = \left[0 \vdots 1\right],\ D_3 = 0$

$\Longrightarrow$

$$F_{ISO3}(s) = \left(s^2-1\right)^{-1}\left[s-1 \vdots 0 \vdots s-1\right] \Longrightarrow$$

$$F_{ISO3nd}(s) = (s+1)^{-1}\left[1 \vdots 0 \vdots 1\right],$$

$$G_{ISO3}(s) = \frac{s-1}{s^2-1} \Longrightarrow G_{ISO3nd}(s) = \frac{1}{s+1}.$$

Both $F_{ISO3}(s)$ *and* $G_{ISO3}(s)$ *are degenerate.*

The conditions for the IO realization of the degenerate $F_{ISO3}(s)$ *read*

$$\nu_3 = 2,\ \mu_3 = 1,\ N_3 = 1,\ M_3 = 1,$$
$$A_3^{(2)} = \left[-1 \vdots 0 \vdots 1\right],\ B_3^{(1)} = \left[-1 \vdots 1\right],$$

and

$$A_3^{(2)} S_1^{(2)}(s) = \left[-1 \vdots 0 \vdots 1\right]\begin{bmatrix} 1 \\ s \\ s^2 \end{bmatrix} = \left[-1+s^2\right] = \left[s^2-1\right],$$

$$B_3^{(1)} S_1^{(1)}(s) = \left[-1 \vdots 1\right]\begin{bmatrix} 1 \\ s \end{bmatrix} = [-1+s] = [s-1],$$

$$-B_3^{(1)} Z_1^{(1-1)}(s) = -\left[-1 \vdots 1\right]\begin{bmatrix} 0 \\ 1 \end{bmatrix} = [-1] \neq [0],$$

$$A_3^{(2)} Z_1^{(1-1)}(s) = \left[-1 \vdots 0 \vdots 1\right]\begin{bmatrix} 0 \vdots & 0 \\ 1 \vdots & 0 \\ s \vdots & 1 \end{bmatrix} = \left[s \vdots 1\right] \neq\ [s-1].$$

We determine now

$$F_{ISOIO3}(s) = (s^2-1)^{-1}\left[s-1 \vdots\ -1 \vdots s \vdots 1\right] = F_{IOISO3nd}(s).$$

The last two conditions for the IO realizability of $F_{ISO3}(s)$ *are not fulfilled (Theorem 237, in Section 12.3). The degenerate* $F_{ISO3}(s)$ *is not IO realizable*

$$F_{ISO3}(s) = \left(s^2-1\right)^{-1}\left[s-1 \vdots 0 \vdots s-1\right] \neq$$

$$\neq (s^2-1)^{-1}\left[s-1 \vdots\ -1 \vdots s \vdots 1\right] = F_{ISOIO3}(s).$$

However, the quadruple obtained from $G_{ISO3}(s)$ determines the exact full transfer function matrix $F_{ISOIO3}(s)$ of the corresponding IO model.

The minimal IO realization $G_{ISO3nd}(s)$ of $G_{ISO3}(s)$ is determined by (12.6), (12.7), Example 239 (in Section 12.3).

Let us verify the conditions for the IO realization of the nondegenerate form $F_{ISO3nd}(s)$ of $F_{ISO3}(s)$,

$$F_{ISO3nd}(s) = (s+1)^{-1} \begin{bmatrix} 1 \vdots 0 \vdots 1 \end{bmatrix}.$$

The conditions read

$$\nu_{3\min} = 1,\ \mu_{3\min} = 0,\ N_{3\min} = 1,\ M_{3\min} = 1,$$
$$A^{(1)}_{3\min} = \begin{bmatrix} 1 \vdots 1 \end{bmatrix},\ B^{(0)}_{3\min} = [1\ ],$$

and

$$A^{(1)}_{3\min} S^{(1)}_1(s) = \begin{bmatrix} 1 \vdots 1 \end{bmatrix} \begin{bmatrix} 1 \\ s \end{bmatrix} = [1+s] = [s+1],$$
$$B^{(0)}_{3\min} S^{(0)}_1(s) = [1]\,[1] = [1] = [1],$$
$$B^{(0)}_{3\min} Z^{(0-1)}_1(s) \text{ does not exist},$$
$$A^{(1)}_{3\min} Z^{(1-1)}_1(s) = \begin{bmatrix} 1 \vdots 1 \end{bmatrix} \begin{bmatrix} 0 \\ 1 \end{bmatrix} = [1] = [1].$$

Hence,

$$F_{IO3nd}(s) = (s+1)^{-1} \begin{bmatrix} 1 \vdots 1 \end{bmatrix} = F_{IO1-3rd}(s).$$

In this case, the nondegenerate form $F_{ISO3nd}(s)$ of $F_{ISO3}(s)$ might seem IO realizable at first glance. Is that correct? Let us check this result by comparing $F_{ISO3nd}(s)$ with $F_{ISOIO3nd}(s)$,

$$F_{ISO3nd}(s) = (s+1)^{-1} \begin{bmatrix} 1 \vdots 0 \vdots 1 \end{bmatrix} \neq$$
$$\neq (s+1)^{-1} \begin{bmatrix} 1 \vdots 1 \end{bmatrix} = F_{ISOIO3nd}(s).$$

The nondegenerate form of the full transfer function matrix $F_{ISO3nd}(s)$ obtained from the ISO system corresponds to the input action vector $[i(t) \vdots x_{10}\delta(t) \vdots x_{20}\delta(t)]^T$, while the full transfer function matrix $F_{ISOIO3nd}(s)$

obtained from the reduced IO system is related to the input action vector $\left[i(t) \vdots y_0\delta(t)\right]^T$. *This explains why their dimensions are different,*

$$F_{ISO3}(s) \in \mathfrak{C}^{1x3},\ F_{IO}(s) \in \mathfrak{C}^{1x2}.$$

The condition of Theorem 237, in Section 12.3, is not satisfied.

4) $A_4 = [-1]\,,\ B_4 = [1]\,,\ C_4 = [1]\,,\ D_4 = 0 \Longrightarrow$

$$F_{ISO4}(s) = (s+1)^{-1}\left[1 \vdots 1\right] = F_{ISO4nd}(s),$$

$$G_{ISO4}(s) = \frac{1}{s+1} = G_{ISO4nd}(s).$$

Both $F_{ISO4}(s)$ *and* $G_{ISO4}(s)$ *are nondegenerate. In this case*

$$\nu_4 = \nu_{4\min} = 1,\ \mu_4 = \mu_{4\min} = 0,\ N_4 = N_{4\min} = 1,\ M_4 = M_{4\min} = 1,$$

$$A_4^{(1)} = A_{4\min}^{(1)} = \left[1 \vdots 1\right],\ B_4^{(0)} = B_{4\min}^{(0)} = [1\ ],$$

and the IO conditions (Theorem 237, in Section 12.3) read

$$A_{4\min}^{(1)} S_1^{(1)}(s) = \left[1 \vdots 1\right]\left[\begin{array}{c} 1 \\ s \end{array}\right] = [1+s] = [s+1]\,,$$

$$B_{4\min}^{(0)} S_1^{(0)}(s) = [1]\,[1] = [1] = [1]\,,$$

$$B_{4\min}^{(0)} Z_1^{(0-1)}(s) \text{ does not exist},$$

$$A_{4\min}^{(1)} Z_1^{(1-1)}(s) = \left[1 \vdots 1\right]\left[\begin{array}{c} 0 \\ 1 \end{array}\right] = [1] = \ [1] \Longrightarrow$$

$$F_{IO4}(s) = (s+1)^{-1}\left[1 \vdots 1\right] = F_{ISO4}(s),$$

$$F_{IO4nd}(s) = (s+1)^{-1}\left[1 \vdots 1\right] = F_{ISO4nd}(s).$$

The full transfer function matrix $F_{ISO4}(s)$ *obtained from the ISO model is IO realizable in this case and its IO realization is minimal.*

Equations (12.6) and (12.7), Example 239 (in Section 12.3), determine the minimal IO model of both $F_{ISO4}(s)$ *and* $G_{ISO4}(s)$. *It is the minimal IO model that realizes* $F_{ISO4}(s)$.

Conclusion 249 *We conclude that the (non)degenerate form of $F_{ISO}(s)$ can be IO unrealizable although the (non)degenerate form of $G_{ISO}(s)$ is IO realizable, respectively. We should have in mind that $F_{ISO}(s)$ follows from the ISO system model, for which the initial conditions are the initial values of the state variables rather than the initial values of output and its derivatives that are related to the IO model. In other words, $F_{ISO}(s)$ obtained from the ISO model can be different from $F_{ISOIO}(s)$ obtained from the IO model (induced by the ISO model), in which case the former can be IO unrealizable although the latter is IO realizable. This is clearly expressed by necessity of the condition of Theorem 237.*

Example 250 *For the ISO system by Kalman (presented in Example 169, in Section 8.2) [80, Example 8, pp. 188, 189]:*

$$A = \begin{bmatrix} 0 \vdots 1 \vdots & 0 \\ 5 \vdots 0 \vdots & 2 \\ -2 \vdots 0 \vdots & -2 \end{bmatrix}, \; B = \begin{bmatrix} 0 \\ 0 \\ 0.5 \end{bmatrix}, \; C = \begin{bmatrix} -2 \vdots 1 \vdots 0 \end{bmatrix}, \; D = O,$$

we found

$$F_{ISO}(s) = [det(sI_n - A)]^{-1} \bullet$$
$$\bullet \left[Cadj(sI_n - A)B + Ddet(sI_n - A) \vdots Cadj(sI_n - A) \right] =$$
$$= \frac{s-2}{s-2} \left[(s+1)(s+3) \right]^{-1} \left[1 \vdots -2s-3 \vdots \; s+2 \vdots 2 \right] \Longrightarrow$$
$$F_{ISOnd}(s) = \frac{\left[1 \vdots -2s-3 \vdots \; s+2 \vdots 2 \right]}{(s+1)(s+3)},$$

$$G_{ISO}(s) = [det(sI_n - A)]^{-1} \left[Cadj(sI_n - A)B + Ddet(sI_n - A) \right] =$$
$$= \left(s^3 + 2s^2 - 5s - 6 \right)^{-1} (s-2) = \frac{s-2}{s-2} \left[(s+1)(s+3) \right]^{-1} \Longrightarrow$$
$$G_{ISOnd}(s) = \frac{1}{(s+1)(s+3)}.$$

Since $G_{ISO}(s)$ is reducible to

$$G_{ISOnd}(s) = \frac{1}{(s+1)(s+3)}$$

then the given ISO realization is not the minimal ISO realization of $G_{ISO}(s)$. Its minimal ISO realization reads

$$A_{\min G} = \begin{bmatrix} 0 \vdots & 1 \\ -3 \vdots & -4 \end{bmatrix}, \quad B_{\min G} = \begin{bmatrix} 0 \\ 1 \end{bmatrix},$$

$$C_{\min G} = \begin{bmatrix} 0 \vdots 1 \end{bmatrix}, \quad D_{\min G} = O.$$

It yields the following fictitious full transfer function matrix $F_{ISO\min}(s)$

$$\frac{\begin{bmatrix} C_{\min G} adj(sI_n - A_{\min G}) B_{\min G} + D_{\min G} det(sI_n - A_{\min G}) \vdots \\ \vdots C_{\min G} adj(sI_n - A_{\min G}) \end{bmatrix}}{det(sI_n - A_{\min G})} =$$

$$= \frac{\begin{bmatrix} 1 \vdots s+4 \vdots 1 \end{bmatrix}}{(s+1)(s+3)} = F_{ISO\min}(s),$$

which is different from the nondegenerate form $F_{ISOnd}(s)$ of $F_{ISO}(s)$,

$$F_{ISO\min}(s) = \frac{\begin{bmatrix} 1 \vdots s+4 \vdots 1 \end{bmatrix}}{(s+1)(s+3)} \neq \frac{\begin{bmatrix} 1 \vdots & -2s-3 \vdots & s+2 \vdots 2 \end{bmatrix}}{(s+1)(s+3)} = F_{ISOnd}(s).$$

The minimal ISO realization $(A_{\min G}, B_{\min G}, C_{\min G}, D_{\min G})$ of $G_{ISO}(s)$ is not the minimal ISO realization of $F_{ISO}(s)$.

Does the nondegenerate form $F_{ISOnd}(s)$ of $F_{ISO}(s)$ imply an ISO realization of $F_{ISO}(s)$ with a lower dimension than that of the given ISO realization (A, B, C, D)?

We start with

$$F_{ISOnd}(s) =$$

$$= \frac{\begin{bmatrix} C_{ISOnd} adj(sI_n - A_{ISOnd}) B_{ISOnd} + D_{ISOnd} det(sI_n - A_{ISOnd}) \vdots \\ \vdots C_{ISOnd} adj(sI_n - A_{ISOnd}) \end{bmatrix}}{det(sI_n - A_{ISOnd})} =$$

$$= \frac{\begin{bmatrix} 1 \vdots & -2s-3 \vdots & s+2 \vdots 2 \end{bmatrix}}{(s+1)(s+3)} \Longrightarrow \deg \begin{bmatrix} 1 \vdots & -2s-3 \vdots & s+2 \vdots 2 \end{bmatrix} = 1,$$

$$\deg (s+1)(s+3) = 2.$$

This implies

$$\deg \begin{bmatrix} C_{ISOnd} adj(sI_n - A_{ISOnd}) B_{ISOnd} + D_{ISOnd} det(sI_n - A_{ISOnd}) \vdots \\ \vdots \, C_{ISOnd} adj(sI_n - A_{ISOnd}) \end{bmatrix} = 1,$$

$$\deg det(sI_n - A_{ISOnd}) = 2$$

$$\Longrightarrow$$

$$D_{ISOnd} = 0, \;\; n = 2,$$

$$\Longrightarrow$$

$$C_{ISOnd} adj(sI_n - A_{ISOnd}) B_{ISOnd} = \begin{bmatrix} 1 \vdots -2s-3 \end{bmatrix},$$

$$C_{ISOnd} adj(sI_n - A_{ISOnd}) = \begin{bmatrix} s+2 \vdots 2 \end{bmatrix}.$$

$$\det(sI_n - A_{ISOnd}) = (s+1)(s+3)$$

$$A_{ISOnd} = \begin{bmatrix} a_{11} & a_{12} \\ a_{21} & a_{22} \end{bmatrix}, \; B_{ISOnd} = \begin{bmatrix} b_1 \\ b_2 \end{bmatrix}, \; C_{ISOnd} = \begin{bmatrix} c_1 \vdots c_2 \end{bmatrix}$$

$$\Longrightarrow$$

$$\begin{bmatrix} c_1 \vdots c_2 \end{bmatrix} \begin{bmatrix} s - a_{22} & a_{12} \\ a_{21} & s - a_{11} \end{bmatrix} \begin{bmatrix} b_1 \\ b_2 \end{bmatrix} = \begin{bmatrix} 1 \vdots -2s-3 \end{bmatrix},$$

$$\begin{bmatrix} c_1 \vdots c_2 \end{bmatrix} \begin{bmatrix} s - a_{22} & -a_{12} \\ -a_{21} & s - a_{11} \end{bmatrix} = \begin{bmatrix} s+2 \vdots 2 \end{bmatrix},$$

$$\det \begin{bmatrix} s - a_{11} & -a_{12} \\ -a_{21} & s - a_{22} \end{bmatrix} = (s+1)(s+3) = s^2 + 4s + 3.$$

The last three equations have many solutions. There exist many ISO system realizations that imply their full transfer function matrixes equal to the nondegenerate form $F_{ISOnd}(s)$ of the given $F_{ISO}(s)$. There exist many ISO realizations (A, B, C, D) of the nondegenerate form $F_{ISOnd}(s)$ of the given $F_{ISO}(s)$. The dimension n of the given ISO realization (A, B, C, D) is not the minimal dimension of the ISO realization of the given $F_{ISO}(s)$.

Conclusion 251 *The preceding analysis shows that the minimal ISO realization of the transfer function matrix $G_{ISO}(s)$ can be, but need not be, the minimal ISO realization of the full transfer function matrix $F_{ISO}(s)$.*

Conclusion 252 *The nondegenerate form $F_{ISOnd}(s)$ of the full transfer function matrix $F_{ISO}(s)$ can have an ISO realization with a lower dimension than that of the given ISO realization (A, B, C, D).*

12.5 Realizations of $F(s)$ of IIO systems

Definitions 220 (in Section 12.1), 230 and 232 (in Section 12.2) are the basis for the following:

Definition 253 *If and only if there is* $(\alpha,\ \beta,\ \mu,\ \nu,\ E^{(\nu)},\ P^{(\beta)},\ Q^{(\alpha)},\ R^{(\alpha)},\ T^{(\mu)})$ *that determines an IIO system realization (3.90) (in Subsection 3.5.4) such that its full transfer function matrix is equal to the nondegenerate form* $F_{IIOnd}(s)$ *of a given* $F_{IIO}(s)$*, then the realization is* ***the minimal (irreducible) IIO realization of the given full transfer function matrix*** $F_{IIO}(s)$.

Theorem 254 *In order for* $\left(\alpha,\ \beta,\ \mu,\ \nu,\ E^{(\nu)},\ P^{(\beta)},\ Q^{(\alpha)},\ R^{(\alpha)},\ T^{(\mu)}\right)$ *to determine the (minimal) IIO system realization (3.90) (in Subsection 3.5.4) it is necessary and sufficient that*

$$\left(E^{(\nu)}S_N^{(\nu)}(s)\right)^{-1}\bullet$$
$$\bullet\left(R^{(\alpha)}S_\rho^{(\alpha)}(s)\left(Q^{(\alpha)}S_\rho^{(\alpha)}(s)\right)^{-1}P^{(\beta)}S_M^{(\beta)}(s)+T^{(\mu)}S_M^{(\mu)}(s)\right)=$$
$$=G_{IIO}(s),$$

$$\left(E^{(\nu)}S_N^{(\nu)}(s)\right)^{-1}\bullet$$
$$\bullet\left(\begin{array}{c}\left[R^{(\alpha)}S_\rho^{(\alpha)}(s)\left(Q^{(\alpha)}S_\rho^{(\alpha)}(s)\right)^{-1}B^{(\beta)}Z_M^{(\beta-1)}(s)\vdots O_{N(\gamma-\beta)}\right]+\\+\left[T^{(\mu)}Z_M^{(\mu-1)}(s)\vdots O_{N(\gamma-\mu)}\right]\end{array}\right)=$$
$$=-G_{IIOi_0}(s),$$

$$\left(E^{(\nu)}S_N^{(\nu)}(s)\right)^{-1}\bullet$$
$$\bullet\left(\begin{array}{c}R^{(\alpha)}S_\rho^{(\alpha)}(s)\left(Q^{(\alpha)}S_\rho^{(\alpha)}(s)\right)^{-1}Q^{(\alpha)}Z_\rho^{(\alpha-1)}(s)-\\-R^{(\alpha)}Z_\rho^{(\alpha-1)}(s)\end{array}\right)=$$
$$=-G_{IIOr_0}(s),$$

$$\left(E^{(\nu)}S_N^{(\nu)}(s)\right)^{-1}E^{(\nu)}Z_N^{(\nu-1)}(s)=G_{IIOy_0}(s),\qquad(12.17)$$

and that they determine the (nondegenerate form $F_{IIOnd}(s)$ of the) given full IIO system transfer function matrix $F_{IIO}(s)$ as follows, respectively,

$$\left[G_{IIO}(s) \vdots G_{IIOi_0}(s) \vdots G_{IIOr_0}(s) \vdots G_{IIOy_0}(s)\right] = F_{IIO}(s) \quad (12.18)$$

$$\left(\left[G_{IIO}(s) \vdots G_{IIOi_0}(s) \vdots G_{IIOr_0}(s) \vdots G_{IIOy_0}(s)\right] = F_{IIOnd}(s)\right). \quad (12.19)$$

Proof. *Necessity.* Let (α, β, μ, ν, $E^{(\nu)}$, $P^{(\beta)}$, $Q^{(\alpha)}$, $R^{(\alpha)}$, $T^{(\mu)}$) determine the (minimal) IIO system realization (3.90). Then Equations 12.17 through 12.19 hold due to Definition 113 (in Chapter 6), Definition 217 through Definition 220, Proposition 224 (in Section 12.1), Definition 230 (in Section 12.2), Definition 232 (in Section 12.2), Definition 253, (8.62) and (8.66) (Theorem 171, in Section 8.3), respectively, which proves necessity of the conditions.

Sufficiency. Let the conditions hold. Hence, (α, β, μ, ν, $E^{(\nu)}$, $P^{(\beta)}$, $Q^{(\alpha)}$, $R^{(\alpha)}$, $T^{(\mu)}$) determines (the minimal) an input-output realization of $F_{IIO}(s)$ in view of (8.62) and (8.66) (Theorem 171, in Section 8.3), which is also (the minimal) an IIO realization of the IIO system (3.90) due to Definition 230 (in Section 12.2), Definition 232, (in Section 12.2), and Definition 253, respectively. ■

Example 255 *Let the full IIO system transfer function matrix be*

$$F_{IIO}(s) = \left[(s-1)^2(s+2)(s+5)\right]^{-1} \bullet \begin{bmatrix} (s-1)^2(s-6)(17s+10) \vdots \\ \vdots -(s-1)(s-7)(17s+10) \vdots \\ \vdots -(s-1)(13s-10) \vdots \\ \vdots\ 10(s-1) \vdots \\ \vdots -5(s-1)(2s+5) \vdots \\ \vdots\ (s-1)(s^2+6s+5) \vdots \\ \vdots\ (s-1)(s+5) \end{bmatrix}^T,$$

$$F_{IIO}(s) \in \mathfrak{C}^{1\times 7}. \quad (12.20)$$

Its IIO system realization $\left(\alpha,\ \beta,\ \mu,\ \nu,\ E^{(\nu)},\ P^{(\beta)},\ Q^{(\alpha)},\ R^{(\alpha)},\ T^{(\mu)}\right)$ *is determined (Example 174, in Section 8.3) by:*

$$\alpha = 2,\ \beta = 2, \mu = 2,\ \nu = 2,$$

$$E^{(\nu)} = E^{(2)} = \begin{bmatrix} -2 \vdots 1 \vdots 1 \end{bmatrix}, \; P^{(\beta)} = P^{(2)} = \begin{bmatrix} 18 \vdots -21 \vdots 3 \end{bmatrix},$$

$$Q^{(\alpha)} = Q^{(2)} = \begin{bmatrix} -5 \vdots 4 \vdots 1 \end{bmatrix}, \; R^{(\alpha)} = R^{(2)} = \begin{bmatrix} 0 \vdots -5 \vdots 5 \end{bmatrix},$$

$$T^{(\mu)} = T^{(2)} = \begin{bmatrix} 12 \vdots -14 \vdots 2 \end{bmatrix}.$$

These data determine the IIO system realization in the vector form:

$$\begin{aligned} Q^{(\alpha)}\mathbf{r}^{\alpha}(t) &= P^{(\beta)}\mathbf{i}^{\beta}(t), \\ E^{(\nu)}\mathbf{y}^{\nu}(t) &= R^{(\alpha)}\mathbf{r}^{\alpha}(t) + T^{(\mu)}\mathbf{i}^{\mu}(t), \end{aligned} \tag{12.21}$$

and, equivalently, in the scalar form:

$$\begin{gathered} r^{(2)}(t) + 4r^{(1)}(t) - 5r(t) = 3i^{(2)}(t) - 21i^{(1)}(t) + 18i(t), \\ y^{(2)}(t) + y^{(1)}(t) - 2y = \\ = 5r^{(2)}(t) - 5r^{(1)}(t) + 2i^{(2)}(t) - 14i^{(1)}(t) + 12i(t). \end{gathered} \tag{12.22}$$

Part III

STABILITY STUDY

Chapter 13

Lyapunov stability

13.1 Lyapunov stability concept

The stability concept of Lyapunov [88] was established and mostly studied for the *ISO* systems. It has the following main qualitative characteristics:

- It concerns the internal dynamical behavior, i.e., the state variation that is the motion, of the system.
- It concerns the system behavior in the nominal regime in terms of total coordinates (e.g., $\mathbf{I}$), i.e., in the free regime in terms of deviations (e.g., $\mathbf{i}$, $\mathbf{i} = \mathbf{I} - \mathbf{I}_N$), Subsection 3.5.1,

$$e.g., \ \mathbf{I}(t) = \mathbf{I}_N(t), \text{ i.e., } \mathbf{i}(t) = \mathbf{0}_M, \ \forall t \in \mathfrak{T}_0.$$

- It treats the influence of nonzero initial conditions on the system dynamical behavior.
- It concerns the system dynamical behavior over the unbounded time interval $\mathfrak{T}_0$. If the system dynamical behavior is satisfactory over $\mathfrak{T}_0$, then it is satisfactory also on any subinterval of $\mathfrak{T}_0$.
- It allows any permitted upper bound ε, $\varepsilon \in \mathfrak{R}^+$, of the instantaneous deviation $\mathbf{x}(t)$, or $\mathbf{y}(t)$, of a real total systems behavior $\mathbf{X}(t)$, or $\mathbf{Y}(t)$, from the total nominal behavior $\mathbf{X}_d(t)$, or $\mathbf{Y}_d(t)$, respectively.
- It demands the existence of the appropriate upper bound δ, $\delta = \delta(\varepsilon) \in \mathfrak{R}^+$, of the initial deviation $\mathbf{x}_0$, or $\mathbf{y}_0$, of the total initial system behavior $\mathbf{X}_0$, or $\mathbf{Y}_0$, from the total initial desired behavior $\mathbf{X}_{d0}$, or $\mathbf{Y}_{d0}$.

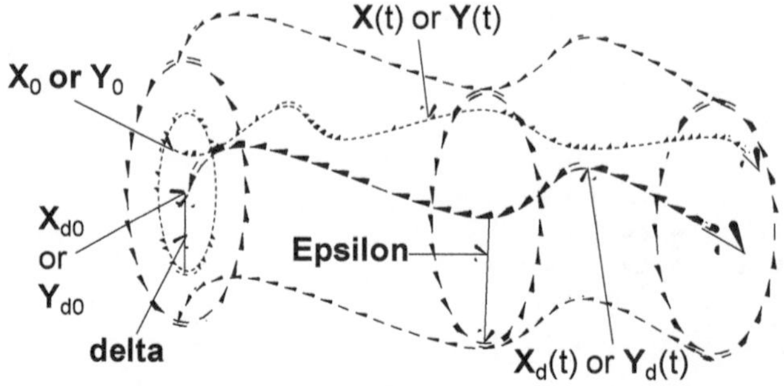

Figure 13.1: A stable total desired behavior $\mathbf{X}_d(t)$, or $\mathbf{Y}_d(t)$.

- Analogously, a positive real number α specifies the arbitrarily requested lower bound of the instantaneous closeness of a real system behavior $\mathbf{X}(t)$, or $\mathbf{Y}(t)$, to the desired behavior $\mathbf{X}_d(t)$, or $\mathbf{Y}_d(t)$, where $\alpha = \varepsilon^{-1} \in \mathfrak{R}^+$.

- A positive real number $\beta = \beta(\alpha)$ specifies the lower bound of the initial closeness of $\mathbf{X}_0$, or $\mathbf{Y}_0$, to $\mathbf{X}_{d0}$, or $\mathbf{Y}_{d0}$, which corresponds to α, where $\beta = \delta^{-1} \in \mathfrak{R}^+$.

- If and only if for arbitrarily chosen permitted upper bound ε, there exists an appropriate upper bound δ of the initial deviation $\mathbf{x}_0$, or $\mathbf{y}_0$, such that the norm $\|\mathbf{x}_0\|$, or $\|\mathbf{y}_0\|$ less than δ guarantees that the norm $\|\mathbf{x}(t)\|$, or $\|\mathbf{y}(t)\|$, is less than the chosen permitted upper bound ε at every $t \in \mathfrak{T}_0$, then **the desired behavior $\mathbf{X}_d(t)$**, or $\mathbf{Y}_d(t)$, respectively, is **stable** (**the linear system is limiting stable**, in other words **the linear system is on the boundary of stability,** equivalently: **it is critically stable**). See Fig. 13.1. If and only if additionally, $\delta \longrightarrow \infty$ as $\varepsilon \longrightarrow \infty$, then the stability is **global (in the whole**). Equivalently, **the zero vector deviation $\mathbf{x} = \mathbf{0}_n$**, or **$\mathbf{y} = \mathbf{0}_N$, is stable**; see Fig. 13.2.

- If and only if there exists a Δ-neighborhood of the initial desired behavior $\mathbf{X}_d(0)$, or $\mathbf{Y}_d(0)$, such that for every initial condition $\mathbf{X}_0$, or $\mathbf{Y}_0$, from the Δ-neighborhood, the corresponding system dynamical behavior asymptotically approaches the desired behavior $\mathbf{X}_d(t)$, or $\mathbf{Y}_d(t)$, as $t \to \infty$, then the desired behavior $\mathbf{X}_d(t)$, or $\mathbf{Y}_d(t)$, is **attractive**; see Fig. 13.3. Equivalently, then, and only then, **the zero vector deviation $\mathbf{x} = \mathbf{0}_n$**, or **$\mathbf{y} = \mathbf{0}_N$, is attractive**; see Fig. 13.4. If and only if this

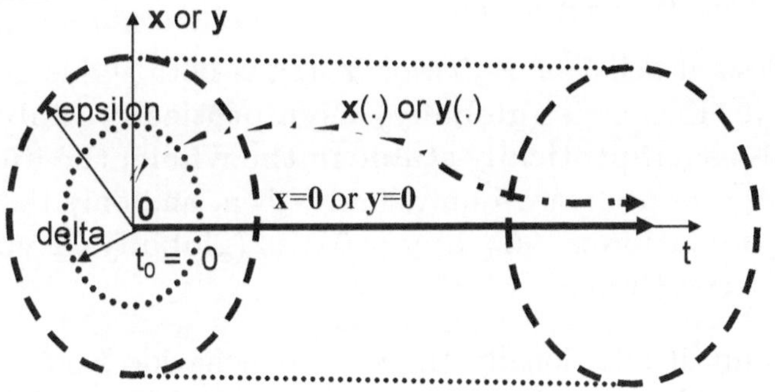

Figure 13.2: The zero vector deviation $\mathbf{x} = \mathbf{0}_n$, or $\mathbf{y} = \mathbf{0}_N$, is stable.

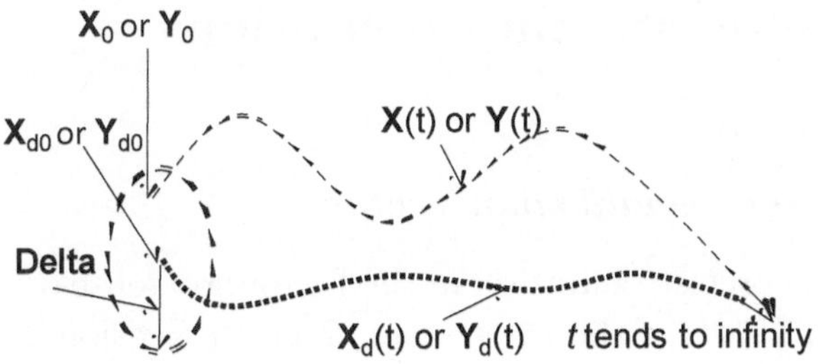

Figure 13.3: The total desired behavior $\mathbf{X}_d(t)$, or $\mathbf{Y}_d(t)$, is attractive.

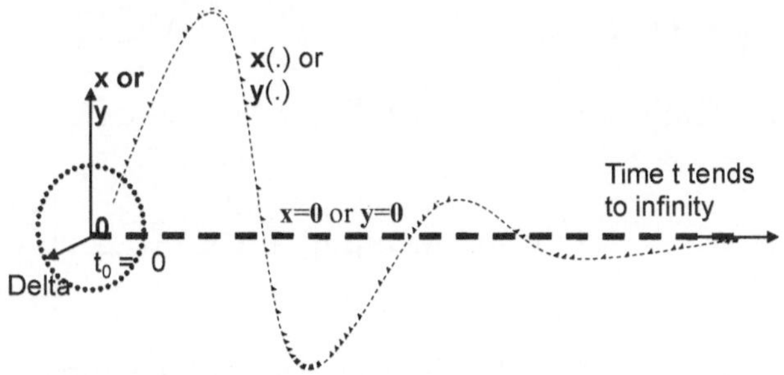

Figure 13.4: The zero vector deviation $\mathbf{x} = \mathbf{0}_n$, or $\mathbf{y} = \mathbf{0}_N$, is attractive.

holds for any initial conditions, i.e., for $\Delta = \infty$, then the attraction is **global**, i.e., **in the whole.**

- If the desired behavior $\mathbf{X}_d(t)$, or $\mathbf{Y}_d(t)$, is both stable and (globally) attractive, then it is **(globally) asymptotically stable** (which is called also **asymptotically stable in the whole, the linear system is stable**), respectively. Equivalently, then, and only then, **the zero vector deviation $\mathbf{x} = \mathbf{0}_n$**, or **$\mathbf{y} = \mathbf{0}_N$, is (globally) asymptotically stable**, respectively.

- If and only if additionally the desired behavior $\mathbf{X}_d(t)$, or $\mathbf{Y}_d(t)$, is globally stable, then the desired behavior $\mathbf{Y}_d(t)$ is **strictly globally asymptotically stable.** Equivalently, then, and only then, **the zero vector deviation $\mathbf{x} = \mathbf{0}_n$**, or **$\mathbf{y} = \mathbf{0}_N$, is strictly globally asymptotically stable.**

13.2 Lyapunov stability definitions

13.2.1 *IO* systems

Stability of the zero equilibrium vector

We present now the precise definition in the Lyapunov sense of *stability* of the desired behavior $\mathbf{Y}_d^{\nu-1}(t)$ of the IO system (2.1), (in Section 2.1), in terms of the total coordinates, i.e., of the desired behavior deviation $\mathbf{y}_d^{\nu-1}(t) \equiv \mathbf{0}_{\nu N}$ of the system (3.77), (in Subsection 3.5.2), although Lyapunov gave it only for the general nonlinear systems in Cauchy (normal) form (i.e., in the form of IS systems), [88].

Let $||.|| : \mathfrak{R}^k \rightarrow \mathfrak{R}_+$ be *an accepted norm on* $\mathfrak{R}^k$,

$$k \in \{1, 2, 3, .., M, \ ...., N, .., n, ..., \ \nu N, ...\},$$

which can be the *Euclidean norm on* $\mathfrak{R}^k$ in a special case:

$$\mathbf{w} = [w_1 \ w_2 \ ... \ w_k]^T \in \mathfrak{R}^k \Longrightarrow ||\mathbf{w}|| = \sqrt{\mathbf{w}^T\mathbf{w}} = \sqrt{\sum_{i=1}^{i=k} w_i^2},$$

Let $\mathfrak{B}_\xi(\mathbf{z})$ be an open hyperball with the radius ξ centered at the point $\mathbf{z}$ also in the space $\mathfrak{R}^k$,

$$\mathfrak{B}_\xi(\mathbf{z}) = \{\mathbf{w} : \ ||\mathbf{w} - \mathbf{z}|| < \xi\}.$$

If and only if $\mathbf{z} = \mathbf{0}_k$, then we omit $(\mathbf{0})$ from the notation $\mathfrak{B}_\xi(\mathbf{0})$,

$$\mathfrak{B}_\xi = \mathfrak{B}_\xi(\mathbf{0}).$$

In this in subsection, $k = \nu N$, $\mathfrak{R}^k = \mathfrak{R}^{\nu N}$.

Lyapunov defined stability properties of the desired behavior by assuming that the input vector $\mathbf{I}$ is nominal,

$$\mathbf{I}(t)= \mathbf{I}_N(t),\ \forall t \in \mathfrak{T}_0.$$

This means that in terms of the input deviation vector $\mathbf{i}(t)$,

$$\mathbf{i}(t)= \mathbf{I}(t) - \mathbf{I}_N(t)\ \forall t \in \mathfrak{T}_0,$$

we consider the system in the free regime,

$$\mathbf{i}(t)= \mathbf{I}_N(t) - \mathbf{I}_N(t) = \mathbf{0}_M,\ \forall t \in \mathfrak{T}_0.$$

Definition 256 *a) The desired behavior $\mathbf{Y}_d^{\nu-1}(t)$ of the IO system (2.1) is* ***stable*** *if and only if for every positive real number ε there exists a positive real number δ, the value of which depends on ε such that*

$$\begin{gathered}\forall \varepsilon \in \mathfrak{R}^+,\ \exists \delta \in \mathfrak{R}^+,\ \delta = \delta(\varepsilon), \left\| \mathbf{Y}_{d0}^{\nu-1} - \mathbf{Y}_0^{\nu-1} \right\| < \delta \Longrightarrow \\ \left\| \mathbf{Y}_d^{\nu-1}(t) - \mathbf{Y}^{\nu-1}(t; \mathbf{Y}_0^{\nu-1}; \mathbf{I}_N) \right\| < \varepsilon,\ \forall t \in \mathfrak{T}_0. \end{gathered} \tag{13.1}$$

The zero equilibrium vector $\mathbf{y}_e^{\nu-1} = \mathbf{0}_{\nu N}$ of the IO system (3.77) is ***stable*** *if and only if*

$$\begin{gathered}\forall \varepsilon \in \mathfrak{R}^+,\ \exists \delta \in \mathfrak{R}^+,\ \delta = \delta(\varepsilon), \\ \left\| \mathbf{y}_0^{\nu-1} \right\| < \delta \Longrightarrow \left\| \mathbf{y}^{\nu-1}(t; \mathbf{y}_0^{\nu-1}; \mathbf{0}_M) \right\| < \varepsilon,\ \forall t \in \mathfrak{T}_0. \end{gathered} \tag{13.2}$$

b) The desired behavior $\mathbf{Y}_d^{\nu-1}(t)$ of the IO system (2.1) is ***globally stable*** *(i.e.,* ***stable in the whole****) if and only if a) holds so that the maximal $\delta(\varepsilon)$, denoted by $\delta_{\max}(\varepsilon)$, diverges to infinity as ε goes to infinity,*

$$\varepsilon \to \infty \Longrightarrow \delta_{\max}(\varepsilon) \to \infty. \tag{13.3}$$

The zero equilibrium vector $\mathbf{y}_e^{\nu-1} = \mathbf{0}_{\nu N}$ of the IO system (3.77) is ***globally stable*** *(i.e.,* ***stable in the whole)*** *if and only if a) holds so that the maximal $\delta(\varepsilon)$, denoted by $\delta_{\max}(\varepsilon)$, diverges to infinity as ε goes to infinity, that is that (13.3) is valid.*

Conclusion 257 ***Stability of*** $\mathbf{Y}_d^{\nu-1}(t)$ ***and of*** $\mathbf{y}_e^{\nu-1} = \mathbf{0}_{\nu N}$

The desired behavior $\mathbf{Y}_d^{\nu-1}(t)$ *of the IO system (2.1) is (globally) stable if and only if the zero equilibrium vector* $\mathbf{y}_e^{\nu-1} = \mathbf{0}_{\nu N}$ *of the IO system (3.77) is (globally) stable, respectively.*

Note 258 ***Geometrical explanation***

From the geometrical point of view this definition means that the zero equilibrium vector $\mathbf{y}_e^{\nu-1} = \mathbf{0}_{\nu N}$ *of the IO system (3.77) is* ***stable*** *if and only if for every open hyperball* $\mathfrak{B}_\varepsilon \in \mathfrak{R}^{N\nu}$ *centered at the origin with the radius* ε*, there exists an open hyperball* $\mathfrak{B}_\delta \in \mathfrak{R}^{N\nu}$ *centered at the origin with the radius* δ*, which depends on* $\mathfrak{B}_\varepsilon$*, such that* $\mathbf{y}_0^{\nu-1} \in \mathfrak{B}_\delta$ *guarantees that* $\mathbf{y}^{\nu-1}(t;\mathbf{y}_0^{\nu-1};\mathbf{0}_M)$ *stays in* $\mathfrak{B}_\varepsilon$ *all the time from* $t=0$ *on, Fig. 13.2, (in Section 13.1),*

$$\begin{gathered}\forall \mathfrak{B}_\varepsilon \subseteq \mathfrak{R}^{N\nu},\ \exists \mathfrak{B}_\delta \subseteq \mathfrak{R}^{N\nu},\ \mathfrak{B}_\delta = \mathfrak{B}_\delta(\mathfrak{B}_\varepsilon),\ \Longrightarrow \\ \mathbf{y}_0^{\nu-1} \in \mathfrak{B}_\delta \Longrightarrow \mathbf{y}^{\nu-1}(t;\mathbf{y}_0^{\nu-1};\mathbf{0}_M) \in \mathfrak{B}_\varepsilon,\ \forall t \in \mathfrak{T}_0.\end{gathered} \tag{13.4}$$

If and only if additionally $\mathfrak{B}_\varepsilon \longrightarrow \mathfrak{R}^n$ *implies* $\mathfrak{B}_\delta \longrightarrow \mathfrak{R}^n$*, then* $\mathbf{x}_e = \mathbf{0}_n$ *is* ***globally stable*** *(i.e.,* ***stable in the whole****).*

The response of the *IO* system (3.77) in the free regime is given by (Equation 10.4, in Section 10.1),

$$\mathbf{y}(t;\mathbf{y}_0^{\nu-1};\mathbf{0}_M) = \Gamma_{IOy_0}(t)\mathbf{y}_0^{\nu-1}. \tag{13.5}$$

This yields

$$\mathbf{y}^{(k)}(t;\mathbf{y}_0^{\nu-1};\mathbf{0}_M) = \Gamma_{IOy_0}^{(k)}(t)\mathbf{y}_0^{\nu-1},\ k = 0, 1, ..., \nu,$$

and

$$\mathbf{y}^{\nu-1}(t;\mathbf{y}_0^{\nu-1};\mathbf{0}_M) = \Gamma_{IOy_0}^{\nu-1}(t)\mathbf{y}_0^{\nu-1}, \tag{13.6}$$

where

$$\Gamma_{IOy_0}^{\nu-1}(t) = \begin{bmatrix} \Gamma_{IOy_0}(t) \\ \Gamma_{IOy_0}^{(1)}(t) \\ \cdots\cdots \\ \Gamma_{IOy_0}^{(\nu-1)}(t) \end{bmatrix} \in \mathfrak{R}^{\nu N \mathrm{x} \nu N}. \tag{13.7}$$

Theorem 259 *In order for the zero equilibrium vector* $\mathbf{y}_e^{\nu-1} = \mathbf{0}_{\nu N}$ *of the IO system (3.77) to be globally stable, it is necessary and sufficient to be stable.*

Proof. *Necessity.* Let the zero equilibrium vector $\mathbf{y}_e^{\nu-1} = \mathbf{0}_{\nu N}$ of the *IO* system (3.77) be globally stable. Then it is stable due to b) of Definition 256.

Sufficiency. Let the zero equilibrium vector $\mathbf{y}_e^{\nu-1} = \mathbf{0}_{\nu N}$ of the *IO* system (3.77) be stable. Then, Definition 256 holds. Let $\varepsilon \in \mathfrak{R}^+$ be arbitrarily chosen. Let $\delta = \delta(\varepsilon) \in \mathfrak{R}^+$ be the corresponding δ obeying (256) so that, due to Equation 13.5 through Equation 13.7,

$$\left\|\mathbf{y}_0^{\nu-1}\right\| < \delta \Longrightarrow \left\|\mathbf{y}^{\nu-1}(t;\mathbf{y}_0^{\nu-1};\mathbf{0}_M)\right\| = \left\|\Gamma_{IOy_0}^{\nu-1}(t)\mathbf{y}_0^{\nu-1}\right\| < \varepsilon,\ \forall t \in \mathfrak{T}_0.$$

Let δ_{ij} be the Kronecker delta, $\delta_{ij} = 0$ for $i \neq j$, and $\delta_{ij} = 1$ for $i = j$, and let $\mathbf{w}_k^{\nu N} = \left[\delta_{1k} \vdots \delta_{2k} \vdots \ldots \vdots \delta_{\nu N,k}\right]^T \in \mathfrak{R}^{\nu N}$, $k \in \{1, 2, ..., \nu N\}$, and

$$\begin{gathered}
\mathbf{y}_{0k}^{\nu-1} = \frac{\delta}{2}\mathbf{w}_k^{\nu N} \Longrightarrow \left\|\mathbf{y}_{0k}^{\nu-1}\right\| = \frac{\delta}{2} < \delta \Longrightarrow \\
\left\|\mathbf{y}^{\nu-1}(t;\mathbf{y}_{0k}^{\nu-1};\mathbf{0}_M)\right\| = \left\|\Gamma_{IOy_0}^{\nu-1}(t)\mathbf{y}_{0k}^{\nu-1}\right\| = \\
= \frac{\delta}{2}\left\|\Gamma_{IOy_0}^{\nu-1}(t)\mathbf{w}_k^{\nu N}\right\| < \varepsilon,\ \forall t \in \mathfrak{T}_0, \\
\forall k \in \{1, 2, ..., \nu N\} \Longrightarrow
\end{gathered}$$

$$\begin{gathered}
\exists \gamma \in \mathfrak{R}^+ \Longrightarrow \left\|\Gamma_{IOy_0}^{\nu-1}(t)\right\| < \gamma,\ \forall t \in \mathfrak{T}_0 \Longrightarrow \\
0 \leq \left\|\mathbf{y}^{\nu-1}(t;\mathbf{y}_0^{\nu-1};\mathbf{0}_M)\right\| = \left\|\Gamma_{IOy_0}^{\nu-1}(t)\mathbf{y}_0^{\nu-1}\right\| \leq \gamma\left\|\mathbf{y}_0^{\nu-1}\right\|.
\end{gathered}$$

Hence,

$$\begin{gathered}
\forall t \in \mathfrak{T}_0,\ \forall \mathbf{y}_0^{\nu-1} \in \mathfrak{R}^{\nu N},\ \left\|\mathbf{y}_0^{\nu-1}\right\| < \delta\ for\ \delta = \delta(\varepsilon) = \frac{\varepsilon}{2\gamma} \Longrightarrow \\
\left\|\mathbf{y}^{\nu-1}(t;\mathbf{y}_0^{\nu-1};\mathbf{0}_M)\right\| = \left\|\Gamma_{IOy_0}^{\nu-1}(t)\mathbf{y}_0^{\nu-1}\right\| \leq \gamma\left\|\mathbf{y}_0^{\nu-1}\right\| < \\
< \gamma\delta(\varepsilon) = \frac{\varepsilon}{2} < \varepsilon,\ \forall t \in \mathfrak{T}_0, \\
\delta_{\max}(\varepsilon) \geq \delta(\varepsilon) = \frac{\varepsilon}{2\gamma} \to \infty\ as\ \varepsilon \to \infty.
\end{gathered}$$

This proves global stability of the zero equilibrium vector $\mathbf{y}_e^{\nu-1} = \mathbf{0}_{\nu N}$ of the *IO* system (3.77). ∎

Conclusion 257 and this theorem imply the following:

Theorem 260 *In order for the desired behavior $\mathbf{Y}_d^{\nu-1}(t)$ of the IO system (2.1) to be globally stable it is necessary and sufficient to be stable.*

Definition 261 *The desired behavior* $\mathbf{Y}_d^{\nu-1}(t)$ *of the IO system (2.1) is* ***unstable*** *if and only if there exists a positive real number* ε *such that for every positive real number* δ *there exist an initial vector* $\mathbf{Y}_0^{\nu-1}$ *satisfying*

$$\left\|\mathbf{Y}_{d0}^{\nu-1} - \mathbf{Y}_0^{\nu-1}\right\| < \delta,$$

and a moment $\tau \in \mathfrak{T}_0$*, which imply*

$$\left\|\mathbf{Y}^{\nu-1}(t) - \mathbf{Y}^{\nu-1}(\tau; \mathbf{y}_0^{\nu-1}; \mathbf{I}_N)\right\| \geq \varepsilon.$$

The zero equilibrium vector $\mathbf{y}_e^{\nu-1} = \mathbf{0}_{\nu N}$ *of the IO system (3.77) is* ***unstable*** *if and only if*

$$\exists \varepsilon \in \mathfrak{R}^+ \longrightarrow \forall \delta \in \mathfrak{R}^+, \ \exists (\mathbf{y}_0^{\nu-1}, \tau) \in \mathfrak{R}^{N\nu} \times \mathfrak{T}_0,$$
$$\left\|\mathbf{y}_0^{\nu-1}\right\| < \delta \Longrightarrow \left\|\mathbf{y}^{\nu-1}(\tau; \mathbf{y}_0^{\nu-1}; \mathbf{0}_M)\right\| \geq \varepsilon.$$

Conclusion 262 ***Instability of*** $\mathbf{Y}_d^{\nu-1}(t)$ ***and of*** $\mathbf{y}_e^{\nu-1} = \mathbf{0}_{\nu N}$

The desired behavior $\mathbf{Y}_d^{\nu-1}(t)$ *of the IO system (2.1) is unstable if and only if the zero equilibrium vector* $\mathbf{y}_e^{\nu-1} = \mathbf{0}_{\nu N}$ *of the IO system (3.77) is unstable.*

Note 263 *Geometrical explanation*

The zero equilibrium vector $\mathbf{y}_e^{\nu-1} = \mathbf{0}_{\nu N}$ *of the IO system (3.77) is* ***unstable*** *if and only if there exists an open hyperball* $\mathfrak{B}_\varepsilon \subset \mathfrak{R}^{N\nu}$ *centered at the origin with the radius* ε*, such that for every open hyperball* $\mathfrak{B}_\delta \subset \mathfrak{R}^{N\nu}$ *there are a moment* $\tau \in \mathfrak{T}_0$ *and* $\mathbf{y}_0^{\nu-1} \in \mathfrak{B}_\delta$ *implying in the free regime that* $\mathbf{y}^{\nu-1}(t; \mathbf{y}_0^{\nu-1}; \mathbf{0}_M)$ *escapes* $\mathfrak{B}_\varepsilon$ *at the instant* τ*,*

$$\exists \mathfrak{B}_\varepsilon \subset \mathfrak{R}^{N\nu}, \ \forall \mathfrak{B}_\delta \subset \mathfrak{R}^{N\nu} \Longrightarrow \exists (\mathbf{y}_0^{\nu-1}, \tau) \in \mathfrak{B}_\delta \times \mathfrak{T}_0 \Longrightarrow$$
$$\mathbf{y}^{\nu-1}(\tau; \mathbf{y}_0^{\nu-1}; \mathbf{0}_M) \in \mathfrak{R}^{N\nu} \backslash \mathfrak{B}_\varepsilon,$$
$$\mathfrak{R}^{N\nu} \backslash \mathfrak{B}_\varepsilon = \left\{\mathbf{y}^{\nu-1} : \ \mathbf{y}^{\nu-1} \in \mathfrak{R}^{N\nu}, \ \mathbf{y}^{\nu-1} \notin \mathfrak{B}_\varepsilon\right\}.$$

Attraction of the zero equilibrium vector

Another qualitative dynamical system property is *attraction* of the equilibrium vector. We present its definition now in the framework of the *IO* systems.

Definition 264 *a) The desired behavior* $\mathbf{Y}_d^{\nu-1}(t)$ *of the IO system (2.1) is* ***attractive*** *if and only if there exists a positive real number* Δ*, or* $\Delta = \infty$*, such that*

$$\left\|\mathbf{Y}_{d0}^{\nu-1} - \mathbf{Y}_0^{\nu-1}\right\| < \Delta$$

implies

$$\lim_{t\to\infty} \left\| \mathbf{Y}_d^{\nu-1}(t) - \mathbf{Y}^{\nu-1}(t; \mathbf{Y}_0^{\nu-1}; \mathbf{I}_N) \right\| = 0. \tag{13.8}$$

The zero equilibrium vector $\mathbf{y}_e^{\nu-1} = \mathbf{0}_{\nu N}$ *of the IO system (3.77) is* ***attractive*** *if and only if*

$$\exists \Delta > 0 \Longrightarrow$$

$$\left\| \mathbf{y}_0^{\nu-1} \right\| < \Delta \Longrightarrow \lim_{t\to\infty} \mathbf{y}^{\nu-1}(t; \mathbf{y}_0^{\nu-1}; \mathbf{0}_M) = \mathbf{0}_{\nu N}. \tag{13.9}$$

b) If and only if a) holds for $\Delta = \infty$ *then the desired behavior* $\mathbf{Y}_d^{\nu-1}(t)$ *of the IO system (2.1), i.e., the zero equilibrium vector* $\mathbf{y}_e^{\nu-1} = \mathbf{0}_{\nu N}$ *of the IO system (3.77), is* ***globally attractive (attractive in the whole)***.

Conclusion 265 ***Attraction of*** $\mathbf{Y}_d^{\nu-1}(t)$ ***and of*** $\mathbf{y}_e^{\nu-1} = \mathbf{0}_{\nu N}$

The desired behavior $\mathbf{Y}_d^{\nu-1}(t)$ *of the IO system (2.1) is (globally) attractive if and only if the zero equilibrium vector* $\mathbf{y}_e^{\nu-1} = \mathbf{0}_{\nu N}$ *of the IO system (3.77) is (globally) attractive, respectively.*

Note 266 ***Geometrical explanation***

The zero equilibrium vector $\mathbf{y}_e^{\nu-1} = \mathbf{0}_{\nu N}$ *of the IO system (3.77) is* ***attractive*** *if and only if there exists an open neighborhood* $\mathfrak{B}_\Delta \subseteq \mathfrak{R}^{N\nu}$ *of* $\mathbf{0}_{\nu N}$ *such that every initial vector* $\mathbf{y}_0^{\nu-1}$ *in* $\mathfrak{B}_\Delta$, $\mathbf{y}_0^{\nu-1} \in \mathfrak{B}_\Delta$, *ensures that the system response* $\mathbf{y}^{\nu-1}(t; \mathbf{y}_0^{\nu-1}; \mathbf{0}_M)$ *approaches asymptotically the equilibrium state* $\mathbf{y}_e^{\nu-1} = \mathbf{0}_{\nu N}$ *as* $t \to \infty$, *Fig. 13.4, (in Section 13.1).*

The global attraction of the zero equilibrium vector $\mathbf{y}_e^{\nu-1} = \mathbf{0}_{\nu N}$ of the *IO* system (3.77) means that for every initial vector $\mathbf{y}_0^{\nu-1} \in \mathfrak{R}^{N\nu}$ the system response $\mathbf{y}^{\nu-1}(t; \mathbf{y}_0^{\nu-1}; \mathbf{0}_M)$ approaches asymptotically the equilibrium state $\mathbf{y}_e^{\nu-1} = \mathbf{0}_{\nu N}$ as $t \to \infty$.

Global attraction of the zero equilibrium vector $\mathbf{y}_e^{\nu-1} = \mathbf{0}_{\nu N}$ of the *IO* system (3.77) implies its attraction. The inverse also holds as the following theorem explains.

Theorem 267 *In order for the zero equilibrium vector* $\mathbf{y}_e^{\nu-1} = \mathbf{0}_{\nu N}$ *of the IO system (3.77) to be globally attractive, it is necessary and sufficient to be attractive.*

Proof. *Necessity.* Let the zero equilibrium vector $\mathbf{y}_e^{\nu-1} = \mathbf{0}_{\nu N}$ of the *IO* system (3.77) be globally attractive. Let Δ be any positive real number. Since $\mathbf{y}^{\nu-1}(t; \mathbf{y}_0^{\nu-1}; \mathbf{0}_M)$ approaches asymptotically the equilibrium state $\mathbf{y}_e^{\nu-1} =$

$\mathbf{0}_{\nu N}$ as $t \to \infty$ for every $\mathbf{y}_0^{\nu-1} \in \mathfrak{R}^{\nu N}$, b) of Definition 264, then $\left\|\mathbf{y}_0^{\nu-1}\right\| < \Delta < \infty$ implies in the free regime $\lim_{t\to\infty} \mathbf{y}^{\nu-1}(t;\mathbf{y}_0^{\nu-1};\mathbf{0}_M) = \mathbf{0}_{\nu N}$. The zero equilibrium vector $\mathbf{y}_e^{\nu-1} = \mathbf{0}_{\nu N}$ of the IO system (3.77) is attractive in view of a) of Definition 264.

Sufficiency. Let the zero equilibrium vector $\mathbf{y}_e^{\nu-1} = \mathbf{0}_{\nu N}$ of the IO system (3.77) be attractive. Its response in the free regime is given by Equation 13.5 through Equation 13.7. Since

$$\mathbf{y}^{\nu-1}(t;\mathbf{y}_0^{\nu-1};\mathbf{0}_M) = \Gamma_{IOy_0}^{\nu-1}(t)\mathbf{y}_0^{\nu-1} \to \mathbf{0}_{\nu N} \ as\ t \to \infty,$$
$$\forall \mathbf{y}_0^{\nu-1} \in \mathfrak{R}^{\nu N} \ obeying\ \left\|\mathbf{y}_0^{\nu-1}\right\| < \Delta,\ a)\ of\ Definition\ 264,$$

then

$$\Gamma_{IOy_0}^{\nu-1}(t) \to O_{\nu N,\nu N} \ as\ t \to \infty.$$

This implies

$$\lim_{t\to\infty} \mathbf{y}^{\nu-1}(t;\mathbf{y}_0^{\nu-1};\mathbf{0}_M) = \mathbf{0}_{\nu N},\ \forall \mathbf{y}_0^{\nu-1} \in \mathfrak{R}^{\nu N},$$

which shows that the zero equilibrium vector $\mathbf{y}_e^{\nu-1} = \mathbf{0}_{\nu N}$ of the IO system (3.77) is globally attractive (part b) of Definition 264. ■

Theorem 267 enables us to establish the relation between the uniqueness of the zero equilibrium vector and its attraction.

Theorem 268 *In order for the zero equilibrium vector* $\mathbf{y}_e^{\nu-1} = \mathbf{0}_{\nu N}$ *of the IO system (3.77) to be attractive, it is necessary (but not sufficient) to be the unique equilibrium vector of the system.*

Proof. *Necessity.* Let the zero equilibrium vector $\mathbf{y}_e^{\nu-1} = \mathbf{0}_{\nu N}$ of the IO system (3.77) be nonunique and attractive. Then there is another equilibrium vector $\mathbf{y}_{e2}^{\nu-1} \neq \mathbf{0}_{\nu N}$ of the IO system (3.77). The zero equilibrium vector $\mathbf{y}_e^{\nu-1} = \mathbf{0}_{\nu N}$ of the IO system is globally attractive due to Theorem 267. However,

$$\lim_{t\to\infty} \mathbf{y}^{\nu-1}(t;\mathbf{y}_{e2}^{\nu-1};\mathbf{0}_M) = \mathbf{y}_{e2}^{\nu-1} \neq \mathbf{0}_{\nu N}$$

due to Definition 97 (in Subsection 3.7.2), of the equilibrium vector of the IO system (3.77), which contradicts global attraction of $\mathbf{y}_e^{\nu-1} = \mathbf{0}_{\nu N}$. Hence, there does not exist another equilibrium vector $\mathbf{y}_{e2}^{\nu-1} \neq \mathbf{0}_{\nu N}$ of the IO system (3.77), i.e., the zero equilibrium vector $\mathbf{y}_e^{\nu-1} = \mathbf{0}_{\nu N}$ of the IO system is unique.

Insufficiency. Let the IO system (3.77) be defined by

$$y^{(1)}(t) - y(t) = i(t).$$

Its equilibrium point $y_e = 0$ is unique. Its response in the free regime reads $y(t; y_0; 0) = e^t y_0$, $\forall t \in \mathfrak{T}_0$, so that $\lim_{t \to \infty} y(t; y_0; 0) = \infty \; sign \; y_0$, $\forall (y_0 \neq 0) \in \mathfrak{R}$. The equilibrium point $y_e = 0$ is not attractive. ■

This theorem and Theorem 99, (in Subsection 3.7.2), yield directly

Theorem 269 *In order for the zero equilibrium vector $\mathbf{y}_e^{\nu-1} = \mathbf{0}_{\nu N}$ of the IO system (3.77) to be attractive, it is necessary (but not sufficient) that the matrix A_0 is nonsingular.*

Asymptotic stability of the zero equilibrium vector

Definition 270 *The desired behavior $\mathbf{Y}_d^{\nu-1}(t)$ of the IO system (2.1), i.e., the zero equilibrium vector $\mathbf{y}_e^{\nu-1} = \mathbf{0}_{\nu N}$ of the IO system (3.77) is:*

a) ***asymptotically stable*** *if and only if it is both stable and attractive,*

b) ***globally asymptotically stable*** *if and only if it is both stable and globally attractive,*

c) ***strictly (completely, fully) globally asymptotically stable*** *if and only if it is both globally stable and globally attractive.*

This definition, Theorem 259, and Theorem 267 imply directly,

Theorem 271 *In order for the zero equilibrium vector $\mathbf{y}_e^{\nu-1} = \mathbf{0}_{\nu N}$ of the IO system (3.77) to be strictly globally asymptotically stable, it is necessary and sufficient that it is asymptotically stable.*

Exponential stability of the zero equilibrium vector

The exponential stability ensures a higher convergence rate of system behaviors to the equilibrium vector than its asymptotic stability.

Definition 272 *a) The desired behavior $\mathbf{Y}_d^{\nu-1}(t)$ of the IO system (2.1) is* ***exponentially stable*** *if and only if there exist positive real numbers $\alpha \geq 1$, β, and Δ, or $\Delta = \infty$, such that*

$$\left\|\mathbf{Y}_{d0}^{\nu-1} - \mathbf{Y}_0^{\nu-1}\right\| < \Delta$$

implies

$$\left\|\mathbf{Y}_d^{\nu-1}(t) - \mathbf{Y}^{\nu-1}(t; \mathbf{Y}_0^{\nu-1}; \mathbf{I}_N)\right\| \leq \alpha \exp(-\beta t) \left\|\mathbf{Y}_{d0}^{\nu-1} - \mathbf{Y}_0^{\nu-1}\right\|, \quad \forall t \in \mathfrak{T}_0. \tag{13.10}$$

The zero equilibrium vector $\mathbf{y}_e^{\nu-1} = \mathbf{0}_{\nu N}$ *of the IO system (3.77) is* ***exponentially stable*** *if and only if*

$$\begin{gathered}\exists\alpha \in \mathfrak{R}^+,\ \alpha \geq 1,\ \exists\beta \in \mathfrak{R}^+, and\ \exists\Delta > 0 \Longrightarrow \\ \left\|\mathbf{y}_0^{\nu-1}\right\| < \Delta \Longrightarrow \\ \left\|\mathbf{y}^{\nu-1}(t;\mathbf{y}_0^{\nu-1};\mathbf{0}_M)\right\| \leq \alpha\exp(-\beta t)\left\|\mathbf{y}_0^{\nu-1}\right\|,\ \forall t \in \mathfrak{T}_0. \end{gathered} \tag{13.11}$$

b) If and only if a) holds for $\Delta = \infty$, *then the desired behavior* $\mathbf{Y}_d^{\nu-1}(t)$ *of the IO system (2.1), i.e., the zero equilibrium vector* $\mathbf{y}_e^{\nu-1} = \mathbf{0}_{\nu N}$ *of the IO system (3.77), is* ***globally exponentially stable (exponentially stable in the whole)****.*

Conclusion 273 ***Exponential stability of*** $\mathbf{Y}_d^{\nu-1}(t)$ ***and of*** $\mathbf{y}_e^{\nu-1} = \mathbf{0}_{\nu N}$

The desired behavior $\mathbf{Y}_d^{\nu-1}(t)$ *of the IO system (2.1) is (globally) exponentially stable if and only if the zero equilibrium vector* $\mathbf{y}_e^{\nu-1} = \mathbf{0}_{\nu N}$ *of the IO system (3.77) is (globally) exponentially stable, respectively.*

The number α determines the upper bound of $\left\|\mathbf{y}^{\nu-1}(t;\mathbf{y}_0^{\nu-1};\mathbf{0}_M)\right\|$ for fixed $\mathbf{y}_{0^-}^{\nu-1}$, which is a measure of the maximal deviation of $\mathbf{y}^{\nu-1}(t;\mathbf{y}_0^{\nu-1};\mathbf{0}_M)$ from the zero equilibrium vector $\mathbf{y}_e^{\nu-1} = \mathbf{0}_{\nu N}$. The number β signifies the minimal convergence rate of $\left\|\mathbf{y}^{\nu-1}(t;\mathbf{y}_0^{\nu-1};\mathbf{0}_M)\right\|$ to zero, i.e., of the system state $\mathbf{y}^{\nu-1}(t;\mathbf{y}_0^{\nu-1};\mathbf{0}_M)$ to the zero equilibrium vector $\mathbf{y}_e^{\nu-1} = \mathbf{0}_{\nu N}$.

Theorem 274 *In order for the zero equilibrium vector* $\mathbf{y}_e^{\nu-1} = \mathbf{0}_{\nu N}$ *of the IO system (3.77) to be globally exponentially stable, it is necessary and sufficient to be exponentially stable.*

The proof of this theorem is analogous to the proof of Theorem 267.

Theorem 275 *In order for the zero equilibrium vector* $\mathbf{y}_e^{\nu-1} = \mathbf{0}_{\nu N}$ *of the IO system (3.77) to be exponentially stable it is necessary and sufficient to be asymptotically stable.*

Proof. *Necessity.* Let the zero equilibrium vector $\mathbf{y}_e^{\nu-1} = \mathbf{0}_{\nu N}$ of the *IO* system (3.77) be exponentially stable. Then, it is attractive because (Definition 272),

$$\begin{gathered}\left\|\mathbf{y}_0^{\nu-1}\right\| < \Delta \Longrightarrow \\ \left\|\mathbf{y}^{\nu-1}(t;\mathbf{y}_0^{\nu-1};\mathbf{0}_M)\right\| \leq \alpha\exp(-\beta t)\left\|\mathbf{y}_0^{\nu-1}\right\|,\ \forall t \in \mathfrak{T}_0,\end{gathered}$$

implies

$$\mathbf{y}^{\nu-1}(t;\mathbf{y}_0^{\nu-1};\mathbf{0}_M) \to \mathbf{0}_{\nu N} \ as \ t \longrightarrow \infty.$$

This proves attraction of $\mathbf{y}_e^{\nu-1} = \mathbf{0}_{\nu N}$. Let $\delta(\varepsilon) = \varepsilon\alpha^{-1}$, $\forall\varepsilon \in \mathfrak{R}^+$. Hence,

$$\left\|\mathbf{y}_0^{\nu-1}\right\| < \delta(\varepsilon) = \varepsilon\alpha^{-1}$$

and

$$\left\|\mathbf{y}^{\nu-1}(t;\mathbf{y}_0^{\nu-1};\mathbf{0}_M)\right\| \leq \alpha\exp(-\beta t)\left\|\mathbf{y}_0^{\nu-1}\right\|, \ \forall t \in \mathfrak{T}_0,$$

guarantee

$$\begin{array}{c}\left\|\mathbf{y}^{\nu-1}(t;\mathbf{y}_0^{\nu-1};\mathbf{0}_M)\right\| \leq \alpha\exp(-\beta t)\left\|\mathbf{y}_0^{\nu-1}\right\| < \\ < \alpha\exp(-\beta t)\varepsilon\alpha^{-1} \leq \varepsilon, \ \forall\varepsilon \in \mathfrak{R}^+, \ \forall t \in \mathfrak{T}_0.\end{array}$$

This proves stability of $\mathbf{y}_e^{\nu-1} = \mathbf{0}_{\nu N}$. Since it is also attractive, then it is asymptotically stable.

Sufficiency. Notice that the IO system (3.77) can be written in the free regime in the ISO form:

$$\frac{d\mathbf{y}^{\nu-1}}{dt} = \begin{bmatrix} \mathbf{y}^{(1)} \\ \mathbf{y}^{(2)} \\ - \\ \mathbf{y}^{(\nu-1)} \\ \mathbf{y}^{(\nu)} \end{bmatrix} =$$

$$= \underbrace{\begin{bmatrix} O_N & I_N & O_N & \dots & O_N & O_N \\ O_N & O_N & I_N & \dots & O_N & O_N \\ - & - & - & \dots & - & - \\ O_N & O_N & O_N & \dots & O_N & I_N \\ -A_\nu^{-1}A_0 & -A_\nu^{-1}A_1 & -A_\nu^{-1}A_2 & \dots & -A_\nu^{-1}A_{\nu-2} & -A_\nu^{-1}A_{\nu-1} \end{bmatrix}}_{A_{IO}} \mathbf{y}^{\nu-1}, \tag{13.12}$$

i.e.,

$$\frac{d\mathbf{w}}{dt} = A_{IO}\mathbf{w}, \ for \ \mathbf{w} = \mathbf{y}^{\nu-1}, \ A_{IO} \in \mathfrak{R}^{\nu N x \nu N}. \tag{13.13}$$

The solution $\mathbf{w}(t;\mathbf{w}_0)$ of (13.13) reads

$$\mathbf{w}(t;\mathbf{w}_0) = e^{A_{IO}t}\mathbf{w}_0,$$

i.e.,

$$\mathbf{y}^{\nu-1}(t;\mathbf{y}_{0\mp}^{\nu-1};\mathbf{0}_M) = e^{A_{IO}t}\mathbf{y}_{0\mp}^{\nu-1}. \tag{13.14}$$

Let the zero equilibrium vector $\mathbf{y}_e^{\nu-1} = \mathbf{0}_{\nu N}$ of the *IO* system (3.77) be asymptotically stable. Then, Definition 256, Definition 270 and (13.14) yield

$$\forall \varepsilon \in \mathfrak{R}^+,\ \exists \delta \in \mathfrak{R}^+,\ \delta = \delta(\varepsilon),\ \Longrightarrow$$
$$\left\| \mathbf{y}_0^{\nu-1} \right\| < \delta \Longrightarrow \left\| e^{A_{IO}t} \mathbf{y}_0^{\nu-1} \right\| < \varepsilon,\ \forall t \in \mathfrak{T}_0, \tag{13.15}$$

Definition 264 and Definition 270 furnish

$$\exists \Delta > 0 \Longrightarrow$$
$$\left\| \mathbf{y}_0^{\nu-1} \right\| < \Delta \Longrightarrow \lim_{t \to \infty} e^{A_{IO}t} \mathbf{y}_0^{\nu-1} = \mathbf{0}_{\nu N}. \tag{13.16}$$

From (13.15) and (13.16) follows that

$$\exists \gamma \in \mathfrak{R}^+ \Longrightarrow \left\| e^{A_{IO}t} \right\| < \gamma,\ \forall t \in \mathfrak{T}_0, \tag{13.17}$$

and

$$\lim_{t \to \infty} e^{A_{IO}t} = O_{\nu N}, \tag{13.18}$$

which, together with (13.14), implies, for $\alpha \geq \gamma$, $\alpha \in \mathfrak{R}^+$, and for some $\beta \in \mathfrak{R}^+$, which exists due to (13.14),

$$\left\| \mathbf{y}^{\nu-1}(t; \mathbf{y}_0^{\nu-1}; \mathbf{0}_M) \right\| = \left\| e^{A_{IO}t} \mathbf{y}_0^{\nu-1} \right\| \leq$$
$$\leq \alpha \exp(-\beta t) \left\| \mathbf{y}_0^{\nu-1} \right\|,\ \forall \mathbf{y}_0^{\nu-1} \in \mathfrak{R}^{N\nu},\ \forall t \in \mathfrak{T}_0. \tag{13.19}$$

The zero equilibrium vector $\mathbf{y}_e^{\nu-1} = \mathbf{0}_{\nu N}$ of the *IO* system (3.77) is globally exponentially stable (Definition 272), hence exponentially stable (Theorem 274). ■

The end of this proof permits the proof of the following:

Theorem 276 *In order for the zero equilibrium vector $\mathbf{y}_e^{\nu-1} = \mathbf{0}_{\nu N}$ of the IO system (3.77) to be exponentially stable, it is necessary and sufficient to be attractive.*

Proof. *Necessity.* Let the zero equilibrium vector $y_e^{\nu-1} = 0_{\nu N}$ of the *IO* system (3.77) be exponentially stable. Then it is attractive that we have proved in the proof of the necessity of the condition of Theorem 275.

Sufficiency. Let the zero equilibrium vector $y_e^{\nu-1} = 0_{\nu N}$ of the *IO* system (3.77) be attractive. The proof of Theorem 275 provides (13.14) through (13.16), which imply (13.17) through (13.19), hence prove exponential stability of zero equilibrium vector $y_e^{\nu-1} = 0_{\nu N}$ of the *IO* system (3.77). It is also global in view of Theorem 274. ■

This theorem, Theorem 268, Theorem 269 and Definition 270 imply directly the following:

Theorem 277 *In order for the zero equilibrium vector* $\mathbf{y}_e^{\nu-1} = \mathbf{0}_{\nu N}$ *of the IO system (3.77) to be exponentially stable, it is necessary (but not sufficient) that* $\mathbf{y}_e^{\nu-1} = \mathbf{0}_{\nu N}$ *is the unique equilibrium vector of the system, i.e., that the matrix* A_0 *is nonsingular.*

Conclusions 257, 262, 265, and 273 result in:

Conclusion 278 ***Stability properties of the total desired behavior and of the zero deviation vector***

The total desired behavior $\mathbf{Y}_d^{\nu-1}(t)$ *the IO system (2.1) and the zero deviation vector* $\mathbf{y}_e^{\nu-1} = \mathbf{0}_{\nu N}$ *of the IO system (3.77) possess the same stability properties.*

13.2.2 *ISO* systems

Stability of the zero equilibrium state

A. M. Lyapunov [88] defined stability of the desired motion and of the zero equilibrium vector of the nonlinear systems in Cauchy (normal, i.e., *IS* systems) form. We will expose it for the *ISO* system (2.17) and (2.18), (in Section 2.2), i.e., for (3.84) and (3.85), (in Subsection 3.5.3).

Definition 279 *a) The desired motion* $\mathbf{X}_d(t)$ *of the ISO system (2.17) and (2.18) is* ***stable*** *if and only if for every positive real number* ε *there exists a positive real number* δ*, the value of which depends on* $\varepsilon, \delta = \delta(\varepsilon)$*, such that* $\|\mathbf{X}_{d0} - \mathbf{X}_0\| < \delta$ *implies* $\|\mathbf{X}_d(t) - \mathbf{X}(t;\mathbf{X}_0;\mathbf{I}_N)\| < \varepsilon$ *for all* $t \in \mathfrak{T}_0$,

The zero equilibrium vector $\mathbf{x}_e = \mathbf{0}_n$ *of the ISO system (3.84) and (3.85) is* ***stable*** *if and only if*

$$\forall\varepsilon \in \mathfrak{R}^+,\ \exists\delta \in \mathfrak{R}^+,\ \delta = \delta(\varepsilon),\ \Longrightarrow$$
$$\|\mathbf{x}_0\| < \delta \Longrightarrow \|\mathbf{x}(t;\mathbf{x}_0;\mathbf{0}_M)\| < \varepsilon,\ \forall t \in \mathfrak{T}_0. \tag{13.20}$$

b) The desired motion $\mathbf{X}_d(t)$ *of the ISO system (2.17) and (2.18), i.e., the zero equilibrium vector* $\mathbf{x}_e = \mathbf{0}_n$ *of the ISO system (3.84) and (3.85), is* ***globally stable*** *(* ***i.e., stable in the whole*** *) if and only if a) holds so that the maximal* $\delta(\varepsilon)$*, denoted by* $\delta_{\max}(\varepsilon)$*, diverges to infinity as* ε *goes to infinity,*

$$\varepsilon \to \infty \Longrightarrow \delta_{\max}(\varepsilon) \to \infty. \tag{13.21}$$

Conclusion 280 ***Stability of*** $\mathbf{X}_d(t)$ ***and of*** $\mathbf{x}_e = \mathbf{0}_n$

For the desired motion $\mathbf{X}_d(t)$ *of the ISO system (2.17) and (2.18) to be* ***stable,*** *it is necessary and sufficient that the zero equilibrium vector* $\mathbf{x}_e = \mathbf{0}_n$ *of the ISO system (3.84) and (3.85) is stable.*

Note 281 *Geometrical explanation*

This definition means in the geometrical sense that the zero equilibrium vector $\mathbf{x}_e = \mathbf{0}_n$ *of the ISO system (3.84) and (3.85) is* ***stable*** *if and only if for every open hyperball* $\mathfrak{B}_\varepsilon \subseteq \mathfrak{R}^n$ *centered at the origin with the radius* ε*, there exists an open hyperball* $\mathfrak{B}_\delta \subseteq \mathfrak{R}^n$ *centered at the origin* $\mathbf{x} = \mathbf{0}_n$ *with the radius* δ*, which depends on* $\mathfrak{B}_\varepsilon$*, such that* $\mathbf{x}_0 \in \mathfrak{B}_\delta$ *guarantees that the system instantaneous state* $\mathbf{x}(t) = \mathbf{x}(t;\mathbf{x}_0;\mathbf{0}_M)$ *stays in* $\mathfrak{B}_\varepsilon$ *all the time at and after* $t = t_0 = 0$*, Fig. 13.2 (in Section 13.1),*

$$\forall \mathfrak{B}_\varepsilon \subseteq \mathfrak{R}^n,\ \exists \mathfrak{B}_\delta \subseteq \mathfrak{R}^n,\ \mathfrak{B}_\delta = \mathfrak{B}_\delta(\mathfrak{B}_\varepsilon),\ \Longrightarrow$$
$$\mathbf{x}_0 \in \mathfrak{B}_\delta \Longrightarrow \mathbf{x}(t;\mathbf{x}_0;\mathbf{0}_M) \in \mathfrak{B}_\varepsilon,\ \forall t \in \mathfrak{T}_0. \tag{13.22}$$

If and only if additionally $\mathfrak{B}_\varepsilon \longrightarrow \mathfrak{R}^n$ *implies* $\mathfrak{B}_\delta \longrightarrow \mathfrak{R}^n$*, then* $\mathbf{x}_e = \mathbf{0}_n$ *is* ***globally stable*** *(i.e.,* ***stable in the whole****).*

It is well known in the linear systems theory (which is easy to verify or see in Section 10.2) that the motion $\mathbf{x}(.;\mathbf{x}_0;\mathbf{0}_M)$ of the *ISO* system (3.84) and (3.85) in the free regime is given by

$$\mathbf{x}(t;\mathbf{x}_0;\mathbf{0}_M) = e^{At}\mathbf{x}_0,\ \forall t \in \mathfrak{T}_0. \tag{13.23}$$

Theorem 282 *In order for the zero equilibrium vector* $\mathbf{x}_e = \mathbf{0}_n$ *of the ISO system (3.84) and (3.85) to be globally stable, it is necessary and sufficient to be stable.*

Proof. *Necessity.* Let the zero equilibrium vector $\mathbf{x}_e = \mathbf{0}_n$ of the *ISO* system (3.84) and (3.85) be globally stable. Then it is stable due to b) of Definition 279.

Sufficiency. Let the zero equilibrium vector $\mathbf{x}_e = \mathbf{0}_n$ of the *ISO* system (3.84) and (3.85) be stable. Then, Definition 279 holds. Let $\varepsilon \in \mathfrak{R}^+$ be arbitrarily chosen. Let $\delta = \delta(\varepsilon) \in \mathfrak{R}^+$ be the corresponding δ obeying Definition 279 so that, due to Equation 13.23,

$$\|\mathbf{x}_0\| < \delta \Longrightarrow \|\mathbf{x}(t;\mathbf{x}_0;\mathbf{0}_M)\| = \left\|e^{At}\mathbf{x}_0\right\| < \varepsilon,\ \forall t \in \mathfrak{T}_0.$$

Let δ_{ij} be the Kronecker delta,

$$\mathbf{w}_k^n = \left[\delta_{1k} \vdots \delta_{2k} \vdots ... \vdots \delta_{nk}\right]^T \in \mathfrak{R}^n,\ k \in \{1,2,...,n\},$$

and

$$\mathbf{x}_0 = \frac{\delta}{2}\mathbf{w}_k^n \Longrightarrow \|\mathbf{x}_0\| = \frac{\delta}{2} < \delta \Longrightarrow$$
$$\|\mathbf{x}(t;\mathbf{x}_0;\mathbf{0}_M)\| = \left\|e^{At}\mathbf{x}_0\right\| = \frac{\delta}{2}\left\|e^{At}\mathbf{w}_k^n\right\| < \varepsilon,\ \forall t \in \mathfrak{T}_0 \Longrightarrow$$

$$\exists \gamma \in \mathfrak{R}^{+} \Longrightarrow \left\| e^{At} \right\| < \gamma, \ \forall t \in \mathfrak{T}_0.$$

Therefore,

$$\forall t \in \mathfrak{T}_0, \ \forall \mathbf{x}_0 \in \mathfrak{R}^n, \ \|\mathbf{x}_0\| < \delta \ for \ \delta = \delta(\varepsilon) = \frac{\varepsilon}{2\gamma} \Longrightarrow$$
$$\|\mathbf{x}(t;\mathbf{x}_0;\mathbf{0}_M)\| = \left\|e^{At}\mathbf{x}_0\right\| \leq \gamma \|\mathbf{x}_0\| < \gamma\delta = \frac{\varepsilon}{2} < \varepsilon, \ \forall t \in \mathfrak{T}_0, \Longrightarrow$$
$$\delta_{\max}(\varepsilon) \geq \delta(\varepsilon) = \frac{\varepsilon}{2\gamma} \rightarrow \infty \ as \ \varepsilon \rightarrow \infty.$$

This proves global stability of the zero equilibrium vector $\mathbf{x}_e = \mathbf{0}_n$ of the *ISO* system (3.84) and (3.85). ■

Definition 283 *a) The zero equilibrium vector* $\mathbf{x}_e = \mathbf{0}_n$ *of the ISO system (3.84) and (3.85) is* ***unstable*** *if and only if there exists a positive real number* ε *such that for every positive real number* δ *there exist an initial vector* $\mathbf{x}_0$ *satisfying* $\|\mathbf{x}_0\| < \delta$ *and a moment* $\tau \in \mathfrak{T}_0$*, which imply* $\|\mathbf{x}_0(\tau;\mathbf{x}_0;\mathbf{0}_M)\| \geq \varepsilon$ *in the free regime,*

$$\exists \varepsilon \in \mathfrak{R}^{+} \longrightarrow \forall \delta \in \mathfrak{R}^{+}, \ \exists (\mathbf{x}_0, \tau) \in \mathfrak{R}^n x \mathfrak{T}_0,$$
$$\|\mathbf{x}_0\| < \delta \Longrightarrow \|\mathbf{x}_0(\tau;\mathbf{x}_0;\mathbf{0}_M)\| \geq \varepsilon.$$

Note 284 *Geometrically interpreted this means the following: the zero equilibrium vector* $\mathbf{x}_e = \mathbf{0}_n$ *of the ISO system (3.84) and (3.85) is* ***unstable*** *if and only if there exists an open hyperball* $\mathfrak{B}_\varepsilon \subseteq \mathfrak{R}^n$ *centered at the origin with the radius* ε*, such that for every open hyperball* $\mathfrak{B}_\delta \subseteq \mathfrak{R}^n$ *there are a moment* $\tau \in \mathfrak{T}_0$ *and* $\mathbf{x}_0 \in \mathfrak{B}_\delta$ *implying in the free regime that* $\mathbf{x}(t;\mathbf{x}_0;\mathbf{0}_M)$ *escapes* $\mathfrak{B}_\varepsilon$ *at the instant* τ*,*

$$\exists \mathfrak{B}_\varepsilon \subseteq \mathfrak{R}^n, \ \forall \mathfrak{B}_\delta \subseteq \mathfrak{R}^n \Longrightarrow \exists (\mathbf{x}_0, \tau) \in \mathfrak{B}_\delta x \mathfrak{T}_0 \Longrightarrow$$
$$\mathbf{x}_0(\tau;\mathbf{x}_0;\mathbf{0}_M) \notin \mathfrak{B}_\varepsilon.$$

Attraction of the zero equilibrium state

Attraction of the equilibrium state $\mathbf{x} = \mathbf{0}_n$ is another important qualitative dynamical system property. We present its definition now in the framework of the *ISO* systems described by (2.17) and (2.18), i.e., (3.84) and (3.85).

Definition 285 *a) The desired motion* $\mathbf{X}_d(t)$ *of the ISO system (2.17) and (2.18) is* ***attractive*** *if and only if there exists a positive real number* Δ*, or* $\Delta = \infty$*, such that* $\|\mathbf{X}_{d0} - \mathbf{X}_0\| < \Delta$ *implies in the free regime*

$$\lim_{t\rightarrow\infty} \left[\mathbf{X}_d(t) - \mathbf{X}(t;\mathbf{X}_0;\mathbf{I}_N)\right] = \mathbf{0}_n,$$

The zero equilibrium vector $\mathbf{x}_e = \mathbf{0}_n$ *of the ISO system (3.84) and (3.85) is* ***attractive*** *if and only if*

$$\exists \Delta > 0 \Longrightarrow \|\mathbf{x}_0\| < \Delta \Longrightarrow \lim_{t \to \infty} \mathbf{x}(t; \mathbf{x}_0; \mathbf{0}_M) = \mathbf{0}_n.$$

b) If and only if a) holds for $\Delta = \infty$ *then the desired motion* $\mathbf{X}_d(t)$ *of the ISO system (2.17) and (2.18), i.e., the zero equilibrium vector* $\mathbf{x}_e = \mathbf{0}_n$ *of the ISO system (3.84) and (3.85), is* ***globally attractive (attractive in the whole)****.*

Conclusion 286 ***Attraction of*** $\mathbf{X}_d(t)$ ***and of*** $\mathbf{x}_e = \mathbf{0}_n$

For the desired motion $\mathbf{X}_d(t)$ *of the ISO system (2.17) and (2.18) to be* ***(globally) attractive,*** *it is necessary and sufficient that the zero equilibrium vector* $\mathbf{x}_e = \mathbf{0}_n$ *of the ISO system (3.84) and (3.85) is (globally) attractive, respectively.*

Note 287 *Geometrical explanation*

The zero equilibrium vector $\mathbf{x}_e = \mathbf{0}_n$ *of the ISO system (3.84) and (3.85) is* ***attractive*** *if and only if there exists an open hyperball* $\mathfrak{B}_\Delta \subseteq \mathfrak{R}^n$ *such that for every initial vector* $\mathbf{x}_0$ *in* $\mathfrak{B}_\Delta$, $\mathbf{x}_0 \in \mathfrak{B}_\Delta$, *the system state* $\mathbf{x}(t; \mathbf{x}_0; \mathbf{0}_M)$ *approaches asymptotically the equilibrium state* $\mathbf{x} = \mathbf{0}_n$ *as* $t \to \infty$, *Fig. 13.4, (in Section 13.1).*

The global attraction of the zero equilibrium state $\mathbf{x} = \mathbf{0}_n$ *of the ISO system (3.84) and (3.85) means that for every initial vector* $\mathbf{x}_0 \in \mathfrak{R}^n$ *the system state* $\mathbf{x}(t; \mathbf{x}_0; \mathbf{0}_M)$ *approaches asymptotically the equilibrium state* $\mathbf{x}_e = \mathbf{0}_n$ *as* $t \to \infty$.

Global attraction of the zero equilibrium vector $\mathbf{x}_e = \mathbf{0}_n$ of the *ISO* system (3.84) and (3.85) implies its attraction. The inverse also holds as the following theorem explains.

Theorem 288 *In order for the zero equilibrium vector* $\mathbf{x}_e = \mathbf{0}_n$ *of the ISO system (3.84) and (3.85) to be globally attractive, it is necessary and sufficient to be attractive.*

Proof. *Necessity.* Let the zero equilibrium vector $\mathbf{x}_e = \mathbf{0}_n$ of the *ISO* system (3.84) and (3.85) be globally attractive. The equilibrium state $\mathbf{x} = \mathbf{0}_n$ is attractive due to b) of Definition 285.

Sufficiency. Let the zero equilibrium vector $\mathbf{x}_e = \mathbf{0}_n$ of the *ISO* system (3.84) and (3.85) be attractive. Its motion in the free regime is given by

Equation 13.23. Since

$$\mathbf{x}(t;\mathbf{x}_0;\mathbf{0}_M) = e^{At}\mathbf{x}_0 \to 0 \; as \; t \to \infty,$$
$$\forall \mathbf{x}_0 \in \mathfrak{R}^n \; obeying \; \|\mathbf{x}_0\| < \Delta, \; a) \; of \; Definition \; 285,$$

then

$$e^{At} \to O_n \; as \; t \to \infty.$$

This implies

$$\lim_{t\to\infty} \mathbf{x}(t;\mathbf{x}_0;\mathbf{0}_M) = \lim_{t\to\infty} e^{At}\mathbf{x}_0 = \mathbf{0}_n, \; \forall \mathbf{x}_0 \in \mathfrak{R}^n,$$

which shows that the zero equilibrium vector $\mathbf{x}_e = \mathbf{0}_n$ of the ISO system (3.84) and (3.85) is globally attractive (the part # b) of Definition 285). ∎

This theorem permits us to state the relation between the uniqueness of the zero equilibrium vector and its attraction.

Theorem 289 *In order for the zero equilibrium vector* $\mathbf{x}_e = \mathbf{0}_n$ *of the* ISO *system (3.84) and (3.85) to be attractive, it is necessary (but not sufficient) that it is the unique equilibrium vector of the system.*

Proof. *Necessity.* Let the zero equilibrium vector $\mathbf{x}_e = \mathbf{0}_n$ of the ISO system (3.84) and (3.85) be nonunique and attractive. Then there is another equilibrium vector $\mathbf{x}_{e2} \neq \mathbf{0}_n$ of the ISO system (3.84) and (3.85). The zero equilibrium vector $\mathbf{x}_e = \mathbf{0}_n$ of the ISO system is globally attractive due to Theorem 288. However,

$$\lim_{t\to\infty} \mathbf{x}(t;\mathbf{x}_{e2};\mathbf{0}_M) = \mathbf{x}_{e2} \neq \mathbf{0}_n$$

due to Definition 102, (in Subsection 3.7.3), of the equilibrium vector of the ISO system (3.84) and (3.85), which contradicts global attraction of $\mathbf{x}_e = \mathbf{0}_n$. Hence, there does not exist another equilibrium vector $\mathbf{x}_{\mathbf{e2}} \neq \mathbf{0}_n$ of the ISO system (3.84) and (3.85). The zero equilibrium vector $\mathbf{x}_e = \mathbf{0}_n$ of the ISO system is unique.

Insufficiency. Let the ISO system (3.84) and (3.85) be defined by

$$\frac{dx(t)}{dt} = x(t) + i(t), \; y = x.$$

Its equilibrium point $x_e = 0$ is unique. Its motion in the free regime reads $x(t;x_0;0) = e^t x_0$ so that $\lim_{t\to\infty} x(t;x_0;0) = \infty sign x_0, \; \forall (x_0 \neq 0) \in \mathfrak{R}$. The equilibrium point $x_e = 0$ is not attractive. ∎

This theorem and Theorem 104, (in Subsection 3.7.3), yield directly,

Theorem 290 *In order for the zero equilibrium vector* $\mathbf{x}_e = \mathbf{0}_n$ *of the ISO system (3.84) and (3.85) to be attractive, it is necessary (but not sufficient) that the matrix A is nonsingular.*

Asymptotic stability of the zero equilibrium state

Definition 291 *The desired motion* $\mathbf{X}_d(t)$ *of the ISO system (2.17) and (2.18), i.e., the zero equilibrium vector* $\mathbf{x}_e = \mathbf{0}_n$ *of the ISO system (3.84) and (3.85), is:*

a) ***asymptotically stable*** *if and only if it is both stable and attractive,*

b) ***globally asymptotically stable*** *if and only if it is both stable and globally attractive,*

c) ***strictly (completely, fully) globally asymptotically stable*** *if and only if it is both globally stable and globally attractive.*

This definition, Theorem 282, and Theorem 288 imply directly,

Theorem 292 *In order for the zero equilibrium vector* $\mathbf{x}_e = \mathbf{0}_n$ *of the ISO system (3.84) and (3.85) to be strictly globally asymptotically stable, it is necessary and sufficient to be asymptotically stable.*

Exponential stability of the zero equilibrium state

The exponential stability is originally defined in the framework of the *ISO* systems.

Definition 293 *a) The desired motion* $\mathbf{X}_d(t)$ *of the ISO system (2.17) and (2.18) is* ***exponentially stable*** *if and only if there exist positive real numbers* $\alpha \geq 1$, β, *and* Δ, *or* $\Delta = \infty$, *such that* $\|\mathbf{X}_{d0} - \mathbf{X}_0\| < \Delta$ *implies*

$$\|\mathbf{X}_d(t) - \mathbf{X}(t;\mathbf{x}_0;\mathbf{I}_N)\| \leq \alpha \exp(-\beta t) \|\mathbf{X}_{d0} - \mathbf{X}_0\|$$

for all $t \in \mathfrak{T}_0$.

The zero equilibrium state $\mathbf{x}_e = \mathbf{0}_n$ *of the ISO system (3.84) and (3.85) is* ***exponentially stable*** *if and only if there exist positive real numbers* $\alpha \geq 1$, β, *and* Δ, *or* $\Delta = \infty$, *such that* $\|\mathbf{x}_0\| < \Delta$ *implies in the free regime* $\|\mathbf{x}(t;\mathbf{x}_0;\mathbf{0}_M)\| \leq \alpha \exp(-\beta t) \|\mathbf{x}_0\|$ *for all* $t \in \mathfrak{T}_0$,

$$\begin{gathered}\exists \alpha \in \mathfrak{R}^+,\ \alpha \geq 1,\ \exists \beta \in \mathfrak{R}^+,\ \text{and } \exists \Delta > 0 \Longrightarrow \\ \|\mathbf{x}_0\| < \Delta \Longrightarrow \\ \|\mathbf{x}(t;\mathbf{x}_0;\mathbf{0}_M)\| \leq \alpha \exp(-\beta t) \|\mathbf{x}_0\|,\ \forall t \in \mathfrak{T}_0. \end{gathered} \tag{13.24}$$

b) If and only if a) holds for $\Delta = \infty$, then the desired motion $\mathbf{X}_d(t)$ of the ISO system (2.17) and (2.18), i.e., the zero equilibrium state $\mathbf{x}_e = \mathbf{0}_n$ of the ISO system (3.84) and (3.85), is ***globally exponentially stable (exponentially stable in the whole).***

The number α specifies the upper bound of $\|\mathbf{x}(t;\mathbf{x}_0;\mathbf{0}_M)\|$. It serves as a measure of the maximal deviation of $\mathbf{x}(t;\mathbf{x}_0;\mathbf{0}_M)$ from the zero equilibrium state $\mathbf{x}_e = \mathbf{0}_n$. The number β expresses the minimal convergence rate of $\|\mathbf{x}(t;\mathbf{x}_0;\mathbf{0}_M)\|$ to zero, i.e., of $\mathbf{x}(t;\mathbf{x}_0;\mathbf{0}_M)$ to the zero equilibrium state $\mathbf{x}_e = \mathbf{0}_n$.

Conclusion 294 ***Exponential stability of $\mathbf{X}_d(t)$ and of $\mathbf{x}_e = \mathbf{0}_n$***
For the desired motion $\mathbf{X}_d(t)$ of the ISO system (2.17) and (2.18) to be ***(globally) exponentially stable,*** *it is necessary and sufficient that the zero equilibrium vector $\mathbf{x}_e = \mathbf{0}_n$ of the ISO system (3.84) and (3.85) is (globally) exponentially stable, respectively.*

Theorem 295 *In order for the zero equilibrium state $\mathbf{x}_e = \mathbf{0}_n$ of the ISO system (3.84) and (3.85) to be globally exponentially stable, it is necessary and sufficient to be exponentially stable.*

The proof of this theorem is analogous to the proof of Theorem 288.

Theorem 296 *In order for the zero equilibrium state $\mathbf{x}_e = \mathbf{0}_n$ of the ISO system (3.84) and (3.85) to be exponentially stable, it is necessary and sufficient to be asymptotically stable.*

Proof. *Necessity.* Let the zero equilibrium state $\mathbf{x}_e = \mathbf{0}_n$ of the *ISO* system (3.84) and (3.85) be exponentially stable. Then it is attractive because (Definition 293),

$$\|\mathbf{x}_0\| < \Delta \Longrightarrow \|\mathbf{x}(t;\mathbf{x}_0;\mathbf{0}_M)\| \leq \alpha \exp(-\beta t) \|\mathbf{x}_0\|, \ \forall t \in \mathfrak{T}_0,$$

which implies

$$t \longrightarrow \infty \Longrightarrow \mathbf{x}(t;\mathbf{x}_0;\mathbf{0}_M) \longrightarrow \mathbf{0}_n.$$

This proves the attraction of $\mathbf{x}_e = \mathbf{0}_n$. Let $\delta(\varepsilon) = \varepsilon\alpha^{-1}$, $\forall \varepsilon \in \mathfrak{R}^+$. Hence,

$$\|\mathbf{x}_0\| < \delta(\varepsilon) = \varepsilon\alpha^{-1}$$

and

$$\|\mathbf{x}(t;\mathbf{x}_0;\mathbf{0}_M)\| \leq \alpha \exp(-\beta t) \|\mathbf{x}_0\|, \ \forall t \in \mathfrak{T}_0,$$

guarantee

$$\|\mathbf{x}(t;\mathbf{x}_0;\mathbf{0}_M)\| \leq \alpha\exp(-\beta t)\,\|\mathbf{x}_0\| < \\ < \alpha\exp(-\beta t)\varepsilon\alpha^{-1} \leq \varepsilon,\ \forall\varepsilon\in\mathfrak{R}^+,\ \forall t\in\mathfrak{T}_0.$$

This proves stability of $\mathbf{x}_e = \mathbf{0}_n$. Since it is also attractive, then it is asymptotically stable.

Sufficiency. Let the zero equilibrium state $\mathbf{x}_e = \mathbf{0}_n$ of the ISO system (3.84) and (3.85) be asymptotically stable. The system solutions are found in the form

$$\mathbf{x}(t;\mathbf{x}_0;\mathbf{0}_M) = e^{At}\mathbf{x}_0,\ \forall t\in\mathfrak{T}_0.$$

This yields

$$\|\mathbf{x}_0\| < \Delta \Longrightarrow \|\mathbf{x}(t;\mathbf{x}_0;\mathbf{0}_M)\| = \left\|e^{At}\mathbf{x}_0\right\| \leq \\ \leq \left\|e^{At}\right\|\|\mathbf{x}_0\|,\ \forall t\in\mathfrak{T}_0.$$

By repeating the procedure of the proof of sufficiency of Theorem 259, (in Subsection 13.2.1), we prove that

$$\exists\gamma\in\mathfrak{R}^+ \Longrightarrow$$

$$\left\|e^{At}\right\| < \gamma\exp(-\beta t),\ \forall t\in\mathfrak{T}_0.$$

This further implies, for $\alpha\geq\gamma$, $\alpha\in\mathfrak{R}^+$,

$$\|\mathbf{x}(t;\mathbf{x}_0;\mathbf{0}_M)\| = \left\|e^{At}\mathbf{x}_0\right\| \leq \left\|e^{At}\right\|\|\mathbf{x}_0\| \leq \\ \leq \alpha\exp(-\beta t)\,\|\mathbf{x}_0\|,\ \forall\mathbf{x}_0\in\mathfrak{R}^n,\ \forall t\in\mathfrak{T}_0.$$

The zero equilibrium state $\mathbf{x}_e = \mathbf{0}_n$ of the IO system (3.84) and (3.85) is globally exponentially stable (Definition 293), hence exponentially stable. ■

This theorem and Theorem 290 imply directly the following:

Theorem 297 *In order for the zero equilibrium state $\mathbf{x}_e = \mathbf{0}_n$ of the ISO system (3.84) and (3.85) to be exponentially stable, it is necessary (but not sufficient) that $\mathbf{x}_e = \mathbf{0}_n$ is the unique equilibrium state of the system, i.e., that the matrix A is nonsingular.*

We summarize Conclusions 280, 286 and 294:

Conclusion 298 ***Stability properties of $\mathbf{X}_d(t)$ and of $\mathbf{x}_e = \mathbf{0}_n$***

The desired motion $\mathbf{X}_d(t)$ of the ISO system (2.17) and (2.18) and the zero equilibrium state $\mathbf{x}_e = \mathbf{0}_n$ of the ISO system (3.84) and (3.85) possess the same stability and attraction properties.

13.2.3 *IIO* systems

Stability of the zero equilibrium vector

The definitions of the stability properties of $\mathbf{r}_e^{\alpha-1} = \mathbf{0}_{\alpha\rho}$ are the same as the definitions of the corresponding stability properties of $\mathbf{y}_e^{\nu-1} = \mathbf{0}_{\nu N}$ of the *IO* system (in Section 13.2.1).

If $\nu = 0$, then the overall system equilibrium state of the *IIO* system is $\mathbf{r}_e^{\alpha-1} = \mathbf{0}_{\alpha\rho}$ and the overall system dynamics reduces to its internal dynamics, i.e., to its state behavior.

We introduce the precise definition in the Lyapunov sense of stability of the zero equilibrium vector $[(\mathbf{r}_e^{\alpha-1})^T \vdots (\mathbf{y}_e^{\nu-1})^T\,]^T$ of the *IIO* system (3.90), (in Subsection 3.5.4), [88].

Definition 299 *a) The zero equilibrium vector* $[(\mathbf{r}_e^{\alpha-1})^T \vdots (\mathbf{y}_e^{\nu-1})^T\,]^T = \mathbf{0}_{\alpha\rho+\nu N}$ *of the IIO system (3.90), in which* $\nu \geq 1$, *is* ***stable*** *if and only if for every positive real number* ε *there exists a positive real number* δ, *the value of which depends on* ε, $\delta = \delta(\varepsilon)$, *such that* $\left\| [\mathbf{r}_0^{\alpha-1} \vdots\, \mathbf{y}_0^{\nu-1}]^T \right\| < \delta$ *implies in the free regime*

$$\left\| \left[\left(\mathbf{r}^{\alpha-1}(t;\mathbf{r}_0^{\alpha-1};\mathbf{0}_M)\right)^T \vdots \left(\mathbf{y}^{\nu-1}(t;\mathbf{r}_0^{\alpha-1};\mathbf{y}_0^{\nu-1};\mathbf{0}_M)\right)^T \right]^T \right\| < \varepsilon$$

for all $t \in \mathfrak{T}_0$,

$$\begin{gathered} \forall \varepsilon \in \mathfrak{R}^+,\ \exists \delta \in \mathfrak{R}^+,\ \delta = \delta(\varepsilon),\ \Longrightarrow \\ \left\| [(\mathbf{r}_0^{\alpha-1})^T \vdots (\mathbf{y}_0^{\nu-1})^T\,]^T \right\| < \delta \Longrightarrow \\ \left\| \left[\left(\mathbf{r}^{\alpha-1}(t;\mathbf{r}_0^{\alpha-1};\mathbf{0}_M)\right)^T \vdots \left(\mathbf{y}^{\nu-1}(t;\mathbf{r}_0^{\alpha-1};\mathbf{y}_0^{\nu-1};\mathbf{0}_M)\right)^T \right]^T \right\| < \varepsilon,\ \forall t \in \mathfrak{T}_0. \end{gathered} \tag{13.25}$$

b) The zero equilibrium vector $[(\mathbf{r}_e^{\alpha-1})^T \vdots (\mathbf{y}_e^{\nu-1})^T\,]^T = \mathbf{0}_{\alpha\rho+\nu N}$ *of the IIO system (3.90), in which* $\nu \geq 1$, *is* ***globally stable*** *(i.e.,* ***stable in the whole****) if and only if a) holds so that the maximal* $\delta(\varepsilon)$, *denoted by* $\delta_{\max}(\varepsilon)$, *diverges to infinity as* ε *goes to infinity,*

$$\varepsilon \to \infty \Longrightarrow \delta_{\max}(\varepsilon) \to \infty. \tag{13.26}$$

Note 300 ***Geometrical explanation***

The geometrical sense of this definition is the same as for the IO systems and for the ISO systems, i.e., that the zero equilibrium vector $[(\mathbf{r}_e^{\alpha-1})^T \vdots (\mathbf{y}_e^{\nu-1})^T]^T = \mathbf{0}_{\alpha\rho+\nu N}$ *of the IIO system (3.90), in which* $\nu \geq 1$, *is* ***stable*** *if and only if for every open hyperball* $\mathfrak{B}_\varepsilon \subseteq \mathfrak{R}^{\alpha\rho+\nu N}$ *centered at the origin with the radius* ε, *there exists an open hyperball* $\mathfrak{B}_\delta \subseteq \mathfrak{R}^{\alpha\rho+\nu N}$ *centered at the origin with the radius* δ, *which depends on* $\mathfrak{B}_\varepsilon$, *such that* $[(\mathbf{r}_0^{\alpha-1})^T \vdots (\mathbf{y}_0^{\nu-1})^T]^T \in \mathfrak{B}_\delta$ *guarantees that*

$$[(\mathbf{r}^{\alpha-1}(t;\mathbf{r}_0^{\alpha-1};\mathbf{0}_M))^T \vdots (\mathbf{y}^{\nu-1}(t;\mathbf{r}_0^{\alpha-1};\mathbf{y}_0^{\nu-1};\mathbf{0}_M))^T]^T$$

stays in $\mathfrak{B}_\varepsilon$ *all the time from* $t = t_0 = 0$ *on, Fig. 13.2, (in Section 13.1),*

$$\forall\mathfrak{B}_\varepsilon \subseteq \mathfrak{R}^{\alpha\rho+\nu N},\ \exists\mathfrak{B}_\delta \subseteq \mathfrak{R}^{\alpha\rho+\nu N},\ \mathfrak{B}_\delta = \mathfrak{B}_\delta(\mathfrak{B}_\varepsilon),\ \Longrightarrow$$
$$[(\mathbf{r}_0^{\alpha-1})^T \ \vdots\ (\mathbf{y}_0^{\nu-1})^T\ ]^T \in \mathfrak{B}_\delta \Longrightarrow$$
$$[(\mathbf{r}^{\alpha-1}(t;\mathbf{r}_0^{\alpha-1};\mathbf{0}_M))^T \ \vdots\ (\mathbf{y}^{\nu-1}(t;\mathbf{r}_0^{\alpha-1};\mathbf{y}_0^{\nu-1};\mathbf{0}_M))^T]^T \in \mathfrak{B}_\varepsilon,\ \forall t \in \mathfrak{T}_0. \tag{13.27}$$

The dynamic behavior of the *IIO* system (3.90) is given by (10.30), (in Section 10.3), which we apply here for the *IIO* system in the free regime,

$$\mathbf{r}(t;\mathbf{r}_0^{\alpha-1};\mathbf{0}_M) = \Gamma_{IIOroIS}(t)\mathbf{r}_0^{\alpha-1},\ \Gamma_{IIOroIS}(t) \in \mathfrak{R}^{\rho\mathrm{x}\alpha\rho}, \tag{13.28}$$

which yields

$$\mathbf{r}^{(k)}(t;\mathbf{r}_0^{\alpha-1};\mathbf{0}_M) = \Gamma_{IIOroIS}^{(k)}(t)\mathbf{r}_0^{\alpha-1},$$
$$\forall t \in \mathfrak{T}_0,\ k = 0,1,...,\nu-1, \tag{13.29}$$

where

$$\Gamma_{IIOroIS}^{\alpha-1}(t) = \begin{bmatrix} \Gamma_{IIOroIS}(t) \\ \Gamma_{IIOroIS}^{(1)}(t) \\ \cdots\cdots \\ \Gamma_{IIOroIS}^{(\alpha-1)}(t) \end{bmatrix} \in \mathfrak{R}^{\alpha\rho\mathrm{x}\alpha\rho}, \tag{13.30}$$

so that

$$\mathbf{r}^{\alpha-1}(t;\mathbf{r}_0^{\alpha-1};\mathbf{0}_M) = \Gamma_{IIOroIS}^{\alpha-1}(t)\mathbf{r}_0^{\alpha-1} \in \mathfrak{R}^{\alpha\rho}. \tag{13.31}$$

and by Equation 10.42, (in Section 10.3),

$$\mathbf{y}(t;\mathbf{r}_{0^-}^{\alpha-1};\mathbf{y}_0^{\nu-1};\mathbf{i}) = \left(\int_{0^-}^{t} \Gamma_{IIO}(\tau)\mathbf{i}(t-\tau)d\tau\right) + \Gamma_{IIOi_0}(t)\mathbf{i}_0^{\gamma-1} +$$
$$+\Gamma_{IIOr_0}(t)\mathbf{r}_0^{\alpha-1} + \Gamma_{IIOy_0}(t)\mathbf{y}_0^{\nu-1},\ \forall t \in \mathfrak{T}_0, \tag{13.32}$$

which becomes for the free regime

$$\mathbf{y}^{(k)}(t;\mathbf{r}_{0^-}^{\alpha-1};\mathbf{y}_0^{\nu-1};\mathbf{0}_M)=\Gamma_{IIOr_0}^{(k)}(t)\mathbf{r}_0^{\alpha-1}+\Gamma_{IIOy_0}^{(k)}(t)\mathbf{y}_0^{\nu-1},$$
$$\forall t\in\mathfrak{T}_0,\ k=0,1,...,\nu-1,$$

and

$$\mathbf{y}^{\nu-1}(t;\mathbf{r}_{0^-}^{\alpha-1};\mathbf{y}_0^{\nu-1};\mathbf{0}_M)=\Gamma_{IIOr_0}^{\nu-1}(t)\mathbf{r}_0^{\alpha-1}+\Gamma_{IIOy_0}^{\nu-1}(t)\mathbf{y}_0^{\nu-1}=$$
$$=\Gamma_{IIOryo}^{\nu-1}(t)\begin{bmatrix}\mathbf{r}_0^{\alpha-1}\\ \mathbf{y}_0^{\nu-1}\end{bmatrix},\ \forall t\in\mathfrak{T}_0, \tag{13.33}$$

where

$$\Gamma_{IIOr_0}^{\nu-1}(t)=\begin{bmatrix}\Gamma_{IIOr_0}(t)\\ \Gamma_{IIOr_0}^{(1)}(t)\\ .....\\ \Gamma_{IIOr_0}^{(\nu-1)}(t)\end{bmatrix}\in\mathfrak{R}^{\nu N\times\alpha\rho},$$

$$\Gamma_{IIOy_0}^{\nu-1}(t)=\begin{bmatrix}\Gamma_{IIOy_0}(t)\\ \Gamma_{IIOy_0}^{(1)}(t)\\ .....\\ \Gamma_{IIOy_0}^{(\nu-1)}(t)\end{bmatrix}\in\mathfrak{R}^{\nu N\times\nu N},$$

$$\Gamma_{IIOry_0}^{\nu-1}(t)=\left[\Gamma_{IIOr_0}^{\nu-1}(t)\ \vdots\ \Gamma_{IIOy_0}^{\nu-1}(t)\right]\in\mathfrak{R}^{\nu N\times(\alpha\rho+\nu N)}. \tag{13.34}$$

In order to simplify the exposition, let

$$\Gamma_{IIOo}(t)=$$
$$=\begin{bmatrix}\Gamma_{IIOr_0IS}^{\alpha-1}(t)\ \vdots & O_{\rho,\nu N}\\ \Gamma_{IIOr_0}^{\nu-1}(t)\ \vdots & \Gamma_{IIOy_0}^{\nu-1}(t)\end{bmatrix}\in\mathfrak{R}^{(\alpha\rho+\nu N)\times(\alpha\rho+\nu N)}, \tag{13.35}$$

so that

$$\begin{bmatrix}\mathbf{r}^{\alpha-1}(t;\mathbf{r}_0^{\alpha-1};\mathbf{0}_M)\\ \mathbf{y}^{\nu-1}(t;\mathbf{r}_0^{\alpha-1};\mathbf{y}_0^{\nu-1};\mathbf{0}_M)\end{bmatrix}=\Gamma_{IIOo}(t)\begin{bmatrix}\mathbf{r}_0^{\alpha-1}\\ \mathbf{y}_0^{\nu-1}\end{bmatrix}\in\mathfrak{R}^{\alpha\rho+\nu N}. \tag{13.36}$$

Theorem 301 *In order for the zero equilibrium state* $[(\mathbf{r}_e^{\alpha-1})^T \vdots (\mathbf{y}_e^{\nu-1})^T]^T = \mathbf{0}_{\alpha\rho+\nu N}$ *of the IIO system (3.90), in which* $\nu\geq 1$, *to be globally stable, it is necessary and sufficient to be stable.*

Proof. *Necessity.* Let the zero equilibrium vector $[(\mathbf{r}_e^{\alpha-1})^T \vdots (\mathbf{y}_e^{\nu-1})^T\ ]^T = \mathbf{0}_{\alpha\rho+\nu N}$ of the *IIO* system (3.90), in which $\nu \geq 1$, be globally stable. Then it is stable due to b) of Definition 299.

Sufficiency. Let the zero equilibrium vector $[(\mathbf{r}_e^{\alpha-1})^T \vdots (\mathbf{y}_e^{\nu-1})^T\ ]^T = \mathbf{0}_{\alpha\rho+\nu N}$ of the *IIO* system (3.90), in which $\nu \geq 1$, be stable. Then, Definition (299) holds. Let $\varepsilon \in \mathfrak{R}^+$ be arbitrarily chosen. Let $\delta = \delta(\varepsilon) \in \mathfrak{R}^+$ be the corresponding δ obeying Definition 299 so that, due to Equation 13.33 through Equation 13.36,

$$\left\|[(\mathbf{r}_0^{\alpha-1})^T \vdots (\mathbf{y}_0^{\nu-1})^T\ ]^T\right\| < \delta \Longrightarrow$$

$$\left\|[(\mathbf{r}^{\alpha-1}(t;\mathbf{r}_0^{\alpha-1};\mathbf{0}_M))^T \vdots (\mathbf{y}^{\nu-1}(t;\mathbf{r}_0^{\alpha-1};\mathbf{y}_0^{\nu-1};\mathbf{0}_M))^T]^T\right\| =$$

$$= \left\|\Gamma_{IIOo}(t)[(\mathbf{r}_0^{\alpha-1})^T \vdots (\mathbf{y}_0^{\nu-1})^T\ ]^T\right\| < \varepsilon,\ \forall t \in \mathfrak{T}_0.$$

Let δ_{ij} be the Kronecker delta, $L = \alpha\rho+\nu N$, $\mathbf{w}_k^L = \left[\delta_{1k} \vdots \delta_{2k} \vdots ... \vdots \delta_{Lk}\right]^T \in \mathfrak{R}^L$, $k \in \{1,2,...,L\}$, and

$$[(\mathbf{r}_{0k-}^{\alpha-1})^T \vdots (\mathbf{y}_{0k-}^{\nu-1})^T]^T = \frac{\delta}{2}\mathbf{w}_k^L \Longrightarrow$$

$$\left\|[(\mathbf{r}_{0k-}^{\alpha-1})^T \vdots (\mathbf{y}_{0k-}^{\nu-1})^T]^T\right\| = \frac{\delta}{2} < \delta \Longrightarrow$$

$$\left\|\left[(\mathbf{r}^{\alpha-1}(t;\mathbf{r}_{0k}^{\alpha-1};\mathbf{0}_M))^T \vdots (\mathbf{y}^{\nu-1}(t;\mathbf{r}_{0k}^{\alpha-1};\mathbf{y}_{0k}^{\nu-1};\mathbf{0}_M))^T\right]^T\right\| =$$

$$= \left\|\Gamma_{IIOo}\left[(\mathbf{r}_{0k-}^{\alpha-1})^T \vdots (\mathbf{y}_{0k-}^{\nu-1})^T\right]^T\right\| =$$

$$= \frac{\delta}{2}\left\|\Gamma_{IIOo}\mathbf{w}_k^L\right\| < \varepsilon,\ \forall t \in \mathfrak{T}_0 \Longrightarrow$$

$$\exists \gamma \in \mathfrak{R}^+ \Longrightarrow \|\Gamma_{IIOo}(t)\| < \gamma,\ \forall t \in \mathfrak{T}_0 \Longrightarrow$$

$$\forall \left[(\mathbf{r}_{0-}^{\alpha-1})^T \vdots (\mathbf{y}_{0-}^{\nu-1})^T\right]^T \in \mathfrak{R}^{\alpha\rho+\nu N},$$

$$\left\|[(\mathbf{r}_0^{\alpha-1})^T \vdots (\mathbf{y}_0^{\nu-1})^T\ ]^T\right\| < \delta \ for\ \delta = \delta(\varepsilon) = \frac{\varepsilon}{2\gamma} \Longrightarrow$$

$$\left\| \left[\left(\mathbf{r}^{\alpha-1}(t;\mathbf{r}_0^{\alpha-1};\mathbf{0}_M)\right)^T \vdots \left(\mathbf{y}^{\nu-1}(t;\mathbf{r}_0^{\alpha-1};\mathbf{y}_0^{\nu-1};\mathbf{0}_M)\right)^T \right]^T \right\| =$$

$$= \left\| \Gamma_{IIOo}[(\mathbf{r}_0^{\alpha-1})^T \vdots (\mathbf{y}_0^{\nu-1})^T]^T \right\| \leq$$

$$\leq \gamma \left\| [(\mathbf{r}_0^{\alpha-1})^T \vdots (\mathbf{y}_0^{\nu-1})^T]^T \right\| < \delta\gamma = \frac{\varepsilon}{2} < \varepsilon, \ \forall t \in \mathfrak{T}_0,$$

$$\delta_{\max}(\varepsilon) \geq \delta(\varepsilon) = \frac{\varepsilon}{2\gamma} \rightarrow \infty \ as \ \varepsilon \rightarrow \infty.$$

This proves global stability of the zero equilibrium state $[(\mathbf{r}_e^{\alpha-1})^T \vdots (\mathbf{y}_e^{\nu-1})^T]^T = \mathbf{0}_{\alpha\rho+\nu N}$ of the IIO system (3.90), in which $\nu \geq 1$. ■

Definition 302 *a) The zero equilibrium state* $[(\mathbf{r}_e^{\alpha-1})^T \vdots (\mathbf{y}_e^{\nu-1})^T]^T = \mathbf{0}_{\alpha\rho+\nu N}$ *of the IIO system (3.90), in which* $\nu \geq 1$, *is* ***unstable*** *if and only if there exists a positive real number* ε *such that for every positive real number* μ *there exist a moment* $\tau \in \mathfrak{T}_0$ *and an initial vector* $[(\mathbf{r}_e^{\alpha-1})^T \vdots (\mathbf{y}_e^{\nu-1})^T]^T$ *satisfying* $\left\| [(\mathbf{r}_0^{\alpha-1})^T \vdots (\mathbf{y}_0^{\nu-1})^T]^T \right\| < \mu$, *which imply*

$$\left\| \left[\mathbf{r}^{\alpha-1^T}(\tau;\mathbf{r}_0^{\alpha-1};\mathbf{0}_M) \vdots \mathbf{y}^{\nu-1^T}(\tau;\mathbf{r}_0^{\alpha-1};\mathbf{y}_0^{\nu-1};\mathbf{0}_M) \right]^T \right\| \geq \varepsilon$$

in the free regime,

$$\exists \varepsilon \in \mathfrak{R}^+ \longrightarrow \forall \mu \in \mathfrak{R}^+, \ \exists([\mathbf{r}_0^{\alpha-1^T} \vdots \mathbf{y}_0^{\nu-1^T}]^T, \tau) \in \mathfrak{R}^{\alpha\rho+N\nu} x \mathfrak{T}_0, \Longrightarrow$$

$$\left\| [(\mathbf{r}_0^{\alpha-1})^T \vdots (\mathbf{y}_0^{\nu-1})^T]^T \right\| < \mu \Longrightarrow$$

$$\left\| \left[\mathbf{r}^{\alpha-1^T}(\tau;\mathbf{r}_0^{\alpha-1};\mathbf{0}_M) \ \vdots \ \mathbf{y}^{\nu-1^T}(\tau;\mathbf{r}_0^{\alpha-1};\mathbf{y}_0^{\nu-1};\mathbf{0}_M) \right]^T \right\| \geq \varepsilon.$$

Note 303 ***Geometrical explanation***

In the geometrical sense this means the following: the zero equilibrium vector $[(\mathbf{r}_e^{\alpha-1})^T \vdots (\mathbf{y}_e^{\nu-1})^T]^T = \mathbf{0}_{\alpha\rho+\nu N}$ *of the IIO system (3.90), in which* $\nu \geq 1$, *is* ***unstable*** *if and only if there exists an open hyperball* $\mathfrak{B}_\varepsilon \subseteq \mathfrak{R}^{\alpha\rho+N\nu}$ *centered at the origin with the radius* ε, *such that for every open hyperball* $\mathfrak{B}_\mu \subseteq \mathfrak{R}^{2\alpha\rho+N\nu}$ *there are a moment* $\tau \in \mathfrak{T}_0$ *and* $[(\mathbf{r}_0^{\alpha-1})^T \vdots (\mathbf{y}_0^{\nu-1})^T]^T \in \mathfrak{B}_\mu$ *implying in the free regime that*

$$[\mathbf{r}^{\alpha-1}(\tau;\mathbf{r}_0^{\alpha-1};\mathbf{0}_M) \vdots \mathbf{y}^{\nu-1^T}(\tau;\mathbf{r}_0^{\alpha-1};\mathbf{y}_0^{\nu-1};\mathbf{0}_M)]^T$$

escapes $\mathfrak{B}_\varepsilon$ *at the instant* τ,

$$\exists \mathfrak{B}_\varepsilon \subseteq \mathfrak{R}^{\alpha\rho+N\nu},\ \forall \mathfrak{B}_\mu \subseteq \mathfrak{R}^{\alpha\rho+N\nu} \Longrightarrow$$
$$\exists ([(\mathbf{r}_0^{\alpha-1})^T \ \vdots\ (\mathbf{y}_0^{\nu-1})^T\ ]^T, \tau) \in \mathfrak{B}_\mu x \mathfrak{T}_0 \Longrightarrow$$
$$[\mathbf{r}^{\alpha-1^T}(\tau; \mathbf{r}_0^{\alpha-1}; \mathbf{0}_M)\ \vdots\ \mathbf{y}^{\nu-1^T}(\tau; \mathbf{r}_0^{\alpha-1}; \mathbf{y}_0^{\nu-1}; \mathbf{0}_M)]^T \in \mathfrak{R}^{\alpha\rho+\nu N} \backslash \mathfrak{B}_\varepsilon,$$

where

$$\mathfrak{R}^{\alpha\rho+\nu N} \backslash \mathfrak{B}_\varepsilon = \left\{ \begin{array}{c} [(\mathbf{r}_0^{\alpha-1})^T \ \vdots\ (\mathbf{y}_0^{\nu-1})^T\ ]^T : \\ : \left(\begin{array}{c} [(\mathbf{r}_0^{\alpha-1})^T \ \vdots\ (\mathbf{y}_0^{\nu-1})^T\ ]^T \in \mathfrak{R}^{\alpha\rho+\nu N}, \\ [(\mathbf{r}_0^{\alpha-1})^T \ \vdots\ (\mathbf{y}_0^{\nu-1})^T\ ]^T \notin \mathfrak{B}_\varepsilon \end{array} \right) \end{array} \right\}.$$

Attraction of the zero equilibrium vector

We introduce now the definition of attraction of the equilibrium vector in the framework of the *IIO* systems described by (3.90).

Definition 304 *a) The zero equilibrium state* $[(\mathbf{r}_e^{\alpha-1})^T \ \vdots\ (\mathbf{y}_e^{\nu-1})^T\ ]^T = \mathbf{0}_{\alpha\rho+\nu N}$ *of the IIO system (3.90), in which* $\nu \geq 1$, *is* ***attractive*** *if and only if there exists a positive real number* Δ, *or* $\Delta = \infty$, *such that the norm of the initial state less than* Δ, $\left\| [(\mathbf{r}_0^{\alpha-1})^T \vdots (\mathbf{y}_0^{\nu-1})^T\ ]^T \right\| < \Delta$, *implies in the free regime that*

$$\left[\left(\mathbf{r}^{\alpha-1}(t; \mathbf{r}_0^{\alpha-1}; \mathbf{0}_M)\right)^T \ \vdots\ \left(\mathbf{y}^{\nu-1}(t; \mathbf{r}_0^{\alpha-1}; \mathbf{y}_0^{\nu-1}; \mathbf{0}_M)\right)^T \ \right]^T$$

tends to $\mathbf{0}_{\alpha\rho+\nu N}$ *as* t *escapes to infinity,*

$$\exists \Delta > 0 \Longrightarrow \left\| [(\mathbf{r}_0^{\alpha-1})^T \vdots (\mathbf{y}_0^{\nu-1})^T\ ]^T \right\| < \Delta \Longrightarrow$$
$$\lim_{t\to\infty} \left[\left(\mathbf{r}^{\alpha-1}(t; \mathbf{r}_0^{\alpha-1}; \mathbf{0}_M)\right)^T \vdots\ \left(\mathbf{y}^{\nu-1}(t; \mathbf{r}_0^{\alpha-1}; \mathbf{y}_0^{\nu-1}; \mathbf{0}_M)\right)^T \ \right]^T = \mathbf{0}_{\alpha\rho+\nu N}.$$

b) If and only if a) holds for $\Delta = \infty$ *then the zero equilibrium vector* $[(\mathbf{r}_e^{\alpha-1})^T \vdots (\mathbf{y}_e^{\nu-1})^T\ ]^T = \mathbf{0}_{\alpha\rho+\nu N}$ *of the IIO system (3.90), in which* $\nu \geq 1$, *is* ***globally attractive (attractive in the whole)***.

Note 305 ***Geometrical interpretation***

The zero equilibrium vector $[(\mathbf{r}_e^{\alpha-1})^T \vdots (\mathbf{y}_e^{\nu-1})^T]^T = \mathbf{0}_{\alpha\rho+\nu N}$ *of the IIO system (3.90), in which* $\nu \geq 1$, *is* ***attractive*** *if and only if there exists its open neighborhood* $\mathfrak{B}_\Delta \subseteq \mathfrak{R}^{\alpha\rho+N\nu}$ *such that every initial vector* $[(\mathbf{r}_0^{\alpha-1})^T \vdots (\mathbf{y}_0^{\nu-1})^T]^T$ *in* $\mathfrak{B}_\Delta$, $[(\mathbf{r}_0^{\alpha-1})^T \vdots (\mathbf{y}_0^{\nu-1})^T]^T \in \mathfrak{B}_\Delta$, *ensures that the system behavior* $[(\mathbf{r}^{\alpha-1}(t;\mathbf{r}_0^{\alpha-1};\mathbf{0}_M))^T \vdots (\mathbf{y}^{\nu-1}(t;\mathbf{r}_0^{\alpha-1};\mathbf{y}_0^{\nu-1};\mathbf{0}_M))^T]^T$ *in the free regime approaches asymptotically the equilibrium state* $[(\mathbf{r}_e^{\alpha-1})^T \vdots (\mathbf{y}_e^{\nu-1})^T]^T = \mathbf{0}_{\alpha\rho+\nu N}$ *as* $t \to \infty$, *Fig. 13.4, (in Section 13.1).*

Global attraction of the zero equilibrium vector $[(\mathbf{r}_e^{\alpha-1})^T \vdots (\mathbf{y}_e^{\nu-1})^T]^T = \mathbf{0}_{\alpha\rho+\nu N}$ *of the IIO system (3.90) means that for every* $[(\mathbf{r}_0^{\alpha-1})^T \vdots (\mathbf{y}_0^{\nu-1})^T]^T \in \mathfrak{R}^{\alpha\rho+N\nu}$ *the system behavior*

$$[(\mathbf{r}^{\alpha-1}(t;\mathbf{r}_0^{\alpha-1};\mathbf{0}_M))^T \vdots (\mathbf{y}^{\nu-1}(t;\mathbf{r}_0^{\alpha-1};\mathbf{y}_0^{\nu-1};\mathbf{0}_M))^T]^T$$

approaches asymptotically the equilibrium state $[(\mathbf{r}_e^{\alpha-1})^T \vdots (\mathbf{y}_e^{\nu-1})^T]^T = \mathbf{0}_{\alpha\rho+\nu N}$ *as* $t \to \infty$.

Global attraction of the zero equilibrium vector $[(\mathbf{r}_e^{\alpha-1})^T \vdots (\mathbf{y}_e^{\nu-1})^T]^T = \mathbf{0}_{\alpha\rho+\nu N}$ of the *IIO* system (3.90) implies its attraction. The inverse is also true:

Theorem 306 *In order for the zero equilibrium vector* $[(\mathbf{r}_e^{\alpha-1})^T \vdots (\mathbf{y}_e^{\nu-1})^T]^T = \mathbf{0}_{\alpha\rho+\nu N}$ *of the IIO system (3.90), in which* $\nu \geq 1$, *to be globally attractive, it is necessary and sufficient to be attractive.*

Proof. *Necessity.* Let the zero equilibrium vector $[(\mathbf{r}_e^{\alpha-1})^T \vdots (\mathbf{y}_e^{\nu-1})^T]^T = \mathbf{0}_{\alpha\rho+\nu N}$ of the *IIO* system (3.90), in which $\nu \geq 1$, be globally attractive. The zero equilibrium vector $[(\mathbf{r}_e^{\alpha-1})^T \vdots (\mathbf{y}_e^{\nu-1})^T]^T = \mathbf{0}_{\alpha\rho+\nu N}$ of the *IIO* system (3.90), in which $\nu \geq 1$, is attractive in view of a) of Definition 304.

Sufficiency. Let the zero equilibrium vector $[(\mathbf{r}_e^{\alpha-1})^T \vdots (\mathbf{y}_e^{\nu-1})^T]^T = \mathbf{0}_{\alpha\rho+\nu N}$ of the *IIO* system (3.90), in which $\nu \geq 1$, be attractive. Its response in the free regime is given by Equation (13.28) through Equation (13.36).

Since

$$\left[\left(\mathbf{r}^{\alpha-1}(t;\mathbf{r}_0^{\alpha-1};\mathbf{0}_M)\right)^T\vdots\ \left(\mathbf{y}^{\nu-1}(t;\mathbf{r}_0^{\alpha-1};\mathbf{y}_0^{\nu-1};\mathbf{0}_M)\right)^T\ \right]^T=$$

$$=\Gamma_{IIOo}(t)\left[\left(\mathbf{r}_0^{\alpha-1}\right)^T\ \vdots\ \left(\mathbf{y}_0^{\nu-1}\right)^T\ \right]^T\rightarrow\mathbf{0}_{\alpha\rho+\nu N}\ as\ t\rightarrow\infty,$$

$$\forall[(\mathbf{r}_0^{\alpha-1})^T\vdots(\mathbf{y}_0^{\nu-1})^T\ ]^T\in\mathfrak{R}^{\alpha\rho+N\nu}\ obeying\ \left\|[\mathbf{r}_0^{\alpha-1^T}\vdots\ \mathbf{y}_0^{\nu-1^T}]^T\right\|<\Delta,$$

$$due\ to\ a)\ of\ Definition\ 304,$$

then

$$\Gamma_{IIOo}(t)\rightarrow O_{(\alpha\rho+\nu N)\times(\alpha\rho+\nu N)}\ as\ t\rightarrow\infty.$$

This implies

$$\lim_{t\rightarrow\infty}[(\mathbf{r}^{\alpha-1}(t;\mathbf{r}_0^{\alpha-1};\mathbf{0}_M))^T\vdots\ (\mathbf{y}^{\nu-1}(t;\mathbf{r}_0^{\alpha-1};\mathbf{y}_0^{\nu-1};\mathbf{0}_M))^T]^T=$$

$$=\mathbf{0}_{\alpha\rho+N\nu},\ \forall[\mathbf{r}_0^{\alpha-1^T}\vdots\ \mathbf{y}_0^{\nu-1^T}]^T\in\mathfrak{R}^{\alpha\rho+N\nu},$$

which shows that the zero equilibrium vector $[(\mathbf{r}_e^{\alpha-1})^T\ \vdots\ (\mathbf{y}_e^{\nu-1})^T\ ]^T=\mathbf{0}_{\alpha\rho+\nu N}$ of the *IIO* system (3.90), in which $\nu\geq 1$, is globally attractive (part b) of Definition 304. ■

This theorem enables us to discover the link between the uniqueness of the zero equilibrium vector and its attraction in the framework of the *IIO* system described by (3.90).

Theorem 307 *In order for the zero equilibrium vector* $[(\mathbf{r}_e^{\alpha-1})^T\ \vdots\ (\mathbf{y}_e^{\nu-1})^T\ ]^T=\mathbf{0}_{\alpha\rho+\nu N}$ *of the IIO system (3.90), in which* $\nu\geq 1$, *to be attractive, it is necessary (but not sufficient) to be the unique equilibrium vector of the system.*

Proof. *Necessity.* Let the zero equilibrium vector $[(\mathbf{r}_e^{\alpha-1})^T\ \vdots\ (\mathbf{y}_e^{\nu-1})^T\ ]^T=\mathbf{0}_{\alpha\rho+\nu N}$ of the *IIO* system (3.90), in which $\nu\geq 1$, be nonunique and attractive. Then there is another equilibrium vector $[(\mathbf{r}_{e2}^{\alpha-1})^T\ \vdots\ (\mathbf{y}_{e2}^{\nu-1})^T\ ]^T\neq\mathbf{0}_{\alpha\rho+\nu N}$ of the *IIO* system (3.90), in which $\nu\geq 1$. The zero equilibrium vector $[(\mathbf{r}_e^{\alpha-1})^T\ \vdots\ (\mathbf{y}_e^{\nu-1})^T\ ]^T=\mathbf{0}_{\alpha\rho+\nu N}$ of the *IIO* system is globally attractive due to Theorem 306. However,

$$\lim_{t\to\infty}\left[\left(\mathbf{r}^{\alpha-1}(t;\mathbf{r}_{e2}^{\alpha-1};\mathbf{0}_M)\right)^T \vdots \left(\mathbf{y}^{\nu-1}(t;\mathbf{r}_{e2}^{\alpha-1};\mathbf{y}_{e2}^{\nu-1};\mathbf{0}_M)\right)^T\right]^T =$$

$$= [(\mathbf{r}_{e2}^{\alpha-1})^T \vdots (\mathbf{y}_{e2}^{\nu-1})^T\ ]^T \neq \mathbf{0}_{\alpha\rho+\nu N}$$

due to Definition 108, (in Subsection 3.7.4), of the equilibrium vector of the *IIO* system (3.90), in which $\nu \geq 1$. This contradicts global attraction of $[(\mathbf{r}_e^{\alpha-1})^T \vdots (\mathbf{y}_e^{\nu-1})^T\ ]^T = \mathbf{0}_{\alpha\rho+\nu N}$. Hence, there does not exist another equilibrium vector $[(\mathbf{r}_{e2}^{\alpha-1})^T \vdots (\mathbf{y}_{e2}^{\nu-1})^T\ ]^T \neq \mathbf{0}_{\alpha\rho+\nu N}$ of the *IIO* system (3.90), in which $\nu \geq 1$, i.e., the zero equilibrium vector $[(\mathbf{r}_e^{\alpha-1})^T \vdots (\mathbf{y}_e^{\nu-1})^T\ ]^T = \mathbf{0}_{\alpha\rho+\nu N}$ of the *IIO* system is unique.

Insufficiency. Let the *IIO* system (3.90) be defined by

$$\begin{aligned} r^{(1)}(t) - r(t) &= i(t) \\ y^{(1)}(t) - y(t) &= r(t) + i(t). \end{aligned}$$

Its equilibrium point $\begin{bmatrix} r_e & \vdots\ y_e \end{bmatrix} = \mathbf{0}_2^T$ is unique. Its dynamic behavior in the free regime reads for $|r_0| > 0$ and $|y_0| > 0$:

$$\begin{bmatrix} r(t;|r_0|;0) \\ y(t;|r_0|;|y_0|;0) \end{bmatrix} =$$

$$= \begin{bmatrix} e^t & 0 \\ te^t & e^t \end{bmatrix} \begin{bmatrix} |r_0| \\ |y_0| \end{bmatrix} \longrightarrow \begin{bmatrix} \infty \\ \infty \end{bmatrix},\ t \longrightarrow \infty.$$

The equilibrium vector $\begin{bmatrix} r_e^T & \vdots\ y_e^T \end{bmatrix}^T = \mathbf{0}_2$ is not attractive. ■

This theorem and Theorem 110, (in Subsection 3.7.4), yield directly the following:

Theorem 308 *In order for the zero equilibrium vector* $[(\mathbf{r}_e^{\alpha-1})^T \vdots (\mathbf{y}_e^{\nu-1})^T\ ]^T = \mathbf{0}_{\alpha\rho+\nu N}$ *of the IIO system (3.90), in which* $\nu \geq 1$*, to be attractive, it is necessary (but not sufficient) that the matrix*

$$\begin{bmatrix} Q_0 & \vdots & O_{\rho N} \\ -R_0 & \vdots & E_0 \end{bmatrix}$$

is nonsingular.

Asymptotic stability of the zero equilibrium vector

Definition 309 *The zero equilibrium state* $[(\mathbf{r}_e^{\alpha-1})^T \vdots (\mathbf{y}_e^{\nu-1})^T\ ]^T = \mathbf{0}_{\alpha\rho+\nu N}$ *of the IIO system (3.90), in which* $\nu \geq 1$, *is*

a) ***asymptotically stable*** *if and only if it is both stable and attractive,*

b) ***globally asymptotically stable*** *if and only if it is both stable and globally attractive,*

c) ***strictly (completely, fully) globally asymptotically stable*** *if and only if it is both globally stable and globally attractive.*

This definition, Theorem 301 and Theorem 306 imply directly the following:

Theorem 310 *In order for the zero equilibrium vector* $[(\mathbf{r}_e^{\alpha-1})^T \vdots (\mathbf{y}_e^{\nu-1})^T\ ]^T = \mathbf{0}_{\alpha\rho+\nu N}$ *of the IIO system (3.90), in which* $\nu \geq 1$, *to be strictly globally asymptotically stable, it is necessary and sufficient that it is asymptotically stable.*

Definition 311 *The zero equilibrium vector* $\mathbf{r}_e^{\alpha-1} = \mathbf{0}_{\alpha\rho}$ *of the first subsystem of the IIDO system (3.94) is*

a) ***asymptotically stable*** *if and only if it is both stable and attractive,*

b) ***globally asymptotically stable*** *if and only if it is both stable and globally attractive,*

c) ***strictly (completely, fully) globally asymptotically stable*** *if and only if it is both globally stable and globally attractive.*

Exponential stability of the zero equilibrium vector

We will extend the property of exponential stability to the *IIO* systems described by (3.90).

Definition 312 *a) The zero equilibrium vector* $[(\mathbf{r}_e^{\alpha-1})^T \vdots (\mathbf{y}_e^{\nu-1})^T\ ]^T = \mathbf{0}_{\alpha\rho+\nu N}$ *of the IIO system (3.90), in which* $\nu \geq 1$, *is* ***exponentially stable*** *if and only if there exist positive real numbers* $\xi \geq 1$, β, *and* Δ, *or* $\Delta = \infty$, *such that* $\left\| [(\mathbf{r}_0^{\alpha-1})^T \vdots (\mathbf{y}_0^{\nu-1})^T\ ]^T \right\| < \Delta$ *implies in the free regime*

$$\left\| [(\mathbf{r}^{\alpha-1}(t;\mathbf{r}_0^{\alpha-1};\mathbf{0}_M))^T \vdots (\mathbf{y}^{\nu-1}(t;\mathbf{r}_0^{\alpha-1};\mathbf{y}_0^{\nu-1};\mathbf{0}_M))^T]^T \right\| \leq$$

$$\leq \xi \exp(-\beta t) \left\| [(\mathbf{r}_0^{\alpha-1})^T \vdots (\mathbf{y}_0^{\nu-1})^T\ ]^T \right\| \ \textit{for all}\ t \in \mathfrak{T}_0,$$

i.e.,

$$\exists\xi \in \mathfrak{R}^{+},\ \xi \geq 1, \exists\beta \in \mathfrak{R}^{+},\ and\ \exists\Delta > 0 \Longrightarrow$$
$$\left\| [(\mathbf{r}_0^{\alpha-1})^T \ \vdots\ (\mathbf{y}_0^{\nu-1})^T\ ]^T \right\| < \Delta \Longrightarrow$$

$$\left\| [(\mathbf{r}^{\alpha-1}(t;\mathbf{r}_0^{\alpha-1};\mathbf{0}_M))^T \ \vdots\ (\mathbf{y}^{\nu-1}(t;\mathbf{r}_0^{\alpha-1};\mathbf{y}_0^{\nu-1};\mathbf{0}_M))^T]^T \right\| \leq$$
$$\leq \xi \exp(-\beta t) \left\| [(\mathbf{r}_0^{\alpha-1})^T \ \vdots\ (\mathbf{y}_0^{\nu-1})^T\ ]^T \right\|,\ \forall t \in \mathfrak{T}_0.$$

b) If and only if a) holds for $\Delta = \infty$ *then the zero equilibrium vector* $[(\mathbf{r}_e^{\alpha-1})^T \ \vdots\ (\mathbf{y}_e^{\nu-1})^T\ ]^T = \mathbf{0}_{\alpha\rho+\nu N}$ *of the IIO system (3.90), in which* $\nu \geq 1$*, is* ***globally exponentially stable (exponentially stable in the whole)***.

The number ξ determines the upper bound of

$$\left\| \left[(\mathbf{r}^{\alpha-1}(t;\mathbf{r}_0^{\alpha-1};\mathbf{0}_M))^T \vdots\ (\mathbf{y}^{\nu-1}(t;\mathbf{r}_0^{\alpha-1};\mathbf{y}_0^{\nu-1};\mathbf{0}_M))^T \right]^T \right\|,$$

which is a measure of the maximal deviation of

$$\left[(\mathbf{r}^{\alpha-1}(t;\mathbf{r}_0^{\alpha-1};\mathbf{0}_M))^T \vdots\ (\mathbf{y}^{\nu-1}(t;\mathbf{r}_0^{\alpha-1};\mathbf{y}_0^{\nu-1};\mathbf{0}_M))^T \right]^T$$

from the zero equilibrium vector $[(\mathbf{r}_e^{\alpha-1})^T \ \vdots\ (\mathbf{y}_e^{\nu-1})^T\ ]^T = \mathbf{0}_{\alpha\rho+\nu N}$. The number β signifies the minimal convergence rate of

$$\left\| [(\mathbf{r}^{\alpha-1}(t;\mathbf{r}_0^{\alpha-1};\mathbf{0}_M))^T \ \vdots\ (\mathbf{y}^{\nu-1}(t;\mathbf{r}_0^{\alpha-1};\mathbf{y}_0^{\nu-1};\mathbf{0}_M))^T]^T \right\|$$

to zero, i.e., of

$$[(\mathbf{r}^{\alpha-1}(t;\mathbf{r}_0^{\alpha-1};\mathbf{0}_M))^T \ \vdots\ (\mathbf{y}^{\nu-1}(t;\mathbf{r}_0^{\alpha-1};\mathbf{y}_0^{\nu-1};\mathbf{0}_M))^T]^T$$

to the zero equilibrium vector $[(\mathbf{r}_e^{\alpha-1})^T \ \vdots\ (\mathbf{y}_e^{\nu-1})^T\ ]^T = \mathbf{0}_{\alpha\rho+\nu N}$.

Theorem 313 *In order for the zero equilibrium state* $[(\mathbf{r}_e^{\alpha-1})^T \vdots (\mathbf{y}_e^{\nu-1})^T\ ]^T = \mathbf{0}_{\alpha\rho+\nu N}$ *of the IIO system (3.90) with* $\nu \geq 1$ *to be globally exponentially stable, it is necessary and sufficient to be exponentially stable.*

The proof of this theorem is analogous to the proof of Theorem 306.

Theorem 314 *In order for the zero equilibrium state* $[(\mathbf{r}_e^{\alpha-1})^T \vdots (\mathbf{y}_e^{\nu-1})^T]^T = \mathbf{0}_{\alpha\rho+\nu N}$ *of the IIO system (3.90), in which* $\nu \geq 1$, *to be exponentially stable, it is necessary and sufficient to be asymptotically stable.*

Proof. *Necessity.* Let the zero equilibrium vector $[(\mathbf{r}_e^{\alpha-1})^T \vdots (\mathbf{y}_e^{\nu-1})^T]^T = \mathbf{0}_{\alpha\rho+\nu N}$ of the *IIO* system (3.90) with $\nu \geq 1$ be exponentially stable. Then it is attractive because (Definition 312)

$$\left\|[(\mathbf{r}_0^{\alpha-1})^T \vdots (\mathbf{y}_0^{\nu-1})^T]^T\right\| < \Delta \Longrightarrow$$

$$\left\|\left[\left(\mathbf{r}^{\alpha-1}(t;\mathbf{r}_0^{\alpha-1};\mathbf{0}_M)\right)^T \vdots \left(\mathbf{y}^{\nu-1}(t;\mathbf{r}_0^{\alpha-1};\mathbf{y}_0^{\nu-1};\mathbf{0}_M)\right)^T\right]^T\right\| \leq$$

$$\leq \xi \exp(-\beta t)\left\|[(\mathbf{r}_0^{\alpha-1})^T \vdots (\mathbf{y}_0^{\nu-1})^T]^T\right\|,\ \forall t \in \mathfrak{T}_0,\ \xi \in \mathfrak{R}^+,\ \xi \geq 1$$

implies

$$[(\mathbf{r}^{\alpha-1}(t;\mathbf{r}_0^{\alpha-1};\mathbf{0}_M))^T \vdots (\mathbf{y}^{\nu-1}(t;\mathbf{r}_0^{\alpha-1};\mathbf{y}_0^{\nu-1};\mathbf{0}_M))^T]^T \longrightarrow \mathbf{0}_{\alpha\rho+\nu N},\ t \to \infty.$$

Let $\delta(\varepsilon) = \varepsilon\xi^{-1}$, $\forall\varepsilon \in \mathfrak{R}^+$. Hence,

$$\left\|[(\mathbf{r}_0^{\alpha-1})^T \vdots (\mathbf{y}_0^{\nu-1})^T]^T\right\| < \delta(\varepsilon) = \varepsilon\xi^{-1}$$

and

$$\left\|[(\mathbf{r}^{\alpha-1}(t;\mathbf{r}_0^{\alpha-1};\mathbf{0}_M))^T \vdots (\mathbf{y}^{\nu-1}(t;\mathbf{r}_0^{\alpha-1};\mathbf{y}_0^{\nu-1};\mathbf{0}_M))^T]^T\right\| \leq$$

$$\leq \xi \exp(-\beta t)\left\|[(\mathbf{r}_0^{\alpha-1})^T \vdots (\mathbf{y}_0^{\nu-1})^T]^T\right\|,\ \forall t \in \mathfrak{T}_0,$$

guarantee

$$\left\|[(\mathbf{r}^{\alpha-1}(t;\mathbf{r}_0^{\alpha-1};\mathbf{0}_M))^T \vdots (\mathbf{y}^{\nu-1}(t;\mathbf{r}_0^{\alpha-1};\mathbf{y}_0^{\nu-1};\mathbf{0}_M))^T]^T\right\| \leq$$

$$\leq \xi \exp(-\beta t)\left\|[(\mathbf{r}_0^{\alpha-1})^T \vdots (\mathbf{y}_0^{\nu-1})^T]^T\right\| <$$

$$< \xi \exp(-\beta t)\varepsilon\xi^{-1} \leq \varepsilon,\ \forall\varepsilon \in \mathfrak{R}^+,\ \forall t \in \mathfrak{T}_0.$$

This proves stability of $[(\mathbf{r}_e^{\alpha-1})^T \vdots (\mathbf{y}_e^{\nu-1})^T\ ]^T = \mathbf{0}_{\alpha\rho+\nu N}$. Since it is also attractive, then it is asymptotically stable.

Sufficiency. Let the zero equilibrium vector $[(\mathbf{r}_e^{\alpha-1})^T \vdots (\mathbf{y}_e^{\nu-1})^T\ ]^T = \mathbf{0}_{\alpha\rho+\nu N}$ of the IIO system (3.90), in which $\nu \geq 1$, be asymptotically stable. In the free regime, the system takes the following form:

$$Q^{(\alpha)}\mathbf{r}^{\alpha}(t) = \mathbf{0}_{\rho},\ t \in \mathfrak{T},$$
$$R^{(\alpha)}\mathbf{r}^{\alpha}(t) - E^{(\nu)}\mathbf{y}^{\nu}(t) = \mathbf{0}_N,\ t \in \mathfrak{T}. \tag{13.37}$$

or equivalently

$$\frac{d\mathbf{w}}{dt} = A_w\mathbf{w}, \tag{13.38}$$

where

$$p = \alpha\rho + \nu N,\ \mathbf{w} = [\mathbf{r}^{\alpha-1} \ \vdots \mathbf{y}^{\nu-1}\ ]^T \in \mathfrak{R}^p,$$

and A_w is defined by (13.39) through (13.42),

$$A_{IIO1} = \begin{bmatrix} O_\rho & I_\rho & O_\rho & \dots & O_\rho & O_\rho \\ O_\rho & O_\rho & I_\rho & \dots & O_\rho & O_\rho \\ - & - & - & \dots & - & - \\ O_\rho & O_\rho & O_\rho & \dots & O_\rho & I_\rho \\ -Q_\alpha^{-1}Q_0 & -Q_\alpha^{-1}Q_1 & -Q_\alpha^{-1}Q_2 & \dots & -Q_\alpha^{-1}Q_{\alpha-2} & -Q_\alpha^{-1}Q_{\alpha-1} \end{bmatrix}$$
$$\in \mathfrak{R}^{\alpha\rho \mathrm{x} \alpha\rho}, \tag{13.39}$$

$$A_{IIO2} = \begin{bmatrix} O_{N,\rho} & O_{N,\rho} & O_{N,\rho} & \dots & O_{N,\rho} & O_{N,\rho} \\ O_{N,\rho} & O_{N,\rho} & O_{N,\rho} & \dots & O_{N,\rho} & O_{N,\rho} \\ - & - & - & \dots & - & - \\ O_{N,\rho} & O_{N,\rho} & O_{N,\rho} & \dots & O_{N,\rho} & O_{N,\rho} \\ E_\nu^{-1}R_0 & E_\nu^{-1}R_1 & E_\nu^{-1}R_2 & \dots & E_\nu^{-1}R_{\alpha-1} & E_\nu^{-1}R_\alpha \end{bmatrix}$$
$$\in \mathfrak{R}^{\nu N \mathrm{x} (\alpha+1)\rho}, \tag{13.40}$$

$$A_{IIO3} = \begin{bmatrix} O_N & I_N & O_N & \dots & O_N & O_N \\ O_N & O_N & I_N & \dots & O_N & O_N \\ - & - & - & \dots & - & - \\ O_N & O_N & O_N & \dots & O_N & I_N \\ -E_\nu^{-1}E_0 & -E_\nu^{-1}E_1 & -E_\nu^{-1}E_2 & \dots & -E_\nu^{-1}E_{\nu-2} & -E_\nu^{-1}E_{\nu-1} \end{bmatrix}$$
$$\in \mathfrak{R}^{\nu N \mathrm{x} \nu N}, \tag{13.41}$$

$$A_w = A_{IIOry} = \begin{bmatrix} A_{IIO1} & O_{\alpha\rho,\nu N} \\ A_{IIO2} & A_{IIO3} \end{bmatrix}, \ \nu \geq 1,. \tag{13.42}$$

The solution of (13.38) reads for $\nu > 0$:

$$\mathbf{w}(t;\mathbf{w}_0) = \begin{bmatrix} \mathbf{r}^{\alpha-1}(t;\mathbf{r}_0^{\alpha-1};\mathbf{0}_M) \\ \mathbf{y}^{\nu-1}(t;\mathbf{r}_0^{\alpha-1};\mathbf{y}_0^{\nu-1};\mathbf{0}_M) \end{bmatrix} = e^{A_w t} \begin{bmatrix} \mathbf{r}_0^{\alpha-1} \\ \mathbf{y}_0^{\nu-1} \end{bmatrix}.$$

Asymptotic stability of $[(\mathbf{r}_e^{\alpha-1})^T \vdots (\mathbf{y}_e^{\nu-1})^T]^T = \mathbf{0}_{\alpha\rho+\nu N}$ implies both

$$\exists \gamma \in \mathfrak{R}^+ \Longrightarrow \left\| e^{A_w t} \right\| < \gamma, \ \forall t \in \mathfrak{T}_0,$$

and

$$t \to \infty \Longrightarrow e^{A_w t} \to O_p.$$

Hence,

$$\exists (\zeta > 1, \xi) \in \mathfrak{R}^+ \times \mathfrak{R}^+$$

such that

$$\left\| \begin{bmatrix} \mathbf{r}^{\alpha-1}(t;\mathbf{r}_0^{\alpha-1};\mathbf{0}_M) \\ \mathbf{y}^{\nu-1}(t;\mathbf{r}_0^{\alpha-1};\mathbf{y}_0^{\nu-1};\mathbf{0}_M) \end{bmatrix} \right\| = \left\| e^{A_w t} \begin{bmatrix} \mathbf{r}_0^{\alpha-1} \\ \mathbf{y}_0^{\nu-1} \end{bmatrix} \right\| \leq$$

$$\leq \zeta e^{-\xi t} \left\| \begin{bmatrix} \mathbf{r}_0^{\alpha-1} \\ \mathbf{y}_0^{\nu-1} \end{bmatrix} \right\|, \ \forall t \in \mathfrak{T}, \ \forall \begin{bmatrix} \mathbf{r}_0^{\alpha-1} \\ \mathbf{y}_0^{\nu-1} \end{bmatrix} \in \mathfrak{R}^{\alpha\rho+\nu N}.$$

The zero equilibrium vector $[(\mathbf{r}_e^{\alpha-1})^T \vdots (\mathbf{y}_e^{\nu-1})^T]^T = \mathbf{0}_{\alpha\rho+\nu N}$ of the *IIO* system (3.90) is (globally) exponentially stable. ∎

This theorem, Theorem 308 and Definition 309 imply directly the following:

Theorem 315 *In order for the zero equilibrium vector* $[(\mathbf{r}_e^{\alpha-1})^T \vdots (\mathbf{y}_e^{\nu-1})^T]^T = \mathbf{0}_{\alpha\rho+\nu N}$ *of the IIO system (3.90) with* $\nu \geq 1$ *to be exponentially stable, it is necessary (but not sufficient) that* $[(\mathbf{r}_e^{\alpha-1})^T \vdots (\mathbf{y}_e^{\nu-1})^T]^T = \mathbf{0}_{\alpha\rho+\nu N}$ *is the unique equilibrium state of the system, i.e., that the matrix*

$$\begin{bmatrix} Q_0 & O_{\rho,N} \\ -R_0 & E_0 \end{bmatrix}$$

is nonsingular.

13.3 Lyapunov method and theorems

13.3.1 Outline of Lyapunov's original theory

Lyapunov's method, which Lyapunov originally called *the second method*, is also known as *the direct Lyapunov method*. It means the study of properties of the system nominal motion under the influence of arbitrary initial conditions via the sign properties of a subsidiary function $V(.)$ and of its total *time* derivative along system motions without using any information about the motions themselves, hence without solving the system mathematical model (the system differential equation).

Lyapunov theorems determine the conditions on the subsidiary function $V(.)$ called *Lyapunov function* and on its total *time* derivative along system motions in order for the system motions to have some qualitative properties.

Lyapunov methodology results from Lyapunov theorems and it determines how to apply Lyapunov method: should we start with a choice of the subsidiary function $V(.)$ or with its total *time* derivative along system motions. Lyapunov himself determined one methodology for *time*-invariant linear systems and another one for *time*-varying linear and all nonlinear systems.

For *time*-invariant linear systems, his methodology demands to start with a choice of the total *time* derivative of the subsidiary function $V(.)$ along system motions, and then to test the properties of the function $V(.)$. This methodology resolves stability problems completely.

For *time*-varying linear systems and for all nonlinear systems, Lyapunov's methodology requires to start with a choice of the function $V(.)$ and then to test the properties of its total *time* derivative along system motions, which is inverse to his methodology for *time*-invariant linear systems. This inverse Lyapunov methodology leaves unsolved the crucial stability problems in the framework of the nonlinear systems:

- determining the necessary and sufficient conditions for asymptotic stability, which are *not expressed in terms of the existence of a Lyapunov function*,
- identifying the direct generation of a system Lyapunov function,
- determining the exact domain of the asymptotic stability.

We cite Lyapunov's original theorems for *time*-invariant continuous-*time* linear systems, [88, p. 67 in the Russian edition], which A. M. Lyapunov himself described by (13.43),

$$\frac{dx_s}{dt} = p_{s1}x_1 + p_{s2}x_2 + ... + p_{sn}x_n, \; s = 1, 2, ..., n, \tag{13.43}$$

or in the vector form (13.44),

$$\frac{d\mathbf{x}}{dt} = P_L\mathbf{x},\ P_L = \begin{bmatrix} p_{11} & .... & p_{1n} \\ -- & .... & -- \\ p_{n1} & & p_{nn} \end{bmatrix}. \tag{13.44}$$

The zeros (the roots) of the system characteristic equation, denoted by $\varkappa_i$ in them, and the natural numbers m and m_i also in them, obey

$$m_1 + m_2 + ... + m_n = m.$$

Theorem 316 ***Original Lyapunov's Theorem I on the linear systems in the free regime [88, English translation of Lyapunov's original text published on pages 75 and 76 of the Russian edition]***

When the roots $\varkappa_1, \varkappa_2, \dots, \varkappa_n$, of the characteristic equation are such that for a given positive integer m the relationships of the following forms:

$$m_1\varkappa_1 + m_2\varkappa_2 + ... + m_n\varkappa_n = 0,$$

are impossible for them, in which all m_s are nonnegative integers, the sum of which is m, then it is possible to find and then unique, fully homogeneous, function V of the power m of the variables x_s, which satisfies the equation

$$\sum_{s=1}^{n} (p_{s1}x_1 + p_{s2}x_2 + ... + p_{sn}x_n) \frac{\partial V}{\partial x_s} = U \tag{13.45}$$

for an arbitrarily given fully homogeneous function U of quantities x_s with the same power m.

The vector form of (13.45) reads

$$(gradV)^T P_L\mathbf{x} = \mathbf{U}. \tag{13.46}$$

Remark 317 ***Lyapunov's remark on his Theorem I 316, [88, English translation of Lyapunov's original text published on page 76 in the Russian edition]***

The conditions, considered in the theorem, will be, for example, fulfilled and then for every m, when the real parts of all quantities $\varkappa_s$ are different from zero and have the same sign.

Comment 318 *Lyapunov's theorem I 316 for the linear systems establishes only necessary conditions for the existence of the unique solution function $V(.)$ of (13.45), but not sufficiency for roots $\varkappa_1, \varkappa_2, \dots, \varkappa_n$, of the characteristic equation to obey the conditions of the theorem.*

Theorem 319 ***Original Lyapunov's Theorem II on the linear systems in the free regime [88, English translation of Lyapunov's original text published on page 76 in the Russian edition]***

When the real parts of all roots $\varkappa_s$ are negative and when in (13.45) the function U is sign definite form of any even power m, then the form V of the power m satisfying that equation is also sign definite and additionally with the sign opposite to the sign of U.

Comment 320 *The conditions in this Lyapunov Theorem 319 are presented as necessary, but not as both necessary and sufficient for the real parts of all roots $\varkappa_s$ to be negative. The sufficient conditions follow from Lyapunov's following statements related to stability properties of the unperturbed motion of the nonlinear systems.*

Theorem 321 ***Original Lyapunov's Theorem I on the nonlinear systems in the free regime [88, English translation of Lyapunov's original text published on page 59 in the Russian edition]***

If the differential equations of the perturbed motions are such, that it is possible to find a sign definite function V, the derivative V' of which due to those equations is either a sign semidefinite function with the sign inverse to the sign of V, or identically equal to zero, the unperturbed motion is stable.

Note 322 ***Original Lyapunov's Note II on the nonlinear systems in the free regime [88, English translation of Lyapunov's original text published on page 61 in the Russian edition]***

If the function V, which satisfies the conditions of Theorem 321, and simultaneously permits the infinitesimally small upper bound, and its derivative is sign definite function, then it is possible to prove, that every perturbed motion, sufficiently close to the unperturbed one, will converge to it asymptotically.

Comment 323 *Every positive definite time-independent function permits the infinitesimally small upper bound.*

13.3.2 Lyapunov method, theorems and methodology for the linear systems

We will broaden Lyapunov's original theory to be directly applicable in the same, unified, manner to all classes of the systems treated herein.

Let $\mathbf{w}$ be *a subsidiary real-valued vector*, $\mathbf{w} = [w_1 \ w_2 \ ... \ w_p]^T \in \mathfrak{R}^p$,

$$\mathbf{w} \in \left\{ \left[\left(\mathbf{r}^{\alpha-1}\right)^T \vdots \left(\mathbf{y}^{\nu-1}\right)^T \right]^T, \ \mathbf{r}^{\alpha-1}, \ \mathbf{x}, \ \mathbf{y}^{\nu-1} \right\},$$

$$p \in \{ \ n, \ \nu N, \ \alpha\rho, \ \alpha\rho + \nu N\}. \tag{13.47}$$

The vector $\mathbf{w}$ becomes

- $\mathbf{y}^{\nu-1}$,

$$\mathbf{w} = \mathbf{y}^{\nu-1} \in \mathfrak{R}^{\nu N} , \tag{13.48}$$

  for the *IO* systems (3.77),(in Subsection 3.5.2),

- $\mathbf{x}$,

$$\mathbf{w} = \mathbf{x} \in \mathfrak{R}^n, \tag{13.49}$$

  for the *ISO* systems (3.84),(3.85), (in Subsection 3.5.3),

- $\left[\left(\mathbf{r}^{\alpha-1}\right)^T \vdots \left(\mathbf{y}^{\nu-1}\right)^T \right]^T$ if $\nu \geq 1$, or $\mathbf{r}^{\alpha-1}$ if $\nu = 0$,

$$\mathbf{w} = \left\{ \begin{array}{c} \left[\left(\mathbf{r}^{\alpha-1}\right)^T \vdots \left(\mathbf{y}^{\nu-1}\right)^T \right]^T \in \mathfrak{R}^{\alpha\rho+\nu N}, \ \nu \geq 1, \\ \mathbf{r}^{\alpha-1} \in \mathfrak{R}^{\alpha\rho}, \ \nu = 0 \end{array} \right\} \tag{13.50}$$

  for the *IIO* systems (3.89), (in Subsection 3.5.4).

The properties of the symmetric matrix W of the quadratic form $v(\mathbf{w})$,

$$v(\mathbf{w}) = \mathbf{w}^T W \mathbf{w}, \ W = W^T \in \mathfrak{R}^{pxp}, \tag{13.51}$$

determine some particular properties of the quadratic form itself, where

$$W = W^T \in \left\{ G = G^T \in \mathfrak{R}^{qxq}, \ H = H^T \in \mathfrak{R}^{qxq} \right\}, \tag{13.52}$$

and

$$q \in \{ \ n, \ N, \ \nu N, \ \alpha\rho, \ \rho + N, \ \alpha\rho + \nu N\}. \tag{13.53}$$

The following is well known in the matrix theory:

Claim 324 *Let $\lambda_m(W)$ and $\lambda_M(W)$ be the minimal and the maximal eigenvalue of the matrix W. They are real numbers because the matrix W is symmetric. The quadratic form $v(\mathbf{w}) = \mathbf{w}^T W \mathbf{w}$ obeys the following estimates:*

$$\lambda_m(W) \left\| \mathbf{w} \right\|^2 \leq \mathbf{w}^T W \mathbf{w} \leq \lambda_M(W) \left\| \mathbf{w} \right\|^2, \forall \mathbf{w} \in \mathfrak{R}^p. \tag{13.54}$$

It is easy to verify the following properties of the quadratic form

Property 325 *The quadratic form (13.51)*

$$v(\mathbf{w}) = \mathbf{w}^T W \mathbf{w}, \ W = W^T \in \mathfrak{R}^{p\mathrm{x}p}$$

is

1. everywhere defined, continuous, infinitely many times continuously differentiable on $\mathfrak{R}^p$,

$$v(\mathbf{w}) \in \mathfrak{C}^{\infty}\left(\mathfrak{R}^p\right),$$

and

2. zero valued at the origin,

$$v(\mathbf{0}_p) = 0.$$

Definition 326 *The quadratic form (13.51):*

*- is **positive definite**, denoted by $v > 0$, if and only if it is both*

a) zero valued only at the origin of $\mathfrak{R}^p$,

$$v(\mathbf{w}) = 0 \Leftrightarrow \mathbf{w} = \mathbf{0}_p,$$

and

b) positive valued out of the origin on $\mathfrak{R}^p$,

$$v(\mathbf{w}) > 0, \ \forall\left(\mathbf{w} \neq \mathbf{0}_p\right) \in \mathfrak{R}^p,$$

*- is **negative definite** if and only if $-v(\mathbf{w})$ is positive definite.*

Definition 327 *The matrix W, $W = W^T$, of the quadratic form $v(\mathbf{w})$ is positive (negative) definite, denoted respectively by $W > O$ ($W < O$), if and only if the quadratic form itself is positive (negative) definite, respectively.*

The well-known **Sylvester criterion,** [23], [24], represents the necessary and sufficient conditions for positive definiteness of the matrix W, $W = [w_{ij}]$. It reads:

Criterion 328 ***Sylvester criterion***

In order for the symmetric matrix W, $W = W^T = [w_{ij}] \in R^{p\mathrm{x}p}$, to be positive definite it is:

- necessary (but not sufficient) that

$$w_{k,k} > 0, \ \forall k = 1, \ 2, \ \ldots, \ p,$$

- *both necessary and sufficient that*

$$\begin{vmatrix} w_{11} & w_{12} & \dots & w_{1,k} \\ w_{21} & w_{22} & \dots & w_{2,k} \\ - & - & \dots & - \\ w_{k,1} & w_{k,2} & \dots & w_{k,k} \end{vmatrix} > 0, \ \forall k = 1, \ 2, \ \dots, \ p.$$

The gradient *grad* $v(\mathbf{w})$ of the quadratic form (13.51) is well defined everywhere on $\mathfrak{R}^p$ because the form $v(\mathbf{w})$ is continuously differentiable on $\mathfrak{R}^p$,

$$grad\ v(\mathbf{w}) = \left[\frac{\partial v(\mathbf{w})}{\partial w_1} \ \frac{\partial v(\mathbf{w})}{\partial w_2} \dots \frac{\partial v(\mathbf{w})}{\partial w_p}\right]^T = 2W\mathbf{w}, \tag{13.55}$$

so that the total *time* derivative of $v(\mathbf{w})$ can be expressed in terms of *grad* $v(\mathbf{w})$ and $\mathbf{w}^{(1)}$,

$$\frac{dv(\mathbf{w})}{dt} = [grad\ v(\mathbf{w})]^T \frac{d\mathbf{w}}{dt} = 2(W\mathbf{w})^T \frac{d\mathbf{w}}{dt}. \tag{13.56}$$

Since

$$(W\mathbf{w})^T \frac{d\mathbf{w}}{dt}$$

is scalar, then

$$2(W\mathbf{w})^T \frac{d\mathbf{w}}{dt} = \mathbf{w}^T W^T \frac{d\mathbf{w}}{dt} + \frac{d\mathbf{w}}{dt}^T W\mathbf{w}.$$

and, due to (13.56),

$$\frac{dv(\mathbf{w})}{dt} = \mathbf{w}^T W^T \frac{d\mathbf{w}}{dt} + \frac{d\mathbf{w}}{dt}^T W\mathbf{w}. \tag{13.57}$$

Let

$$\frac{d\mathbf{w}}{dt} = A_w \mathbf{w}. \tag{13.58}$$

This, and $W = W^T$ (13.52) transform (13.57) into

$$\frac{dv(\mathbf{w})}{dt} = \mathbf{w}^T \left(W A_w + A_w^T W\right) \mathbf{w}. \tag{13.59}$$

Definition 329 *A square matrix $A_w \in \mathfrak{R}^{p \times p}$ is* ***stable*** *(or* ***stability,*** *or* ***Hurwitz) matrix*** *if and only if the real parts of all its eigenvalues are negative.*

The following is well known in the matrix theory [23], [24]:

Claim 330 *If and only if W is positive definite, then its minimal and maximal eigenvalues are positive,*

$$W > O \Leftrightarrow \lambda_M(W) \geq \lambda_m(W) > 0. \tag{13.60}$$

We will present the complete proof of the following theorem, well-known as *Lyapunov theorem* in the stability theory.

Theorem 331 ***Lyapunov theorem for the system (13.58)***

In order for the zero equilibrium state $\mathbf{w}_e = \mathbf{0}_p$ of the system (13.58) to be asymptotically stable, it is necessary and sufficient that for an arbitrary positive definite quadratic form $\mathbf{w}^T G\mathbf{w}$, $G = G^T \in \mathfrak{R}^{pxp}$, $G > O$, the solution function $v(.)$, $v(.) : \mathfrak{R}^p \to \mathfrak{R}$, of the differential equation

$$\frac{dv(\mathbf{w})}{dt} = -\mathbf{w}^T G\mathbf{w} \tag{13.61}$$

is also positive definite, and unique, quadratic form,

$$v(\mathbf{w}) = \mathbf{w}^T H\mathbf{w}, \ H = H^T \in \mathfrak{R}^{pxp}. \ H > O. \tag{13.62}$$

Proof. *Necessity.* Let the zero equilibrium state $\mathbf{w}_e = \mathbf{0}_p$ of the system (13.58) be asymptotically stable. Then, it is globally exponentially stable, due to Theorem 292, Theorem 295, and Theorem 296, (in Subsection 13.2.2). Theorem 289 and Definition 291, (in Subsection 13.2.2), ensure that the zero equilibrium state $\mathbf{w}_e = \mathbf{0}_p$ is unique.

N. N. Krasovskii [83, inequality 4.110], [84], proved the following for the motions of the system (13.58):

$$\exists\zeta \in \mathfrak{R}^+, \ \exists\eta \in \mathfrak{R}^+ \Longrightarrow \ \zeta e^{-\eta t} \|\mathbf{w}_0\| \leq \|\mathbf{w}(t;\mathbf{w}_0;\mathbf{0}_M)\|, \ \forall t \in \mathfrak{T}, \tag{13.63}$$

Let $v(\mathbf{w})$ be the solution of (13.61), which can be integrated to lead to

$$v[\mathbf{w}(t;\mathbf{w}_0;\mathbf{0}_M)]_0^\infty = -\int_0^\infty \mathbf{w}^T(t;\mathbf{w}_0;\mathbf{0}_M) G\mathbf{w}(t;\mathbf{w}_0;\mathbf{0}_M)\, dt. \tag{13.64}$$

Global exponential stability of $\mathbf{w}_e = \mathbf{0}_p$ implies $\mathbf{w}(\infty;\mathbf{w};\mathbf{0}_M) = \mathbf{0}_p$, hence

$$v[\mathbf{w}(\infty;\mathbf{w}_0;\mathbf{0}_M)] = 0, \ \forall \mathbf{w}_0 \in \mathfrak{R}^p,$$

due to the quadratic form of $v(\mathbf{w}) = \mathbf{w}^T H\mathbf{w}$. This reduces (13.64) to

$$v(\mathbf{w}_0) = \int_0^\infty \mathbf{w}^T(t;\mathbf{w}_0;\mathbf{0}_M) G\mathbf{w}(t;\mathbf{w}_0;\mathbf{0}_M)\, dt. \tag{13.65}$$

The estimates (13.24), (in Subsection 13.2.2), and (13.63), together with (13.54) (Claim 324) and (13.65) allow

$$\lambda_m(G)\zeta^2 \|\mathbf{w}_0\|^2 \int_0^\infty e^{-2\eta t}dt \leq v(\mathbf{w}_0) \leq \lambda_M(G)\alpha^2 \|\mathbf{w}_0\|^2 \int_0^\infty e^{-2\beta t}dt.$$

After carrying out the integration, these inequalities become

$$\frac{\lambda_m(G)\zeta^2}{2\eta} \|\mathbf{w}_0\|^2 \leq v(\mathbf{w}_0) \leq \frac{\lambda_M(G)\alpha^2}{2\beta} \|\mathbf{w}_0\|^2, \ \forall \mathbf{w}_0 \in \mathfrak{R}^p.$$

Since $\mathbf{w}_0 \in \mathfrak{R}^p$ is arbitrary, then we may write

$$k_1 \|\mathbf{w}\|^2 \leq v(\mathbf{w}) \leq k_2 \|\mathbf{w}\|^2, \ \forall \mathbf{w} \in \mathfrak{R}^p,$$
$$k_1 = \frac{\lambda_m(G)\zeta^2}{2\eta} \in \mathfrak{R}^+, \ k_2 = \frac{\lambda_M(G)\alpha^2}{2\beta} \in \mathfrak{R}^+.$$

These results prove positive definiteness of the solution function $v(.)$ of (13.61) in view of (13.60) (Claim 330). Linearity of (13.61) in $v(.)$ implies its uniqueness, as proved by Lyapunov in Theorem 316.

Sufficiency. Let all the conditions of the theorem statement hold. Positive definiteness of the solution $v(.)$ of (13.61) ensures (Claim 330) $\lambda_M(H) \geq \lambda_m(H) > 0$ in (13.66),

$$0 < \lambda_m(H) \|\mathbf{w}\|^2 \leq v(\mathbf{w}) = \mathbf{w}^T H \mathbf{w} \leq \lambda_M(H) \|\mathbf{w}\|^2, \quad \forall (\mathbf{w} \neq \mathbf{0}_p) \in \mathfrak{R}^p, \tag{13.66}$$

or

$$\lambda_M^{-1}(H) v(\mathbf{w}) \leq \|\mathbf{w}\|^2 \leq \lambda_m^{-1}(H) v(\mathbf{w}), \ \forall (\mathbf{w} \neq \mathbf{0}_p) \in \mathfrak{R}^p. \tag{13.67}$$

This, positive definiteness of G, hence (13.60) (Claim 330) holds for G that replaces therein W in view of (13.52), and (13.61) lead to

$$\frac{dv(\mathbf{w})}{dt} = -\mathbf{w}^T G \mathbf{w} \leq -\lambda_m(G) \|\mathbf{w}\|^2 \leq -\lambda_m(G)\lambda_M^{-1}(H) v(\mathbf{w}),$$

the solution of which obeys

$$v[\mathbf{w}(t; \mathbf{w}_0; \mathbf{0}_M)] \leq e^{-\lambda_m(G)\lambda_M^{-1}(H)t} v(\mathbf{w}_0), \ \forall t \in \mathfrak{T}, \forall \mathbf{w} \in \mathfrak{R}^p. \tag{13.68}$$

We apply (13.66) to these inequalities,

$$\lambda_m(H) \|\mathbf{w}(t; \mathbf{w}_0; \mathbf{0}_M)\|^2 \leq v[\mathbf{w}(t; \mathbf{w}_0; \mathbf{0}_M)] \leq e^{-\lambda_m(G)\lambda_M^{-1}(H)t} v(\mathbf{w}_0) \leq$$
$$\leq \lambda_M(H) e^{-\lambda_m(G)\lambda_M^{-1}(H)t} \|\mathbf{w}_0\|^2, \ \forall t \in \mathfrak{T}, \forall \mathbf{w} \in \mathfrak{R}^p,$$

or

$$\|\mathbf{w}(t;\mathbf{w}_0;\mathbf{0}_M)\| \leq \alpha e^{-\beta t}\|\mathbf{w}_0\|,\ \forall t \in \mathfrak{T},\ \forall \mathbf{w}_0 \in \mathfrak{R}^p,$$
$$\alpha = \sqrt{\lambda_m^{-1}(H)\lambda_M(H)} \in \mathfrak{R}^+,\ \beta = \frac{\lambda_m(G)\lambda_M^{-1}(H)}{2} \in \mathfrak{R}^+.$$

This proves global exponential stability of the zero equilibrium state $\mathbf{w}_e = \mathbf{0}_p$ of the system (13.58), (Definition 295 in Subsection 13.2.2), and its global asymptotic stability, (Theorem 296 in Subsection 13.2.1). ■

Definition 332 ***Lyapunov function of the system (13.58)***
The quadratic form (13.62), which is the solution of (13.61), is ***a Lyapunov function of the system (13.58)****.*

Comment 333 ***The physical meaning of the Lyapunov theorem***
Let a mathematical model of a physical system be the linear time-invariant continuous-time system of the form (13.58). If the energy e(.) *and power* p(.) *of the physical system are quadratic forms such that the power is negative definite quadratic form,*

$$p(\mathbf{w}) = \mathbf{w}^T P\mathbf{w} = -\mathbf{w}^T G\mathbf{w} \ for\ P = P^T = -G,\ G > O,$$

then for the zero equilibrium state of the system to be asymptotically (hence, exponentially) stable it is both necessary and sufficient that the system energy e(.), $e(w) = w^T E w$, $E = E^T$, *is positive definite quadratic form since the power* p(.) *is the first derivative of the energy,*

$$\frac{de(\mathbf{w})}{dt} = p(\mathbf{w}).$$

For $H = E = E^T = H^T$,

$$\frac{de(\mathbf{w})}{dt} = \mathbf{w}^T\left(A_w^T E + E^T A_w\right)\mathbf{w} = \mathbf{w}^T\left(A_w^T H + H A_w\right)\mathbf{w} =$$
$$= -\mathbf{w}^T G\mathbf{w} = \mathbf{w}^T P\mathbf{w} = p(\mathbf{w}),\ -G = P = \left(A_w^T H + H A_w\right) \Longrightarrow$$
$$\frac{de(\mathbf{w})}{dt} = -\mathbf{w}^T G\mathbf{w}, \tag{13.69}$$

which is (13.61).

Theorem 334 ***Lyapunov matrix theorem for the system (13.58)***

In order for the matrix A_w to be stable matrix, it is necessary and sufficient that for any positive definite symmetric matrix G_w, $G_w = G_w^T \in \mathfrak{R}^{p\times p}$, the matrix solution H_w of Lyapunov matrix equation

$$A_w^T H_w + H_w A_w = -G_w \tag{13.70}$$

is also positive definite symmetric matrix and the unique solution to (13.70).

Proof. It is well known that for the zero equilibrium vector $\mathbf{w}_e = \mathbf{0}_p$ of the system (13.58) to be asymptotically stable, it is both necessary and sufficient that the real parts of all eigenvalues of the system matrix A_w (13.58) are negative. This, and Definition 329 imply that for the zero equilibrium vector $\mathbf{w}_e = \mathbf{0}_p$ of the system (13.58) to be asymptotically stable, it is both necessary and sufficient that the system matrix A_w (13.58) is stable (i.e., Hurwitz) matrix.

Let $G_w = G_w^T \in \mathfrak{R}^{p\times p}$ be any positive definite symmetric matrix, which determines the quadratic form

$$\mathbf{w}^T G_w \mathbf{w}.$$

Let $H_w = H_w^T \in \mathfrak{R}^{p\times p}$ be a symmetric matrix of the quadratic form

$$v(\mathbf{w}) = \mathbf{w}^T H_w \mathbf{w},$$

which is the solution of Lyapunov differential Equation (13.61),

$$\frac{dv(\mathbf{w})}{dt} = -\mathbf{w}^T G_w \mathbf{w}. \tag{13.71}$$

Lyapunov Theorem 331 implies that the matrix $H_w = H_w^T$ is also positive definite and that its quadratic form $v(\mathbf{w}) = \mathbf{w}^T H_w \mathbf{w}$ is the unique solution of (13.71) if and only if the zero equilibrium vector $\mathbf{w}_e = \mathbf{0}_p$ of the system (13.58) is asymptotically stable. The derivative of $v(\mathbf{w})$ along motions of the system (13.58) reads

$$\frac{dv(\mathbf{w})}{dt} = \frac{d\left(\mathbf{w}^T H_w \mathbf{w}\right)}{dt} = \frac{d\mathbf{w}^T}{dt} H_w \mathbf{w} + \mathbf{w}^T H_w^T \frac{d\mathbf{w}}{dt}.$$

This, and (13.58) permit

$$\frac{dv(\mathbf{w})}{dt} = \mathbf{w}^T A_w^T H_w \mathbf{w} + \mathbf{w}^T H_w^T A_w \mathbf{w}.$$

We replace $dv(\mathbf{w})/dt$ by $\mathbf{w}^T A_w^T H_w \mathbf{w} + \mathbf{w}^T H_w^T A_w \mathbf{w}$ in (13.71),

$$\mathbf{w}^T A_w H_w \mathbf{w} + \mathbf{w}^T H_w^T A_w \mathbf{w} = -\mathbf{w}^T G_w \mathbf{w},$$

i.e.,

$$\mathbf{w}^T \left(A_w H_w + H_w^T A_w \right) \mathbf{w} = -\, \mathbf{w}^T G_w \mathbf{w}.$$

Since this holds for every $\mathbf{w} \in \mathfrak{R}^p$, and since $H_w = H_w^T$, then

$$A_w^T H_w + H_w A_w = -G_w.$$

This completes the proof. ■

Note 335 ***The fundamental matrix theorem***

Theorem 334 is the well known fundamental matrix theorem of the stability theory of the linear time-invariant continuous-time systems.

In order to show how we can effectively solve Lyapunov matrix equation (13.70) for H_w we explain the Kronecker matrix product of two matrices $M = [m_{ij}] \in \mathfrak{R}^{\mu x s}$ and $U = [u_{ij}] \in \mathfrak{R}^{\upsilon x \sigma}$, which is denoted by $\otimes$,

$$M \otimes U = \begin{bmatrix} m_{11}U & m_{12}U & ... & m_{1s}U \\ m_{21}U & m_{21}U & ... & m_{2s}U \\ - & - & ... & - \\ m_{\mu 1}U & m_{\mu 2}U & ... & m_{\mu s}U \end{bmatrix} \in \mathfrak{R}^{\mu \upsilon x s \sigma}. \tag{13.72}$$

We define the vectors $\mathbf{h}$ and $\mathbf{g}$ induced by the matrices $H = H^T = [h_{ij}] \in \mathfrak{R}^{pxp}$ and $G = G^T = [g_{ij}] \in \mathfrak{R}^{pxp}$, respectively,

$$\mathbf{h} = [h_{11}\ h_{12} ...\ h_{1p}\ h_{21}\ h_{22} ...\ h_{2p}\ ...\ h_{p1}\ ...\ h_{p2} ...\ h_{pp}]^T \in \mathfrak{R}^{pp}, \tag{13.73}$$

$$\mathbf{g} = [g_{11}\ g_{12} ...\ g_{1p}\ g_{21}\ g_{22} ...\ g_{2p}\ ...\ g_{p1}\ ...\ g_{p2} ...\ g_{pp}]^T \in \mathfrak{R}^{pp}. \tag{13.74}$$

These vectors enable us

- to determine in the straightforward procedure Lyapunov function of the system (13.58),

and

- to set Lyapunov matrix Equation (13.70) in the vector form (13.75) by using the Kronecker matrix product between the matrix A_w and the identity matrix I_p of the dimension pxp,

$$\left(A_w^T \otimes I_p + I_p \otimes A_w \right) \mathbf{h} = -\mathbf{g}. \tag{13.75}$$

If the eigenvalues $\lambda_i(A_w)$ of the matrix A_w obey

$$\lambda_j(A_w) + \lambda_k(A_w) \neq 0, \ \forall j,k = 1,2,...,p, \tag{13.76}$$

then Equation (13.75) is solvable in $\mathbf{h}$,

$$\mathbf{h} = -\left(A_w^T \otimes I_p + I_p \otimes A_w\right)^{-1} \mathbf{g}. \tag{13.77}$$

This equation determines the vector $\mathbf{h}$, which, together with (13.73), (13.74), defines completely the matrix H, $H = H^T \in \mathfrak{R}^{p\times p}$, and its quadratic form $v(\mathbf{w}) = \mathbf{w}^T H\mathbf{w}$ being Lyapunov function of the system (13.58).

The condition (13.76) means that:

- the matrix A_w does not have an eigenvalue with the zero real part (i.e., on the imaginary axis of the complex plane),

- the complex eigenvalues $\lambda_j(A_w)$ of the matrix A_w with positive real parts satisfying

$$Re\lambda_j(A_w) = -Re\lambda_k(A_w)$$

must have imaginary parts obeying

$$Im\lambda_j(A_w) \neq -Im\lambda_k(A_w),$$

and vice versa, and

- the real eigenvalues of the matrix A_w with the equal absolute values may not have the opposite signs.

If the matrix A_w is stability matrix, then it satisfies all these conditions, i.e., then Equation (13.75) is solvable in $\mathbf{h}$. Equation (13.77) determines the solution $\mathbf{h}$ that induces directly the matrix H_w via (13.73).

13.3.3 Lyapunov theorem for the *IO* systems

In the framework of the *IO* systems described by (3.77), (in Subsection 3.5.2),

$$p = \nu N, \ \mathbf{w} = \mathbf{y}^{\nu-1}, \ W = G_{IO} = G_{IO}^T \in \mathfrak{R}^{\nu N\times\nu N},$$
$$H_w = H_{IO} = H_{IO}^T \in \mathfrak{R}^{\nu N\times\nu N}, \ v(\mathbf{w}) = v_{IO}\left(\mathbf{y}^{\nu-1}\right). \tag{13.78}$$

In the sequel the total *time* derivative, (13.56), of a function $v_{IO}(.)$, $v_{IO}(.) : \mathfrak{R}^{\nu N} \to \mathfrak{R}$, along motions of the *IO* system (3.77), i.e., (13.79)

$$A^{(\nu)}\mathbf{y}^{\nu}(t) = B^{(\mu)}\mathbf{i}^{\mu}(t), \ t \in \mathfrak{T}, \tag{13.79}$$

in the free regime,

$$A^{(\nu)}\mathbf{y}^{\nu}(t) = \sum_{k=0}^{k=\nu} A_k \mathbf{y}^{(k)}(t) = \mathbf{0}_{\nu N},\ t \in \mathfrak{T}, \tag{13.80}$$

is denoted by $dv_{IO}\left(\mathbf{y}^{\nu-1}\right)/dt$,

$$\frac{dv_{IO}\left(\mathbf{y}^{\nu-1}\right)}{dt} = \left[gradv_{IO}\left(\mathbf{y}^{\nu-1}\right)\right]^T \frac{d\mathbf{y}^{\nu-1}}{dt}. \tag{13.81}$$

Theorem 336 ***Lyapunov theorem for the IO system (13.79) in the free regime***

In order for the zero equilibrium state $\mathbf{y}_e^{\nu-1} = \mathbf{0}_{\nu N}$ of the IO system (13.79) to be asymptotically stable, it is necessary and sufficient that for an arbitrary positive definite quadratic form ${\mathbf{y}^{\nu-1}}^T G_{IO}\mathbf{y}^{\nu-1}$, $G_{IO} = G_{IO}^T \in \mathfrak{R}^{\nu N x \nu N}$, $G_{IO} > O$, the solution function $v_{IO}(.)$, $v_{IO}(.) : \mathfrak{R}^{\nu N} \rightarrow \mathfrak{R}$, of the differential equation

$$\frac{dv_{IO}\left(\mathbf{y}^{\nu-1}\right)}{dt} = -{\mathbf{y}^{\nu-1}}^T G_{IO}\mathbf{y}^{\nu-1} \tag{13.82}$$

is also positive definite, and unique, quadratic form,

$$v_{IO}\left(\mathbf{y}^{\nu-1}\right) = {\mathbf{y}^{\nu-1}}^T H_{IO}\mathbf{y}^{\nu-1},\ H_{IO} = H_{IO}^T \in \mathfrak{R}^{\nu N x \nu N},\ H_{IO} > O. \tag{13.83}$$

Proof. We use $\mathbf{w}$ defined by (13.78) so that along the behavior of the IO system (13.79) in the free regime, i.e., (13.80), we find the following since A_ν^{-1} exists due to Condition 25, (in Section 2.1):

$$\frac{d\mathbf{w}}{dt} = \frac{d\mathbf{y}^{\nu-1}}{dt} = \left[\mathbf{y}^{(1)} \vdots \mathbf{y}^{(2)} \vdots ... \vdots \mathbf{y}^{(\nu-1)} \vdots \mathbf{y}^{(\nu)}\right]^T = A_w\mathbf{w}, \tag{13.84}$$

$$\underbrace{A_w = \begin{bmatrix} O_N & I_N & O_N & ... & O_N & O_N \\ O_N & O_N & I_N & ... & O_N & O_N \\ - & - & - & ... & - & - \\ O_N & O_N & O_N & ... & O_N & I_N \\ -A_\nu^{-1}A_0 & -A_\nu^{-1}A_1 & -A_\nu^{-1}A_2 & ... & -A_\nu^{-1}A_{\nu-2} & -A_\nu^{-1}A_{\nu-1} \end{bmatrix}}. \tag{13.85}$$

Equations (13.84), (13.85) and (13.78) give

$$\frac{d\mathbf{w}}{dt} = A_w\mathbf{w},\ for\ p = \nu N,\ A_w = A_{IO} \in \mathfrak{R}^{\nu N x \nu N}. \tag{13.86}$$

Equation (13.86) is the *ISO* form of the *IO* system (13.79) in the free regime, i.e., the *ISO* form of (13.80). Theorem 331 is therefore applicable to the *IO* system (13.79) via (13.78), and (13.86) that is (13.58). This proves the statement of the theorem. ■

From this theorem we will deduce in in Subsection 13.4.2, in the framework of the *IO* system (13.79), the well-known Lyapunov matrix theorem.

Definition 337 ***Lyapunov function of the IO system (13.79)***

The quadratic form (13.83) that is the solution of (13.82) is ***Lyapunov function of the IO system (13.79).***

When we set

$$A_w = A_{IO},\ \mathbf{h} = \mathbf{h}_{IO},\ \mathbf{g} = \mathbf{g}_{IO},$$

in Equation (13.77) then it determines the vector $\mathbf{h}_{IO}$, which, together with (13.73) and (13.74), defines completely the matrix H_{IO}, $H_{IO} = H_{IO}^T \in \mathfrak{R}^{pxp}$, and its quadratic form $v_{IO}\left(\mathbf{y}^{\nu-1}\right) = \mathbf{y}^{\nu-1^T} H_{IO} \mathbf{y}^{\nu-1}$ being Lyapunov function of the *IO* system (13.79).

13.3.4 Lyapunov theorem for the *ISO* systems

Theorem 338 ***Lyapunov theorem for the ISO system (3.84) and (3.85), (in Subsection 3.5.3), in the free regime***

In order for the zero equilibrium state $\mathbf{x}_e = \mathbf{0}_n$ of the ISO system (3.84) and (3.85) to be asymptotically stable, it is necessary and sufficient that for an arbitrary positive definite quadratic form $\mathbf{x}^T G \mathbf{x}$, $G = G^T \in \mathfrak{R}^{nxn}$, $G > O$, the solution function $v(.)$, $v(.) : \mathfrak{R}^n \to \mathfrak{R}$, of the differential equation

$$\frac{dv(\mathbf{x})}{dt} = -\mathbf{x}^T G \mathbf{x} \tag{13.87}$$

is also positive definite, and unique, quadratic form,

$$v(\mathbf{x}) = \mathbf{x}^T H \mathbf{x},\ H = H^T \in \mathfrak{R}^{nxn},\ H > O. \tag{13.88}$$

Proof. When we set

$$p = n,\ \mathbf{w} = \mathbf{x}, \tag{13.89}$$

in (13.58), (in Subsection 13.3), then it becomes (3.84) and (3.85) in the free regime, and vice versa. Besides we set

$$v(\mathbf{w}) = v(\mathbf{x}) \tag{13.90}$$

in (13.61) and (13.77), so that Theorem 331 takes the form of this theorem. ■

Definition 339 ***Lyapunov function of the*** *ISO* ***system (3.84) and (3.85)***

The quadratic form (13.88) that is the solution of (13.87) is ***Lyapunov function of the*** *ISO* ***system (3.84) and (3.85).***

When we set

$$v(\mathbf{w}) = v_{ISO}(\mathbf{x}), \ H = H_{ISO}, G = G_{ISO},$$

in (13.61) and (13.77), and

$$A_w = A,$$

in Equation (13.77) then it determines the vector $\mathbf{h}$, which, together with (13.73), (13.74), defines completely the matrix H, $H = H^T \in \mathfrak{R}^{n \times n}$, and its quadratic form $v_{ISO}(\mathbf{x}) = \mathbf{x}^T H \mathbf{x}$ that is Lyapunov function of the ISO system (3.84) and (3.85).

13.3.5 Lyapunov theorem for the *IIO* systems

Theorem 340 ***Lyapunov theorem for the*** *IIO* ***system*** *2.35*

a) Let $\nu \geq 1$ in (2.35). In order for the zero equilibrium state

$$\left[\left(\mathbf{r}_e^{\alpha-1}\right)^T \ \vdots \ \left(\mathbf{y}_e^{\nu-1}\right)^T \right]^T = \mathbf{0}_{\alpha\rho+\nu N}$$

of the IIO system (2.35) to be asymptotically stable, it is necessary and sufficient that for an arbitrary positive definite quadratic form

$$\left[\left(\mathbf{r}^{\alpha-1}\right)^T \ \vdots \ \left(\mathbf{y}^{\nu-1}\right)^T \right] G_{IIO} \left[\left(\mathbf{r}^{\alpha-1}\right)^T \vdots \left(\mathbf{y}^{\nu-1}\right)^T \right]^T,$$
$$G_{IIO} = G_{IIO}^T \in \mathfrak{R}^{(\alpha\rho+\nu N) \times (\alpha\rho+\nu N)},$$

the solution function $v(.)$, $v(.) : \mathfrak{R}^{\alpha\rho+\nu N} \to \mathfrak{R}$, of the differential equation

$$\frac{dv\left(\left[\left(\mathbf{r}^{\alpha-1}\right)^T \ \vdots \ \left(\mathbf{y}^{\nu-1}\right)^T \right]^T\right)}{dt} =$$
$$= -\left[\left(\mathbf{r}^{\alpha-1}\right)^T \ \vdots \ \left(\mathbf{y}^{\nu-1}\right)^T \right] G_{IIO} \left[\left(\mathbf{r}^{\alpha-1}\right)^T \ \vdots \ \left(\mathbf{y}^{\nu-1}\right)^T \right]^T \qquad (13.91)$$

is also positive definite, and unique, quadratic form,

$$v\left(\left[\left(\mathbf{r}^{\alpha-1}\right)^T \vdots \left(\mathbf{y}^{\nu-1}\right)^T\right]^T\right) =$$

$$= \left[\left(\mathbf{r}^{\alpha-1}\right)^T \vdots \left(\mathbf{y}^{\nu-1}\right)^T\right] H_{IIO} \left[\left(\mathbf{r}^{\alpha-1}\right)^T \vdots \left(\mathbf{y}^{\nu-1}\right)^T\right]^T,$$

$$H_{IIO} = H_{IIO}^T \in \mathfrak{R}^{(\alpha\rho+\nu N)x(\alpha\rho+\nu N)},\ H_{IIO} > O. \tag{13.92}$$

b) Let $\nu = 0$ in (2.35). In order for the zero equilibrium state $\mathbf{r}_e^{\alpha-1} = \mathbf{0}_\rho$ of the IIO system (2.35), the PMD system (2.36), the GISO system (2.39), (in Section 2.3), to be asymptotically stable, it is necessary and sufficient that for an arbitrary positive definite quadratic form $\mathbf{r}^T G_{IIOr}\mathbf{r}$, $G_{IIOr} = G_{IIOr}^T \in \mathfrak{R}^{\rho x \rho}$, the solution function $v(.)$, $v(.) : \mathfrak{R}^\rho \to \mathfrak{R}$, of the differential equation

$$\frac{dv(\mathbf{r})}{dt} = -\mathbf{r}^T G_{IIOr}\mathbf{r} \tag{13.93}$$

is also positive definite, and unique, quadratic form,

$$v(\mathbf{r}) = \mathbf{r}^T H_{IIOr}\mathbf{r},\ H_{IIOr} = H_{IIOr}^T \in \mathfrak{R}^{\rho x \rho},\ H_{IIOr} > O. \tag{13.94}$$

Proof. a) Let $\nu \geq 1$ in (2.35). The matrices Q_α and E_ν are nonsingular. They have inverses Q_α^{-1} and E_ν^{-1}, respectively. When we set

$$p = \alpha\rho + \nu N,\ \mathbf{w} = \left[\left(\mathbf{r}^{\alpha-1}\right)^T \vdots \left(\mathbf{y}^{\nu-1}\right)^T\right]^T,$$

$$v(\mathbf{w}) = v_{IIO}\left(\left[\left(\mathbf{r}^{\alpha-1}\right)^T \vdots \left(\mathbf{y}^{\nu-1}\right)^T\right]^T\right),$$

$$\begin{aligned} H &= H_{IIO} = H_{IIO}^T \in \mathfrak{R}^{(\alpha\rho+\nu N)x(\alpha\rho+\nu N)}, \\ G &= G_{IIO} = G_{IIO}^T \in \mathfrak{R}^{(\alpha\rho+\nu N)x(\alpha\rho+\nu N)}, \end{aligned}$$

and set (13.39)-(13.42), (in Subsection 13.2.3), in (13.58) then it becomes 2.35), and vice versa, so that Theorem 331 takes the form of the statement under a) of this theorem.

b) Let $\nu = 0$ in(2.35). We set

$$p = \alpha\rho,\ \mathbf{w} = \mathbf{r}^{\alpha-1},\ v(\mathbf{w}) = v_{IIOr}\left(\mathbf{r}^{\alpha-1}\right),$$

$$H = H_{IIOr} = H_{IIOr}^T \in \mathfrak{R}^{\alpha\rho x \alpha\rho},\ G = G_{IIOr} = G_{IIOr}^T \in \mathfrak{R}^{\alpha\rho x \alpha\rho},$$

$$A_w = A_{IIO1} = A_{IIOr} =$$

$$= \begin{bmatrix} O_\rho & I_\rho & O_\rho & \dots & O_\rho & O_\rho \\ O_\rho & O_\rho & I_\rho & \dots & O_\rho & O_\rho \\ - & - & - & \dots & - & - \\ O_\rho & O_\rho & O_\rho & \dots & O_\rho & I_\rho \\ -Q_\alpha^{-1}Q_0 & -Q_\alpha^{-1}Q_1 & -Q_\alpha^{-1}Q_2 & \dots & -Q_\alpha^{-1}Q_{\alpha-2} & -Q_\alpha^{-1}Q_{\alpha-1} \end{bmatrix} \in$$

$$\in \mathfrak{R}^{\alpha\rho x \alpha\rho}, \tag{13.95}$$

in (13.58) that takes the form of (2.35), i.e., (2.39), in the free regime, and vice versa. Hence, Theorem 331 becomes the statement under b) of this theorem. ∎

Definition 341 ***Lyapunov function of the IIO system (2.35)***
The quadratic form (13.92) that is the solution of (13.91) is ***Lyapunov function of the IIO system (2.35).***

13.4 Lyapunov stability conditions via *F(s)*

13.4.1 Generating theorem

A complex valued matrix function $F(.) : \mathfrak{C} \to \mathfrak{C}^{mxn}$ is real rational matrix function if and only if every entry is a quotient of two polynomials in s and it becomes a real-valued matrix for the real value of the complex variable s, i.e., for $s = \sigma \in \mathfrak{R}$.

Let $F(s)$ have μ different poles denoted by s_k^*, $k = 1, 2, ..., \mu$. The multiplicity of the pole s_k^* is designated by ν_k. We denote its real and imaginary part by $\text{Re}\, s_k^*$ and $\text{Im}\, s_k^*$, respectively.

Theorem 342 ***Generating theorem***
Let $F(.) : \mathfrak{R} \to \mathfrak{R}^{pxn}$, $F(t) = [F_{ij}(t)]$, have the Laplace transform $F(.) : \mathfrak{C} \to \mathfrak{C}^{pxn}$, which is real rational matrix function, $F(s) = [F_{ij}(s)]$.

- *a) In order for the norm $\|F(t)\|$ of the original $F(t)$ to be* ***bounded****, i.e.,*

  $$\exists \alpha \in \mathfrak{R}^+ \Longrightarrow \|F(t)\| < \alpha, \ \forall t \in \mathfrak{T}_0,$$

  it is necessary and sufficient that

  1. the real parts of all poles of $F(s)$ are nonpositive, Fig. 13.5,

  $$\text{Re}\, s_i^* \leq 0, \ \forall i = 1, 2, ..., \ \mu,$$

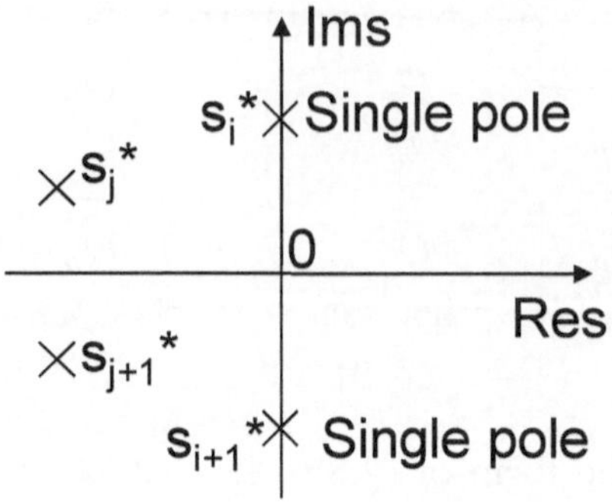

Figure 13.5: Poles with zero or negative real parts.

2. all imaginary poles of $F(s)$ are simple (i.e., with the multiplicity ν_i that is equal to one),

$$\operatorname{Re} s_i^* = 0, \nu_i = 1,$$

3. $F(s)$ is the zero matrix in the infinity, i.e., it is strictly proper,

$$F(\infty) = O_{pn} \in \mathfrak{R}^{p \times n},$$

- *b) In order for $\|F(t)\|$ to* ***vanish asymptotically****, i.e., in order for the following condition to hold:*

$$\lim[\|F(t)\| : t \longrightarrow \infty] = 0,$$

it is necessary and sufficient that

1. the real parts of all poles of $F(s)$ are negative, Fig. 13.6,

$$\operatorname{Re} s_i^* < 0, \ \forall i = 1, 2, ..., \ \mu,$$

and

2. $F(s)$ is the zero matrix in the infinity, i.e., it is strictly proper,

$$F(\infty) = O_{pn} \in \mathfrak{R}^{p \times n}.$$

Proof. Let the Laplace transform $F(s)$ of $F(t)$ have μ different poles denoted by s_k^* with the multiplicity ν_k^*, $k = 1, 2, \ldots, \mu$. We know (from the Heaviside expansion of $F(s)$) that the original $F(t)$ and its Laplace transform $F(s)$ are interrelated by the following formulae:

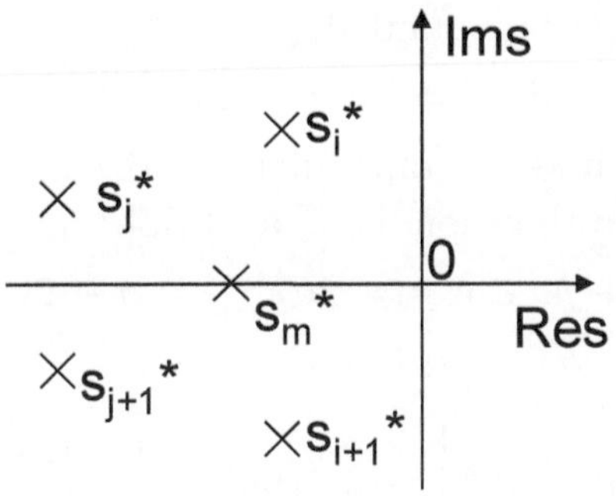

Figure 13.6: Poles with negative real parts.

- in the matrix form (13.96) and (13.97),

$$F(t) = \mathcal{L}^{-1}\left\{F(s)\right\} = \delta^{-}(t)R_0 + \sum_{k=1}^{k=\mu} e^{s_k^* t}\left[\sum_{r=1}^{r=\nu_k^*} \frac{1}{(r-1)!} t^{r-1} R_{kr}\right],$$
$$R_{kr} \in \mathfrak{R}^{pxn}, \tag{13.96}$$

$$F(s) = L^{-}\left\{F(t)\right\} =$$
$$= \mathcal{L}^{-}\left\{\delta^{-}(t)R_0 + \sum_{k=1}^{k=\mu} e^{s_k^* t}\left[\sum_{r=1}^{r=\nu_k^*} \frac{1}{(r-1)!} t^{r-1} R_{kr}\right]\right\}, \tag{13.97}$$

- in the scalar form (13.98) and (13.99), where s_k^{im*} is one of poles of the $\mu_{im} - th$ entry $F_{im}(s)$ of $F(s)$, $k \in \{1, 2, \ldots, \mu\}$, the multiplicity of which is denoted by ν_k^{im*}:

$$F_{im}(t) = \mathcal{L}^{-1}\left\{F_{im}(s)\right\} = \delta^{-}(t)R_0^{im} +$$
$$+ \sum_{k=1}^{k=\mu_{im}} e^{s_k^{im*} t}\left[\sum_{r=1}^{r=\nu_k^{im*}} \frac{1}{(r-1)!} t^{r-1} R_{kr}^{im}\right], \; R_{kr}^{im} \in \mathfrak{R}, \tag{13.98}$$

$$F_{im}(s) = \mathcal{L}^{-}\left\{F_{im}(t)\right\} =$$
$$= \mathcal{L}^{-}\left\{\delta^{-}(t)R_0^{im} + \sum_{k=1}^{k=\mu_{im}} e^{s_k^{im*} t}\left[\sum_{r=1}^{r=\nu_k^{im*}} \frac{1}{(r-1)!} t^{r-1} R_{kr}^{im}\right]\right\}. \tag{13.99}$$

Necessity. a) Let $F(t)$ be bounded, i.e.,

$$\exists \alpha \in \mathfrak{R}^+ \Longrightarrow \|F(t)\| < \alpha, \ \forall t \in \mathfrak{T}_0. \tag{13.100}$$

We will apply the method of contradiction to complete the proof of the necessity. Let us assume that condition a-1) does not hold, i.e.,

$$\exists s_k^{im*} = \sigma + j\omega \in \mathfrak{C} \Longrightarrow \operatorname{Re} s_k^{im*} = \sigma \in \mathfrak{R}^+. \tag{13.101}$$

This, $\delta(t)R_0^{im} = O$ for $t \neq 0$, and (13.98) imply

$$\left| e^{s_k^{im*}t} \left[\sum_{r=1}^{r=\nu_k^{im*}} \frac{1}{(r-1)!} t^{r-1} R_{kr}^{im} \right] \right| = e^{\sigma t} \left| e^{j\omega t} \right| \left| \left[\sum_{r=1}^{r=\nu_k^{im*}} \frac{1}{(r-1)!} t^{r-1} R_{kr}^{im} \right] \right| =$$

$$= e^{\sigma t} \left| \left[\sum_{r=1}^{r=\nu_k^{im*}} \frac{1}{(r-1)!} t^{r-1} R_{kr}^{im} \right] \right| \longrightarrow \infty \ as \ t \to \infty,$$

$$\Longrightarrow$$

$$\lim_{t\to\infty} |F_{im}(t)| =$$

$$= \lim_{t\to\infty} \left| \delta^-(t) R_0^{im} + \sum_{k=1}^{k=\mu_{im}} e^{s_k^{im*}t} \left[\sum_{r=1}^{r=\nu_k^{im*}} \frac{1}{(r-1)!} t^{r-1} R_{kr}^{im} \right] \right| = \infty \Longrightarrow$$

$$\lim_{t\to\infty} \|F(t)\| = \infty \ .$$

It follows that $F(t)$ is not bounded, which contradicts (13.100). The contradiction is a consequence of (13.101) implying that (13.101) is incorrect. This proves necessity of a-1).

We continue with the method of contradiction. Let us suppose that the condition a-2) does not hold, i.e.,

$$\exists s_k^{im*} = \sigma + j\omega \in \mathfrak{C} \Longrightarrow \operatorname{Re} s_k^{im*} = \sigma = 0 \ and \ \nu_k^{im*} \geq 2. \tag{13.102}$$

Now, $\delta(t)R_0^{im} = O$ for $t \neq 0$, (13.98) and (13.102) imply:

$$\lim_{t\to\infty} \left| e^{s_k^{im*}t} \left[\sum_{r=1}^{r=\nu_k^{im*}\geq 2} \frac{1}{(r-1)!} t^{r-1} R_{kr}^{im} \right] \right| =$$

$$= \lim_{t\to\infty} \left| e^{j\omega t} \right| \left| \left[\sum_{r=1}^{r=\nu_k^{im*}\geq 2} \frac{1}{(r-1)!} t^{r-1} R_{kr}^{im} \right] \right| =$$

$$= \lim_{t\to\infty} \left| \left[\sum_{r=1}^{r=\nu_k^{im*}\geq 2} \frac{1}{(r-1)!} t^{r-1} R_{kr}^{im} \right] \right| = \infty \Longrightarrow$$

$$\lim_{t\to\infty} |F_{im}(t)| =$$

$$= \lim_{t\to\infty} \left| \delta^-(t) R_0^{im} + \sum_{k=1}^{k=\mu_{im}} e^{s_k^{im*} t} \left[\sum_{r=1}^{r=\nu_k^{im*}} \frac{1}{(r-1)!} t^{r-1} R_{kr}^{im} \right] \right| = \infty \Longrightarrow$$

$$\lim_{t\to\infty} \|F(t)\| = \infty.$$

It follows that $F(t)$ is unbounded, which contradicts (13.100). The contradiction is a consequence of (13.102), which implies that (13.102) is not correct. This proves necessity of the condition a-2).

We continue further with the contradiction method. We assume that a-3) is not valid, i.e.,

$$\lim_{s\to\infty} F(s) \neq O \Longrightarrow \exists i \in \{1,2,...,p\},\ \exists m \in \{1,2,...,n\} \Longrightarrow R_0^{im} \neq 0. \tag{13.103}$$

Hence, (13.98) and (13.103) yield for $t = 0$:

$$|F_{im}(t)|_{t=0} = \left| \delta^-(t) R_0^{im} + \sum_{k=1}^{k=\mu_{ij}} e^{s_k^{im*} t} \left[\sum_{r=1}^{r=\nu_k^{im*}} \frac{1}{(r-1)!} t^{r-1} R_{kr}^{im} \right] \right|_{t=0} =$$

$$= \left| \delta^-(0) R_0^{im} + \sum_{k=1}^{k=\mu_{im}} R_{k1}^{im} \right|_{t=0} \in [0,\infty]$$

$$\Longrightarrow$$

$$\|F(t)\|_{t=0} \in [0,\infty]$$

This means that $F(t)$ is not bounded, which contradicts (13.100). The contradiction is a consequence of (13.103), which shows that (13.103) is incorrect. Hence,

$$R_0^{im} = 0,\ \forall i \in \{1,2,...,p\},\ \forall m \in \{1,2,...,n\}.$$

implying $R_0 = O_{pn}$ and $F(t)_{t=0} = O_{pn}$, or equivalently: $\lim_{s\to\infty} F(s) = O_{pn}$. This proves necessity of the condition a-3).

b) We keep on using the contradiction method. Let

$$\lim[\|F(t)\| : t \to \infty] = 0. \tag{13.104}$$

be true and let us suppose that the condition b-1) does not hold. If (13.101) were valid, then $F(t)$ would be unbounded as shown above in the proof of necessity of a-1), which would contradict (13.104). If (13.102) were valid, then $F(t)$ would be unbounded as shown above in the proof of necessity of a-2), which would again contradict (13.104). If

$$\exists s_k^{im*} = \sigma + j\omega \in \mathfrak{C} \Longrightarrow \operatorname{Re} s_k^{im*} = \sigma = 0 \; and \; \nu_k^{im*} = 1 \tag{13.105}$$

then

$$\lim_{t\to\infty} |F_{im}(t)| =$$
$$= \lim_{t\to\infty} \left| \delta^-(t) R_0^{im} + \sum_{k=1}^{k=\mu_{im}} e^{s_k^{im*}t} \left[\sum_{r=1}^{r=\nu_k^{im*}=1} \frac{1}{(r-1)!} t^{r-1} R_{kr}^{im} \right] \right| \geq$$
$$\geq \lim_{t\to\infty} \left| e^{j\omega t} \right| \left| [R_{kr}^{im}] \right| \in \mathfrak{R}^+ ,$$

which would also contradict (13.104). Altogether, the validity of (13.104) proves the validity of the condition b-1). Necessity of the condition b-2) is proved under a-3).

Sufficiency. a) Let the conditions under a) hold. Then $F(s)$ is the zero matrix in the infinity,

$$F(\infty) = O_{pn} \in \mathfrak{R}^{pxn}.$$

Hence,

$$R_0^{im} = 0, \forall i \in \{1, 2, ..., p\}, \forall m \in \{1, 2, ..., n\}. \tag{13.106}$$

We recall the following facts:

1) If (13.105) holds, then

$$\lim_{t\to\infty} |F_{im}(t)| \in \mathfrak{R}^+. \tag{13.107}$$

2) If $\operatorname{Re} s_k^{im*} = \sigma_{im} < 0$, then

$$\lim_{t\to\infty} |F_{im}(t)| = \lim_{t\to\infty} \left| \sum_{k=1}^{k=\mu_{im}} e^{s_k^{im*}t} \left[\sum_{r=1}^{r=\nu_k^{im*}=1} \frac{1}{(r-1)!} t^{r-1} R_{kr}^{im} \right] \right| =$$
$$\lim_{t\to\infty} \left| e^{-|\sigma_{im}|t} e^{j\omega t} \left[\sum_{r=1}^{r=\nu_k^{im*}} \frac{1}{(r-1)!} t^{r-1} R_{kr}^{im} \right] \right| =$$
$$\lim_{t\to\infty} \left| e^{-|\sigma_{im}|t} \left[\sum_{r=1}^{r=\nu_k^{im*}} \frac{1}{(r-1)!} t^{r-1} R_{kr}^{im} \right] \right| = 0. \tag{13.108}$$

3) The results (13.106) through (13.108) prove boundedness of $\|F(t)\|$, i.e.,

$$\exists \alpha \in \mathfrak{R}^{+} \Longrightarrow \|F(t)\| < \alpha, \ \forall t \in \mathfrak{T}_0.$$

b) Let the conditions under b) hold. Now, $\operatorname{Re} s_k^{im*} = \sigma < 0$, $\forall i \in \{1, 2, ..., p\}$, $\forall m \in \{1, 2, ..., n\}$, so that (13.108) holds $\forall i \in \{1, 2, ..., p\}$, $\forall m \in \{1, 2, ..., n\}$, which proves that $\|F(t)\|$ vanishes asymptotically, i.e.,

$$\lim[\|F(t)\| : \ t \to \infty] = 0.$$

This completes the proof. ■

Comment 343 ***Importance of the Generating theorem***

Qualitative system properties (e.g., controllability, observability, optimality, stability, trackability) concern families of dynamical behaviors of a dynamical system, which are caused by sets of initial conditions and/or by sets of external actions. They take place in time. Their definitions are given in the time domain. It is impractical in the framework of linear systems (practically impossible in the framework of nonlinear systems) to use their definitions directly in order to test whether a given system possesses a requested qualitative dynamical property. It is preferable to establish conditions and criteria for them in the algebraic and/or in the complex domain, which enable us to test them without knowing individual system behaviors, i.e., without solving system mathematical models for every initial condition and for every external action. The Generating Theorem 342 is the basis to establish such conditions and criteria in the complex domain for stability properties of the continuous-time time-invariant linear dynamical systems.

13.4.2 *IO* systems

We usually say that the negativeness of the real parts of all poles of the transfer function matrix $G_{IO}(s)$ of the *IO* system (3.77), (in Subsection 3.5.2), is necessary and sufficient for asymptotic stability of the zero equilibrium vector. The following counterexample illustrates the need to refine this statement.

Example 344 *The SISO IO system of Example 154 (in Section 8.1),*

$$y^{(2)}(t) + y^{(1)}(t) - 2y(t) = 2i^{(2)}(t) - 14i^{(1)}(t) + 12i(t).$$

has the following transfer function $G_{IO}(s)$*:*

$$G_{IO}(s) = \frac{2s^2 - 14s + 12}{s^2 + s - 2} = \frac{(s-1)(s-6)}{(s-1)(s+2)}.$$

It has the same zero s^o and pol s^*, $s^o = s^* = 1$. They do not influence the impulse response of the system under all zero initial conditions. The nondegenerate form $G_{IOnd}(s)$ of $G_{IO}(s)$ reads:

$$G_{IOnd}(s) = \frac{s-6}{s+2}.$$

It has only one pole $s^* = -2 < 0$. It is real and negative. If we used it to conclude about the stability properties of the system, then we would conclude (wrongly) that its zero state is asymptotically stable equilibrium. Let us verify this. The system differential equation yields

$$G_{IOyo}(s) = \left[\frac{s+1}{(s-1)(s+2)} \vdots \frac{1}{(s-1)(s+2)}\right],$$

so that in the free regime

$$\mathbf{y}^1(t;\mathbf{y}_0^1;0) = \frac{1}{3}\begin{bmatrix} 2e^t + e^{-2t} & e^t - e^{-2t} \\ 2e^t - 2e^{-2t} & e^t + 2e^{-2t} \end{bmatrix} \mathbf{y}_0^1 \Longrightarrow$$
$$\mathbf{y}_0^1 \neq \mathbf{0}_2 \Longrightarrow \left\|\mathbf{y}^1(t;\mathbf{y}^1;0)\right\| \rightharpoonup \infty \text{ as } t \rightharpoonup \infty.$$

The zero equilibrium state is unstable. If we concluded that it is asymptotically stable by referring to the poles of the nondegenerate form $G_{IOnd}(s)$ of $G_{IO}(s)$, then we would make a cardinal mistake. It would be the consequence of using the nondegenerate form $G_{IOnd}(s)$ of $G_{IO}(s)$ without testing the system full transfer function matrix $F_{IO}(s)$ for the cancellation of the positive pole $s^* = 1$.

The system full transfer function matrix $F_{IO}(s)$ (Example 184 in Section 10.1),

$$F_{IO}(s) = \frac{\left[2(s-1)(s-6) \vdots 2(7-s) \vdots -2 \vdots s+1 \vdots 1\right]}{(s-1)(s+2)}$$

shows that it is nondegenerate. The cancellation of $(s-1)$ in the numerator and in the denominator is neither possible in all entries of $F_{IO}(s)$ nor in all entries of any row of $G_{IOy_0}(s)$, which means in the single row of $G_{IOy_0}(s)$ because it is 1x2 matrix, i.e., the row vector,

$$G_{IOy_0}(s) = \left[-2\frac{s-7}{(s-1)(s+2)} \vdots -2\frac{1}{(s-1)(s+2)}\right],$$

so that $(s-1)$ may not be cancelled in $G_{IO}(s)$ when it is to be used for the Lyapunov stability test.

Comment 345 *There was discussion in [6] on the problem of the correctness of the pole-zero cancellation in the system transfer function related to the internal, i.e., Lyapunov stability of the system. The preceding example (Example 344) gives the clear, correct and unique explanation of the problem, the answer and the solution to the problem. The pole-zero cancellation from Lyapunov stability point of view is permitted only in the IO system transfer function (matrix)* $G_{IOy_0}(s)$, *and from the point of view of the IO system complete response it is allowed only in the system full transfer function matrix* $F_{IO}(s)$.

Lemma 346 *In order for* $\mathbf{y}^{\nu-1}(t;\mathbf{y}_0^{\nu-1};\mathbf{0}_M)$ *of the IO system (3.77) in the free regime*

a) to be bounded on $\mathfrak{T}_0$,

$$\exists\zeta \in \mathfrak{R}^+ \Longrightarrow \left\|\mathbf{y}^{\nu-1}(t;\mathbf{y}_0^{\nu-1};\mathbf{0}_M)\right\| < \zeta,\ \forall t \in \mathfrak{T}_0,$$

it is necessary and sufficient that $\mathbf{y}(t;\mathbf{y}_0^{\nu-1};\mathbf{0}_M)$ *is bounded on* $\mathfrak{T}_0$,

$$\exists\xi \in \mathfrak{R}^+ \Longrightarrow \left\|\mathbf{y}(t;\mathbf{y}_0^{\nu-1};\mathbf{0}_M)\right\| < \xi,\ \forall t \in \mathfrak{T}_0,$$

b) to vanish as $t \to \infty$,

$$t \to \infty \Longrightarrow \mathbf{y}^{\nu-1}(t;\mathbf{y}_0^{\nu-1};\mathbf{0}_M) \to \mathbf{0}_{\nu N},$$

it is necessary and sufficient that $\mathbf{y}(t;0;\mathbf{y}_0^{\nu-1})$ *vanishes as* $t \to \infty$,

$$t \to \infty \Longrightarrow \mathbf{y}(t;\mathbf{y}_0^{\nu-1};\mathbf{0}_M) \to \mathbf{0}_N.$$

Proof. *Necessity.* a) In order for $\mathbf{y}^{\nu-1}(t;\mathbf{y}_0^{\nu-1};\mathbf{0}_M)$ to be bounded on $\mathfrak{T}_0$, it is necessary that every entry is bounded on $\mathfrak{T}_0$. Hence, it is necessary that $\mathbf{y}(t;\mathbf{y}_0^{\nu-1};\mathbf{0}_M)$ is bounded on $\mathfrak{T}_0$.

b) In order for $\mathbf{y}^{\nu-1}(t;\mathbf{y}_0^{\nu-1};\mathbf{0}_M)$ to vanish as $t \to \infty$, it is necessary that every entry of $\mathbf{y}^{\nu-1}(t;\mathbf{y}_0^{\nu-1};\mathbf{0}_M)$ vanishes as $t \to \infty$. Hence, it is necessary that $\mathbf{y}(t;\mathbf{y}_0^{\nu-1};\mathbf{0}_M)$ vanishes as $t \to \infty$.

Sufficiency. a) Let the condition of the statement of Lemma under a) be valid. This, (3.77) in the free regime, the property of the Laplace transform and the Generating Theorem 342 (in Subsection 13.4.1) yield

$$\mathbf{y}(t;\mathbf{y}_0^{\nu-1};\mathbf{0}_M) = \left\{R_0\delta^-(t) + \sum_{k=1}^{k=\nu N} e^{s_k^* t}\left[\sum_{r=1}^{r=\nu_k} \frac{1}{(r-1)!}t^{r-1}R_{kr}\right]\right\}\mathbf{y}_0^{\nu-1},$$
$$R_0 \in \mathfrak{R}^{N\mathrm{x}\nu N},\ R_{kr} \in \mathfrak{C}^{N\mathrm{x}\nu N},\ \mathrm{Re}\, s_k^* \leq 0,$$
$$\forall k = 1,2,..,\nu N,\ \ \mathrm{Re}\, s_k^* = 0 \Longrightarrow \nu_k = 1, \tag{13.109}$$

where ν_k is the multiplicity of s_k^*. This shows that $\mathbf{y}(t;\mathbf{y}_0^{\nu-1};\mathbf{0}_M)$ is infinitely times continuously differentiable at every $t \in \mathfrak{T}$. All its derivatives rest bounded on $\mathfrak{T}_0$. This and the Generating Theorem 342 guarantee that $\mathbf{y}^{\nu-1}(t;\mathbf{y}_0^{\nu-1};\mathbf{0}_M)$ is also bounded on $\mathfrak{T}_0$.

b) Let the condition of the statement of Lemma under b) hold. This, (3.77) in the free regime, the properties of the Laplace transform and (13.109) imply

$$\mathbf{y}(t;\mathbf{y}_0^{\nu-1};\mathbf{0}_M) = \left\{ \sum_{k=1}^{k=\nu N} e^{s_k^* t} \left[\sum_{r=1}^{r=\nu_k} \frac{1}{(r-1)!} t^{r-1} R_{kr} \right] \right\} \mathbf{y}_0^{\nu-1}.$$

This shows that $\mathbf{y}(t;\mathbf{y}_0^{\nu-1};\mathbf{0}_M)$ is infinitely times continuously differentiable at every $t \in \mathfrak{T}$. All its derivatives rest bounded on $\mathfrak{T}_0$ and vanish as $t \to \infty$ due to the Generating Theorem 342. Every entry of $\mathbf{y}^{\nu-1}(t;\mathbf{y}_0^{\nu-1};\mathbf{0}_M)$ is bounded and vanishes as $t \to \infty$. ■

Example 347 *Let us consider the second order IO system described by*

$$y^{(2)}(t) + 3y^{(1)}(t) + 2y(t) = 2i^{(2)}(t) - 14i^{(1)}(t) + 12i(t).$$

The system differential equation yields

$$G_{IOyo}(s) = \left[\frac{s+3}{(s+1)(s+2)} \vdots \frac{1}{(s+1)(s+2)} \right] \Longrightarrow$$
$$Y(s) = Y\left(s;\mathbf{y}_0^1;0\right) = G_{IOyo}(s)\mathbf{y}_0^1,$$

$G_{IOyo}(s)$ is strictly proper, but $F_{IO}(s)$ and $G_{IO}(s)$,

$$F_{IO}(s) = \frac{\left[2(s-1)(s-6) \vdots -2(s+7) \vdots -2 \vdots s+3 \vdots 1 \right]}{(s+1)(s+2)} \Longrightarrow$$
$$G_{IO}(s) = \frac{2(s-1)(s-6)}{(s+1)(s+2)} = 2 - \frac{28}{s+1} + \frac{48}{s+2},$$

are only proper. The unit impulse response $y(t;\mathbf{0}_2;\delta)$ is the inverse Laplace transform of $G_{IO}(s)$,

$$y(t;\mathbf{0}_2;\delta) = \mathcal{L}^{-1}\left\{G_{IO}(s)\right\} = 2\delta(t) - 28e^{-t} + 48e^{-2t} \Longrightarrow y(0;\mathbf{0}_2;\delta) \in [0,\infty].$$

It is unbounded at the initial moment due to its unit impulse component $2\delta(t)$. Should we conclude that the zero equilibrium state is unstable? No, *because if we test $G_{IOyo}(s)$, then we conclude that it is strictly proper and the real parts of all its poles are negative, i.e., -1 and -2. Hence, the zero equilibrium state is globally asymptotically stable. Let us verify this in the time domain. The time evolutions of $y(t;\mathbf{y}^1;0)$ and $y^{(1)}(t;\mathbf{y}^1;0)$ describe completely the system output behavior in the free regime*

$$y(t;\mathbf{y}_0^1;0) = \left(2e^{-t} - e^{-2t}\right) y_0 \; + \; \left(e^{-t} - e^{-2t}\right) y_0^{(1)} \;,\; y(0;\mathbf{y}_0^1;0) = y_0.$$

The output response $y(t;\mathbf{y}_0^1;0)$ is bounded on $\mathfrak{T}_0$ for every bounded $\mathbf{y}_0^1$, and vanishes as $t \to \infty$. We determine easily its first derivative as

$$y^{(1)}(t;\mathbf{y}_0^1;0) = \left(-2e^{-t} + 2e^{-2t}\right) y_0 \; + \; \left(-e^{-t} + 2e^{-2t}\right) y_0^{(1)} \;,$$
$$y^{(1)}(0;0;\mathbf{y}_0^1) = y_0^{(1)}.$$

Altogether, $\mathbf{y}^1(t;\mathbf{y}_0^1;0)$,

$$\mathbf{y}^1(t;0;\mathbf{y}_0^1) = \begin{bmatrix} 2e^{-t} - e^{-2t} & e^{-t} - e^{-2t} \\ -2e^{-t} + 2e^{-2t} & -e^{-t} + 2e^{-2t} \end{bmatrix} \mathbf{y}_0^1,\; \mathbf{y}^1(0;0;\mathbf{y}_0^1) = \mathbf{y}_0^1,$$

is also bounded on $\mathfrak{T}_0$ for every bounded $\mathbf{y}_0^1$, and vanishes as $t \to \infty$. The zero equilibrium state is globally asymptotically, hence globally exponentially, stable.

Note 348 *Equation (13.109), $Res_k^* < 0$, $\forall k = 1, 2, ..., p = \nu N$, together with its derivatives, and (13.14) (in the proof of sufficiency of Theorem 275, in Subsection 13.2.1) imply the existence of positive numbers α and β such that*

$$\left\| \mathbf{y}^{\nu-1}(t;\mathbf{y}_0^{\nu-1};\mathbf{0}_M) \right\| \leq \beta e^{-\alpha t} \left\| \mathbf{y}_0^{\nu-1} \right\|,\; \forall t \in \mathfrak{T}_0,\; \forall \mathbf{y}_0^{\nu-1} \in \mathfrak{R}^{\nu N}. \quad (13.110)$$

Lemma 349 *If $G_{IOy_0}(s)$ is strictly proper, then in order to test stability, asymptotic stability or exponential stability of the zero equilibrium vector $\mathbf{y}_e^{\nu-1} = \mathbf{0}_{\nu N}$ of the IO system (3.77) via the transfer function matrices of the system it is necessary and sufficient to use the denominator polynomial $f_{IOy_0rnd}(s)$ of the row nondegenerate form $G_{IOy_0rnd}(s)$ of the submatrix $G_{IOy_0}(s)$ of $F_{IO}(s)$.*

Proof. The output response of the *IO* system (3.77) in the free regime on $\mathfrak{T}_0$ is determined, due to Lemma 123 (Chapter 6) and Lemma 346, by

$$\mathbf{y}(t;\mathbf{y}_0^{\nu-1};\mathbf{0}_M) = \mathcal{L}^{-1}\left\{G_{IOy_{0rnd}}(s)\right\} \mathbf{y}_0^{\nu-1},\; \forall t \in \mathfrak{T}_0. \quad (13.111)$$

This implies, since $G_{IOy_0}(s)$, hence $G_{IOy_0rnd}(s)$, is strictly proper, that the zeros of the denominator polynomial $f_{IOy_0rnd}(s) = \det\left[DenG_{IOy_0rnd}(s)\right]$ of the row nondegenerate form $G_{IOy_0rnd}(s)$ of the submatrix $G_{IOy_0}(s)$ of $F_{IO}(s)$, which are the poles of $G_{IOy_0rnd}(s)$, determine completely the character of $\mathbf{y}(t;\mathbf{y}_0^{\nu-1};\mathbf{0}_M)$. ■

The preceding results enable the simple form of the proof of the conditions for stability and for asymptotic stability of the equilibrium vector.

Theorem 350 *In order for the zero equilibrium state $\mathbf{y}_e^{\nu-1} = \mathbf{0}_{\nu N}$ of the IO system (3.77) to be*

a) stable it is necessary and sufficient that

1. $G_{IOy_0}(s)$ is strictly proper,

and

2. that the real parts of all the poles of the row nondegenerate form $G_{IOy_0rnd}(s)$ of $G_{IOy_0}(s)$ are nonpositive and the multiplicity of its poles with zero real parts should be equal to one (i.e., the poles should be simple),

b) asymptotically stable it is necessary and sufficient that

1. $G_{IOy_0}(s)$ is strictly proper,

and that

2. the real parts of all poles of the row nondegenerate form $G_{IOy_0rnd}(s)$ of $G_{IOy_0}(s)$ are negative.

Proof. Lemma 123, the Generating Theorem 342, Lemma 346 and Lemma 349 prove this theorem. ■

Note 351 *This theorem discovers that the competent transfer function matrix of the IO system (3.77) for the test of a Lyapunov stability property is the row nondegenerate form $G_{IOy_0rnd}(s)$ of $G_{IOy_0}(s)$, and not the system transfer function matrix $G_{IO}(s)$. This confirms the statement in Comment 345.*

Corollary 352 *In order for the cancellation of the same zero and pole of the equal order and with nonnegative real parts in the transfer function matrix $G_{IO}(s)$ of the IO system (3.77) to be without any influence on the results on any Lyapunov stability property of the equilibrium vector, it is necessary and sufficient that their same cancellation is possible in all rows of the transfer function matrix $G_{IOy_0}(s)$ of the system with respect to $\mathbf{y}_0^{\nu-1}$.*

The matrix A_{IO} (13.85) permits the transformation of the IO system (13.79) in the free regime, i.e., (13.80), into its ISO equivalent system (13.86), (in Section 13.3).

Lemma 353 *The eigenvalues of the matrix A_{IO} (13.85) of the IO system (13.79) are the poles of the system transfer function matrix $G_{IOy_0}(s)$, and vice versa.*

Proof. Equation (8.12), (in Section 8.1), together with (2.2), (in Section 2.1), (8.4), and (8.5), (in Section 8.1), show that the denominator polynomial $f_{IOyo}(s)$ of $G_{IOy_0}(s)$ is given by

$$f_{IOyo}(s) = det\left(A^{(\nu)}S_N^{(\nu)}(s)\right) = det\left(\sum_{k=0}^{k=\nu} s^k A_k)\right) = det\left(sI_{\nu N} - A_{IO}\right). \tag{13.112}$$

Its zeros are the poles of $G_{IOy_0}(s)$.

The characteristic polynomial $f_{IOA}(s)$ of the matrix A_{IO} reads

$$f_{IOA}(s) = det\left(s^\nu I_N + A_\nu^{-1}\sum_{k=0}^{k=\nu-1} s^k A_k)\right) = detA_\nu^{-1} det\left(\sum_{k=0}^{k=\nu} s^k A_k)\right).$$

The eigenvalues of the matrix A_{IO} are the zeros of $f_{IOA}(s)$, which are the zeros of

$$det\left(\sum_{k=0}^{k=\nu} s^k A_k)\right) = det\left(sI_{\nu N} - A_{IO}\right) = f_{IOyo}(s),$$

and vice versa. This, and (13.112) prove that the eigenvalues of the matrix A_{IO} (13.85) of the IO system (13.79) are the poles of the system transfer function matrix $G_{IOy_0}(s)$, and vice versa. ■

Theorem 354 ***The Lyapunov matrix theorem for the IO system (3.77)***

In order for the zero equilibrium vector $\mathbf{y}_e^{\nu-1} = \mathbf{0}_{\nu N}$ of the IO system (3.77) to be asymptotically stable, equivalently for the matrix A_{IO} (13.85) to be stable matrix, it is necessary and sufficient that for any positive definite symmetric matrix G_{IO}, $G_{IO} = G_{IO}^T \in \mathfrak{R}^{\nu N x \nu N}$, the matrix solution H_{IO} of the Lyapunov matrix equation

$$A_{IO}^T H_{IO} + H_{IO} A_{IO} = -G_{IO} \tag{13.113}$$

is also positive definite symmetric matrix and the unique solution to (13.113).

Proof. Theorem 350, and Lemma 353 show the equivalence between asymptotic stability of the zero equilibrium vector $\mathbf{y}_e^{\nu-1} = \mathbf{0}_{\nu N}$ of the IO system (3.77) and stability of the matrix A_{IO}, (13.85), (in Section 13.3). We apply the replacements determined by (13.78), (in Section 13.3), to Theorem 334 that then becomes this theorem. ■

Note 355 *This theorem, called* **the Lyapunov matrix theorem**, *is the fundamental matrix theorem for stability of the IO system (3.77). It enables the following estimate of* $\left\|\mathbf{y}^{\nu-1}(t;\mathbf{0}_M;\mathbf{y}_0^{\nu-1})\right\|$ *of* $\mathbf{y}^{\nu-1}(t;\mathbf{0}_M;\mathbf{y}_0^{\nu-1})$ *of the IO system (3.77) due to (13.78), (13.82), (13.84), (13.85) and (13.86) (in Section 13.3) and (13.113):*

$$v(\mathbf{y}^{\nu-1}) = \mathbf{y}^{\nu-1^T} H_{IO}\mathbf{y}^{\nu-1} \Longrightarrow \frac{dv_{IO}\left(\mathbf{y}^{\nu-1}\right)}{dt} = -\mathbf{y}^{\nu-1^T} G_{IO}\mathbf{y}^{\nu-1} \Longrightarrow$$

$$\frac{dv_{IO}\left(\mathbf{y}^{\nu-1}\right)}{dt} \leq -\lambda_m(G_{IO})\lambda_M^{-1}(H_{IO})v_{IO}\left(\mathbf{y}^{\nu-1}\right) \Longrightarrow$$

$$v_{IO}\left(\mathbf{y}^{\nu-1}\right) \leq \exp\left[-\lambda_m(G_{IO})\lambda_M^{-1}(H_{IO})t\right] v_{IO}\left(\mathbf{y}_0^{\nu-1}\right) \Longrightarrow$$

$$\lambda_m(H_{IO})\left\|\mathbf{y}^{\nu-1}(t)\right\|^2 \leq$$
$$\leq \lambda_M(H_{IO})e^{\left[-\lambda_m(G_{IO})\lambda_M^{-1}(H_{IO})t\right]}\left\|\mathbf{y}_0^{\nu-1}\right\|^2,\ \forall t\in\mathfrak{T},\ \forall \mathbf{y}_0^{\nu-1}\in\mathfrak{R}^{\nu N} \Longrightarrow$$

$$\left\|\mathbf{y}^{\nu-1}(t;\mathbf{0}_M;\mathbf{y}_0^{\nu-1})\right\| \leq \beta e^{-\alpha t}\left\|\mathbf{y}_0^{\nu-1}\right\|,\ \forall t\in\mathfrak{T},\ \forall \mathbf{y}_0^{\nu-1}\in\mathfrak{R}^{\nu N},$$
$$\alpha = \frac{1}{2}\lambda_m(G_{IO})\lambda_M^{-1}(H_{IO}),\ \beta = \left[\lambda_m^{-1/2}(H_{IO})\lambda_M^{1/2}(H_{IO})\right].$$

Comment 356 ***The physical meaning of the Lyapunov matrix theorem for the IO system (3.77)***

Let the system power p(.) be a negative definite quadratic form, $p(\mathbf{w}) = \mathbf{w}^T P\mathbf{w}$, $P = P^T < O$. *It has the properties determined in Comment 333 (in Section 13.3). Equations (13.69) of Comment 333 imply*

$$A_{IO}^T E + E^T A_{IO} = P,\ P = P^T < O.$$

In order for the zero equilibrium state of the system to be asymptotically (hence, exponentially) stable, it is both necessary and sufficient that the system energy e(.) is positive definite quadratic form, $e(\mathbf{w}) = \mathbf{w}^T E\mathbf{w}$ *with* $E = E^T > O$.

13.4.3 *ISO* systems

The motion $\mathbf{x}(.;\mathbf{x}_0;\mathbf{0}_M)$ of the *ISO* system (3.84) and (3.85), (in Subsection 3.5.3), in the free regime is determined by the matrix A and the initial state vector $\mathbf{x}_0$, (10.11) and (10.13), (in Section 10.2),

$$\mathbf{x}(t;\mathbf{x}_0;\mathbf{0}_M) = e^{At}\mathbf{x}_0. \quad (13.114)$$

The well-known Lyapunov stability conditions are naturally expressed in terms of the eigenvalues of the matrix A. They are in the complex domain usually stated in terms of the poles of the system transfer function matrix $G_{ISO}(s)$. If it is obtained from the state-space system description (3.84) and (3.85), then it reads

$$G_{ISO}(s) = [det(sI_n - A)]^{-1} [Cadj(sI_n - A)B + Ddet(sI_n - A)] .$$

Its denominator polynomial $f_{ISO}(s) = \det(sI_n - A)$ is the characteristic polynomial of the matrix A, if $G_{ISO}(s)$ is nondegenerate matrix. However, if it is degenerate matrix and the same poles and zeros were cancelled, and if among them there were poles with nonnegative real parts, then we should not use such nondegenerate form of $G_{ISO}(s)$ for the Lyapunov stability tests.

Example 357 *We analyze again the ISO system by Kalman (Example 169 in in Section 8.2) [80, Example 8, pp. 188, 189]:*

$$\frac{d\mathbf{x}}{dt} = \underbrace{\begin{bmatrix} 0 & 1 & 0 \\ 5 & 0 & 2 \\ -2 & 0 & -2 \end{bmatrix}}_{A} \mathbf{x} + \begin{bmatrix} 0 \\ 0 \\ 0.5 \end{bmatrix} i,$$

$$y = \begin{bmatrix} -2 \vdots 1 \vdots 0 \end{bmatrix} \mathbf{x}.$$

The eigenvalues of the matrix A are

$$\lambda_1 = -1,\ \lambda_2 = 2,\ \lambda_3 = -3.$$

Since $\lambda_2 = 2$ is positive real eigenvalue of the matrix A, then we conclude that the equilibrium state $\mathbf{x}_e = \mathbf{0}_n$ of the system is unstable. Its transfer function $G_{ISO}(s)$,

$$G_{ISO}(s) = \frac{s-2}{s^3 + 2s^2 - 5s - 6} = \frac{s-2}{(s+1)(s-2)(s+3)},$$

is degenerate. If we used its nondegenerate form $G_{ISOnd}(s)$,

$$G_{ISOnd}(s) = \frac{1}{(s+1)(s+3)},$$

for Lyapunov stability test, then we would conclude wrongly that the equilibrium state $\mathbf{x}_e = \mathbf{0}_n$ of the system is asymptotically stable because both poles $s_1^* = -1$ and $s_2^* = -3$ of $G_{ISOnd}(s)$ are real and negative.

Besides, either the transfer function matrix in its original form $G_{ISO}(s)$, or in its nondegenerate form $G_{ISOnd}(s)$, does not determine the system motion in the free regime. They are not adequate for the Lyapunov stability tests.

The system transfer function matrix $G_{ISOSS}(s)$ with respect to the initial state determines the Laplace transform of the system motion in the free regime, (7.31), (in Section 7.3),

$$\mathbf{X}^{\mp}(s) = G_{ISOSS}(s)\mathbf{x}_0, \ \mathbf{i}(t) \equiv \mathbf{0}_M, \tag{13.115}$$

where $G_{ISOSS}(s)$ is in fact the system resolvent matrix $(sI_n - A)^{-1}$, (8.34), (8.2),

$$G_{ISOSS}(s) = (sI_n - A)^{-1} = \frac{adj(sI_n - A)}{\det(sI_n - A)}. \tag{13.116}$$

Lemma 358 *In order to test a Lyapunov stability property of the zero equilibrium vector $\mathbf{x}_e = \mathbf{0}_n$ of the ISO system (3.84) and (3.85) via the transfer function matrices of the system, it is necessary and sufficient to use the denominator polynomial of the row nondegenerate form of the system SS transfer function matrix $G_{ISOSS}(s)$.*

Proof. Equation (13.116) shows that $G_{ISOSS}(s)$ is strictly proper. The dynamic behavior of the *ISO* system (3.84) and (3.85) in the free regime is determined by (13.115) in the complex domain. This equation shows that the same poles and zeros of any row of $G_{ISOSS}(s)$ do not influence $\mathbf{x}(t;\mathbf{x}_0;\mathbf{0}_M)$,

$$\mathbf{x}(t;\mathbf{x}_0;\mathbf{0}_M) = L^{-1}\{G_{ISOSS}(s)\mathbf{x}_0\}.$$

Hence, Lemma 123, (Chapter 6), shows that we should use the row nondegenerate form of $G_{ISOSS}(s)$ for any Lyapunov stability test, which leads to the use of its denominator polynomial. ∎

Theorem 359 *In order for the zero equilibrium state $\mathbf{x}_e = \mathbf{0}_n$ of the ISO system (3.84) and (3.85) to be*

a) stable, it is necessary and sufficient that the real parts of all poles of the row nondegenerate form of $G_{ISOSS}(s)$ are nonpositive, and the multiplicity of its poles with zero real parts should be equal to one (i.e., the poles should be simple),

b) asymptotically stable, it is necessary and sufficient that the real parts of all poles of the row nondegenerate form of $G_{ISOSS}(s)$ are negative.

Proof. The Generating Theorem 342, (in Subsection 13.4.1), and Lemma 358 imply directly the statement of the theorem. ∎

Note 360 *The preceding theorem discovers that the competent transfer function matrix of the ISO system (3.84) and (3.85) for the test of a Lyapunov stability property is the row nondegenerate form* $G_{ISOSSnd}(s)$ *of* $G_{ISOSS}(s)$*, and not the system transfer function matrix* $G_{ISO}(s)$*.*

Corollary 361 *In order for the cancellation of the same zero and pole of the equal order and with nonnegative real parts in the transfer function matrix* $G_{ISO}(s)$ *of the ISO system (3.84) and (3.85) to be without any influence on the results on any Lyapunov stability property of the equilibrium vector, it is necessary and sufficient that their same cancellation is possible in all rows of the transfer function matrix* $G_{ISOSS}(s)$ *of the system with respect to* $\mathbf{x}_0$*.*

Example 362 *We continue to use the system of Example 357. We compute the resolvent matrix* $(sI_n - A)^{-1}$ *that is* $G_{ISOSS}(s)$*, (13.116),*

$$G_{ISOSS}(s) = (sI_n - A)^{-1} =$$

$$= \begin{bmatrix} \frac{2s+s^2}{s^3+2s^2-5s-6} & \frac{s+2}{s^3+2s^2-5s-6} & \frac{2}{s^3+2s^2-5s-6} \\ \frac{5s+6}{s^3+2s^2-5s-6} & \frac{2s+s^2}{s^3+2s^2-5s-6} & 2\frac{s}{s^3+2s^2-5s-6} \\ -2\frac{s}{s^3+2s^2-5s-6} & -\frac{2}{s^3+2s^2-5s-6} & \frac{s^2-5}{s^3+2s^2-5s-6} \end{bmatrix}.$$

It is row nondegenerate. Its poles are all eigenvalues of the resolvent matrix $(sI_n - A)^{-1}$*,*

$$s_1^* = \lambda_1 = -1,\ s_2^* = \lambda_2 = 2,\ s_3^* = \lambda_3 = -3.$$

This shows that we should not cancel the positive real pole with the same positive real zero, $s_2^* = s_2^o = 2$*, in* $G_{ISO}(s)$ *if we use* $G_{ISO}(s)$ *for the Lyapunov stability test.*

Lemma 363 *The eigenvalues of the matrix A of the ISO system (3.84) and (3.85) are the poles of the system transfer function matrix* $G_{ISOSS}(s)$*, and vice versa.*

Proof. Equation (8.34), (in Section 8.2), shows that the denominator polynomial $f_{ISOSS}(s)$ of $G_{ISOSS}(s)$ obeys

$$f_{ISOSS}(s) = det\,(sI_n - A)\,. \tag{13.117}$$

Its zeros are the poles of $G_{ISOSS}(s)$.

The characteristic polynomial $f_{IOA}(s)$ of the matrix A is $det\,(sI_n - A)$,

$$f_{ISOA}(s) = det\,(sI_n - A) = f_{ISOSS}(s).$$

Hence, the eigenvalues of the matrix A are the zeros of $f_{ISOSS}(s)$ and the poles of $G_{ISOSS}(s)$, and vice versa. ■

Theorem 364 ***The Lyapunov matrix theorem for the ISO system (3.84) and (3.85)***

In order for the zero equilibrium state $\mathbf{x}_e = \mathbf{0}_n$ *of the ISO system (3.84) and (3.85) to be asymptotically stable, equivalently for its matrix A to be stable matrix, it is necessary and sufficient that for any positive definite symmetric matrix G,* $G = G^T \in \mathfrak{R}^{nxn}$*, the matrix solution H of the Lyapunov matrix equation*

$$A^T H + HA = -G \tag{13.118}$$

is also positive definite symmetric matrix and the unique solution to (13.118).

Proof. Theorem 359, and Lemma 363 show the equivalence between asymptotic stability of the zero equilibrium vector $\mathbf{x}_e = \mathbf{0}_n$ of the *ISO* system (3.84) and (3.85) and stability of its matrix A. We apply the replacements determined by (13.89), (13.90), (in Section 13.3), to Theorem 334 that then becomes this theorem. ■

Note 365 *Theorem 364 is the fundamental matrix theorem for stability of the ISO system (3.84) and (3.85).*

Comment 366 ***The physical meaning of Lyapunov matrix theorem for the ISO system (3.84) and (3.85)***

Let the system energy $\mathrm{e}(.)$ *and power* $\mathrm{p}(.)$ *have the properties determined in Comment 333 (in Section 13.3). Equations (13.69) of Comment 333 imply*

$$A^T E + E^T A = P,\ P = P^T < O.$$

The matrix P of the system power $\mathrm{p}(\mathbf{w}) = \mathbf{w}^T P\mathbf{w}$ *is negative definite and symmetric. In order for the zero equilibrium state of the ISO system (3.84) and (3.85) to be asymptotically (hence, exponentially) stable, it is both necessary and sufficient that the matrix E of the system energy* $\mathrm{e}(\mathbf{w}) = \mathbf{w}^T E\mathbf{w}$ *is positive definite and symmetric,* $E = E^T > O$.

13.4.4 *IIO* systems

We will treat simultaneously the case $\nu \geq 1$ and $\nu = 0$ of the *IIO* systems (3.90), (in Subsection 3.5.4). The latter incorporates both the Rosenbrock systems (3.92) and the *GISO* systems (3.94), (in Subsection 3.5.4).

Lemma 367 *In order for*

$$\left[\left(\mathbf{r}^{\alpha-1}(t;\mathbf{r}_0^{\alpha-1};\mathbf{0}_M)\right)^T \vdots \left(\mathbf{y}^{\nu-1}(t;\mathbf{r}_0^{\alpha-1};\mathbf{y}_0^{\nu-1};\mathbf{0}_M)\right)^T\right]^T, \ \nu \geq 1,$$
$$\mathbf{r}^{\alpha-1}(t;\mathbf{r}_0^{\alpha-1};\mathbf{0}_M), \ \nu = 0,$$

of the IIO system (3.90) in the free regime

a) to be bounded on $\mathfrak{T}_0$,

$$\exists \zeta \in \mathfrak{R}^+ \Longrightarrow \left\| \begin{bmatrix} \mathbf{r}^{\alpha-1}(t;\mathbf{r}_0^{\alpha-1};\mathbf{0}_M) \\ \mathbf{y}^{\nu-1}(t;\mathbf{r}_0^{\alpha-1};\mathbf{y}_0^{\nu-1};\mathbf{0}_M) \end{bmatrix} \right\| < \zeta, \ \forall t \in \mathfrak{T}_0, \ \nu \geq 1,$$
$$\exists \theta \in \mathfrak{R}^+ \Longrightarrow \left\|\mathbf{r}^{\alpha-1}(t;\mathbf{r}_0^{\alpha-1};\mathbf{0}_M)\right\| < \theta, \ \forall t \in \mathfrak{T}_0, \ \nu = 0,$$

it is necessary and sufficient that

$$\left[\mathbf{r}^T(t;\mathbf{r}_0^{\alpha-1};\mathbf{0}_M) \vdots \mathbf{y}^T(t;\mathbf{r}_0^{\alpha-1};\mathbf{y}_0^{\nu-1};\mathbf{0}_M)\right]^T, \ \nu \geq 1,$$
$$\mathbf{r}(t;\mathbf{r}_0^{\alpha-1};\mathbf{0}_M), \ \nu = 0,$$

is bounded on $\mathfrak{T}_0$,

$$\exists \xi \in \mathfrak{R}^+ \Longrightarrow \left\| \begin{bmatrix} \mathbf{r}(t;\mathbf{r}_0^{\alpha-1};\mathbf{0}_M) \\ \mathbf{y}(t;\mathbf{r}_0^{\alpha-1};\mathbf{y}_0^{\nu-1};\mathbf{0}_M) \end{bmatrix} \right\| < \xi, \ \forall t \in \mathfrak{T}_0, \ \nu \geq 1,$$
$$\exists \sigma \in \mathfrak{R}^+ \Longrightarrow \left\|\mathbf{r}(t;\mathbf{r}_0^{\alpha-1};\mathbf{0}_M)\right\| < \sigma, \ \forall t \in \mathfrak{T}_0, \ \nu = 0,$$

b) to vanish as $t \to \infty$,

$$t \to \infty \Longrightarrow$$
$$\left[\left(\mathbf{r}^{\alpha-1}(t;\mathbf{r}_0^{\alpha-1};\mathbf{0}_M)\right)^T \vdots \left(\mathbf{y}^{\nu-1}(t;\mathbf{r}_0^{\alpha-1};\mathbf{y}_0^{\nu-1};\mathbf{0}_M)\right)^T\right]^T \to \mathbf{0}_{\alpha\rho+\nu N}, \ \nu \geq 1,$$
$$\mathbf{r}^{\alpha-1}(t;\mathbf{r}_0^{\alpha-1};\mathbf{0}_M) \to \mathbf{0}_{\alpha\rho}, \ \nu = 0,$$

it is necessary and sufficient that

$$\left[\mathbf{r}^T(t;\mathbf{r}_0^{\alpha-1};\mathbf{0}_M) \vdots \mathbf{y}^T(t;\mathbf{r}_0^{\alpha-1};\mathbf{y}_0^{\nu-1};\mathbf{0}_M)\right]^T, \ \nu \geq 1,$$
$$\mathbf{r}(t;\mathbf{r}_0^{\alpha-1};\mathbf{0}_M), \ \nu = 0,$$

vanishes as $t \to \infty$,

$$t \to \infty \Longrightarrow \begin{bmatrix} \mathbf{r}(t;\mathbf{r}_0^{\alpha-1};\mathbf{0}_M) \\ \mathbf{y}(t;\mathbf{r}_0^{\alpha-1};\mathbf{y}_0^{\nu-1};\mathbf{0}_M) \end{bmatrix} \to \mathbf{0}_{\rho+N},\ \nu \geq 1,$$
$$t \to \infty \Longrightarrow \mathbf{r}(t;\mathbf{r}_0^{\alpha-1};\mathbf{0}_M) \to \mathbf{0}_\rho,\ \nu = 0.$$

Proof. *Necessity.* a) Let

$$\left[\left(\mathbf{r}^{\alpha-1}(t;\mathbf{r}_0^{\alpha-1};\mathbf{0}_M)\right)^T \vdots \left(\mathbf{y}^{\nu-1}(t;\mathbf{r}_0^{\alpha-1};\mathbf{y}_0^{\nu-1};\mathbf{0}_M)\right)^T\right]^T,\ \nu \geq 1,$$
$$\mathbf{r}^{\alpha-1}(t;\mathbf{r}_0^{\alpha-1};\mathbf{0}_M),\ \nu = 0,$$

be bounded on $\mathfrak{T}_0$. Each of its vector entries is bounded on $\mathfrak{T}_0$. Its subvector

$$\left[\mathbf{r}^T(t;\mathbf{r}_0^{\alpha-1};\mathbf{0}_M) \vdots \mathbf{y}^T(t;\mathbf{r}_0^{\alpha-1};\mathbf{y}_0^{\nu-1};\mathbf{0}_M)\right]^T,\ \nu \geq 1,$$
$$\mathbf{r}(t;\mathbf{r}_0^{\alpha-1};\mathbf{0}_M),\ \nu = 0,$$

is therefore bounded on $\mathfrak{T}_0$.

b) Let

$$\left[\left(\mathbf{r}^{\alpha-1}(t;\mathbf{r}_0^{\alpha-1};\mathbf{0}_M)\right)^T \vdots \left(\mathbf{y}^{\nu-1}(t;\mathbf{r}_0^{\alpha-1};\mathbf{y}_0^{\nu-1};\mathbf{0}_M)\right)^T\right]^T,\ \nu \geq 1,$$
$$\mathbf{r}^{\alpha-1}(t;\mathbf{r}_0^{\alpha-1};\mathbf{0}_M),\ \nu = 0,$$

vanish as $t \to \infty$. Every subvector of it vanishes as $t \to \infty$, which holds also for its first subvector,

$$t \to \infty \Longrightarrow \begin{bmatrix} \mathbf{r}(t;\mathbf{r}_0^{\alpha-1};\mathbf{0}_M) \\ \mathbf{y}(t;\mathbf{r}_0^{\alpha-1};\mathbf{y}_0^{\nu-1};\mathbf{0}_M) \end{bmatrix} \to \mathbf{0}_{\rho+N},\ \nu \geq 1,$$
$$t \to \infty \Longrightarrow \mathbf{r}(t;\mathbf{r}_0^{\alpha-1};\mathbf{0}_M) \to \mathbf{0}_\rho,\ \nu = 0.$$

Sufficiency. Let the conditions of the statement of Lemma hold.

a) The properties of the Laplace transform and the conditions of the Generating Theorem 342 (in Subsection 13.4.1) permit the following:

$$\begin{bmatrix} \begin{bmatrix} \mathbf{r}(t;\mathbf{r}_0^{\alpha-1};\mathbf{0}_M) \\ \mathbf{y}(t;\mathbf{r}_0^{\alpha-1};\mathbf{y}_0^{\nu-1};\mathbf{0}_M) \end{bmatrix},\ \nu \geq 1, \\ \mathbf{r}(t;\mathbf{r}_0^{\alpha-1};\mathbf{0}_M),\ \nu = 0, \end{bmatrix} =$$

$$= \left[\begin{array}{c} \left[\begin{array}{c} \left\{ \delta^-(t) R_{0r} + \sum_{k=1}^{k=\alpha\rho} e^{s_k^* t} \left[\sum_{r=1}^{r=\nu_k} \frac{1}{(r-1)!} t^{r-1} R_{krr} \right] \right\} \mathbf{r}_0^{\alpha-1} \\ \left\{ \delta^-(t) R_{0y} + \sum_{k=1}^{k=\nu N} e^{s_k^* t} \left[\sum_{r=1}^{r=\nu_k} \frac{1}{(r-1)!} t^{r-1} R_{kry} \right] \right\} \left[\begin{array}{c} \mathbf{r}_0^{\alpha-1} \\ \mathbf{y}_0^{\nu-1} \end{array} \right] \end{array} \right], \\ iff\ \nu \geq 1, \\ \left\{ \delta^-(t) R_{0r} + \sum_{k=1}^{k=\alpha\rho} e^{s_k^* t} \left[\sum_{r=1}^{r=\nu_k} \frac{1}{(r-1)!} t^{r-1} R_{krr} \right] \right\} \mathbf{r}_0^{\alpha-1},\ iff\ \nu = 0. \end{array} \right]$$

$$\operatorname{Re} s_k^* \leq 0,\ \forall k = 1, 2, .., \nu N,\ \ \operatorname{Re} s_k^* = 0 \Longrightarrow \nu_k = 1, \tag{13.119}$$

where ν_k is the multiplicity of s_k^* and the constant matrices R_{0r}, R_{krr}, R_{0y}, and R_{kry} are of the appropriate dimensions. All derivatives of

$$\left[\begin{array}{c} \left[\begin{array}{c} \mathbf{r}(t; \mathbf{r}_0^{\alpha-1}; \mathbf{0}_M) \\ \mathbf{y}(t; \mathbf{r}_0^{\alpha-1}; \mathbf{y}_0^{\nu-1}; \mathbf{0}_M) \end{array} \right],\ \nu \geq 1, \\ \mathbf{r}(t; \mathbf{r}_0^{\alpha-1}; \mathbf{0}_M),\ \nu = 0, \end{array} \right]$$

are bounded and infinitely times continuously differentiable, which guarantees the same properties to

$$\left[\left(\mathbf{r}^{\alpha-1}(t; \mathbf{r}_0^{\alpha-1}; \mathbf{0}_M) \right)^T \vdots \left(\mathbf{y}^{\nu-1}(t; \mathbf{r}_0^{\alpha-1}; \mathbf{y}_0^{\nu-1}; \mathbf{0}_M) \right)^T \right]^T,\ \nu \geq 1,$$
$$\mathbf{r}^{\alpha-1}(t; \mathbf{r}_0^{\alpha-1}; \mathbf{0}_M),\ \nu = 0.$$

b) Since

$$\left[\begin{array}{c} \left[\begin{array}{c} \mathbf{r}(t; \mathbf{r}_0^{\alpha-1}; \mathbf{0}_M) \\ \mathbf{y}(t; \mathbf{r}_0^{\alpha-1}; \mathbf{y}_0^{\nu-1}; \mathbf{0}_M) \end{array} \right],\ \nu \geq 1, \\ \mathbf{r}(t; \mathbf{r}_0^{\alpha-1}; \mathbf{0}_M),\ \nu = 0, \end{array} \right]$$

vanishes as $t \to \infty$ then, in view of (13.119), all its derivatives also vanish as $t \to \infty$. This guarantees that

$$\left[\left(\mathbf{r}^{\alpha-1}(t; \mathbf{r}_0^{\alpha-1}; \mathbf{0}_M) \right)^T \vdots \left(\mathbf{y}^{\nu-1}(t; \mathbf{r}_0^{\alpha-1}; \mathbf{y}_0^{\nu-1}; \mathbf{0}_M) \right)^T \right]^T,\ \nu \geq 1,$$
$$\mathbf{r}^{\alpha-1}(t; \mathbf{r}_0^{\alpha-1}; \mathbf{0}_M),\ \nu = 0,$$

vanishes as $t \to \infty$. ■

Note 368 *$Res_k^* < 0$, $\forall k = 1, 2, ..., p = \alpha\rho + \nu N$, (13.119), Generating Theorem 342, (in Subsection 13.4.1), and its derivatives imply the existence* *of positive numbers* $\varsigma \geq 1$ *and* ξ *such that*

$$\nu \geq 1 \Longrightarrow \left\| \left[\left(\mathbf{r}^{\alpha-1}(t;\mathbf{r}_0^{\alpha-1};\mathbf{0}_M)\right)^T \vdots \left(\mathbf{y}^{\nu-1}(t;\mathbf{r}_0^{\alpha-1};\mathbf{y}_0^{\nu-1};\mathbf{0}_M)\right)^T \right]^T \right\| \leq$$

$$\leq \varsigma e^{-\xi t} \left\| \begin{bmatrix} \mathbf{r}_0^{\alpha-1} \\ \mathbf{y}_0^{\nu-1} \end{bmatrix} \right\|, \ \forall t \in \mathfrak{T}_0, \ \forall \begin{bmatrix} \mathbf{r}_0^{\alpha-1} \\ \mathbf{y}_0^{\nu-1} \end{bmatrix} \in \mathfrak{R}^{\alpha\rho+\nu N},$$

$$\nu = 0 \Longrightarrow \left\|\mathbf{r}^{\alpha-1}(t;\mathbf{r}_0^{\alpha-1};\mathbf{0}_M)\right\| \leq \varsigma e^{-\xi t} \left\| \left[\mathbf{r}_0^{\alpha-1}\right] \right\|, \ \forall t \in \mathfrak{T}_0, \ \forall \mathbf{r}_0^{\alpha-1} \in \mathfrak{R}^{\alpha\rho}. \tag{13.120}$$

Lemma 369 *If the IIO system overall transfer function matrix* $G_{IIOr_0y_0}(s)$ *(13.121) with respect to the initial conditions,*

$$G_{IIOr_0y_0}(s) = \left\{ \begin{array}{c} \begin{bmatrix} G_{IIOr_0IS}(s) & O_{\rho,\nu N} \\ G_{IIOr_0}(s) & G_{IOy_0}(s) \end{bmatrix} \ iff \ \nu \geq 1, \\ G_{IIOr_0IS}(s) \ iff \ \nu = 0, \end{array} \right\} \tag{13.121}$$

is strictly proper, then in order to test stability, asymptotic stability or exponential stability of the zero equilibrium vector

- $\left[\mathbf{r}_e^{\alpha-1^T} \vdots \mathbf{y}_e^{\nu-1^T}\right]^T = \mathbf{0}_{\alpha\rho+\nu N}$ *of the IIO system (3.90), iff* $\nu \geq 1$,

- $\mathbf{r}_e^{\alpha-1} = \mathbf{0}_{\alpha\rho}$ *of the IIO system (3.90), iff* $\nu = 0$, *or of the GISO systems (3.94),*

via the system transfer function matrices it is necessary and sufficient to use the denominator polynomial $f_{IIOrnd}(s)$,

$$f_{IIOrnd}(s) = \{\det [DenG_{IIOr_0ISrnd}(s)]\} \bullet$$

$$\bullet \left\{ \begin{array}{c} \{\det [DenG_{IIOr_0rnd}(s)]\} \{\det [DenG_{IOy_0rnd}(s)]\}, \ \nu \geq 1, \\ 1, \ \nu = 0 \end{array} \right\}, \tag{13.122}$$

of the row nondegenerate form $G_{IIOr_0y_0rnd}(s)$,

$$G_{IIOr_0y_0rnd}(s) = \left\{ \begin{array}{c} \begin{bmatrix} G_{IIOr_0ISrnd}(s) & O_{\rho,\nu N} \\ G_{IIOr_0rnd}(s) & G_{IOy_0rnd}(s) \end{bmatrix} \ iff \ \nu \geq 1, \\ G_{IIOr_0ISrnd}(s) \ iff \ \nu = 0 \end{array} \right\}, \tag{13.123}$$

of $G_{IIOr_0y_0}(s)$ *(13.121).*

Proof. The dynamical behavior of the *IIO* system (3.90) in the free regime is determined by (3.11) due to (10.30), (Definition 193), (10.32),

(Theorem 194), i.e., (10.44), (10.42), (Definition 195), (10.45), (Theorem 197), (in Section 10.3), and Lemma 123 (in Chapter 6):

$$\left[\begin{array}{c} \mathbf{r}^{\alpha-1}(t;\mathbf{r}_0^{\alpha-1};\mathbf{0}_M) \\ \mathbf{y}^{\nu-1}(t;\mathbf{r}_0^{\alpha-1};\mathbf{y}_0^{\nu-1};\mathbf{0}_M) \end{array}\right] =$$

$$= \mathcal{L}^{-1}\left\{ G_{IIOr_0y_0rnd}(s) \left[\begin{array}{c} \mathbf{r}_0^{\alpha-1} \\ \left\{ \begin{array}{c} \mathbf{y}_0^{\nu-1} \ iff \ \nu \geq 1, \\ \mathbf{0}_{\nu N} \ iff \ \nu = 0 \end{array} \right\} \end{array}\right] \right\} \tag{13.124}$$

This, and Lemma 367 imply, because $G_{IIOr_0y_0}(s)$, hence $G_{IIOr_0y_0rnd}(s)$, is strictly proper, that the zeros of the denominator polynomial $f_{IIOy_0rnd}(s)$ of the row nondegenerate form $G_{IIOr_0y_0rnd}(s)$ (13.123) of the submatrix $G_{IIOr_0y_0}(s)$ (13.121) of $F_{IO}(s)$, which are the poles of $G_{IIOr_0y_0rnd}(s)$ (13.123), determine completely the character of

$$\left[\mathbf{r}^T(t;\mathbf{r}_0^{\alpha-1};\mathbf{0}_M) \vdots \mathbf{y}^T(t;\mathbf{r}_0^{\alpha-1};\mathbf{y}_0^{\nu-1};\mathbf{0}_M)\right]^T \quad iff \ \nu \geq 1,$$

$$\mathbf{r}(t;\mathbf{r}_0^{\alpha-1};\mathbf{0}_M) \ iff \ \nu = 0$$

The proof is completed. ■

Theorem 370 *In order for the zero equilibrium state*

- $\left[\mathbf{r}_e^{\alpha-1^T} \vdots \mathbf{y}_e^{\nu-1^T}\right]^T = \mathbf{0}_{\alpha\rho+\nu N}$ *of the IIO system (3.90), iff* $\nu \geq 1$,

- $\mathbf{r}_e^{\alpha-1} = \mathbf{0}_{\alpha\rho}$ *of the IIO system (3.90), iff* $\nu = 0$, *or of the GISO systems (3.94),*

to be

a) stable, it is necessary and sufficient that

1. the matrix $G_{IIOr_0y_0}(s)$ *(13.121) is strictly proper,*

and that

2. the real parts of all poles of the row nondegenerate $G_{IIOr_0y_0rnd}(s)$ *(13.123) of the IIO system combined transfer function matrix* $G_{IIOr_0y_0}(s)$ *(13.121), i.e., that the real parts of all zeros of the characteristic polynomial* $f_{IIOrnd}(s)$ *(13.122), are nonpositive, and the multiplicity of each of them with zero real part should be equal to one (i.e., they should be simple),*

b) asymptotically stable, it is necessary and sufficient that

1. the matrix $G_{IIOr_0y_0}(s)$ *(13.121) is strictly proper,*

and that

2. the real parts of all poles of the row nondegenerate $G_{IIOr_0y_0rnd}(s)$ *(13.123) of the IIO combined transfer function matrix* $G_{IIOr_0y_0}(s)$ *(13.121), i.e., that the real parts of all zeros of its characteristic polynomial* $f_{IIOrnd}(s)$ *(13.122), are negative.*

Proof. Lemma 123, the Generating Theorem 342, Lemma 367 and Lemma 369 prove this theorem. ■

Note 371 *This theorem discovers that the competent transfer function matrix of the IIO system (3.90) for the test of a Lyapunov stability property is the row nondegenerate form* $G_{IIOr_0y_0rnd}(s)$ *(13.123) of* $G_{IIOr_0y_0}(s)$ *(13.121), but not the system transfer function* $G_{IIO}(s)$.

Corollary 372 *In order for the cancellation of the same zero and pole of the equal order and with nonnegative real parts in the transfer function matrix* $G_{IIO}(s)$ *of the IIO system (3.90) to be without any influence on the results on any Lyapunov stability property of the equilibrium vector, it is necessary and sufficient that their same cancellation is possible in the IIO system overall transfer function matrix* $G_{IIOr_0y_0}(s)$ *(13.121) with respect to the initial conditions.*

Theorem 373 ***Lyapunov matrix theorem for the** IIO **system (3.90)***

In order for the zero equilibrium vector

- $\left[\mathbf{r}_e^{\alpha-1^T} \vdots \mathbf{y}_e^{\nu-1^T}\right]^T = \mathbf{0}_{\alpha\rho+\nu N}$ *of the IIO system (3.90), iff* $\nu \geq 1$,

- $\mathbf{r}_e^{\alpha-1} = \mathbf{0}_{\alpha\rho}$ *of the IIO system (3.90), iff* $\nu = 0$*, or of the IIDO systems (3.94),*

to be asymptotically stable it is necessary and sufficient that for any positive definite symmetric matrix G_{IIO},

$$G_{IIO} = G_{IIO}^T \in \left\{ \begin{array}{c} \mathfrak{R}^{(\alpha\rho+\nu N)\text{x}(\alpha\rho+\nu N)},\ iff\ \nu \geq 1, \\ \mathfrak{R}^{\alpha\rho\text{x}\alpha\rho},\ iff\ \nu = 0 \end{array} \right\},$$

the matrix solution H_{IIO},

$$H_{IIO} \in \left\{ \begin{array}{c} \mathfrak{R}^{(\alpha\rho+\nu N)\text{x}(\alpha\rho+\nu N)},\ iff\ \nu \geq 1, \\ \mathfrak{R}^{\alpha\rho\text{x}\alpha\rho},\ iff\ \nu = 0 \end{array} \right\},$$

of the Lyapunov matrix equation

$$A_{IIO}^T H_{IIO} + H_{IIO} A_{IIO} = -G_{IIO} \qquad (13.125)$$

is also positive definite symmetric matrix and the unique solution to (13.125).

Proof. The application of the matrix A_{IIO} defined by (13.39) through (13.42), (in Subsection 13.2.3), and by

$$A = A_{IIO} = \left\{ \begin{array}{c} A_{IIOry} = \left[\begin{array}{cc} A_{IIOr} & O_{\alpha\rho,\nu N} \\ A_{IIO2} & A_{IIO3} \end{array} \right] \ iff\ \nu \geq 1, \\ A_{IIOr}\ iff\ \nu = 0 \end{array} \right\}. \tag{13.126}$$

permits the transformation of the IIO system (3.90) in the free regime into the ISO system (13.58) (in Section 13.3) also in the free regime. This, and the replacements

$$p = \alpha\rho + \nu N,\ \mathbf{w} = \left\{ \begin{array}{c} [(\mathbf{r}^{\alpha-1})^T \vdots (\mathbf{y}^{\nu-1})^T]^T\ iff\ \nu \geq 1, \\ \mathbf{r}^{\alpha-1},\ iff\ \nu = 0 \end{array} \right\},$$

$$v(\mathbf{w}) = v_{IIO}\left(\left\{ \begin{array}{c} [(\mathbf{r}^{\alpha-1})^T \vdots (\mathbf{y}^{\nu-1})^T]^T\ iff\ \nu \geq 1, \\ \mathbf{r}^{\alpha-1},\ iff\ \nu = 0 \end{array} \right\} \right),$$

$$G_{IIO} = G_{IIO}^T \in \left\{ \begin{array}{c} \mathfrak{R}^{(\alpha\rho+\nu N)\times(\alpha\rho+\nu N)},\ iff\ \nu \geq 1, \\ \mathfrak{R}^{\alpha\rho\times\alpha\rho},\ iff\ \nu = 0 \end{array} \right\},$$

$$H_{IIO} = H_{IIO}^T \in \left\{ \begin{array}{c} \mathfrak{R}^{(\alpha\rho+\nu N)\times(\alpha\rho+\nu N)},\ iff\ \nu \geq 1, \\ \mathfrak{R}^{\alpha\rho\times\alpha\rho},\ iff\ \nu = 0 \end{array} \right\},$$

transform Theorem 364, (in Subsection 13.4.3) into this theorem. ■

Note 374 *This theorem is the fundamental matrix theorem for stability of the IIO system (3.90) and for stability of the IIDO systems (3.94).*

Comment 375 ***The physical meaning of the Lyapunov matrix theorem for the IIO system (3.90)***

Let the system power $p(.)$ be negative definite quadratic form $p(\mathbf{w}) = \mathbf{w}^T P \mathbf{w}$, $P = P^T < O$. Theorem 334, (in Section 13.3), implies

$$A_{IIO}^T E + E^T A_{IIO} = P,\ P = P^T < O.$$

In order for the zero equilibrium state of the system to be asymptotically (hence, exponentially) stable, it is both necessary and sufficient that the system energy $e(.)$ is positive definite quadratic form, $e(\mathbf{w}) = \mathbf{w}^T E \mathbf{w}$ with $E = E^T > O$.

Chapter 14

Bounded Input stability

14.1 BI stability and initial conditions

The **Bounded Input** (BI) **stability** concept is essentially different from Lyapunov's (LY) stability concept for its following characteristics:

- The BI stability concept concerns the system behavior under actions of input variables, while the original LY stability concept treats the system behavior in the free regime considered in the framework of deviations, i.e., in the total nominal regime in terms of the total coordinates [88].

- The original BI stability concept of the linear dynamical systems demands boundedness of the system behaviors for bounded input variables and *for all zero initial conditions* [114, p. 311], which we will broaden to the boundedness of the system behaviors for bounded input variables *and any bounded initial conditions*, while the LY stability concept demands Lyapunov's ε-δ closeness of the system behaviors to the zero equilibrium under all zero input deviations. This means that various new BI stability properties will be introduced and defined, and complex domain criteria will be discovered for them. E. D. Sontag et al. [115], [117] established the theory of "**input to state stability**" that concerns both the bounded input vector and arbitrary initial state conditions. It is valid in the general setting of nonlinear systems. Their results, which hold in *time* domain, exploit the Lyapunov method in combination with the comparison functions of W. Hahn [76].

- The BI stability concept characterizes various stability properties of a dynamical system, while the LY stability concept determines various stability properties of a desired (nominal) system behavior in terms of total coordinates, i.e., of an equilibrium vector in terms of deviations.

- The BI stability concept does not demand asymptotic convergence (as $t \to \infty$) of the system behaviors to the equilibrium vector, while LY stability properties-attraction, asymptotic and exponential stability of the equilibrium vector, do.

Both the LY stability concept and the BI stability concept treat system behaviors over the infinite *time* interval $[0, \infty]$, i.e., over $\mathfrak{T}_0$.

Since we consider only *time*-invariant systems, then the initial instant is zero moment, $t_0 = 0$, for both the BI stability concept and the LY stability concept.

The BI stability concept that demands boundedness of the system behaviors for bounded input variables and for zero or arbitrary initial conditions incorporates a great variety of BI stability properties among which are the following :

- **Bounded Input-Bounded Internal State** ($BIBIS$) **stability** that demands the bounded system internal state, for all $t \in \mathfrak{T}_0$ and for every bounded input vector function under all zero initial conditions.

- **Bounded Input-Bounded Output** ($BIBO$) **stability** that demands the bounded system output response for all $t \in \mathfrak{T}_0$ and for every bounded input vector function under all zero initial conditions [4], [13], [77].

- **Bounded Input-Bounded Internal State and Output** ($BIBISO$) **stability** that demands the bounded system internal state and output, for all $t \in \mathfrak{T}_0$ and for every bounded input vector function under all zero initial conditions.

- **Bounded Input-Bounded Output State** ($BIBOS$) **stability** that demands the bounded system output state for all $t \in \mathfrak{T}_0$ and for every bounded input vector function under all zero initial conditions.

- **Bounded Input-Bounded State** ($BIBS$) **stability** that demands the bounded system state for all $t \in \mathfrak{T}_0$ and for every bounded input vector function under all zero initial conditions.

- **Bounded Input-Bounded State and Output** ($BIBSO$) **stability** that demands the bounded system state and its output response for all $t \in \mathfrak{T}_0$, and for every bounded input vector function under all zero initial conditions.

These BI stability properties assume *all zero initial conditions*, and they are valid only under this assumption. In reality, at least one initial condition is, most often, different from zero. It is natural to consider system behaviors under the bounded input vector function and arbitrary initial conditions. This leads to the following generalizations of the BI stability concept and properties that hold for arbitrary initial conditions:

- **Bounded Input and Initial Internal State-Bounded Internal State** (*BIISBIS*) **stability** demands the bounded system internal state for all $t \in \mathfrak{T}_0$ and for every bounded the input vector function and the initial internal state.

- **Bounded Input and Initial Internal State-Bounded Output** (*BIISBO*) **stability** demands the bounded system output response for all $t \in \mathfrak{T}_0$ and for every bounded the input vector function and the initial internal state.

- **Bounded Input and Initial Internal State-Bounded Output State** (*BIISBOS*) **stability** demands the bounded system output state for all $t \in \mathfrak{T}_0$ and for every bounded the input vector function and the initial internal state.

- **Bounded Input and Initial Internal State-Bounded Internal State and Output** (*BIISBISO*) **stability** demands the bounded system internal state and the system output response, for all $t \in \mathfrak{T}_0$ and for every bounded the input vector function and the initial internal state.

- **Bounded Input and Initial Internal State-Bounded State** (*BIISBS*) **stability** demands the bounded system state for all $t \in \mathfrak{T}_0$ and for every bounded the input vector function and the initial internal state.

- **Bounded Input and Initial Internal State-Bounded State and Output** (*BIISBSO*) **stability** demands the bounded system state and the system output response for all $t \in \mathfrak{T}_0$ and for every the bounded input vector function and the initial internal state.

- **Bounded Input and Initial State-Bounded Output** (*BISBO*) **stability** demands the bounded system output response, for all $t \in \mathfrak{T}_0$ and for every bounded the input vector function and the initial state.

- **Bounded Input and Initial State-Bounded Output State** (*BISBOS*) **stability** demands the bounded system output state for all $t \in \mathfrak{T}_0$ and for every bounded the input vector function and the initial state.

- **Bounded Input and Initial State-Bounded State and Output** (*BISBSO*) **stability** demands the bounded system output state for all $t \in \mathfrak{T}_0$ and for every bounded the input vector function and the initial state.

- **Bounded Input and Initial State-Bounded State** (*BISBS*) **stability** demands the bounded system state for all $t \in \mathfrak{T}_0$ and for every bounded the input vector function and the initial state.

Comment 376 *Every BI stability property valid under arbitrary initial conditions implies the corresponding BI stability property valid only under all zero initial conditions.*

We will prove the conditions for each of the BI stability properties for input vector functions $\mathbf{i}(.) : \mathfrak{T}_0 \to \mathfrak{R}^M$ belonging to the class $\mathfrak{I}^\mu$ (2.14), (in Section 2.1).

The conditions will discover other relationships among BI stability properties. They will enable us also to compare them with those for Lyapunov stability properties.

14.2 BI stability definitions

14.2.1 *IO* systems

It might seem at first glance that $\mathbf{y}(t; \mathbf{y}_0^{\nu-1}; \mathbf{i})$ expresses both the internal dynamical behavior of the IO system (3.77), (in Subsection 3.5.2), and its output dynamics. It is not correct. The *time* evolution of $\mathbf{y}^{\nu-1}(t; \mathbf{y}_0^{\nu-1}; \mathbf{i})$ expresses both the internal dynamical behavior and the output dynamical behavior of the IO system. Hence, it represents the IO system dynamics. The *time* evolution $\mathbf{y}(t; \mathbf{y}_0^{\nu-1}; \mathbf{i})$ is the ordinary temporal output response of the IO system and it expresses only the output variation in *time*. We can say that $\mathbf{y}^{\nu-1}(t; \mathbf{y}_0^{\nu-1}; \mathbf{i})$ is also *the system output response of the order* $\nu - 1$, or the *system full output response*. It is also the system internal state, the system output state and the system state at the moment t.

Definition 377 *The IO system (3.77) is*

a) ***Bounded Input-Bounded State*** *(BIBS)* ***stable,*** *if and only if for every positive real number κ there exists a positive real number μ, the value of which depends on κ, $\mu = \mu(\kappa)$, such that for every input vector function obeying $\|\mathbf{i}(t)\| < \mu$ for all $t \in \mathfrak{T}_0$, under all zero initial conditions the system state satisfies $\left\|\mathbf{y}^{\nu-1}(t; \mathbf{0}_{\nu N}; \mathbf{i})\right\| < \kappa$ for all $t \in \mathfrak{T}_0$,*

$$\forall \kappa \in \mathfrak{R}^+, \ \exists \mu \in \mathfrak{R}^+, \ \mu = \mu(\kappa), \ \|\mathbf{i}(t)\| < \mu, \ \forall t \in \mathfrak{T}_0, \Longrightarrow \left\|\mathbf{y}^{\nu-1}(t; \mathbf{0}_{\nu N}; \mathbf{i})\right\| < \kappa, \ \forall t \in \mathfrak{T}_0. \tag{14.1}$$

b) ***Bounded Input-Bounded Output*** *(BIBO)* ***stable,*** *if and only if for every positive real number γ there exists a positive real number μ, the value of which depends on γ, $\mu = \mu(\gamma)$, such that for every input vector function obeying $\|\mathbf{i}(t)\| < \mu$ for all $t \in \mathfrak{T}_0$, the output response under all zero initial conditions satisfies $\|\mathbf{y}(t; \mathbf{0}_{\nu N}; \mathbf{i})\| < \gamma$ for all $t \in \mathfrak{T}_0$,*

$$\forall \gamma \in \mathfrak{R}^+, \ \exists \mu \in \mathfrak{R}^+, \ \mu = \mu(\gamma), \ \|\mathbf{i}(t)\| < \mu, \ \forall t \in \mathfrak{T}_0, \Longrightarrow \|\mathbf{y}(t; \mathbf{0}_{\nu N}; \mathbf{i})\| < \gamma, \ \forall t \in \mathfrak{T}_0. \tag{14.2}$$

c) ***Bounded Input and Initial State-Bounded State*** *(BISBS)* ***stable****, if and only if for every positive real number* $\varkappa$ *there exist positive real numbers* μ *and* υ*, the values of which depend on* $\varkappa$*,* $\mu = \mu(\varkappa)$*,* $\upsilon = \upsilon(\varkappa)$*, such that for every input vector function obeying* $\|\mathbf{i}(t)\| < \mu$ *for all* $t \in \mathfrak{T}_0$*, and for every initial state satisfying* $\|\mathbf{y}_0^{\nu-1}\| < \upsilon$ *the state satisfies* $\|\mathbf{y}^{\nu-1}(t;\mathbf{y}_0^{\nu-1};\mathbf{i})\| < \varkappa$ *for all* $t \in \mathfrak{T}_0$*,*

$$\begin{gathered}\forall \varkappa \in \mathfrak{R}^+,\ \exists \mu \in \mathfrak{R}^+,\ \mu = \mu(\kappa),\ \exists \upsilon \in \mathfrak{R}^+,\ \upsilon = \upsilon(\kappa),\\ \|\mathbf{i}(t)\| < \mu,\ \forall t \in \mathfrak{T}_0,\ \|\mathbf{y}_0^{\nu-1}\| < \upsilon \Longrightarrow\\ \|\mathbf{y}^{\nu-1}(t;\mathbf{y}_0^{\nu-1};\mathbf{i})\| < \varkappa,\ \forall t \in \mathfrak{T}_0.\end{gathered} \tag{14.3}$$

d) ***Bounded Input and Initial State-Bounded Output*** *(BISBO)* ***stable****, if and only if for every positive real number* γ *there exist positive real numbers* μ *and* υ*, the values of which depend on* γ*,* $\mu = \mu(\gamma)$*,* $\vartheta = \vartheta(\gamma)$*, such that for every input vector function obeying* $\|\mathbf{i}(t)\| < \mu$ *for all* $t \in \mathfrak{T}_0$*, and for every initial state satisfying* $\|\mathbf{y}_0^{\nu-1}\| < \vartheta$ *the output response satisfies* $\|\mathbf{y}(t;\mathbf{y}_0^{\nu-1};\mathbf{i})\| < \gamma$ *for all* $t \in \mathfrak{T}_0$*,*

$$\begin{gathered}\forall \gamma \in \mathfrak{R}^+,\ \exists \mu \in \mathfrak{R}^+,\ \mu = \mu(\gamma),\ \exists \vartheta \in \mathfrak{R}^+,\ \vartheta = \vartheta(\gamma),\\ \|\mathbf{i}(t)\| < \mu,\ \forall t \in \mathfrak{T}_0,\ \|\mathbf{y}_0^{\nu-1}\| < \vartheta \Longrightarrow\\ \|\mathbf{y}(t;\mathbf{y}_0^{\nu-1};\mathbf{i})\| < \gamma,\ \forall t \in \mathfrak{T}_0.\end{gathered} \tag{14.4}$$

Note 378 *The boundedness condition on the input vector that*

$$\|\mathbf{i}(t)\| < \mu \quad \textit{for all } t \in \mathfrak{T}_0$$

implies the boundedness by μ *of the initial input vector,*

$$\|\mathbf{i}(0)\| = \|\mathbf{i}_0\| < \mu.$$

Note 379 *Since the initial state vector* $\mathbf{y}_0^{\nu-1}$ *of the IO system (3.77) is also the vector of all its initial conditions, then the following BI stability properties coincide: BIBIS, BIBOS, BIBS and BIBSO; BIISBIS, BIISBOS, BIISBSO and BIISBS; BISBSO , BISBOS and BISBS.*

Note 380 *These BI stability properties, like Lyapunov stability of the equilibrium vector, do not demand that system behaviors converge to the equilibrium vector. Moreover, the BI stability properties do not demand the existence of an equilibrium vector in general.*

14.2.2 *ISO* systems

A characteristic of the *ISO* system (3.84) and (3.85), (in Subsection 3.5.3), is the linear algebraic dependence of the system output vector on both the system state vector and on the input vector. This simplifies their study. This is clearer from the definition of the *BI* stability properties of these systems.

Definition 381 *a) The ISO system (3.84) and (3.85) is*

a) **Bounded Input-Bounded State** *(BIBS)* **stable** *if and only if for every positive real number* κ *there exists a positive real number* η*, the value of which depends on* κ*,* $\eta = \eta(\kappa)$*, such that for every input vector function obeying* $\|\mathbf{i}(t)\| < \eta$ *for all* $t \in \mathfrak{T}_0$*, the system state vector under all zero initial conditions satisfies* $\|\mathbf{x}(t;\mathbf{0}_n;\mathbf{i})\| < \kappa$ *for all* $t \in \mathfrak{T}_0$*,*

$$\forall \kappa \in \mathfrak{R}^+,\ \exists \eta \in \mathfrak{R}^+,\ \eta = \eta(\kappa),\ \ \|\mathbf{i}(t)\| < \eta,\ \forall t \in \mathfrak{T}_0 \Longrightarrow \\ \|\mathbf{x}(t;\mathbf{0}_n;\mathbf{i})\| < \kappa,\ \forall t \in \mathfrak{T}_0. \tag{14.5}$$

b) **Bounded Input-Bounded Output** *(BIBO)* **stable** *if and only if for every positive real number* γ *there exists a positive real number* μ*, the value of which depends on* γ*,* $\mu = \mu(\gamma)$*, such that for every input vector function obeying* $\|\mathbf{i}(t)\| < \gamma$ *for all* $t \in \mathfrak{T}_0$*, the system output vector under all zero initial conditions obeys* $\|\mathbf{y}(t;\mathbf{0}_n;\mathbf{i})\| < \gamma$ *for all* $t \in \mathfrak{T}_0$*,*

$$\forall \gamma \in \mathfrak{R}^+,\ \exists \mu \in \mathfrak{R}^+,\ \mu = \mu(\gamma),\ \ \|\mathbf{i}(t)\| < \mu,\ \forall t \in \mathfrak{T}_0 \Longrightarrow \\ \|\mathbf{y}(t;\mathbf{0}_n;\mathbf{i})\| < \gamma,\ \forall t \in \mathfrak{T}_0. \tag{14.6}$$

c) **Bounded Input-Bounded State and Output** *(BIBSO)* **stable** *if and only if for every positive real number* κ *there exist positive real number* η *and* θ*, the values of which depend on* κ*,* $\eta = \eta(\kappa)$ *and* $\theta = \theta(\kappa)$*, such that for every input vector function obeying* $\|\mathbf{i}(t)\| < \eta$ *for all* $t \in \mathfrak{T}_0$*, the system motion and the system output response under all zero initial conditions satisfy* $\|\mathbf{x}(t;\mathbf{0}_n;\mathbf{i})\| < \kappa$ *and* $\|\mathbf{y}(t;\mathbf{0}_n;\mathbf{i})\| < \theta$ *for all* $t \in \mathfrak{T}_0$*,*

$$\forall \kappa \in \mathfrak{R}^+,\ \exists \eta, \theta \in \mathfrak{R}^+,\ \eta = \eta(\kappa),\ \theta = \theta(\kappa),\ \|\mathbf{i}(t)\| < \eta,\ \forall t \in \mathfrak{T}_0 \Longrightarrow \\ \|\mathbf{x}(t;\mathbf{0}_n;\mathbf{i})\| < \kappa\ \ \text{and}\ \ \|\mathbf{y}(t;\mathbf{0}_n;\mathbf{i})\| < \theta,\ \forall t \in \mathfrak{T}_0. \tag{14.7}$$

d) **Bounded Input and Bounded Initial State - Bounded State** *(BISBS)* **Stable** *if and only if for every positive real number* κ *there exist positive real numbers* η *and* θ*, the values of which depend on* κ*,* $\eta = \eta(\kappa)$ *and* $\theta = \theta(\kappa)$*, such that for every input vector function obeying* $\|\mathbf{i}(t)\| < \eta$

for all $t \in \mathfrak{T}_0$, and for every initial state $\mathbf{x}_0$ satisfying $\|\mathbf{x}_0\| < \theta$, the system motion obeys $\|\mathbf{x}(t;\mathbf{x}_0;\mathbf{i})\| < \kappa$ for all $t \in \mathfrak{T}_0$,

$$\forall \kappa \in \mathfrak{R}^+,\ \exists \eta \in \mathfrak{R}^+,\ \eta = \eta(\kappa),\ \exists \theta \in \mathfrak{R}^+,\ \theta = \theta(\kappa),$$
$$\|\mathbf{i}(t)\| < \eta,\ \forall t \in \mathfrak{T}_0,\ and\ \|\mathbf{x}_0\| < \theta \Longrightarrow$$
$$\|\mathbf{x}(t;\mathbf{x}_0;\mathbf{i})\| < \kappa,\ \forall t \in \mathfrak{T}_0. \tag{14.8}$$

e) ***Bounded Input and Initial State-Bounded Output*** *(BISBO)* ***Stable*** *if and only if for every positive real number γ there exist positive real number μ and ν, the values of which depend on γ, $\mu = \mu(\gamma)$ and $\nu = \nu(\gamma)$, such that for every input vector function obeying $\|\mathbf{i}(t)\| < \mu$ for all $t \in \mathfrak{T}_0$, and for every initial state $\mathbf{x}_0$ satisfying $\|\mathbf{x}_0\| < \nu$, the system output response obeys $\|\mathbf{y}(t;\mathbf{x}_0;\mathbf{i})\| < \gamma$ for all $t \in \mathfrak{T}_0$,*

$$\forall \gamma \in \mathfrak{R}^+,\ \exists \mu \in \mathfrak{R}^+,\ \mu = \mu(\gamma),\ \exists \nu \in \mathfrak{R}^+,\ \nu = \nu(\gamma),$$
$$\|\mathbf{i}(t)\| < \mu,\ \forall t \in \mathfrak{T}_0,\ and\ \|\mathbf{x}_0\| < \nu \Longrightarrow$$
$$\|\mathbf{y}(t;\mathbf{x}_0;\mathbf{i})\| < \gamma,\ \forall t \in \mathfrak{T}_0. \tag{14.9}$$

f) ***Bounded Input and Initial State-Bounded State and Output*** *(BISBSO)* ***Stable*** *if and only if for every positive real number κ there exist positive real numbers ν, μ and ρ, the values of which depend on κ, $\nu = \nu(\kappa)$, $\mu = \mu(\kappa)$ and $\rho = \rho(\kappa)$, such that for every input vector function obeying $\|\mathbf{i}(t)\| < \mu$ for all $t \in \mathfrak{T}_0$, and for every initial state $\mathbf{x}_0$ satisfying $\|\mathbf{x}_0\| < \nu$, the system motion and the system output response obey, respectively, $\|\mathbf{x}(t;\mathbf{x}_0;\mathbf{i})\| < \kappa$ and $\|\mathbf{y}(t;\mathbf{x}_0;\mathbf{i})\| < \rho$ for all $t \in \mathfrak{T}_0$,*

$$\forall k \in \mathfrak{R}^+,\ \exists \nu \in \mathfrak{R}^+,\ \nu = \nu(\kappa),\ \exists \mu \in \mathfrak{R}^+,\ \mu = \mu(\kappa),$$
$$\exists \rho \in \mathfrak{R}^+,\ \rho = \rho(\kappa),$$
$$\|\mathbf{i}(t)\| < \mu,\ \forall t \in \mathfrak{T}_0,\ and\ \|\mathbf{x}_0\| < \nu \Longrightarrow$$
$$\|\mathbf{x}(t;\mathbf{x}_0;\mathbf{i})\| < \kappa\ and\ \|\mathbf{y}(t;\mathbf{x}_0;\mathbf{i})\| < \rho,\ \forall t \in \mathfrak{T}_0. \tag{14.10}$$

Note 382 *The condition that $\|\mathbf{i}(t)\|$ is bounded for all $t \in \mathfrak{T}_0$ guarantees the boundedness of the norm $\|\mathbf{i}_0\|$ of the initial input vector $\mathbf{i}_0$.*

Note 383 *These BI stability properties do not demand that system motions converge to the equilibrium vector $\mathbf{x}_e = \mathbf{0}_n$.*

Claim 384 *The fact that the initial state vector $\mathbf{x}_0$ is the vector of all system initial conditions implies that the following BI stability properties are the same for the ISO system (3.84) and (3.85): BIBIS, BIBOS, BIBS and BIBSO; BIISBIS, BIISBOS, BIISBSO and BIISBS; BISBSO, BISBOS and BISBS.*

The following two Theorems and their proofs explain and verify Claim 384. Besides, they show how we can easily prove Claim 379 in Subsection 14.2.1.

Theorem 385 *In order for the ISO system (3.84) and (3.85) to be $BIBSO$ stable, it is necessary and sufficient to be $BIBS$ stable.*

Proof. *Necessity.* Let the ISO system (3.84) and (3.85) be $BIBSO$ stable. Then, c) of Definition 381 shows that for every input vector function obeying $\|\mathbf{i}(t)\| < \mu$ for all $t \in \mathfrak{T}_0$, the system motion and the system output response under all zero initial conditions satisfy $\|\mathbf{x}(t; \mathbf{0}_n; \mathbf{i})\| < \kappa_1$ and $\|\mathbf{y}(t; \mathbf{0}_n; \mathbf{i})\| < \kappa_2$ for all $t \in \mathfrak{T}_0$,. The system is $BIBS$ stable in view of c) of Definition 381.

Sufficiency. Let the ISO system (3.84) and (3.85) be $BIBS$ stable. From a) of Definition 381, i.e., (14.5), follows

$$\begin{gathered}\forall \kappa \in \mathfrak{R}^+,\ \exists \eta \in \mathfrak{R}^+,\ \eta = \eta(\kappa), \\ \|\mathbf{i}(t)\| < \eta,\ \forall t \in \mathfrak{T}_0 \Longrightarrow \|\mathbf{x}(t; \mathbf{0}_n; \mathbf{i})\| < \kappa,\ \forall t \in \mathfrak{T}_0,\end{gathered}$$

which, together with (3.85), implies

$$\begin{gathered}\forall \kappa \in \mathfrak{R}^+,\ \eta = \eta(\kappa), \\ \exists \rho \in \mathfrak{R}^+,\ \rho = \rho(\kappa) \Longrightarrow \|C\| k + \|D\| \eta < \rho,\ \rho = \rho\left[k, \eta(\kappa)\right] = \rho(\kappa), \\ \Longrightarrow \\ \\ \|\mathbf{i}(t)\| < \eta,\ \forall t \in \mathfrak{T}_0,\ \Longrightarrow \\ \|\mathbf{x}(t; \mathbf{0}_n; \mathbf{i})\| < k,\ \forall t \in \mathfrak{T}_0 \Longrightarrow \\ \|\mathbf{y}(t; \mathbf{0}_n; \mathbf{i})\| \leq \|C\| \|\mathbf{x}(t; \mathbf{0}_n; \mathbf{i})\| + \|D\| \|\mathbf{i}(t)\| < \\ < \|C\| k + \|D\| \eta < \rho < \infty,\ \forall t \in \mathfrak{T}_0.\end{gathered}$$

The system is $BIBSO$ stable in view of c) of Definition 381. ■

By following the proof of Theorem 385, we can easily prove the following claim.

Claim 386 *Variations of the state vector and the input vector determine completely the output dynamical behavior of the ISO system (3.84) and (3.85). This comes out from the algebraic nature of the system output equation. The BI stability properties hold for bounded input vectors. They and various Bounded Input-Bounded State stability properties, which guarantee* *the boundedness of the system state, imply the boundedness of both the output behavior and the full output behavior because they coincide for the ISO system (3.84) and (3.85). This explains that BIBS stability guarantees BIBO stability, BIBOS stability and BIBSO stability.*

Theorem 387 *In order for the ISO system (3.84) and (3.85) to be BISBSO stable, it is necessary and sufficient to be BISBS stable.*

Proof. *Necessity.* Let the *ISO* system (3.84) and (3.85) be *BISBSO* stable. Then, f) of Definition 381 shows that for every input vector function obeying $\|\mathbf{i}(t)\| < \mu$ for all $t \in \mathfrak{T}_0$, and for any initial state satisfying $\|\mathbf{x}_0\| < \theta$, the system motion obeys $\|\mathbf{x}(t;\mathbf{x}_0;\mathbf{i})\| < \kappa$ for all $t \in \mathfrak{T}_0$. The system is *BISBS* stable in view of d) of Definition 381.

Sufficiency. Let the *ISO* system (3.84) and (3.85) be *BISBS* stable. From d) of Definition 381, i.e., (14.8), follows

$$\begin{gathered}\forall\kappa \in \mathfrak{R}^+,\ \exists\eta \in \mathfrak{R}^+,\ \eta = \eta(\kappa),\ \exists\theta \in \mathfrak{R}^+,\ \theta = \theta(\kappa),\\ \|\mathbf{i}(t)\| < \eta,\ \forall t \in \mathfrak{T}_0,\ and\ \|\mathbf{x}_0\| < \theta \Longrightarrow\\ \|\mathbf{x}(t;\mathbf{x}_0;\mathbf{i})\| < \kappa,\ \forall t \in \mathfrak{T}_0,\end{gathered}$$

which, together with (3.85), implies

$$\begin{gathered}\forall\kappa \in \mathfrak{R}^+,\ \exists\psi \in \mathfrak{R}^+,\ \psi = \psi(\kappa) \leq k,\\ \exists\eta \in \mathfrak{R}^+,\ \eta = \eta(\psi),\ \exists\theta \in \mathfrak{R}^+,\ \theta = \theta(\psi) \Longrightarrow\\ \|C\|\,\psi + \|D\|\,\eta < \kappa\\ \Longrightarrow\\ \|\mathbf{i}(t)\| < \eta,\ \forall t \in \mathfrak{T}_0,\ and\ \|\mathbf{x}_0\| < \theta \Longrightarrow\\ \|\mathbf{x}(t;\mathbf{x}_0;\mathbf{i})\| < \psi,\ \forall t \in \mathfrak{T}_0 \Longrightarrow\\ \|\mathbf{y}(t;\mathbf{x}_0;\mathbf{i})\| \leq \|C\|\,\|\mathbf{x}(t;\mathbf{x}_0;\mathbf{i})\| + \|D\|\,\|\mathbf{i}(t)\| <\\ < \|C\|\,\psi + \|D\|\,\eta < \kappa < \infty,\ \forall t \in \mathfrak{T}_0.\end{gathered}$$

The system is *BISBSO* stable in view of f) of Definition 381. ∎

With trivial appropriate adjustments of this proof we prove the following

Claim 388 *Variations of the state vector and the input vector determine completely the output dynamical behavior of the ISO system (3.84) and (3.85). This comes out from the algebraic nature of the system output equation. The BI stability properties hold for bounded input vectors. They and*

various Bounded Input-Bounded State stability properties, which guarantee the boundedness of the system state, imply the boundedness of both the output behavior and the full output behavior because they coincide for the ISO system (3.84) and (3.85). This explains that $BISBS$ stability guarantees $BISBO$ stability and $BISBSO$ stability.

We will establish the conditions for the BI stability properties of the ISO system (3.84) and (3.85) in in Subsection 14.3.2.

14.2.3 *IIO* systems

The IIO system (3.90), (in Subsection 3.5.4), differently than the IO system (3.77), (in Subsection 3.5.2), and differently than the ISO system (3.84), (3.85), (in Subsection 3.5.3), has both *the internal dynamics* expressed by *its state behavior,* which is described by the left-hand side of its first differential equation, and *the output dynamics,* which is defined by the left-hand side of its second differential equation if and only if $\nu > 0$ (Section 2.3). This enriches the Bounded Input (BI) stability concept with new and more complex BI stability properties than those of the IO system (3.77) and of the ISO system (3.84), (3.85). In this regard we recall Note 80, (in Subsection 3.5.4).

The fact that the output behavior depends on the output dynamics (if and only if $\nu > 0$) implies that the following BI stability properties are not significant for the IIO system (3.90): $BIISBIS$, $BIISBO$, $BIISBOS$, $BIISBISO$, and $BIISBSO$.

Definition 389 *a) The IIO system (3.90) is*

a) ***Stable-Bounded Input-Bounded Internal State*** *($BIBIS$) if and only if for every positive real number κ there exists a positive real number η, the value of which depends on κ, $\eta = \eta(\kappa)$, such that for every input vector function obeying $\|\mathbf{i}(t)\| < \eta$ for all $t \in \mathfrak{T}_0$, the system state vector under all zero initial conditions satisfies $\left\|\mathbf{r}^{\alpha-1}(t;\mathbf{0}_{\alpha\rho};\mathbf{i})\right\| < \kappa$ for all $t \in \mathfrak{T}_0$,*

$$\forall\kappa \in \mathfrak{R}^+,\ \exists\eta \in \mathfrak{R}^+,\ \eta = \eta(\kappa),$$
$$\|\mathbf{i}(t)\| < \eta,\ \forall t \in \mathfrak{T}_0, \Longrightarrow \left\|\mathbf{r}^{\alpha-1}(t;\mathbf{0}_{\alpha\rho};\mathbf{i})\right\| < \kappa,\ \forall t \in \mathfrak{T}_0. \quad (14.11)$$

b) ***Bounded Input-Bounded Output*** *($BIBO$)* ***stable*** *if and only if for every positive real number γ there exists a positive real number η, the value of which depends on γ, $\eta = \eta(\gamma)$, such that for every input vector* *function obeying* $\|\mathbf{i}(t)\| < \eta$ *for all* $t \in \mathfrak{T}_0$*, the system output vector in the free regime satisfies* $\|\mathbf{y}(t; \mathbf{0}_{\alpha\rho}; \mathbf{0}_{\nu N}; \mathbf{i})\| < \gamma$ *for all* $t \in \mathfrak{T}_0$,

$$\forall \gamma \in \mathfrak{R}^+, \ \exists \eta \in \mathfrak{R}^+, \ \eta = \eta(\gamma), \ \|\mathbf{i}(t)\| < \eta, \ \forall t \in \mathfrak{T}_0, \implies \|\mathbf{y}(t; \mathbf{0}_{\alpha\rho}; \mathbf{0}_{\nu N}; \mathbf{i})\| < \gamma, \ \forall t \in \mathfrak{T}_0, \tag{14.12}$$

c) ***Bounded Input-Bounded Output State*** *(BIBOS)* ***stable*** *if and only if for every positive real number* γ *there exists a positive real number* η*, the value of which depends on* γ*,* $\eta = \eta(\gamma)$*, such that for every input vector function obeying* $\|\mathbf{i}(t)\| < \eta$ *for all* $t \in \mathfrak{T}_0$*, and for all zero initial conditions, the system output dynamics behavior fulfills* $\left\|\mathbf{y}^{\nu-1}(t; \mathbf{0}_{\alpha\rho}; \mathbf{0}_{\nu N}; \mathbf{i})\right\| < \gamma$ *for all* $t \in \mathfrak{T}_0$,

$$\forall \gamma \in \mathfrak{R}^+, \ \exists \eta \in \mathfrak{R}^+, \ \eta = \eta(\gamma), \ \|\mathbf{i}(t)\| < \eta, \ \forall t \in \mathfrak{T}_0, \implies \left\|\mathbf{y}^{\nu-1}(t; \mathbf{0}_{\alpha\rho}; \mathbf{0}_{\nu N}; \mathbf{i})\right\| < \gamma, \ \forall t \in \mathfrak{T}_0. \tag{14.13}$$

d) ***Bounded Input-Bounded Internal State and Output*** *(BIBISO)* ***stable*** *if and only if for every positive real number* κ *there exist positive real numbers* η *and* θ*, the values of which depend on* κ*,* $\eta = \eta(\kappa)$ *and* $\theta = \theta(\kappa)$*, such that for every input vector function obeying* $\|\mathbf{i}(t)\| < \eta$ *for all* $t \in \mathfrak{T}_0$*, and for all zero initial conditions, the system state vector fulfills* $\left\|\mathbf{r}^{\alpha-1}(t; \mathbf{0}_{\alpha\rho}; \mathbf{i})\right\| < \kappa$ *and its output vector obeys* $\|\mathbf{y}(t; \mathbf{0}_{\alpha\rho}; \mathbf{0}_{\nu N}; \mathbf{i})\| < \theta$ *for all* $t \in \mathfrak{T}_0$,

$$\begin{gathered}\forall \kappa \in \mathfrak{R}^+, \ \exists \eta, \theta \in \mathfrak{R}^+, \ \eta = \eta(\kappa), \ \theta = \theta(\kappa), \\ \|\mathbf{i}(t)\| < \eta, \ \forall t \in \mathfrak{T}_0 \implies \\ \left\|\mathbf{r}^{\alpha-1}(t; \mathbf{0}_{\alpha\rho}; \mathbf{i})\right\| < \kappa, \|\mathbf{y}(t; \mathbf{0}_{\alpha\rho}; \mathbf{0}_{\nu N}; \mathbf{i})\| < \theta, \ \forall t \in \mathfrak{T}_0.\end{gathered} \tag{14.14}$$

e) ***Bounded Input and Initial Internal State-Bounded Internal State*** *(BIISBIS)* ***stable*** *if and only if for every positive real number* κ *there exist positive real numbers* η *and* θ*, the values of which depend on* κ*,* $\eta = \eta(\kappa)$ *and* $\theta = \theta(\kappa)$*, such that for every input vector function obeying* $\|\mathbf{i}(t)\| < \eta$ *for all* $t \in \mathfrak{T}_0$*, and for every initial state vector satisfying* $\left\|\mathbf{r}_0^{\alpha-1}\right\| < \theta$*, the system internal state vector obeys* $\left\|\mathbf{r}^{\alpha-1}(t; \mathbf{r}_0^{\alpha-1}; \mathbf{i})\right\| < \kappa$ *for all* $t \in \mathfrak{T}_0$,

$$\begin{gathered}\forall \kappa \in \mathfrak{R}^+, \ \exists \eta \in \mathfrak{R}^+, \ \eta = \eta(\kappa), \ \exists \theta \in \mathfrak{R}^+, \ \theta = \theta(\kappa), \\ \|\mathbf{i}(t)\| < \eta, \ \forall t \in \mathfrak{T}_0, \text{ and } \left\|\mathbf{r}_0^{\alpha-1}\right\| < \theta \implies \\ \left\|\mathbf{r}^{\alpha-1}(t; \mathbf{r}_0^{\alpha-1}; \mathbf{i})\right\| < \kappa, \ \forall t \in \mathfrak{T}_0.\end{gathered} \tag{14.15}$$

f) ***Bounded Input and Initial State-Bounded Output*** *(BISBO)* ***stable*** *if and only if for every positive real number γ there exist positive real numbers η and ξ, the values of which depend on γ, $\eta = \eta(\gamma)$ and $\xi = \xi(\gamma)$, such that for every input vector function obeying $\|\mathbf{i}(t)\| < \eta$ for all $t \in \mathfrak{T}_0$, and for every initial state satisfying $\left\| \left[\left(\mathbf{r}_0^{\alpha-1}\right)^T \ \vdots \ \left(\mathbf{y}_0^{\nu-1}\right)^T \right]^T \right\| < \xi$, the system response obeys $\left\|\mathbf{y}(t; \mathbf{r}_0^{\alpha-1}; \mathbf{y}_0^{\nu-1}; \mathbf{i})\right\| < \gamma$ for all $t \in \mathfrak{T}_0$,*

$$\forall \gamma \in \mathfrak{R}^+,\ \exists \eta \in \mathfrak{R}^+,\ \eta = \eta(\gamma),\ \exists \xi \in \mathfrak{R}^+,\ \xi = \xi(\gamma),$$
$$\|\mathbf{i}(t)\| < \eta,\ \forall t \in \mathfrak{T}_0,\ and\ \left\| \left[\left(\mathbf{r}_0^{\alpha-1}\right)^T \ \vdots \ \left(\mathbf{y}_0^{\nu-1}\right)^T \right]^T \right\| < \xi \Longrightarrow$$
$$\left\|\mathbf{y}(t; \mathbf{r}_0^{\alpha-1}; \mathbf{y}_0^{\nu-1}; \mathbf{i})\right\| < \gamma,\ \forall t \in \mathfrak{T}_0. \tag{14.16}$$

g) ***Bounded Input and Initial State-Bounded Output State*** *(BISBOS)* ***stable*** *if and only if for every positive real number γ there exist positive real numbers η and ξ, the values of which depend on γ, $\eta = \eta(\gamma)$ and $\xi = \xi(\gamma)$, such that for every input vector function obeying $\|\mathbf{i}(t)\| < \eta$ for all $t \in \mathfrak{T}_0$, and for every initial state satisfying $\left\| \left[\left(\mathbf{r}_0^{\alpha-1}\right)^T \ \vdots \ \left(\mathbf{y}_0^{\nu-1}\right)^T \right]^T \right\| < \xi$, the system output state fulfills $\left\|\mathbf{y}^{\nu-1}(t; \mathbf{r}_0^{\alpha-1}; \mathbf{y}_0^{\nu-1}; \mathbf{i})\right\| < \gamma$ for all $t \in \mathfrak{T}_0$,*

$$\forall \gamma \in \mathfrak{R}^+,\ \exists \eta \in \mathfrak{R}^+,\ \eta = \eta(\gamma),\ \exists \xi \in \mathfrak{R}^+,\ \xi = \xi(\gamma),$$
$$\|\mathbf{i}(t)\| < \eta,\ \forall t \in \mathfrak{T}_0,\ and\ \left\| \left[\left(\mathbf{r}_0^{\alpha-1}\right)^T \ \vdots \ \left(\mathbf{y}_0^{\nu-1}\right)^T \right]^T \right\| < \xi \Longrightarrow$$
$$\left\|\mathbf{y}^{\nu-1}(t; \mathbf{r}_0^{\alpha-1}; \mathbf{y}_0^{\nu-1}; \mathbf{i})\right\| < \gamma,\ \forall t \in \mathfrak{T}_0. \tag{14.17}$$

h) ***Bounded Input and Initial State-Bounded State and Output*** *(BISBSO)* ***stable*** *if and only if for every positive real number κ and γ there exist positive real numbers η and ξ, the values of which depend on κ and γ, $\eta = \eta(\gamma, \kappa)$ and $\xi = \xi(\gamma, \kappa)$, such that for every input vector function obeying $\|\mathbf{i}(t)\| < \eta$ for all $t \in \mathfrak{T}_0$, and for every overall initial vector satisfying $\left\| \left[\left(\mathbf{r}_0^{\alpha-1}\right)^T \ \vdots \ \left(\mathbf{y}_0^{\nu-1}\right)^T \right]^T \right\| < \xi$, the system state vector fulfills $\left\|\mathbf{r}^{\alpha-1}(t; \mathbf{r}_0^{\alpha-1}; \mathbf{i})\right\| < \kappa$ and its output vector obeys $\left\|\mathbf{y}(t; \mathbf{r}_0^{\alpha-1}; \mathbf{y}_0^{\nu-1}; \mathbf{i})\right\| < \gamma$*

for all $t \in \mathfrak{T}_0$,

$$\forall \kappa \in \mathfrak{R}^+,\ \forall \gamma \in \mathfrak{R}^+,\ \exists \eta \in \mathfrak{R}^+,\ \eta = \eta(\gamma,\kappa),\ \exists \xi \in \mathfrak{R}^+,\ \xi = \xi(\gamma,\kappa),$$
$$\|\mathbf{i}(t)\| < \eta,\ \forall t \in \mathfrak{T}_0,\ and\ \left\| \left[\left(\mathbf{r}_0^{\alpha-1}\right)^T \ \vdots\ \left(\mathbf{y}_0^{\nu-1}\right)^T \right]^T \right\| < \xi \Longrightarrow$$
$$\left\|\mathbf{r}^{\alpha-1}(t;\mathbf{r}_0^{\alpha-1};\mathbf{i})\right\| < \kappa, \left\|\mathbf{y}(t;\mathbf{r}_0^{\alpha-1};\mathbf{y}_0^{\nu-1};\mathbf{i})\right\| < \gamma,\ \forall t \in \mathfrak{T}_0. \tag{14.18}$$

i) ***Bounded Input and Initial State-Bounded State*** *(BISBS)* ***stable*** *if and only if for every positive real number* κ *and* γ *there exist positive real numbers* η *and* ξ*, the values of which depend on* κ *and* γ*,* $\eta = \eta(\gamma,\kappa)$ *and* $\xi = \xi(\gamma,\kappa)$*, such that for every input vector function obeying* $\|\mathbf{i}(t)\| < \eta$ *for all* $t \in \mathfrak{T}_0$*, and for every overall initial vector satisfying* $\left\| \left[\left(\mathbf{r}_0^{\alpha-1}\right)^T \ \vdots\ \left(\mathbf{y}_0^{\nu-1}\right)^T \right]^T \right\| < \xi$*, the system state vector fulfills* $\left\|\mathbf{r}^{\alpha-1}(t;\mathbf{r}_0^{\alpha-1};\mathbf{i})\right\| < \kappa$ *and its output state obeys* $\left\|\mathbf{y}^{\nu-1}(t;\mathbf{r}_0^{\alpha-1};\mathbf{y}_0^{\nu-1};\mathbf{i})\right\| < \gamma$ *for all* $t \in \mathfrak{T}_0$,

$$\forall \kappa \in \mathfrak{R}^+,\ \forall \gamma \in \mathfrak{R}^+,\ \exists \eta \in \mathfrak{R}^+,\ \eta = \eta(\gamma,\kappa),\ \exists \xi \in \mathfrak{R}^+,\ \xi = \xi(\gamma,\kappa),$$
$$\|\mathbf{i}(t)\| < \eta,\ \forall t \in \mathfrak{T}_0,\ and\ \left\| \left[\left(\mathbf{r}_0^{\alpha-1}\right)^T \ \vdots\ \left(\mathbf{y}_0^{\nu-1}\right)^T \right]^T \right\| < \xi \Longrightarrow$$
$$\left\|\mathbf{r}^{\alpha-1}(t;\mathbf{r}_0^{\alpha-1};\mathbf{i})\right\| < \kappa, \left\|\mathbf{y}^{\nu-1}(t;\mathbf{r}_0^{\alpha-1};\mathbf{y}_0^{\nu-1};\mathbf{i})\right\| < \gamma,\ \forall t \in \mathfrak{T}_0. \tag{14.19}$$

Note 390 *The norm* $\|\mathbf{i}_0\|$ *of the initial input vector* $\mathbf{i}_0$ *is bounded due to boundedness of* $\|\mathbf{i}(t)\|$ *for all* $t \in \mathfrak{T}_0$.

Note 391 *Positive real numbers* κ *and* γ *can be replaced by a positive real number* ζ *so that (14.19) takes a more compact equivalent form:*

$$\forall \zeta \in \mathfrak{R}^+,\ \exists \eta \in \mathfrak{R}^+,\ \eta = \eta(\zeta),\ \exists \xi \in \mathfrak{R}^+,\ \xi = \xi(\zeta),$$
$$\|\mathbf{i}(t)\| < \eta,\ \forall t \in \mathfrak{T}_0,\ and\ \left\| \left[\left(\mathbf{r}_0^{\alpha-1}\right)^T \ \vdots\ \left(\mathbf{y}_0^{\nu-1}\right)^T \right]^T \right\| < \xi \Longrightarrow$$
$$\left\| \left[\left(\mathbf{r}^{\alpha-1}(t;\mathbf{r}_0^{\alpha-1};\mathbf{i})\right)^T \ \vdots\ \left(\mathbf{y}^{\nu-1}(t;\mathbf{r}_0^{\alpha-1};\mathbf{y}_0^{\nu-1};\mathbf{i})\right)^T \right]^T \right\| < \zeta,\ \forall t \in \mathfrak{T}_0. \tag{14.20}$$

Note 392 *Positive real numbers* κ *and* γ *can be replaced by a positive real* *number* ζ *so that (14.20) takes a more compact equivalent form:*

$$\forall \zeta \in \mathfrak{R}^+,\ \exists \eta \in \mathfrak{R}^+,\ \eta = \eta(\zeta),$$
$$\|\mathbf{i}(t)\| < \eta,\ \forall t \in \mathfrak{T}_0,\ \Longrightarrow$$
$$\left\| \left[\left(\mathbf{r}^{\alpha-1}(t;\mathbf{0}_{\alpha\rho};\mathbf{i})\right)^T \ \vdots\ \left(\mathbf{y}^{\nu-1}(t;\mathbf{0}_{\alpha\rho};\mathbf{0}_{\nu N};\mathbf{i})\right)^T \right]^T \right\| < \zeta,\ \forall t \in \mathfrak{T}_0. \quad (14.21)$$

Note 393 *These BI stability properties allow that the system (reduced internal, output or full) dynamics behavior does not converge to the corresponding zero vector.*

Comment 394 *Every BI stability property valid under nonzero initial conditions guarantees the corresponding BI stability property under all zero initial conditions.*

Subsection 14.3.3 contains the conditions for the main *BI* stability properties of the *IIO* system (3.90).

14.3 BI stability conditions

14.3.1 *IO* systems

We will explore complex domain conditions for various *BI* stability properties of the *IO* system (3.77) under arbitrary all initial (internal and output) conditions (in Subsection 3.5.2) repeated as

$$A^{(\nu)}\mathbf{y}^{\nu}(t) = B^{(\mu)}\mathbf{i}^{\mu}(t),\ \forall t \in \mathfrak{T}_0,\ \nu \geq \mu. \quad (14.22)$$

Lemma 395 *Let the input vector function* $\mathbf{i}(.)$ *belong to the family* $\mathfrak{I}^{\mu}$, *(2.13) (in Section 2.1). In order for* $\mathbf{y}^{\nu-1}(t;\mathbf{i}_0^{\mu-1};\mathbf{y}_0^{\nu-1};\mathbf{i})$ *of the IO system (14.22) in the forced regime*

a) to be bounded on $\mathfrak{T}_0$,

$$\exists \zeta = \zeta\left(\mathbf{i}_0^{\mu-1}, \mathbf{y}_0^{\nu-1}, \mathbf{i},\right) \in \mathfrak{R}^+ \Longrightarrow \left\|\mathbf{y}^{\nu-1}(t;\mathbf{i}_0^{\mu-1};\mathbf{y}_0^{\nu-1};\mathbf{i})\right\| < \zeta,\ \forall t \in \mathfrak{T}_0,$$

it is necessary and sufficient that $\mathbf{y}(t;\mathbf{i}_0^{\mu-1};\mathbf{y}_0^{\nu-1};\mathbf{i})$ *is bounded on* $\mathfrak{T}_0$,

$$\exists \xi = \xi\left(\mathbf{i}_0^{\mu-1}, \mathbf{y}_0^{\nu-1}, \mathbf{i},\right) \in \mathfrak{R}^+ \Longrightarrow \left\|\mathbf{y}(t;\mathbf{i}_0^{\mu-1};\mathbf{y}_0^{\nu-1};\mathbf{i})\right\| < \xi,\ \forall t \in \mathfrak{T}_0,$$

b) to vanish as $t \to \infty$,

$$t \to \infty \Longrightarrow \mathbf{y}^{\nu-1}(t;\mathbf{i}_0^{\mu-1};\mathbf{y}_0^{\nu-1};\mathbf{i}) \to \mathbf{0}_{\nu N},$$

it is necessary and sufficient that $\mathbf{y}(t;\mathbf{i}_0^{\mu-1};\mathbf{y}_0^{\nu-1};\mathbf{i})$ *vanishes as* $t \to \infty$,

$$t \to \infty \Longrightarrow \mathbf{y}(t;\mathbf{i}_0^{\mu-1};\mathbf{y}_0^{\nu-1};\mathbf{i}) \to \mathbf{0}_N,$$

Proof. Let the input vector function $\mathbf{i}(.)$ be arbitrarily chosen from the family $\mathfrak{I}^{\mu}$, (2.13), (in Section 2.1) and be fixed. Let all the initial conditions be also arbitrarily chosen so that they are bounded and fixed.

Necessity. a) In order for $\mathbf{y}^{\nu-1}(t;\mathbf{i}_0^{\mu-1};\mathbf{y}_0^{\nu-1};\mathbf{i})$ to be bounded on $\mathfrak{T}_0$ it is necessary its every entry is bounded on $\mathfrak{T}_0$. Hence, it is necessary that $\mathbf{y}(t;\mathbf{i}_0^{\mu-1};\mathbf{y}_0^{\nu-1};\mathbf{i})$ is bounded on $\mathfrak{T}_0$.

b) In order for $\mathbf{y}^{\nu-1}(t;\mathbf{i}_0^{\mu-1};\mathbf{y}_0^{\nu-1};\mathbf{i})$ to vanish as $t \to \infty$ it is necessary that every entry of $\mathbf{y}^{\nu-1}(t;\mathbf{i}_0^{\mu-1};\mathbf{y}_0^{\nu-1};\mathbf{i})$ vanishes as $t \to \infty$. Hence, it is necessary that $\mathbf{y}(t;\mathbf{i}_0^{\mu-1};\mathbf{y}_0^{\nu-1};\mathbf{i})$ vanishes as $t \to \infty$.

Sufficiency. The Laplace transform of $\mathbf{y}(t;\mathbf{i}_0^{\mu-1};\mathbf{y}_0^{\nu-1};\mathbf{i})$ for $\mu \geq 1$ is determined by (8.6) and (8.8):

$$\mathbf{Y}^{\mp}(s) = \left[G_{IO}(s) \vdots G_{IOi_0}(s) \vdots\ G_{IOy_0}(s)\right]\left[\mathbf{I}^{\mp^T}(s) \vdots \left(\mathbf{i}_{0\mp}^{\mu-1}\right)^T \vdots \left(\mathbf{y}_{0\mp}^{\nu-1}\right)^T\right]^T.$$

The Equations (2.12) and (2.13) transform the preceding equation into:

$$\mathbf{Y}^{\mp}(s) = H_{IO}(s)\left[\mathbf{1}_M^T \vdots \left(\mathbf{i}_{0\mp}^{\mu-1}\right)^T \vdots \left(\mathbf{y}_{0\mp}^{\nu-1}\right)^T\right]^T,\ \mathbf{1}_M^T = [1\ 1\ ...\ 1] \in \mathfrak{R}^{1\times M},$$

$$H_{IO}(s) = \left[G_{IO}(s)I(s) \vdots G_{IOi_0}(s) \vdots\ G_{IOy_0}(s)\right]. \tag{14.23}$$

a) Let the conditions of the statement of Lemma under a) be valid. This, $\mathbf{i}(.) \in \mathfrak{I}^{\mu}$, the boundedness of the initial conditions, the Heaviside expansion of (14.23) and its inverse Laplace transform yield

$$\mathbf{y}(t;\mathbf{i}_0^{\mu-1};\mathbf{y}_0^{\nu-1};\mathbf{i}) = \left\{\begin{array}{c} \delta^-(t)\,R_0 + \\ +\sum\limits_{k=1}^{k=\nu N} e^{s_k^* t}\left[\sum\limits_{r=1}^{r=\nu_k}\frac{1}{(r-1)!}t^{r-1}R_{kr}\right]\end{array}\right\}\left[\begin{array}{c}\mathbf{1}_M \\ \mathbf{i}_0^{\mu-1} \\ \mathbf{y}_0^{\nu-1}\end{array}\right],$$

$$R_0 \in \mathfrak{R}^{N\times((\mu+1)M+\nu N)},\ R_{kr} \in \mathfrak{C}^{N\times((\mu+1)M+\nu N)},$$

$$R_0 = CONST.,\ R_{kr} = CONST.,$$

$$\operatorname{Re} s_k^* \leq 0,\ \forall k = 1,2,..,\nu N,\ \ \operatorname{Re} s_i^* = 0 \Longrightarrow \nu_i = 1, \tag{14.24}$$

where ν_k is the multiplicity of s_k^*. This shows that $\mathbf{y}(t;\mathbf{i}_0^{\mu-1};\mathbf{y}_0^{\nu-1};\mathbf{i})$ is infinitely times continuously differentiable at every $t \in \mathfrak{T}$. All its derivatives rest bounded on $\mathfrak{T}_0$ due to the Generating Theorem 342, which guarantees that $\mathbf{y}^{\nu-1}(t;\mathbf{i}_0^{\mu-1};\mathbf{y}_0^{\nu-1};\mathbf{i})$ is also bounded on $\mathfrak{T}_0$ if $\mu \geq 1$. In case $\mu = 0$ the submatrix $G_{IOi_0}(s)$ is to be replaced by the zero matrix and the vector $\mathbf{i}_0^{\mu-1}$ by the zero vector so that (14.24) becomes

$$\mathbf{y}(t;\mathbf{i}_0^{\mu-1};\mathbf{y}_0^{\nu-1};\mathbf{i}) = \left\{ \begin{array}{c} \delta^-(t)\, R_0 + \\ + \sum_{k=1}^{k=\nu N} e^{s_k^* t} \left[\sum_{r=1}^{r=\nu_k} \frac{1}{(r-1)!} t^{r-1} R_{kr} \right] \end{array} \right\} \left[\begin{array}{c} \mathbf{1}_M \\ \mathbf{y}_0^{\nu-1} \end{array} \right],$$

$$R_0 \in \mathfrak{R}^{N\times(M+\nu N)},\ R_{kr} \in \mathfrak{C}^{N\times(M+\nu N)},$$

$$R_0 = CONST.,\ R_{kr} = CONST.,$$

$$\operatorname{Re} s_k^* \leq 0,\ \forall k = 1, 2, .., \nu N,\ \ \operatorname{Re} s_i^* = 0 \Longrightarrow \nu_i = 1, \tag{14.25}$$

The boundedness of the output response $\mathbf{y}(t;\mathbf{i}_0^{\mu-1};\mathbf{y}_0^{\nu-1};\mathbf{i})$, the condition under a), and the Equations (14.24) and (14.25) imply that the output response $\mathbf{y}(t;\mathbf{i}_0^{\mu-1};\mathbf{y}_0^{\nu-1};\mathbf{i})$ is infinitely times continuously differentiable and bounded on $\mathfrak{T}_0$ due to Generating Theorem 342).

b) Let the condition of the statement of Lemma under b) hold. It, and the Equations (14.24) and (14.25) imply that all derivatives of the output response $\mathbf{y}(t;\mathbf{i}_0^{\mu-1};\mathbf{y}_0^{\nu-1};\mathbf{i})$ are continuously differentiable, bounded on $\mathfrak{T}_0$ and vanish as $t \to \infty$ due to the Generating Theorem 342. ■

This Lemma reduces largely the study of the *BI* stability properties of the *IO* system (14.22).

In view of the fact that the internal state $\mathbf{S}_I$, the output state $\mathbf{S}_O$ and the (full) state $\mathbf{S}$ of the *IO* system (14.22) coincide, (2.8), the *BIISBIS*, *BIISBOS*, *BIISBISO*, and *BIISBSO* stability coincide with *BISBS*, and *BIISBO* with *BISBO*.

Theorem 396 *Let the input vector function* $\mathbf{i}(.)$ *belong to the family* $\mathfrak{I}^{\mu}$. *In order for the IO system (14.22) in the forced regime to be:*

a) ***Bounded Input-Bounded State** (BIBS) **stable,** it is necessary and sufficient that*

1. $G_{IO}(.)$ *is at least proper, or strictly proper, real rational matrix function of s,*

and that

2. the real parts $Res_i^*(G_{IOrnd})$ *of all poles* $s_i^*(G_{IOrnd})$ *of the row nondegenerate form* $G_{IOrnd}(s)$ *of the system transfer function matrix* $G_{IO}(s)$ *are negative,*

$$Res_i^*(G_{IOrnd}) < 0,\ \forall i \in \{1, 2,\ \dots\ ,\ \nu N\}, \tag{14.26}$$

b) **Bounded Input-Bounded Output** *(BIBO)* **stable,** *it is necessary and sufficient that the conditions under a) hold,*

c) **Bounded Input and Initial State-Bounded State** *(BISBS)* **stable,** *it is necessary and sufficient that*

1. $G_{IO}(.)$ *is at least proper, or strictly proper, real rational matrix function of s,* $G_{IOio}(.)$ *and* $G_{IOyo}(.)$ *are strictly proper real rational matrix functions of s,*

and that

2. the real parts $Res_k^*(G_{IOrnd})$ *of all poles* $s_k^*(G_{IOrnd})$ *of the row nondegenerate form* $G_{IOrnd}(s)$ *of the system transfer function matrix* $G_{IO}(s)$ *are negative, the real parts* $Res_k^*(G_{IOiornd})$ *and* $Res_k^*(G_{IOyornd})$ *of all poles* $s_k^*(G_{IOiornd})$ *and* $s_k^*(G_{IOyornd})$ *of the row nondegenerate forms* $G_{IOiornd}(s)$ *and* $G_{IOyornd}(s)$ *of the system transfer function matrices* $G_{IOio}(s)$ *and of* $G_{IOyo}(s)$ *are nonpositive and those with the zero real part are simple,*

$$Res_k^*(G_{IOrnd}) < 0,\ \forall k \in \{1, 2,\ \dots\ ,\ \nu N\},$$
$$Res_k^*(G_{IOiornd}) \leq 0 \text{ and } Res_k^*(G_{IOyornd}) \leq 0,\ \forall k \in \{1, 2,\ \dots\ ,\ \nu N\},$$
$$Res_j^*(G_{IOiornd}) = 0 \Longrightarrow \nu_j = 1,\ Res_m^*(G_{IOyornd}) = 0 \Longrightarrow \nu_m = 1, \quad (14.27)$$

d) **Bounded Input and Initial State-Bounded Output** *(BISBO)* **stable,** *it is necessary and sufficient that the conditions under c) hold.*

Proof. Let the input vector function $\mathbf{i}(.)$ be arbitrarily chosen from the family $\mathfrak{I}^{\mu}$, (2.14) (in Section 2.1) and be fixed. Let all the initial conditions be also arbitrarily chosen so that they are bounded and fixed.

Necessity. a) Let the IO system (14.22) be $BIBS$ stable. Hence, (14.1), (item a) of Definition 377, in Subsection 14.2.1), holds. The overall output response $\mathbf{y}^{\nu-1}(t; \mathbf{0}_{\nu N}; \mathbf{i})$ of the system does not contain an impulse component. The same holds for $\mathbf{y}(t; \mathbf{0}_{\nu N}; \mathbf{i})$ as a subvector of $\mathbf{y}^{\nu-1}(t; \mathbf{0}_{\nu N}; \mathbf{i})$. Its Laplace transform

$$\mathcal{L}\{\mathbf{y}(t; \mathbf{0}_{\nu N}; \mathbf{i})\} = G_{IO}(s)\mathbf{I}(s),$$

is strictly proper real rational vector function due to the Generating theorem 342, (in Subsection 13.4.1). Let

$$\mathbf{i}(t) = h(t)\mathbf{1}_M,\ \mathbf{1}_M = \begin{bmatrix} 1 \vdots 1 \vdots \dots \vdots 1 \end{bmatrix}^T \in \mathfrak{R}^M \Longrightarrow \mathbf{I}(s) = \frac{1}{s}\mathbf{1}_M \in \mathfrak{C}^M.$$

Hence,

$$\mathcal{L}\{\mathbf{y}(t; \mathbf{0}_{\nu N}; h(t)\mathbf{1}_M)\} = s^{-1}G_{IO}(s)\mathbf{1}_M$$

is a strictly proper real rational vector function. This proves necessity of condition a-1). From this equation and Lemma 123, (in Section 6) follows that only the poles of the row nondegenerate form $G_{IOrnd}(s)$ of $G_{IO}(s)$ and the poles of $\mathbf{I}(s)$ determine the character of $\mathbf{y}^{\nu-1}(t;\mathbf{0}_{\nu N};\mathbf{i})$ for every $\mathbf{i}(.) \in \mathfrak{I}^{\mu}$. The Generating theorem 342 and (14.1) prove necessity of the condition a-2), i.e., of (14.26).

b) The preceding proof of the necessity of the conditions under a) applies to b).

c) Let the IO system (14.22) be $BISBS$ stable. Hence, (14.3), (item f) of Definition 377, in Subsection 14.2.1), holds. The overall output response $\mathbf{y}^{\nu-1}(t;\mathbf{i}_0^{\mu-1};\mathbf{y}_0^{\nu-1};\mathbf{i})$ of the system does not contain an impulse component. The same holds for $\mathbf{y}(t;\mathbf{i}_0^{\mu-1};\mathbf{y}_0^{\nu-1};\mathbf{i})$ as a subvector of $\mathbf{y}^{\nu-1}(t;\mathbf{i}_0^{\mu-1};\mathbf{y}_0^{\nu-1};\mathbf{i})$. Its Laplace transform

$$\mathcal{L}\{\mathbf{y}(t;\mathbf{i}_0^{\mu-1};\mathbf{y}_0^{\nu-1};\mathbf{i})\} = F_{IO}(s)\begin{bmatrix} \mathbf{I}(s) \\ \mathbf{i}_0^{\mu-1} \\ \mathbf{y}_0^{\nu-1} \end{bmatrix},\ k = 1, 2, \ \dots \ , \nu - 1,$$

is strictly proper real rational vector function due to the Generating Theorem 342. Let

$$\mathbf{i}(t) = h(t)\mathbf{1}_M,\ \mathbf{1}_M = \begin{bmatrix} 1 \vdots 1 \vdots \dots \vdots 1 \end{bmatrix}^T \in \mathfrak{R}^M \Longrightarrow \mathbf{I}(s) = \frac{1}{s}\mathbf{1}_M \in \mathfrak{C}^M.$$

This, and (8.6) (in Section 8.1) yield

$$\mathcal{L}\{\mathbf{y}(t;\mathbf{i}_0^{\mu-1};\mathbf{y}_0^{\nu-1};h(t)\mathbf{1}_M)\} = s^{-1}G_{IO}(s)\mathbf{1}_M + G_{IOio}(s)\mathbf{i}_0^{\mu-1} + G_{IOyo}(s)\mathbf{y}_0^{\nu-1},$$

which is a strictly proper real rational vector function. This proves necessity of the condition c-1). From this equation and Lemma 123, (Chapter 6) follows that only the poles of the row nondegenerate forms $G_{IOrnd}(s)$, $G_{IOiornd}(s)$ and $G_{IOyornd}(s)$ of $G_{IO}(s)$, $G_{IOio}(s)$ and $G_{IOyo}(s)$, as well as the poles of $\mathbf{I}(s)$, determine the character of $\mathbf{y}^{\nu-1}(t;\mathbf{i}_0^{\mu-1};\mathbf{y}_0^{\nu-1};\mathbf{i})$ for every $\mathbf{i}(.) \in \mathfrak{I}^{\mu}$. The Generating Theorem 342, (in Subsection 13.4.1), and (14.1) prove necessity of the condition c-2), i.e., (14.27).

d) Let the IO system (14.22) be $BISBO$ stable so that (14.4) (the item g) of Definition 377, in Subsection 14.2.1) is valid. The output response $\mathbf{y}(t;\mathbf{i}_0^{\mu-1};\mathbf{y}_0^{\nu-1};\mathbf{i})$ of the system does not contain an impulse component. The repetition of the proof of the necessity under c) from this point on completes the proof of the necessity of d).

Sufficiency. The boundedness of every $\mathbf{i}(.)$ that belongs to the family $\mathfrak{I}^{\mu}$ guarantees that the real parts of all poles of $\mathbf{I}(s)$ are nonpositive, that its poles with zero real parts are simple, and that $\mathbf{I}(s)$ is strictly proper real vector function.

a) Let the conditions under a) hold. The Laplace transform of the system response $\mathbf{y}(t;\mathbf{0}_{\mu M};\mathbf{0}_{\nu N};\mathbf{i})$,

$$\mathcal{L}\{\mathbf{y}(t;\mathbf{0}_{\mu M};\mathbf{0}_{\nu N};\mathbf{i})\} = G_{IO}(s)\mathbf{I}(s),$$

is strictly proper and does not have either a pole in infinity or a pole with positive real part or a multiple pole with the zero real part for every $\mathbf{i}(.)$ that belongs to the family $\mathfrak{I}^{\mu}$. Lemma 123, and the Generating Theorem 342 imply that

$$\mathbf{y}(t;\mathbf{0}_{\mu M};\mathbf{0}_{\nu N};\mathbf{i}) = \mathcal{L}^{-1}\left\{G_{IO}(s)\mathbf{I}(s)\right\}$$

is bounded for every $\mathbf{i}(.) \in \mathfrak{I}^{\mu}$. Hence,

$$\mathbf{y}^{\nu-1}(t;\mathbf{0}_{\mu M};\mathbf{0}_{\nu N};\mathbf{i}) = \mathcal{L}^{-1}\left\{\left[I_N \;\; sI_N \;\ldots\; s^{\nu-1}I_N\right]^T G_{IO}(s)\mathbf{I}(s)\right\}$$

is also bounded for every $\mathbf{i}(.) \in \mathfrak{I}^{\mu}$ due to Lemma 395.

b) Let the conditions under b) hold. The Laplace transform

$$\mathcal{L}\{\mathbf{y}(t;\mathbf{0}_{\mu M};\mathbf{0}_{\nu N};\mathbf{i})\} = G_{IO}(s)\mathbf{I}(s)$$

of $\mathbf{y}(t;\mathbf{0}_{\mu M};\mathbf{0}_{\nu N};\mathbf{i})$ is strictly proper real rational function and it does not have either a pole in infinity or a pole with positive real part or a multiple pole with the zero real part for every $\mathbf{i}(.)$ that belongs to the family $\mathfrak{I}^{\mu}$. Lemma 123 and the Generating Theorem 342 imply that

$$\mathbf{y}(t;\mathbf{0}_{\mu M};\mathbf{0}_{\nu N};\mathbf{i}) = \mathcal{L}^{-1}\left\{G_{IO}(s)\mathbf{I}(s)\right\},$$

is bounded for every $\mathbf{i}(.) \in \mathfrak{I}^{\mu}$.

c) Let all the conditions under c) hold. The system output response $\mathbf{y}(t;\mathbf{i}_{0^-}^{\mu-1};\mathbf{y}_{0^-}^{\nu-1};\mathbf{i})$ obeys the Equations (14.24) and (14.25). The matrix function $H_{IO}(.)$ is strictly proper real rational function for every $\mathbf{i}(.) \in \mathfrak{I}^{\mu}$, and it does not have either a pole in infinity or a pole with positive real part or a multiple pole with the zero real part for every $\mathbf{i}(.)$ that belongs to the family $\mathfrak{I}^{\mu}$. The conditions under c), Lemma 123, the Generating Theorem 342 imply that the system response $\mathbf{y}(t;\mathbf{i}_{0^-}^{\mu-1};\mathbf{y}_{0^-}^{\nu-1};\mathbf{i})\}$ is bounded for every $\mathbf{i}(.) \in \mathfrak{I}^{\mu}$. This and Lemma 395 prove that $\mathbf{y}^{\nu-1}(t;\mathbf{i}_0^{\mu-1};\mathbf{y}_0^{\nu-1};\mathbf{i})$ is also bounded for every $\mathbf{i}(.) \in \mathfrak{I}^{\mu}$.

d) The statement under c) proves the statement under d) as its special case. ■

The BI stability conditions, like the asymptotic stability conditions, demand that the real parts of all poles of the row nondegenerate form of the appropriate system transfer function matrix are at least nonpositive, or negative, depending on the transfer function matrix. However, the appropriate transfer function matrices are different for different BI stability properties and for the asymptotic stability.

Example 397 *For the IO system of Example 154 (in Section 8.1)*

$$y^{(2)}(t) + y^{(1)}(t) - 2y(t) = 2i^{(2)}(t) - 14i^{(1)}(t) + 12i(t).$$

we determined the following transfer function $G_{IO}(s)$ *and the full transfer function matrix* $F_{IO}(s)$:

$$G_{IO}(s) = \frac{2s^2 - 14s + 12}{s^2 + s - 2} = \frac{2(s-1)(s-6)}{(s-1)(s+2)} \Longrightarrow G_{IOnd}(s) = 2\frac{s-6}{s+2},$$

$$F_{IO}(s) =$$

$$= \left[\underbrace{\frac{2s^2 - 14s + 12}{(s-1)(s+2)}}_{G_{IO}(s)} \vdots \underbrace{\frac{-(2s-14)}{(s-1)(s+2)} \vdots \frac{-2}{(s-1)(s+2)}}_{G_{IOio}(s)} \vdots \underbrace{\frac{s+1}{(s-1)(s+2)} \vdots \frac{1}{(s-1)(s+2)}}_{G_{IOyo}(s)} \right] =$$

$$= F_{IOnd}(s).$$

The system transfer function matrix $G_{IO}(s)$ *has the same zero* $s^0 = 1$ *and pole* $s^* = 1$ *so that they cancel yielding the nondegenerate form* $G_{IOnd}(s)$ *of* $G_{IO}(s)$ *that has only one pole. The system is both* $BIBS$ *and* $BIBO$ *stable.*

Since the pole is negative real, $s^* = -2$, *then we could conclude wrongly from* $G_{IOnd}(s)$ *that the system is* $BISBS$ *and* $BISBO$ *stable. However, the cancellation of the same zero* $s^0 = 1$ *and pole* $s^* = 1$ *is not possible in either the system full transfer function matrix* $F_{IO}(s)$ *or in its submatrices* $G_{IOio}(s)$ *and* $G_{IOyo}(s)$. *Since they have one positive real pole* $s^* = 1$, *then it follows that the system is* $BISBS$ *and* $BISBO$ *unstable for some* $i(.) \in \mathfrak{I}^{\mu}$. *This shows that we should not apply the* BI *stability criteria established under all zero initial conditions to* BI *stability properties in general, i.e., under arbitrary initial conditions.*

Example 398 *We analyze further Antsaklis-Michel systems presented under 2) in Example 167 (in Section 8.2), [4, Example 3.8].*

The state space models under 2) in Example 167 yields the second order IO differential equation, hence the degenerate transfer function matrix $G_{IO2}(s)$ and the nondegenerate full transfer function matrix $F_{IOISO2}(s)$:

$$y^{(2)}(t) - y(t) = -i(t) + i^{(1)}(t),$$
$$G_{IO2}(s) = (s^2-1)^{-1}(s-1) \Longrightarrow G_{IO2nd}(s) = (s+1)^{-1},$$
$$F_{IOISO2}(s) = \left[\underbrace{\frac{s-1}{s^2-1}}_{G_{IOi\_iii}(s)} \vdots \underbrace{-\frac{1}{s^2-1}}_{G_{IOio2}(s)} \vdots \underbrace{\frac{s}{s^2-1} \vdots \frac{1}{s^2-1}}_{G_{IOyo2}(s)} \right].$$

The cancellation of the same zero $s^0 = 1$ and pole $s^* = 1$ of the transfer function $G_{IO2}(s)$ is possible only in it, but it cannot be carried out either in the full transfer function matrix $F_{IOISO2}(s)$, in $G_{IOio2}(s)$, or in $G_{IOyo2}(s)$.

The systems are both $BIBS$ and $BIBO$ stable. They are both $BISBS$ and $BISBO$ unstable for some $i(.) \in \mathfrak{I}^{\mu}$.

Example 399 *Let*

$$y^{(2)}(t) + 5y^{(1)}(t) + 6y(t) = i(t) + 2i^{(1)}(t) + i^{(2)}(t) \Longrightarrow \nu = \mu = 2.$$

Its Laplace transform yields

$$G_{IO}(s) = \frac{s^2+2s+1}{s^2+5s+6} = G_{IOnd}(s),$$
$$F_{IO}(s) = \begin{bmatrix} \frac{s^2+2s+1}{s^2+5s+6} \\ -\frac{s+2}{s^2+5s+6} \\ -\frac{1}{s^2+5s+6} \\ \frac{s+5}{s^2+5s+6} \\ \frac{1}{s^2+5s+6} \end{bmatrix}^T = F_{IOnd}(s).$$

Neither $G_{IO}(s)$ nor $F_{IO}(s)$ is strictly proper, but both are proper and nondegenerate. However, $G_{IOx_0}(s)$ and $G_{IOy_0}(s)$ are strictly proper.

The system transfer function matrices obey all the conditions of Theorem 396. The system is $BIBS$, $BIBO$, $BISBS$ and $BISBO$ stable for every $i(.) \in \mathfrak{I}^{\mu}$.

For example, the unit step response of the system under all zero initial conditions

$$y(t; \mathbf{0}_2; \mathbf{0}_2; h) = \mathcal{L}^{-1}\left\{\frac{s^2+2s+1}{s^2+5s+6}\frac{1}{s}\right\},$$

and the unit step response of the system under nonzero initial conditions,

$$y(t;\mathbf{i}_0^1;\mathbf{y}_0^1;h)=\mathcal{L}^{-1}\left\{\begin{array}{c}\frac{s^2+2s+1}{s^2+5s+6}\frac{1}{s}-\frac{s+2}{s^2+2s+1}i_0-\\-\frac{1}{s^2+2s+1}i_0^{(1)}+\frac{s+5}{s^2+2s+1}y_0+\\+\frac{1}{s^2+2s+1}y_0^{(1)}\end{array}\right\}$$

do not contain the impulse components. They are bounded for the input $i(t)=h(t)$ *and for bounded initial conditions,*

$$i_0\in\mathfrak{R},\ i_0^{(1)}\in\mathfrak{R},\ y_0\in\mathfrak{R},\ y_0^{(1)}\in\mathfrak{R}.$$

The real parts of the poles of $G_{IO}(s)$ *and of* $F_{IO}(s)$ *are negative.*

Example 400 *We reconsider the IO system of the Example 124, (Chapter 6),*

$$\begin{bmatrix}1&1\\1&2\end{bmatrix}\mathbf{y}^{(2)}(t)-\begin{bmatrix}1&1\\1&2\end{bmatrix}\mathbf{y}(t)=$$
$$=\begin{bmatrix}-1&2\\0&-3\end{bmatrix}\mathbf{i}(t)+\begin{bmatrix}1&1\\-1&1\end{bmatrix}\mathbf{i}^{(1)}(t)+\begin{bmatrix}0&0\\1&0\end{bmatrix}\mathbf{i}^{(2)}(t).$$

Its transfer function matrix

$$G_{IO}(s)=\frac{s^2-1}{(s^2-1)^2}\begin{bmatrix}-(s-1)(s-2)&s+7\\(s-1)^2&-5\end{bmatrix}$$

is degenerate. The nondegenerate form $G_{IOnd}(s)$ *of* $G_{IO}(s)$ *reads*

$$G_{IOnd}(s)=\frac{1}{s^2-1}\begin{bmatrix}-(s-1)(s-2)&s+7\\(s-1)^2&-5\end{bmatrix}.$$

It is also the row nondegenerate form $G_{IOrnd}(s)$ *of* $G_{IO}(s)$,

$$G_{IOnd}(s)=G_{IOrnd}(s).$$

The column nondegenerate form $G_{IOcnd}(s)$ *of* $G_{IO}(s)$ *is different from them,*

$$G_{IOcnd}(s)=\frac{1}{s+1}\begin{bmatrix}-(s-2)&s+7\\s-1&-5\end{bmatrix}.$$

If we wish to test whether the system is $BIBS$ *stable and/or* $BIBO$ *stable we should use the row nondegenerate form* $G_{IOrnd}(s)$ *of* $G_{IO}(s)$, *but we may not use its column nondegenerate form* $G_{IOcnd}(s)$ *because the pole* $s^*=1$ *cannot* *be cancelled in the rows of* $G_{IOrnd}(s)$. *If we used the column nondegenerate form* $G_{IOcnd}(s)$ *we would conclude wrongly that the system is* $BIBS$ *stable and/or* $BIBO$ *stable. However, it is* $BIBS$ *unstable and* $BIBO$ *unstable for some* $i(.) \in \mathfrak{I}^{\mu}$ *due to the positive pole* $s^* = 1$ *of the row nondegenerate form* $G_{IOrnd}(s)$ *of* $G_{IO}(s)$.

The full system transfer function matrix

$$F_{IO}(s) = \frac{s^2-1}{\left(s^2-1\right)^2} \bullet$$

$$\bullet \begin{bmatrix} -(s-1)(s-2) & (s-1)^2 \\ s+7 & -5 \\ s-1 & -s \\ -1 & 0 \\ 1 & -1 \\ 0 & 0 \\ s & 0 \\ 0 & s \\ 1 & 0 \\ 0 & 1 \end{bmatrix}^T .$$

is degenerate. Its nondegenerate form

$$F_{IOnd}(s) = \frac{1}{s^2-1} \bullet$$

$$\bullet \begin{bmatrix} -(s-1)(s-2) & (s-1)^2 \\ s+7 & -5 \\ s-1 & -s \\ -1 & 0 \\ 1 & -1 \\ 0 & 0 \\ s & 0 \\ 0 & s \\ 1 & 0 \\ 0 & 1 \end{bmatrix}^T .$$

is also its row nondegenerate form $F_{IOnd}(s)$ *and its column nondegenerate form* $F_{IOcnd}(s)$,

$$F_{IOrnd}(s) = F_{IOnd}(s) = F_{IOcnd}(s).$$

They show that the system is both $BISBS$ *unstable for some* $i(.) \in \mathfrak{I}^{\mu}$ *and* $BISBO$ *unstable for some* $i(.) \in \mathfrak{I}^{\mu}$ *because* $G_{IOiocnd}(s)$ *and* $G_{IOyocnd}(s)$ *have positive real pole* $s^* = 1$.

14.3.2 *ISO* systems

For the *ISO* system (3.84) and (3.85), (Subsection 3.5.3),

$$\frac{d\mathbf{x}(t)}{dt} = A\mathbf{x}(t) + B\mathbf{i}(t), \ \forall t \in \mathfrak{T}_0, \tag{14.28}$$

$$\mathbf{y}(t) = C\mathbf{x}(t) + D\mathbf{i}(t), \ \forall t \in \mathfrak{T}_0, \tag{14.29}$$

we recall the fact that its internal state $\mathbf{S}_I$, output state $\mathbf{S}_O$ and (full) state $\mathbf{S}$ coincide, (2.19). This explains that the *BIISBIS*, *BIISBOS*, *BIISBISO*, and *BIISBSO* stability coincide with *BISBS*, and *BIISBO* with *BISBO*.

Theorem 401 *Let the input vector function* $\mathbf{i}(.)$ *belong to the family* $\mathfrak{I}$, *(2.14) (in Section 2.1). In order for the ISO system (14.28) and (14.29) in the forced regime to be:*

a) **Bounded Input-Bounded State** *(BIBS)* **stable,** *it is necessary and sufficient that:*

1. the system IS transfer function matrix $G_{ISOIS}(s)$ *is at least proper, or strictly proper, real rational matrix function of* s,

and that

2. the real parts $Res_i^*(G_{ISOISrnd})$ *of all poles* $s_i^*(G_{ISOISrnd})$ *of the row nondegenerate form* $G_{ISOISrnd}(s)$ *of the system IS transfer function matrix* $G_{ISOIS}(s)$ *are negative,*

$$Res_i^*(G_{ISOISrnd}) < 0, \ \forall i \in \{1, 2, \ \dots\ , \ n\}. \tag{14.30}$$

b) **Bounded Input-Bounded Output** *(BIBO)* **stable,** *it is necessary and sufficient that:*

1. the system IO transfer function matrix $G_{ISO}(s)$ *is at least proper, or strictly proper, real rational matrix function of* s,

and that

2. the real parts $Res_i^*(G_{ISOrnd})$ *of all poles* $s_i^*(G_{ISOrnd})$ *of the row nondegenerate form* $G_{ISOrnd}(s)$ *of the system IO transfer function matrix* $G_{ISO}(s)$ *are negative,*

$$Res_i^*(G_{ISOrnd}) < 0, \ \forall i \in \{1, 2, \ \dots\ , \ n\}. \tag{14.31}$$

c) **Bounded Input and Initial State-Bounded State** *(BISBS)* **stable,** *it is necessary and sufficient that:*

1. the system IS transfer function matrix $G_{ISOIS}(s)$ *is at least proper, or strictly proper, real rational matrix function of* s, *and the system*

SS transfer function matrix $G_{ISOSS}(s)$ *is strictly proper real rational matrix function of* s,

2. the real parts $Res_i^*(G_{ISOISrnd})$ *and* $Res_i^*(G_{ISOSSrnd})$ *of every pole* $s_i^*(G_{ISOISrnd})$ *and* $s_i^*(G_{ISOSSrnd})$ *of the row nondegenerate forms*

$$G_{ISOISrnd}(s) \text{ and } G_{ISOSSrnd}(s)$$

of the system IS transfer function matrix $G_{ISOIS}(s)$ *are negative, and of* $G_{ISOSS}(s)$ *are nonpositive and those with the zero real part are simple, respectively,*

$$\begin{gathered} Res_i^*(G_{ISOISrnd}) < 0,\ \forall i \in \{1, 2, \ ...\ , n\} \\ Res_i^*(G_{ISOSSrnd}) \leq 0,\ \forall i \in \{1, 2, \ ...\ , n\} \text{ and} \\ Res_k^*(G_{ISOSSrnd}) \leq 0 \Longrightarrow \nu_k = 1. \end{gathered} \tag{14.32}$$

d) ***Bounded Input and Initial State-Bounded Output*** *(BISBO)* ***stable,*** *it is necessary and sufficient that:*

1. the system IO transfer function matrix $G_{ISO}(s)$ *is at least proper, or strictly proper, real rational matrix function of* s*, and the system IISO transfer function matrix* $G_{ISOx_0}(s)$ *is strictly proper real rational matrix function of* s,

2. the real parts $Res_i^*(G_{ISOrnd})$ *and* $Res_i^*(G_{ISOxornd})$ *of every pole* $s_i^*(G_{ISOrnd})$ *and* $s_i^*(G_{ISOxornd})$ *of the row nondegenerate forms* $G_{ISOrnd}(s)$ *and* $G_{ISOxornd}(s)$ *of the system IO transfer function matrix* $G_{ISO}(s)$ *are negative, and of* $G_{ISOxo}(s)$ *are nonpositive and those with the zero real part are simple, respectively,*

$$\begin{gathered} Res_i^*(G_{ISOrnd}) < 0,\ \forall i \in \{1, 2, \ ...\ , n\}, \\ Res_i^*(G_{ISOxornd}) \leq 0,\ \forall i \in \{1, 2, \ ...\ , n\} \\ Res_k^*(G_{ISOxornd}) \leq 0 \Longrightarrow \nu_k = 1. \end{gathered} \tag{14.33}$$

Proof. Let the input vector function $\mathbf{i}(.)$ be arbitrarily chosen from the family $\mathfrak{I}$, (2.14), (in Section 2.1) and be fixed. Let all the initial conditions be also arbitrarily chosen so that they are bounded and fixed. Every $\mathbf{i}(.) \in \mathfrak{I}$ is bounded and guarantees that the real parts of all poles of $\mathbf{I}(s)$ are nonpositive, that its poles with the zero real parts are simple, and that $\mathbf{I}(s)$ is strictly proper real vector function.

a) Since Equation (7.30), (in Section 7.3), implies

$$\mathbf{x}(t; \mathbf{0}_n; \mathbf{i}) = \mathcal{L}^{-1}\left\{G_{ISOIS}(s)\mathbf{I}^{\mp}(s)\right\},\ \mathbf{x}_{0\mp} = \mathbf{0}_n,$$

then a) of Definition 381, (in Subsection 14.2.2), $\mathbf{i}(.) \in \mathfrak{I}$, Lemma 123, (in Chapter 6), and the Generating Theorem 342, (in Subsection 13.4.1), prove necessity and sufficiency of the condition a).

b) Equation (7.26), (in Section 7.3), yields

$$\mathbf{y}(t;\mathbf{0}_n;\mathbf{i}) = \mathcal{L}^{-1}\left\{G_{ISO}(s)\mathbf{I}^{\mp}(s)\right\}, \ \mathbf{x}_{0\mp} = \mathbf{0}_n.$$

This, b) of Definition 381 (in Subsection 14.2.2), $\mathbf{i}(.) \in \mathfrak{I}$, Lemma 123, and the Generating Theorem 342 prove necessity and sufficiency of the condition b).

c) From (7.29), (in Section 7.3), we deduce

$$\mathbf{x}(t;\mathbf{x}_0;\mathbf{i}) = \mathcal{L}^{-1}\left\{F_{ISOIS}(s)\left[\mathbf{I}^{(\mp)^T}(s) \vdots \mathbf{x}_0^T\right]^T\right\}.$$

This, d) of Definition 381, $\mathbf{i}(.) \in \mathfrak{I}$, (7.29) and (7.32) (in Section 7.3), Lemma 123, (in Chapter 6), and the Generating Theorem 342, (in Subsection 13.4.1), prove necessity and sufficiency of the condition c).

d) We write (7.23), (in Section 7.3), in the *time* domain by applying the inverse Laplace transform,

$$\mathbf{y}(t;\mathbf{x}_0;\mathbf{i}) = \mathcal{L}^{-1}\left\{F_{ISO}(s)\left[\mathbf{I}^{(\mp)^T}(s) \vdots \mathbf{x}_0^T\right]^T\right\}.$$

This, e) of Definition 381, $\mathbf{i}(.) \in \mathfrak{I}$, (7.23), (7.28) (in Section 7.3), Lemma 123, and the Generating Theorem 342 prove necessity and sufficiency of the condition d). ■

Note 402 *The conditions under d) are necessary and sufficient for BIBSO stability of the ISO system (14.28) and (14.29) due to Theorem 385, (in Subsection 14.2.2).*

Note 403 *The conditions under d) are necessary and sufficient for BISBSO stability of the ISO system (14.28) and (14.29) due to Theorem 387, (in Subsection 14.2.2).*

Example 404 *We continue to analyze the ISO system by Kalman (Example 169 in in Subsection 8.2, Example 357 in in Subsection 13.4.3) [80,*

Example 8, pp. 188, 189]:

$$\frac{d\mathbf{x}}{dt}=\begin{bmatrix}0 & 1 & 0\\ 5 & 0 & 2\\ -2 & 0 & -2\end{bmatrix}\mathbf{x}+\begin{bmatrix}0\\ 0\\ 0.5\end{bmatrix}i,$$

$$y=\begin{bmatrix}-2 \vdots 1 \vdots 0\end{bmatrix}\mathbf{x}.$$

We determined (8.35):

$$F_{ISO}(s)=\frac{s-2}{(s-2)(s+1)(s+3)}\begin{bmatrix}1 \vdots -2s-3 \vdots\ s+2 \vdots 2\end{bmatrix}$$

$$\Longrightarrow$$

$$G_{ISO}(s)=\frac{s-2}{s^3+2s^2-5s-6}=\frac{s-2}{(s-2)(s+1)(s+3)}.$$

Both, $F_{ISO}(s)$ and $G_{ISO}(s)$ are degenerate. Their nondegenerate forms, which are also their row nondegenerate forms, read

$$F_{ISOrnd}(s)=F_{ISOnd}(s)=\frac{\begin{bmatrix}1 \vdots -2s-3 \vdots\ s+2 \vdots 2\end{bmatrix}}{(s+1)(s+3)},$$

$$G_{ISOrnd}(s)=G_{ISOnd}(s)=\frac{1}{(s+1)(s+3)}.$$

Their poles are negative real numbers,

$$s_1^*(F_{ISOrnd})=s_1^*(G_{ISOrnd})=-1,\ s_2^*(F_{ISOrnd})=s_2^*(G_{ISOrnd})=-3.$$

The system is both BIBO and BISBO stable.

Besides,

$$G_{ISOIS}(s)=(sI_n-A)^{-1}B=$$

$$=\begin{bmatrix}\frac{2s+s^2}{s^3+2s^2-5s-6} & \frac{s+2}{s^3+2s^2-5s-6} & \frac{2}{s^3+2s^2-5s-6}\\ \frac{5s+6}{s^3+2s^2-5s-6} & \frac{2s+s^2}{s^3+2s^2-5s-6} & 2\frac{s}{s^3+2s^2-5s-6}\\ -2\frac{s}{s^3+2s^2-5s-6} & -\frac{2}{s^3+2s^2-5s-6} & \frac{s^2-5}{s^3+2s^2-5s-6}\end{bmatrix}\begin{bmatrix}0\\ 0\\ 0.5\end{bmatrix}=$$

$$=\begin{bmatrix}\frac{1}{s^3+2s^2-5s-6}\\ \frac{s}{s^3+2s^2-5s-6}\\ \frac{0.5\left(s^2-5\right)}{s^3+2s^2-5s-6}\end{bmatrix}.$$

It is row nondegenerate,

$$G_{ISOISrnd}(s) = G_{ISOIS}(s).$$

Its poles are

$$s_1^* = -1, \ s_2^* = 2, \ s_3^* = -3.$$

Since the second pole is positive real, the system is BIBS unstable.

Furthermore, $F_{ISOIS}(s)$,

$$F_{ISOIS}(s) = \left[\underbrace{(sI_n - A)^{-1}B}_{G_{ISOIS}(s)} \vdots \underbrace{(sI_n - A)^{-1}}_{G_{ISOSS}(s)} \right] =$$

$$\left[\underbrace{\begin{matrix} \frac{1}{s^3+2s^2-5s-6} \\ \frac{s}{s^3+2s^2-5s-6} \\ \frac{0.5(s^2-5)}{s^3+2s^2-5s-6} \end{matrix}}_{G_{ISOIS}(s)} \underbrace{\begin{matrix} \frac{2s+s^2}{s^3+2s^2-5s-6} & \frac{s+2}{s^3+2s^2-5s-6} & \frac{2}{s^3+2s^2-5s-6} \\ \frac{5s+6}{s^3+2s^2-5s-6} & \frac{2s+s^2}{s^3+2s^2-5s-6} & 2\frac{s}{s^3+2s^2-5s-6} \\ -2\frac{s}{s^3+2s^2-5s-6} & -\frac{2}{s^3+2s^2-5s-6} & \frac{s^2-5}{s^3+2s^2-5s-6} \end{matrix}}_{G_{ISOSS}(s)} \right],$$

is also row nondegenerate,

$$F_{ISOISrnd}(s) = F_{ISOIS}(s).$$

Its submatrices $G_{ISOIS}(s)$ *and* $G_{ISOSS}(s)$ *are also row nondegenerate. They, all three, have the same poles. Their second pole,* $s_2^* = 2$, *is positive real. The system is both BIBS and BISBS unstable.*

Note 405 *This example illustrated the inadequacy of the system transfer function matrix* $G_{ISO}(s)$ *for testing every BI stability property of the ISO system. We should use the full system transfer function matrix* $F_{ISO}(s)$ *(or* $F_{ISOIS}(s)$*) and its corresponding submatrix(es).*

This example illustrates the following Corollary to Theorem 401:

Corollary 406 *If the ISO system (14.28) and (14.29) is:*

a) Bounded Input and Initial State-Bounded State (BISBS) stable then it is also Bounded Input-Bounded State (BIBS) stable.

b) Bounded Input and Initial State-Bounded Output (BISBO) stable, then it is also Bounded Input-Bounded Output (BIBO) stable.

This Corollary results also from Definition 381 and agrees with Comment 376.

14.3.3 *IIO* systems

We recall the following subclass of the *IIO* systems described by (3.90), (in Subsection 3.5.4):

$$Q^{(\alpha)}\mathbf{r}^{\alpha}(t) = P^{(\beta)}\mathbf{i}^{\beta}(t),\ detQ_{\alpha} \neq 0,\ \forall t \in \mathfrak{T}_0,\ \alpha \geq 1, \alpha \geq \beta \geq 0,$$
$$E^{(\nu)}\mathbf{y}^{\nu}(t) = R^{(\alpha)}\mathbf{r}^{\alpha}(t) + T^{(\mu)}\mathbf{i}^{\mu}(t),\ detE_{\nu} \neq 0,\ \forall t \in \mathfrak{T}_0,$$
$$\alpha,\ \beta,\ \nu,\ \mu \in \mathfrak{R}_+,\ \nu \geq \mu. \tag{14.34}$$

The left-hand side of the first equation (14.34) expresses *the state (dynamical) behavior* of the system (in Section 2.3),

$$Q^{(\alpha)}\mathbf{r}^{\alpha}(t) = P^{(\beta)}\mathbf{i}^{\beta}(t),\ \forall t \in \mathfrak{T}_0. \tag{14.35}$$

The left-hand side of the second equation (14.34) determines *the output dynamics, i.e., the output (dynamical) behavior,* of the system (in Section 2.3),

$$E^{(\nu)}\mathbf{y}^{\nu}(t) = R^{(\alpha)}\mathbf{r}^{\alpha}(t) + T^{(\mu)}\mathbf{i}^{\mu}(t),\ \forall t \in \mathfrak{T}_0. \tag{14.36}$$

The left-hand sides of equations (14.34), i.e., (14.35) and (14.36), equivalently of the following equation, determine *the full dynamics* (the state and the output dynamics) of the *IIO* system (14.34), which is also called *the full (the complete) system dynamics*:

$$\begin{bmatrix} Q^{(\alpha)} & O_{\rho,(\nu+1)N} \\ -R^{(\alpha)} & E^{(\nu)} \end{bmatrix} \begin{bmatrix} \mathbf{r}^{\alpha}(t) \\ \mathbf{y}^{\nu}(t) \end{bmatrix} = \begin{bmatrix} P^{(\gamma)} \\ T^{(\gamma)} \end{bmatrix} \mathbf{i}^{\gamma}(t),\ \forall t \in \mathfrak{T}_0,$$
$$\gamma = \max\{\beta, \mu\},\ \nu \geq 1. \tag{14.37}$$

The *BI* stability properties of the *IIO* system (14.34) in the case $\nu = 0$ can be treated as those of the *ISO* system (3.84) and (3.85), (in Subsection 3.5.3). We will consider in the sequel only the case $\nu \geq 1$.

Lemma 407 *Let the input vector function* $\mathbf{i}(.)$ *belong to the family* $\mathfrak{I}^{\beta}$, *(2.12), (2.13), (in Section 2.1). In order for*

$$\left[\left(\mathbf{r}^{\alpha-1}(t;\mathbf{r}_0^{\alpha-1};\mathbf{i})\right)^T \vdots \left(\mathbf{y}^{\nu-1}(t;\mathbf{r}_0^{\alpha-1};\mathbf{y}_0^{\nu-1};\mathbf{i})\right)^T \right]^T,\ \nu \geq 1,$$

of the IIO system (14.34) in the forced regime:

a) to be bounded on $\mathfrak{T}_0$,

$$\exists \zeta = \zeta(\mathbf{r}_0^{\alpha-1};\mathbf{y}_0^{\nu-1};\mathbf{i}) \in \mathfrak{R}^+ \Longrightarrow \left\| \begin{bmatrix} \mathbf{r}^{\alpha-1}(t;\mathbf{r}_0^{\alpha-1};\mathbf{i}) \\ \mathbf{y}^{\nu-1}(t;\mathbf{r}_0^{\alpha-1};\mathbf{y}_0^{\nu-1};\mathbf{i}) \end{bmatrix} \right\| < \zeta,\ \forall t \in \mathfrak{T}_0,$$

it is necessary and sufficient that

$$\left[\mathbf{r}^T(t;\mathbf{r}_0^{\alpha-1};\mathbf{i}) \vdots\ \mathbf{y}^T(t;\mathbf{r}_0^{\alpha-1};\mathbf{y}_0^{\nu-1};\mathbf{i})\right]^T$$

is bounded on $\mathfrak{T}_0$,

$$\exists\xi = \xi(\mathbf{r}_0^{\alpha-1};\mathbf{y}_0^{\nu-1};\mathbf{i}) \in \mathfrak{R}^+ \Longrightarrow \left\| \begin{bmatrix} \mathbf{r}(t;\mathbf{r}_0^{\alpha-1};\mathbf{i}) \\ \mathbf{y}(t;\mathbf{r}_0^{\alpha-1};\mathbf{y}_0^{\nu-1};\mathbf{i}) \end{bmatrix} \right\| < \xi,\ \forall t \in \mathfrak{T}_0,$$

b) to vanish as $t \to \infty$,

$$t \to \infty \Longrightarrow \left[\left(\mathbf{r}^{\alpha-1}(t;\mathbf{r}_0^{\alpha-1};\mathbf{i})\right)^T \ \vdots\ \left(\mathbf{y}^{\nu-1}(t;\mathbf{r}_0^{\alpha-1};\mathbf{y}_0^{\nu-1};\mathbf{i})\right)^T\right]^T \to \mathbf{0}_{\alpha\rho+\nu N},$$

it is necessary and sufficient that

$$\left[\mathbf{r}^T(t;\mathbf{r}_0^{\alpha-1};\mathbf{i}) \ \vdots\ \mathbf{y}^T(t;\mathbf{r}_0^{\alpha-1};\mathbf{y}_0^{\nu-1};\mathbf{i})\right]^T$$

vanishes as $t \to \infty$,

$$t \to \infty \Longrightarrow \left[\mathbf{r}^T(t;\mathbf{r}_0^{\alpha-1};\mathbf{i}) \ \vdots\ \mathbf{y}^T(t;\mathbf{r}_0^{\alpha-1};\mathbf{y}_0^{\nu-1};\mathbf{i})\right] \to \mathbf{0}_{\rho+N},$$

Proof. Let the input vector function $\mathbf{i}(.)$ be arbitrarily chosen from the family $\mathfrak{I}$, (2.12), (2.13), (in Section 2.1) and be fixed. Let all the initial conditions be also arbitrarily chosen so that they are bounded and fixed.

Necessity. a) In order for

$$\left[\left(\mathbf{r}^{\alpha-1}(t;\mathbf{r}_0^{\alpha-1};\mathbf{i}_0^{\beta-1};\mathbf{i})\right)^T \ \vdots\ \left(\mathbf{y}^{\nu-1}(t;\mathbf{r}_0^{\alpha-1};\mathbf{y}_0^{\nu-1};\mathbf{i})\right)^T\right]^T$$

to be bounded on $\mathfrak{T}_0$, it is necessary that its every entry is bounded on $\mathfrak{T}_0$. Hence, it is necessary that

$$\left[\mathbf{r}^T(t;\mathbf{r}_0^{\alpha-1};\mathbf{i}_0^{\beta-1};\mathbf{i}) \ \vdots\ \mathbf{y}^T(t;\mathbf{r}_0^{\alpha-1};\mathbf{y}_0^{\nu-1};\mathbf{i})\right]^T$$

is bounded on $\mathfrak{T}_0$.

b) In order for

$$\left[\left(\mathbf{r}^{\alpha-1}(t;\mathbf{r}_0^{\alpha-1};\mathbf{i}_0^{\beta-1};\mathbf{i})\right)^T \ \vdots\ \left(\mathbf{y}^{\nu-1}(t;\mathbf{r}_0^{\alpha-1};\mathbf{y}_0^{\nu-1};\mathbf{i})\right)^T\right]^T$$

to vanish as $t \to \infty$ it is necessary that its every entry vanishes as $t \to \infty$. Hence, it is necessary that

$$\left[\mathbf{r}^T(t;\mathbf{r}_0^{\alpha-1};\mathbf{i}_0^{\beta-1};\mathbf{i}) \vdots \mathbf{y}^T(t;\mathbf{r}_0^{\alpha-1};\mathbf{y}_0^{\nu-1};\mathbf{i})\right]^T$$

vanishes as $t \to \infty$.

Sufficiency. a) Let the conditions of the statement of Lemma under # a) be valid. By following the proof of the Equations (14.24) and (14.25) in Subsection 14.3.1 and using $\mathbf{i}(.) \in \mathfrak{I}^\beta$, the properties of the Laplace transform, the boundedness of the initial conditions and (2.40) (in Section 2.3) we derive

$$\mathbf{r}(t;\mathbf{r}_0^{\alpha-1};\mathbf{i}_0^{\beta-1};\mathbf{i}) = \left\{ \begin{array}{c} \delta^-(t)\,R_{0r}+ \\ +\sum\limits_{j=1}^{j=\alpha\rho} e^{s_j^* t}\left[\sum\limits_{r=1}^{r=\nu_j}\frac{1}{(r-1)!}t^{r-1}R_{jr}\right] \end{array}\right\}\begin{bmatrix}\mathbf{1}_M \\ \mathbf{i}_0^{\beta-1} \\ \mathbf{r}_0^{\alpha-1}\end{bmatrix},$$
$$R_{0r} \in \mathfrak{R}^{\rho\times((\beta+1)M+\alpha\rho)},\ R_{jr} \in \mathfrak{C}^{\rho\times((\beta+1)M+\alpha\rho)},$$
$$R_{0r} = CONST.,\ R_{jr} = CONST.,$$
$$\forall r \in \{1,2,..,\ \nu_j\},\ \forall j \in \{1,2,..,\ \alpha\rho\},\ \operatorname{Re} s_j^* \le 0, \tag{14.38}$$

and

$$\mathbf{y}(t;\mathbf{r}_0^{\alpha-1};\mathbf{y}_0^{\nu-1};\mathbf{i}) = \left\{ \begin{array}{c} \delta^-(t)\,R_{0y}+ \\ +\sum\limits_{k=1}^{k=\nu N} e^{s_k^* t}\left[\sum\limits_{r=1}^{r=\nu_k}\frac{1}{(r-1)!}t^{r-1}R_{kry}\right] \end{array}\right\}\begin{bmatrix}\mathbf{1}_M \\ \mathbf{i}_0^{\gamma-1} \\ \mathbf{r}_0^{\alpha-1} \\ \mathbf{y}_0^{\nu-1}\end{bmatrix},$$
$$R_{0y} \in \mathfrak{R}^{N\times((\gamma+1)M+\nu N)},\ R_{kry} \in \mathfrak{C}^{N\times((\gamma+1)M+\nu N)},\ \operatorname{Re} s_k^* \le 0,$$
$$R_{0y} = CONST.,\ R_{kry} = CONST.,$$
$$\forall r \in \{1,2,..,\ \nu_k\},\ \forall k \in \{1,2,..,\ \nu N\},\ \operatorname{Re} s_i^* = 0 \Longrightarrow \nu_i = 1, \tag{14.39}$$

where ν_i is the multiplicity of s_i^*. This shows, in view of the Generating Theorem (342), (in Subsection 13.4.1), that the system substate vector

$$\left[\mathbf{r}^T(t;\mathbf{r}_0^{\alpha-1};\mathbf{i}_0^{\beta-1};\mathbf{i}) \vdots \mathbf{y}^T(t;\mathbf{r}_0^{\alpha-1};\mathbf{y}_0^{\nu-1};\mathbf{i})\right]^T$$

is infinitely times continuously differentiable at every $t \in \mathfrak{T}_0$ and bounded on $\mathfrak{T}_0$. All its derivatives rest bounded on $\mathfrak{T}_0$ due to Generating Theorem (342), which guarantees that the system full state vector

$$\left[\left(\mathbf{r}^{\alpha-1}(t;\mathbf{r}_0^{\alpha-1};\mathbf{i}_0^{\beta-1};\mathbf{i})\right)^T \vdots \left(\mathbf{y}^{\nu-1}(t;\mathbf{r}_0^{\alpha-1};\mathbf{y}_0^{\nu-1};\mathbf{i})\right)^T\right]^T$$

is also bounded on $\mathfrak{T}_0$.

b) Let the conditions of the statement of Lemma under b) hold. This, $\mathbf{i}(.) \in \mathfrak{I}$ and the properties of the Laplace transform and the boundedness of the initial conditions imply

$$\mathbf{r}(t;\mathbf{r}_0^{\alpha-1};\mathbf{i}) = \left\{ \sum_{k=1}^{k=\alpha\rho} e^{s_k^* t} \left[\sum_{r=1}^{r=\nu_k} \frac{1}{(r-1)!} t^{r-1} R_{kr} \right] \right\} \begin{bmatrix} \mathbf{1}_M \\ \mathbf{i}_0^{\beta-1} \\ \mathbf{r}_0^{\alpha-1} \end{bmatrix},$$

$$R_{kr} \in \mathfrak{C}^{\rho \times ((\beta+1)M+\alpha\rho)},\ R_{kr} = R_{kr}(\mathbf{i}_0^{\beta-1};\mathbf{r}_0^{\alpha-1};\mathbf{i})) = CONST.,\ \operatorname{Re} s_k^* < 0,$$

$$\mathbf{y}(t;\mathbf{r}_0^{\alpha-1};\mathbf{y}_0^{\nu-1};\mathbf{i}) = \left\{ \sum_{k=1}^{k=\nu N} e^{s_k^* t} \left[\sum_{r=1}^{r=\nu_k} \frac{1}{(r-1)!} t^{r-1} R_{kry} \right] \right\} \begin{bmatrix} \mathbf{1}_M \\ \mathbf{i}_0^{\gamma-1} \\ \mathbf{r}_0^{\alpha-1} \\ \mathbf{y}_0^{\nu-1} \end{bmatrix},$$

$$R_{kry} \in \mathfrak{C}^{N \times ((\gamma+1)M+\nu N)},\ R_{kry} = CONST.,$$
$$\operatorname{Re} s_k^* < 0,\ \forall k \in \{1, 2, .., \ \nu N\}.$$

This shows that the system substate vector

$$\left[\mathbf{r}^T(t;\mathbf{r}_0^{\alpha-1};\mathbf{i}_0^{\beta-1};\mathbf{i}) \ \vdots\ \mathbf{y}^T(t;\mathbf{r}_0^{\alpha-1};\mathbf{y}_0^{\nu-1};\mathbf{i}) \right]^T$$

is infinitely times continuously differentiable at every $t \in \mathfrak{T}$. All its derivatives rest bounded on $\mathfrak{T}_0$ and vanish as $t \to \infty$ in view of the Generating Theorem 342. Every entry of the system full state vector

$$\left[\left(\mathbf{r}^{\alpha-1}(t;\mathbf{r}_0^{\alpha-1};\mathbf{i}_0^{\beta-1}\mathbf{i}) \right)^T \ \vdots\ (\mathbf{y}^{\nu-1}(t;\mathbf{r}_0^{\alpha-1};\mathbf{y}_0^{\nu-1};\mathbf{i}))^T \right]^T$$

is bounded and vanishes as $t \to \infty$ in view of the Generating Theorem 342, which holds also for the whole system full state vector. ■

Lemma 408 *Let the input vector function* $\mathbf{i}(.)$ *belong to the family* $\mathfrak{I}^\beta$, *(2.12), (2.13), (in Section 2.1). In order for*

$$\left(\mathbf{r}^{\alpha-1}(t;\mathbf{r}_0^{\alpha-1};\mathbf{i}_0^{\beta-1};\mathbf{i}) \right)$$

of the IIO system (14.34) in the forced regime:

a) to be bounded on $\mathfrak{T}_0$,

$$\exists \varsigma = \varsigma(\mathbf{r}_0^{\alpha-1};\mathbf{i}_0^{\beta-1};\mathbf{i}) \in \mathfrak{R}^+ \Longrightarrow \left\| \mathbf{r}^{\alpha-1}(t;\mathbf{r}_0^{\alpha-1};\mathbf{i}_0^{\beta-1};\mathbf{i}) \right\| < \varsigma,\ \forall t \in \mathfrak{T}_0,$$

it is necessary and sufficient that

$$\mathbf{r}(t;\mathbf{r}_0^{\alpha-1};\mathbf{i}_0^{\beta-1};\mathbf{i})$$

is bounded on $\mathfrak{T}_0$,

$$\exists\xi=\xi(\mathbf{r}_0^{\alpha-1};\mathbf{i}_0^{\beta-1};\mathbf{i})\in\mathfrak{R}^+\Longrightarrow\left\|\mathbf{r}(t;\mathbf{r}_0^{\alpha-1};\mathbf{i}_0^{\beta-1};\mathbf{i})\right\|<\xi,\ \forall t\in\mathfrak{T}_0,$$

b) to vanish as $t\to\infty$,

$$t\to\infty\Longrightarrow\mathbf{r}^{\alpha-1}(t;\mathbf{r}_0^{\alpha-1};\mathbf{i})\to\mathbf{0}_{\alpha\rho},$$

it is necessary and sufficient that

$$\mathbf{r}(t;\mathbf{r}_0^{\alpha-1};\mathbf{i}_0^{\beta-1};\mathbf{i})$$

vanishes as $t\to\infty$,

$$t\to\infty\Longrightarrow\mathbf{r}(t;\mathbf{r}_0^{\alpha-1};\mathbf{i}_0^{\beta-1};\mathbf{i})\to\mathbf{0}_{\rho}.$$

By following the proof of Lemma 407 we easily verify and prove this Lemma.

Lemma 409 *Let the input vector function* $\mathbf{i}(.)$ *belong to the family* $\mathfrak{I}^{\beta}$, *(2.12), (2.13), (in Section 2.1). In order for*

$$\left(\mathbf{y}^{\nu-1}(t;\mathbf{r}_0^{\alpha-1};\mathbf{y}_0^{\nu-1};\mathbf{i})\right),\ \nu\geq1,$$

of the IIO system (14.34) in the forced regime:

a) to be bounded on $\mathfrak{T}_0$,

$$\exists\zeta=\zeta(\mathbf{r}_0^{\alpha-1};\mathbf{i}_0^{\gamma-1};\mathbf{y}_0^{\nu-1};\mathbf{i})\in\mathfrak{R}^+\Longrightarrow$$
$$\left\|\mathbf{y}^{\nu-1}(t;\mathbf{r}_0^{\alpha-1};\mathbf{i}_0^{\gamma-1};\mathbf{y}_0^{\nu-1};\mathbf{i})\right\|<\zeta,\ \forall t\in\mathfrak{T}_0,$$

it is necessary and sufficient that

$$\mathbf{y}(t;\mathbf{r}_0^{\alpha-1};\mathbf{i}_0^{\gamma-1};\mathbf{y}_0^{\nu-1};\mathbf{i})$$

is bounded on $\mathfrak{T}_0$,

$$\exists\xi=\xi(\mathbf{r}_0^{\alpha-1};\mathbf{i}_0^{\gamma-1};\mathbf{y}_0^{\nu-1};\mathbf{i})\in\mathfrak{R}^+\Longrightarrow\left\|\mathbf{y}(t;\mathbf{r}_0^{\alpha-1};\mathbf{i}_0^{\gamma-1};\mathbf{y}_0^{\nu-1};\mathbf{i})\right\|<\xi,\ \forall t\in\mathfrak{T}_0,$$

b) to vanish as $t \to \infty$,

$$t \to \infty \Longrightarrow \mathbf{y}^{\nu-1}(t;\mathbf{r}_0^{\alpha-1};\mathbf{i}_0^{\gamma-1};\mathbf{y}_0^{\nu-1};\mathbf{i}) \to \mathbf{0}_{\nu N},$$

it is necessary and sufficient that

$$\mathbf{y}(t;\mathbf{r}_0^{\alpha-1};\mathbf{i}_0^{\gamma-1};\mathbf{y}_0^{\nu-1};\mathbf{i})$$

vanishes as $t \to \infty$,

$$t \to \infty \Longrightarrow \mathbf{y}(t;\mathbf{r}_0^{\alpha-1};\mathbf{i}_0^{\gamma-1};\mathbf{y}_0^{\nu-1};\mathbf{i}) \to \mathbf{0}_N,$$

The proof of the Lemma is a special case of the proof of Lemma 407.

These Lemmae are inherent for the BI stability conditions in the framework of the IIO systems (14.34) and for their proofs.

Theorem 410 *Let the input vector function* $\mathbf{i}(.)$ *belong to the family* $\mathfrak{I}^{\beta}$, *(2.12), (2.13), (in Section 2.1). In order for the IIO system (14.34) in the forced regime to be:*

a) ***Bounded Input-Bounded Internal State*** *(BIBIS)* ***stable,*** *it is necessary and sufficient that 1. and 2. hold:*

1. the system IRIS transfer function matrix $G_{IIOIS}(s)$ is at least proper, or strictly proper, real rational matrix function of s,

and that

2. the real parts $Res_i^*(G_{IIOISrnd})$ of all poles $s_i^*(G_{IIOISrnd})$ of the row nondegenerate form $G_{IIOISrnd}(s)$ of $G_{IIOIS}(s)$ are negative,

$$Res_i^*(G_{IIOISrnd}) < 0, \ \forall i \in \{1, 2, \ \dots \ , \ \alpha\rho\}. \tag{14.40}$$

b) ***Bounded Input-Bounded Output*** *(BIBO)* ***stable,*** *it is necessary and sufficient that 1. and 2. are valid:*

1. the system IO transfer function matrix $G_{IIO}(s)$ is at least proper, or strictly proper, real rational matrix function of s,

and that

2. the real parts $Res_i^*(G_{IIOrnd})$ of all poles $s_i^*(G_{IIOrnd})$ of the row nondegenerate form $G_{IIOrnd}(s)$ of $G_{IIO}(s)$ are negative,

$$Res_i^*(G_{IIOrnd}) < 0, \ \forall i \in \{1, 2, \ \dots \ , \ \alpha\rho + \nu N\}. \tag{14.41}$$

c) ***Bounded Input-Bounded Output State*** *(BIBOS)* ***stable,*** *it is necessary and sufficient that the conditions under b) hold,* *d)* **Bounded Input-Bounded Internal State and Output** *(BIBISO)* **stable** *it is necessary and sufficient that the system is $BIBIS$ stable and $BIBO$ stable.*

e) **Bounded Input and Initial Internal State-Bounded Internal State** *($BIISBIS$)* **stable,** *it is necessary and sufficient that 1. and 2. hold:*

1. the system $IRIS$ transfer function matrix $G_{IIOIS}(s)$ is at least proper, or strictly proper, real rational matrix function of s, the system $IRII$ transfer function matrix $G_{IIOi_0IS}(s)$ and the system $IRIR$ transfer function matrix $G_{IIOr_0IS}(s)$ are strictly proper real rational matrix functions of s,

2. the poles of the row nondegenerate form $G_{IIOISrnd}(s)$ of the system $IRIS$ transfer function matrix $G_{IIOIS}(s)$ have negative real parts, the real parts of the poles of the row nondegenerate forms $G_{IIOi_0ISrnd}(s)$ and $G_{IIOr_0ISrnd}(s)$ of $G_{IIOi_0IS}(s)$ and of $G_{IIOr_0IS}(s)$, respectively, are nonpositive and those with zero real part are simple,

$$\begin{gathered} Res_k^*(G_{IIOISrnd}(s) < 0,\ Res_k^*(G_{IIOi_0ISrnd}(s) \leq 0, \\ Res_k^*(G_{IIOr_0ISrnd}(s) \leq 0,\ \forall k \in \{1, 2,\ \dots\ ,\ \alpha\rho\}, \\ Res_j^*(G_{IIOi_0ISrnd}) = 0 \Longrightarrow \nu_j = 1,\ Res_m^*(G_{IIOi_0ISrnd}) = 0 \Longrightarrow \nu_m = 1. \end{gathered} \tag{14.42}$$

f) **Bounded Input and Initial State-Bounded Output** *($BISBO$)* **stable,** *it is necessary and sufficient that 1. and 2. hold:*

1. the system IO transfer function matrix $G_{IIO}(s)$ is at least proper, or strictly proper, real rational matrix function of s, its $IICO$, $IIRO$ and $IIYO$ transfer function matrices $G_{IIOi_0}(s)$, $G_{IIOr_0}(s)$ and $G_{IIOy_0}(s)$ are strictly proper real rational matrix functions of s,

2. the real parts $Res_k^*(G_{IIOrnd})$ of all poles $s_k^*(G_{IIOrnd})$ of the row nondegenerate form $G_{IIOrnd}(s)$ of $G_{IIO}(s)$ are negative, the real parts of the poles of the row nondegenerate forms $G_{IIOi_0rnd}(s)$, $G_{IIOr_0rnd}(s)$ and $G_{IIOy_0rnd}(s)$ of $G_{IIOi_0}(s)$, $G_{IIOr_0}(s)$ and of $G_{IIOy_0}(s)$ are nonpositive and those with zero real part are simple,

$$\begin{gathered} Res_k^*(G_{IIOrnd}) < 0,\ Res_k^*(G_{IIOi_0rnd}(s) \leq 0,\ . \\ Res_k^*(G_{IIOr_0rnd}(s) \leq 0,\ Res_k^*(G_{IIOy_0rnd}(s) \leq 0, \\ \forall k \in \{1, 2,\ \dots\ ,\ \alpha\rho + \nu N\},\ Res_m^*(G_{IIOi_0IIrnd}) = 0 \Longrightarrow \nu_m = 1, \\ Res_j^*(G_{IIOr_0IIrnd}) = 0 \Longrightarrow \nu_j = 1,\ Res_l^*(G_{IIOy_0IIrnd}) = 0 \Longrightarrow \nu_l = 1. \end{gathered} \tag{14.43}$$

g) **Bounded Input and Initial State-Bounded Output State** *(BISBOS)* **stable,** *it is necessary and sufficient that f) holds.*

h) **Bounded Input and Initial State-Bounded State and Output** *(BISBSO)* **stable** *it is necessary and sufficient that g) holds.*

i) **Bounded Input and Initial State-Bounded State** *(BISBS)* **stable,** *it is necessary and sufficient that e) and g) hold.*

Proof. Let $\mathbf{i}(.) \in \mathfrak{I}^{\beta}$, (2.12), (2.12), (2.13), (in Section 2.1).

Necessity. a) The condition $\mathbf{i}(.) \in \mathfrak{I}^{\beta}$, (2.12), (2.13), and $\alpha \geq \beta$ in (14.34) guarantee that the product $G_{IIOIS}(s)(s)\mathbf{I}(s)$ is strictly proper real rational function of s. Let the IIO system (14.34) be $BIBIS$ stable, i.e., let a) of Definition 389, (in Subsection 14.2.3), hold. All initial conditions are equal to zero. The $k-th$ derivative $\mathbf{r}^{(k)}(t;\mathbf{0}_{\beta M};\mathbf{0}_{\alpha\rho};\mathbf{i})$, $\forall k=0,1,2,\ ...\ ,\alpha-1$, is bounded and it does not contain an impulse component. The state behavior $\mathbf{r}(t;\mathbf{0}_{\beta M};\mathbf{0}_{\alpha\rho};\mathbf{i})$ of the system does not contain an impulse component. This implies that the Laplace transform $\mathcal{L}\{\mathbf{r}(t;\mathbf{0}_{\beta M};\mathbf{0}_{\alpha\rho};\mathbf{i})\}$,

$$\mathcal{L}\{\mathbf{r}(t;\mathbf{0}_{\beta M};\mathbf{0}_{\alpha\rho};\mathbf{i})\} = G_{IIOIS}(s)\mathbf{I}(\mathbf{s}),$$

is strictly proper real rational vector function, for which $G_{IIOIS}(s)$ must be proper due to $\mathbf{i}(.) \in \mathfrak{I}^{\beta}$, (2.12), (2.13). We prove this also as follows. Let

$$\mathbf{i}(.) = h(.)\mathbf{1}_M \in \mathfrak{I},\ \mathbf{1}_M = [1\ \ 1\ ...\ 1]^T \in \mathfrak{R}^M \Longrightarrow \mathbf{I}(s) = \frac{1}{s}\mathbf{1}_M \in \mathfrak{C}^M.$$

Hence,

$$\mathcal{L}\{\mathbf{r}^{\alpha-1}(t;\mathbf{0}_{\beta M};\mathbf{0}_{\alpha\rho};h(t)\mathbf{1}_M)\} = \left[s^{-1}I_N\ \ I_N\ \ sI_N...\ s^{\alpha-2}I_N\right]^T G_{IIOIS}(s)\mathbf{1}_M$$

is strictly proper real rational vector function. This implies that the Laplace transform $\mathcal{L}\{\mathbf{r}(t;\mathbf{0}_{\beta M};\mathbf{0}_{\alpha\rho};\mathbf{i})\}$,

$$\mathcal{L}\{\mathbf{r}(t;\mathbf{0}_{\beta M};\mathbf{0}_{\alpha\rho};\mathbf{i})\} = G_{IIOIS}(s)\frac{1}{s}\mathbf{1}_M,$$

is strictly proper real rational vector function, for which $G_{IIOIS}(s)$ must be proper. This proves necessity of the condition a-1). Lemma 123, (in Chapter 6), $\mathbf{i}(.) \in \mathfrak{I}^{\beta}$, (2.12), (2.13), the Generating Theorem 342, (in Subsection 13.4.1), Lemma 408, and a) of Definition 389 prove necessity of the condition a-2), i.e., of (14.40).

b) Let the IIO system (14.34) be $BIBO$ stable so that b) of Definition 389 is valid. Hence, $G_{IIO}(s)\mathbf{I}(s)$ is strictly proper for every $\mathbf{i}(.) \in \mathfrak{I}^{\beta}$, (2.12),

(2.13), which, with $\mathbf{i}(.) \in \mathfrak{I}^{\beta}$, implies that $G_{IIO}(s)$ is proper. This proves the necessity of the condition b-1. Lemma 123, $\mathbf{i}(.) \in \mathfrak{I}^{\beta}$, the Generating Theorem 342 and

$$\mathbf{y}(t;\mathbf{0}_{\gamma M};\mathbf{0}_{\alpha\rho};\mathbf{0}_{\nu N};\mathbf{i}) = \mathcal{L}^{-1}\left\{G_{IIOrnd}(s)\mathbf{I}(s)\right\},$$

imply the necessity of the condition (14.41), i.e., of b-2).

c) Let the IIO system (14.34) be $BIBOS$ stable, i.e., let c) of Definition 389 hold. Then it is also $BIBO$ stable due to b) and c) of Definition 389. Hence, the conditions under c) are necessary due to Lemma 409, and the conditions under b).

d) Let the IIO system (14.34) be $BIBISO$ stable. This, a), b), c) and d) of Definition 389 imply that the system is $BIBIS$ stable and $BIBO$ stable.

e) Let the IIO system (14.34) be $BIISBIS$ stable, i.e., let (14.15)hold. This guarantees the boundedness of $\left\|\mathbf{r}^{\alpha-1}(t;\mathbf{r}_0^{\alpha-1};\mathbf{i})\right\|$ on $\mathfrak{T}_0$ for every $\mathbf{i}(.) \in \mathfrak{I}^{\beta}$, so that $\left\|\mathbf{r}(t;\mathbf{r}_0^{\alpha-1};\mathbf{i})\right\|$ is bounded on $\mathfrak{T}_0$ for every $\mathbf{i}(.) \in \mathfrak{I}^{\beta}$. This, (8.56) and (8.57) (in Section 8.3), Lemma 123, and the Generating Theorem 342 prove the necessity of the conditions under e).

f) Let the IIO system (14.34) be $BISBO$ stable, i.e., let (14.16)hold. Hence, $G_{IIO}(s)\mathbf{I}(s)$ is strictly proper, which with $\mathbf{i}(.) \in \mathfrak{I}^{\beta}$, implies that $G_{IIO}(s)$ is proper, and together with (8.62) through (8.66), (in Section 8.3), imply the necessity of the condition f-1. $BISBO$ stability of the system,

$$\mathbf{y}(t;\mathbf{i}_0^{\gamma-1};\mathbf{r}_0^{\alpha-1};\mathbf{y}_0^{\nu-1};\mathbf{i}) = \mathcal{L}^{-1}\left\{F_{IIO}(s)\begin{bmatrix}\mathbf{I}(s)\\ \mathbf{i}_0^{\gamma-1}\\ \mathbf{r}_0^{\alpha-1}\\ \mathbf{y}_0^{\nu-1}\end{bmatrix}\right\}$$

(8.62), Lemma 123, and the Generating Theorem 342 imply necessity of the condition (14.43), i.e., of f-2).

g) Let the IIO system (14.34) be $BISBOS$ stable, i.e., let (14.17)be valid. Then the system is $BISBO$ stable due to g) of Definition 389 and the fact that the output vector is subvector of the output state vector, which prove the necessity of g).

h) The definition of $BISBSO$ stability under h of Definition 389 implies the validity of g) of Definition 389 due to Lemma 409, which prove the necessity of the conditions under h).

i) The definition of $BISBS$ stability under i of Definition 389, (in Subsection 14.2.3) and the fact that the state vector is composed of the internal state vector and the output state vector prove that $BISBS$ stability implies both $BIISBIS$ and $BISBOS$ due to e) and g) of Definition 389.

Sufficiency. The boundedness of every $\mathbf{i}(.)$ that belongs to the family $\mathfrak{I}^{\beta}$, (2.12), (2.13), guarantees that the real parts of all poles of $\mathbf{I}(s)$ are nonpositive, that its poles with zero real parts are simple, and that $\mathbf{I}(s)$ is strictly proper real vector function.

a) Let the conditions under a) hold. Then, $\mathbf{i}(.) \in \mathfrak{I}^{\beta}$, (2.12), (2.13), and the condition a-1) guarantee that the Laplace transform

$$\mathcal{L}\{\mathbf{r}(t;\mathbf{0}_{\alpha\rho};\mathbf{i})\} = G_{IIOIS}(s)\mathbf{I}(s),$$

of $\mathbf{r}(t;\mathbf{0}_{\alpha\rho};\mathbf{i})$ does not have a pole in infinity. This, the condition a-2), $\mathbf{i}(.) \in \mathfrak{I}^{\beta}$, Lemma 123, and the Generating Theorem 342 imply that

$$\mathbf{r}(t;\mathbf{0}_{\alpha\rho};\mathbf{i}) = \mathcal{L}^{-1}\left\{G_{IIOIS}(s)\mathbf{I}(s)\right\},$$

is bounded for every $\mathbf{i}(.) \in \mathfrak{I}^{\beta}$. Hence, $\mathbf{r}^{\alpha-1}(t;\mathbf{0}_{\alpha\rho};\mathbf{i})$ is also bounded for every $\mathbf{i}(.) \in \mathfrak{I}^{\beta}$ due to Lemma 408. The system is *BIBIS* stable in view of a) of Definition 389.

b) Let the conditions under b) hold. Then, the condition b-1) and $\mathbf{i}(.) \in \mathfrak{I}^{\beta}$ guarantee that the Laplace transform

$$\mathcal{L}\{\mathbf{y}(t;\mathbf{0}_{\gamma M};\mathbf{0}_{\alpha\rho};\mathbf{0}_{\nu N};\mathbf{i})\} = G_{IIO}(s)\mathbf{I}(s),$$

of $\mathbf{y}(t;\mathbf{0}_{\gamma M};\mathbf{0}_{\alpha\rho};\mathbf{0}_{\nu N};\mathbf{i})$ does not have a pole in infinity. This, the condition b-2), $\mathbf{i}(.) \in \mathfrak{I}^{\beta}$, Lemma 123, and the Generating Theorem 342 imply that

$$\mathbf{y}(t;\mathbf{0}_{\gamma M};\mathbf{0}_{\alpha\rho};\mathbf{0}_{\nu N};\mathbf{i}) = \mathcal{L}^{-1}\left\{G_{IIO}(s)\mathbf{I}(s)\right\},$$

is bounded for every $\mathbf{i}(.) \in \mathfrak{I}^{\beta}$. The system is *BIBO* stable in view of b) of Definition 389.

c) Let the conditions under c) be satisfied. The conditions under b) hold. Therefore, $\mathbf{y}(t;\mathbf{0}_{\gamma M};\mathbf{0}_{\alpha\rho};\mathbf{0}_{\nu N};\mathbf{i})$ is bounded for every $\mathbf{i}(.) \in \mathfrak{I}^{\beta}$. This and Lemma 409 prove that the system is *BIBOS* stable in view of c) of Definition 389.

d) Let the *IIO* system (14.34) be *BIBIS* stable and *BIBO* stable. This, a) and c) prove that d) of Definition 389 is satisfied, i.e., that the system is *BIBISO* stable.

e) Let the conditions under e) be valid. The condition e-1) and $\mathbf{i}(.) \in \mathfrak{I}^{\beta}$, (8.56), (8.57) (in Section 8.3) show that the Laplace transform,

$$\mathcal{L}\{\mathbf{r}(t;\mathbf{r}_0^{\alpha-1};\mathbf{i})\} = F_{IIOIS}(s)\left[\mathbf{I}^T(s) \vdots \left(\mathbf{i}_0^{\beta-1}\right)^T \vdots \left(\mathbf{r}_0^{\alpha-1}\right)^T\right]^T,$$

of $\mathbf{r}(t;\mathbf{r}_0^{\alpha-1};\mathbf{i})$ is strictly proper real rational function. This, the condition e-2), $\mathbf{i}(.)\in\mathfrak{I}^{\beta}$, (8.57), Lemma 123, and the Generating Theorem 342 imply that $\mathbf{r}(t;\mathbf{r}_0^{\alpha-1};\mathbf{i})$ is bounded for every $\mathbf{i}(.)\in\mathfrak{I}^{\beta}$. Now, Lemma 407 proves the boundedness of $\mathbf{r}^{\alpha-1}(t;\mathbf{r}_0^{\alpha-1};\mathbf{i})$ for every $\mathbf{i}(.)\in\mathfrak{I}^{\beta}$. The system is *BIISBIS* stable in view of e) of Definition 389.

f) Let the conditions under f) hold. The Laplace transform

$$\mathcal{L}\left\{\mathbf{y}(t;\mathbf{i}_0^{\gamma-1};\mathbf{r}_0^{\alpha-1};\mathbf{y}_0^{\nu-1};\mathbf{i})\right\}=F_{IIO}(s)\begin{bmatrix}\mathbf{I}(s)\\ \mathbf{i}_0^{\gamma-1}\\ \mathbf{r}_0^{\alpha-1}\\ \mathbf{y}_0^{\nu-1}\end{bmatrix},$$

$$F_{IIO}(s)=\left[G_{IIO}(s)\,\vdots\,G_{IIOi_0}(s)\,\vdots\,G_{IIOr_0}(s)\,\vdots\,G_{IIOy_0}(s)\right],$$

of $\mathbf{y}(t;\mathbf{i}_0^{\gamma-1};\mathbf{r}_0^{\alpha-1};\mathbf{y}_0^{\nu-1};\mathbf{i})$ is strictly proper real rational function and it does not have either a pole in infinity or a pole with positive real part or a multiple imaginary pole for every $\mathbf{i}(.)\in\mathfrak{I}^{\beta}$. Lemma 123, and the Generating Theorem 342 imply that

$$\mathbf{y}(t;\mathbf{i}_0^{\gamma-1};\mathbf{r}_0^{\alpha-1};\mathbf{y}_0^{\nu-1};\mathbf{i})=\mathcal{L}^{-1}\left\{F_{IIO}(s)\begin{bmatrix}\mathbf{I}(s)\\ \mathbf{i}_0^{\gamma-1}\\ \mathbf{r}_0^{\alpha-1}\\ \mathbf{y}_0^{\nu-1}\end{bmatrix}\right\},$$

is bounded for every $\mathbf{i}(.)\in\mathfrak{I}^{\beta}$. Hence,

$$\mathbf{y}^{\nu-1}(t;\mathbf{i}_0^{\gamma-1};\mathbf{r}_0^{\alpha-1};\mathbf{y}_0^{\nu-1};\mathbf{i})$$

is also bounded for every $\mathbf{i}(.)\in\mathfrak{I}^{\beta}$ due to Lemma 409. The system is *BISBO* stable in view of f) of Definition 389 .

g) Let the conditions under g) hold. We have just proved under f) that the conditions 1) and 2) of f) guarantee that the system response $\mathbf{y}(t;\mathbf{i}_0^{\gamma-1};\mathbf{r}_0^{\alpha-1};\mathbf{y}_0^{\nu-1};\mathbf{i})$ is bounded for every $\mathbf{i}(.)\in\mathfrak{I}^{\beta}$, which implies the boundedness of the output state

$$\mathbf{y}^{\nu-1}(t;\mathbf{i}_0^{\gamma-1};\mathbf{r}_0^{\alpha-1};\mathbf{y}_0^{\nu-1};\mathbf{i})$$

due to Lemma 409. The system is *BISBOS* stable in view of g) of Definition 389.

h) Let the condition under h) hold, i.e., that the conditions under e) and g) are valid, hence under e) and f) hold. The system is *BIISBIS* stable and

$BISBOS$ stable, which guarantee that the system is also $BISBSO$ due to e), g) and h) of Definition 389.

i) Let i) hold. Hence, h) is valid. The system is $BISBSO$ stable, i.e., h) of Definition 389 is satisfied, which implies that i) of the same definition is fulfilled. The system is $BISBS$ stable. ∎

Part IV

CONCLUSION

Chapter 15

Motivation for the book

Besides the well-studied linear dynamical systems via their classical mathematical models in the IO form and in the ISO form, we have introduced and studied also the third class of the systems that are *the* IIO *systems.* The subclasses of the IIO systems are the PMD *systems*, *Rosenbrock systems*, and the $GISO$ systems.

We have developed a new, unified, and general approach to study in the same manner all three classes of the systems. The result is the unified general linear dynamical *time*-invariant continuous-*time* systems theory in the complex domain. Its applications are simple and straightforward.

In order to be clear and precise in using different notions such as system regime, stationary vector, equilibrium vector, we have presented their definitions and the procedures on how to determine them. These issues are very simple, but constitute the basis of the dynamical systems theory, which is often ignored in the literature on the linear dynamical systems. They facilitate understanding various qualitative properties of the systems (e.g., stability properties of the equilibrium vector).

Every dynamical physical system transfers and transmits simultaneously actions and influences of both the input vector and of all initial conditions. The system transfer function matrix $G(s)$ does not and cannot express and/or describe how the linear *time*-invariant (continuous-*time*, discrete-*time* or hybrid) system transforms all actions on itself into its internal and output behavior. This lack of $G(s)$ is the consequence of its definition and validity only for zero initial conditions.

The Laplace transform of the *n-th* order (scalar or vector) input-output differential equation contains the double sums of the products of the system parameters and initial conditions. Such complex sums and products were the obstacle to treat, in the complex domain, the influence of nonzero (i.e., of arbitrary) initial conditions on the system behavior in the same effective manner as we treat the input vector influence. In the existing theory, the obstacle has been ignored by unjustifiably assuming and accepting that all initial conditions are equal to zero. Such assumption is too severe, a crucial simplification, unjustifiable and unreal.

This obstacle has been removed herein by following [47], which means by solving this mathematical problem of how to put the double sum in the form equivalent to $G(s)\mathbf{I}(s)$ that characterizes the product of the system transfer function matrix $G(s)$ and the Laplace transform $\mathbf{I}(s)$ of the input vector $\mathbf{I}(t)$ under all zero initial conditions. Once this has been solved, we have become able to determine the system complex domain characteristic independent of the input vector and of all initial conditions, which completely expresses and describes how the system transfers, transmits and transforms influences of both the input vector and of all initial conditions on the system state and output behavior. Such characteristic is *the system full transfer function matrix* $F(s)$. It has the same features as the system transfer function matrix $G(s)$:

- the independence of the input vector,
- the independence of all initial conditions,
- the invariance relative to the input vector and to all initial conditions,
- the system order, dimension, structure and parameters completely determine it.

After presenting the definitions of *the system full transfer function matrix* $F(s)$ and of its submatrices for every type of systems, we presented and proved how we can determine $F(s)$ easily by using *the same mathematical knowledge that we apply to determine the system transfer function matrix* $G(s)$. Besides, the book presents the physical meaning of the system full transfer function matrix $F(s)$ and of its submatrices.

The system full transfer function matrix $F(s)$ **unifies** the complex domain theory of the linear *time*-invariant continuous-*time* systems.

In this context we discovered the existence of row (non)degenerate, column (non)degenerate, and (non)degenerate matrix functions.

Moreover, *the system full transfer function matrix* $F(s)$ expresses the channeling information and the system structure through its submatrices.

This book discovers how the use of *the system full transfer function matrix* $F(s)$ permits new results on:

- Pole-zero cancellation, which is admissible if and only if it is possible in *the system full transfer function matrix* $F(s)$ in general, or at least in its appropriate submatrix in a special case.

- The system realization and the minimal realization, which are different from the realization and the minimal realization of the transfer function matrix $G(s)$, which is valid only under all zero initial conditions. The system realization and the minimal realization are the realization and the minimal realization of $F(s)$ because they hold also for any initial conditions, but not of $G(s)$.

- The generalization of the block diagram technique to *the full block diagram technique.*

We have generalized the notion of the system matrix to *the full system matrix* $P(s)$. This has led to new results on the system equivalence. Besides, the book establishes the link between *the system full transfer function matrix* $F(s)$ and *the full system matrix* $P(s)$.

The Lyapunov method has been established for the ISO systems. We have broadened its direct application to both IO systems and IIO systems. This unifies the study of all three classes of the linear systems via the Lyapunov method.

After showing and explaining the inadequacy and the incompetence of the system transfer function matrix $G(s)$ for the Lyapunov stability tests, we have proved that only *the system full transfer function matrix* $F(s)$, or at least its adequate submatrix, which is its submatrix related to the internal initial conditions, is competent for the Lyapunov stability tests.

Starting with the fact that the initial conditions are seldom equal to zero, we have extended the concept of the system stability under bounded inputs (BI stability) and zero initial conditions to the system stability under bounded inputs (BI stability) and *nonzero* initial conditions. This has led to new BI stability properties and the related criteria. They are expressed in terms of the submatrices of *the system full transfer function matrix* $F(s)$, or directly in terms of *the system full transfer function matrix* $F(s)$. The system transfer function matrix $G(s)$ is inapplicable to BI stability tests in general.

Chapter 16

Summary of the contributions

This book completely resolves the problems, which are induced by the lacunae of the theory, for all three classes of the systems. To be specific:

- It discovers, defines and effectively exploits the state variables, i.e., the state vector, of both *IO* and *IIO* systems. They have the full physical sense. This permitted to extend the (Lyapunov and Bounded-Input, or *BI*) stability concepts and properties directly to these classes of the systems. This enabled their direct stability study rather than to study them formally mathematically, without any physical sense, in their formally mathematically (without a physical sense) transformed form of the *ISO* systems.

- It unifies the study and applications of all three classes of the systems. This is due to the following contributions that hold for all three classes of the systems.

- It discovers a complex domain fundamental dynamical characteristic of the systems, which is their full transfer function matrix $F(s)$, and which in the domain of the complex variable s shows how the system in the course of *time* transmits and transfers the simultaneous influence of the system input vector and of all initial conditions on the system output (or on the system state).

- It presents the definition of $F(s)$ and completely resolves the problem of its determination so that it has the same properties as the well-known system transfer function matrix $G(s)$: its compact matrix form, its independence of the external actions and of initial conditions acting on the system, i.e., its dependence only on the system order, structure and parameters.

- It shows the physical meaning of $F(s)$ and its link with the complete *time* response of the system.
- It establishes the full block diagram technique based on the use of $F(s)$, which incorporates the Laplace transform of the input vector and the vector of all initial conditions, and which generalizes the well-known block diagram technique.
- It exactly and completely resolves the problem of the pole-zero cancellation.
- It introduces the concept of the system full matrix $P(s)$ in the complex domain and establishes its link with the system full transfer function matrix $F(s)$.
- It defines the system equivalence under nonzero initial conditions and proves the related conditions.
- It establishes the direct relationship between the system full transfer function matrix $F(s)$ and the Lyapunov stability concept, definitions and conditions. This shows that the Lyapunov stability test via the system transfer function matrix $G(s)$ can end with crucially wrong result. This refines inherently the complex domain criteria for the Lyapunov stability properties.
- It extends, broadens and generalizes the BI stability concept by introducing new BI stability properties that incorporate nonzero initial conditions, presents their exact definitions and proves the conditions for their validity in terms of the system full transfer function matrix $F(s)$ and its submatrices.

Chapter 17

Future teaching and research

The basic notions, discoveries and the results of this book should, hopefully, become the crucial parts of the first course on linear dynamical *time*-invariant continuous-*time* systems. Other issues and results discovered herein should inherently enrich and/or refine the contents of the advanced courses on the linear dynamical *time*-invariant continuous-*time* systems.

The book contributions offer to engineers new, advanced, more powerful and complete complex domain theory of the linear continuous-*time* *time*-invariant systems.

The discoveries and the new results presented herein open new directions for further research. They are the basis for the observability, controllability, tracking and trackability theories of the linear control systems [70]. Another direction is the optimization in general and the conditional optimization in particular (i.e., the system optimization by satisfying the stability demand).

The conditional optimization has been done in the parameter space for zero initial conditions at first by Šiljak* in [111] and later in [35], [46]. This book opens the avenue to generalize their results to the conditional optimization under simultaneous actions of the input vector and nonzero initial conditions on the system.

All that is presented herein is analogously valid for the *time*-invariant discrete-*time* linear dynamical systems, which is shown in the acompanying volume [9]. This book and *Linear Discrete-Time Systems* [9] constitute the entity.

* Pronounced Shilyak.

Part V

BIBLIOGRAPHY

Bibliography

[1] J. K. Acka, "Finding the transfer function of time invariant linear system without computing $(sI - A)^{-1}$", *Int. J. Control*, Vol. 62, No.6, pp. 1517 - 1522, 1995.

[2] B. D. O. Anderson, "A Note on Transmisson Zeros of a Transfer Function Matrix", *IEEE Transactions on Automatic Control*, Vol. AC-21, No. 4, pp. 589 - 591, August 1976.

[3] B. D. O. Anderson and J. B. Moore, *Linear Optimal Control*, Englewood Cliffs: Prentice Hall, 1971.

[4] P. J. Antsaklis and A. N. Michel, *Linear Systems*, New York: The McGraw-Hill Companies, Inc., 1997, Boston: Birkhauser 2006.

[5] P. Borne, G. Dauphin-Tanguy, J.-P. Richard, F. Rotella and I. Zambettakis, *Commande et Optimisation des Processus*, Paris: Éditions TECHNIP, 1990.

[6] R. D. Braatz, "On Internal Stability and Unstable Pole-Zero Cancellations", *IEEE Control Systems Magazine*, vol. 32, No. 5, October 2012, pp. 15, 16.

[7] W. L. Brogan, *Modern Control Theory*, New York: Quantum Publishers, Inc., 1974.

[8] G. S. Brown and D. P. Campbell, *Principles of Servomechanisms*, New York: Wiley, 1948.

[9] Z. M. Buchevats and Ly. T. Gruyitch, *Linear Discrete-Time Systems*, CRC Press/Taylor & Francis Group, Boca Raton, Florida, USA, 2017.

[10] F. M. Callier and C. A. Desoer, *Multivariable Feedback Systems*, New York: Springer-Verlag, 1982

[11] F. M. Callier and C. A. Desoer, *Linear System Theory*, New York: Springer-Verlag, 1991.

[12] G. E. Carlson, *Signal and Linear Systems Analysis and MATLAB*, second edition, New York: Wiley, 1998.

[13] C.-T. Chen, *Linear System Theory and Design*, New York: Holt, Rinehart and Winston, Inc., 1984.

[14] H. Chestnut and R. W. Mayer, *Servomechanisms and Regulating System Design*, New York: Wiley, 1955.

[15] J. J. D'Azzo and C. H. Houpis, *Linear Control System Analysis & Design*, New York: McGraw-Hill Book Company, 1988.

[16] L. Debnath, *Integral Transformations and Their Applications*, Boca Raton: CRC Press, 1995.

[17] C. A. Desoer, *Notes for A Second Course on Linear Systems*, New York: Van Nostrand Reinhold Company, 1970.

[18] C. A. Desoer and J. D. Schulman, "Zeros and Poles of Matrix Transfer Functions and Their Dynamical Interpretation", *IEEE Transactions on Circuits and Systems*, Vol. CAS-21, No. 1, pp. 3 - 8, January 1974.

[19] C. A. Desoer and M. Vidyasagar, *Feedback Systems: Input - Output Properties*, New York: Academic Press, 1975.

[20] O. I. Elgerd, *Control Systems Theory*, New York: McGraw-Hill Book Company, 1967.

[21] D. R. Fannin, W. H. Tranter, and R. E. Ziemer, *Signals & Systems Continuous and Discrete*, fourth edition, Englewood Cliffs: Prentice Hall, 1998.

[22] T. E. Fortmann and K. L. Hitz, *An Introduction to Linear Control Systems*, New York: Marcel Dekker, Inc., 1977.

[23] F. R. Gantmacher, *The Theory of Matrices*, Vol. 1, New York: Chelsea Publishing Co., 1974.

[24] F. R. Gantmacher, *The Theory of Matrices*, Vol. 2, New York: Chelsea Publishing Co., 1974.

[25] Ly. T. Grouyitch, *Automatique : dynamique linéaire*, Notes de cours, Belfort: Ecole Nationale d'Ingénieurs de Belfort, 1997.

[26] Ly. T. Grouyitch, *Automatique: dynamique linéaire*, Lecture Notes, Belfort: University of Technology Belfort - Montbeliard, 1999 - 2000.

[27] Lj. T. Grujić, *Automatique - Dynamique Linéaire*, Lecture Notes, Belfort: Ecole Nationale d'Ingénieurs de Belfort, 1994 - 1996.

[28] Lj. T. Grujić, *Automatique - Dynamique Linéaire*, Lecture Notes, Belfort: Université de Technologie de Belfort - Montbéliard, 2000 - 2003.

[29] Lj. T. Grujić (Ly. T. Gruyitch), *Continuous Time Control Systems*, Lecture notes for the course "DNEL4CN2: Control Systems", Durban: Department of Electrical Engineering, University of Natal, South Africa, 1993.

[30] Lj. T. Grujić, "Natural Trackability and Tracking Control of Robots", *IMACS-IEEE-SMC multiconference CESA'96: Symposium on Control, Optimization and Supervision*, Vol. 1, Lille, France, pp. 38 - 43, 1996.

[31] Lj. T. Grujić, "Natural Trackability and Control: Multiple Time Scale Systems", *Preprints of the IFAC-IFIP-IMACS Conference: Control of Industrial Systems*, **2**, Pergamon, Elsevier, London, pp. 111 - 116, 1997.

[32] Lj. T. Grujić, "Natural trackability and control: multiple time scale systems", *Proceedings of the IFAC Conference: Control of Industrial Systems*, (Ed's. Lj. T. Grujić, P. Borne, A. El Moudni and M. Ferney), Vol. 2, Pergamon, Elsevier, London, pp. 669 - 674, 1997.

[33] Lj. T. Grujić, "Natural Trackability and Control: Perturbed Robots", *Preprints of the IFAC-IFIP-IMACS Conference: Control of Industrial Systems*, **3**, Belfort, France, pp. 691 - 696, 1997.

[34] Lj. T. Grujić, "Natural trackability and control: perturbed robots", *Proceedings of the IFAC Conference: Control of Industrial Systems*, (Ed's. Lj. T. Grujić, P. Borne, A. El Moudni and M. Ferney), Vol. 3, Pergamon, Elsevier, London, pp. 1641 - 1646, 1997.

[35] Lj. T. Grujić, "Possibilities of Linear System Design on the Basis of Conditional Optimization in Parameter Plane", Part I, *Automatika: Theoretical supplement*, Zagreb (Yugoslavia - Croatia), Vol. 2, No. 1-2, pp. 49 - 60, 1966.

[36] Lj. T. Grujić and W. P. Mounfield, *Natural Tracking Controller*, US Patent No 5,379,210, Jan. 3, 1995.

[37] Lj. T. Grujić and W. P. Mounfield, Jr., "Natural Tracking Control of Linear Systems", *Proceedings of the 13th IMACS World Congress on Computation and Applied Mathematics*, Eds. R. Vichnevetsky and J. J. H. Miller, Trinity College, Dublin, Ireland, Vol. 3, pp. 1269 - 1270, July 22 - 26, 1991.

[38] Lj. T. Grujić and W. P. Mounfield, "Natural Tracking Control of Linear Systems", in *Mathematics of the Analysis and Design of Process Control*, Ed. P. Borne, S.G. Tzafestas and N.E. Radhy, Elsevier Science Publishers B. V., IMACS, pp. 53 - 64, 1992.

[39] Lj. T. Grujić and W. P. Mounfield, "Natural Tracking PID Process Control for Exponential Tracking", *American Institute of Chemical Engineers Journal*, **38**, No. 4, pp. 555 - 562, 1992.

[40] Lj. T. Grujić and W. P. Mounfield, "PD-Control for Stablewise Tracking with Finite Reachability Time: Linear Continuous Time MIMO Systems with State-Space Description", *International Journal of Robust and Nonlinear Control*, England, Vol. 3, pp. 341 - 360, 1993.

[41] Lj. T. Grujić and W. P. Mounfield, "PD Natural Tracking Control of an Unstable Chemical Reaction", *Proc. 1993 IEEE International Conference on Systems, Man and Cybernetics*, Le Touquet, Vol. 2, pp. 730 - 735, 1993.

[42] Lj. T. Grujić and Mounfield W. P., "PID Natural Tracking Control of a Robot: Theory", *Proc. 1993 IEEE International Conference on Systems, Man and Cybernetics*, Le Touquet, Vol. 4, pp. 323 - 327, 1993.

[43] Lj. T. Grujić and W. P. Mounfield, "Ship Roll Stabilization by Natural Tracking Control: Stablewise Tracking with Finite Reachability Time", *Proc. 3rd IFAC Workshop on Control Applications in Marine Systems*, Trondheim, Norway, pp. 202-207, 10-12 May, 1995.

[44] Lj. T. Grujić and W. P. Mounfield, "Stablewise Tracking with Finite Reachability Time: Linear Time-Invariant Continuous-Time MIMO Systems", *Proc. of the 31st IEEE Conference on Decision and Control*, Tucson, Arizona, pp. 834 - 839, 1992.

[45] Lj. T. Grujić and W. P. Mounfield, Jr., "Tracking Control of Time-Invariant Linear Systems Described by IO Differential Equations", *Proceedings of the 30th IMACS Conference on Decision and Control*, Brighton, England, Vol. 3, pp. 2441 - 2446, December 11 - 13, 1991.

[46] Lj. T. Grujić and Z. Novaković, "Conditional Optimization of Multi-Input Multi-Output Continuous-Time Automatic Control Systems (A.C.S.) with Application to a Hydraulic A.C.S.", in Serbo-Croatian, *Proceedings of HIPNEF*, Belgrade (Yugoslavia-Serbia), pp. 55 - 61, November 1980.

[47] Ly. T. Gruyitch, *Advances in the Linear Dynamic Systems Theory*, Llumina Press, Tamarac, FL, USA, 2013.

[48] Ly. T. Gruyitch, "A physical principle and consistent Lyapunov methodology: time-invariant nonlinear systems", *Proc. International Conference on Advances in Systems, Signals, Control and Computers*, 1, Durban, South Africa, pp. 42 - 50, 1998.

[49] Ly. T. Gruyitch, *Conduite des systèmes*, Lecture Notes: Notes de cours SY 98, Belfort: University of Technology Belfort - Montbéliard, 2000, 2001.

[50] Ly. T. Gruyitch, "Consistent Lyapunov Methodology for Exponential Stability: PCUP Approach", in *Advances in Stability Theory at the End of the 20th Century*, Ed. A. A. Martynyuk, London: Taylor & Francis, pp. 107 - 120, 2003.

[51] Ly. T. Gruyitch, *Contrôle commande des processus industriels*, Lecture Notes: Notes de cours SY 51, Belfort: University of Technology Belfort - Montbéliard, 2002, 2003.

[52] Ly. T. Gruyitch, *EINSTEIN'S RELATIVITY THEORY. Correct, Paradoxical, and Wrong*, Trafford, Victoria, Canada, 2006.

[53] Ly. T. Gruyitch, "Exponential stabilizing natural tracking control of robots: theory", *Proceedings of the Third ASCE Specialty Conference on Robotics for Challenging Environments*, held in Albuquerque, New

Mexico, USA, (Ed's. Laura A. Demsetz, Raymond H. Bryne and John P. Wetzel), Reston, Virginia, USA: American Society of Civil Engineers (ASCE), pp. 286 - 292, April 26 - 30, 1998.

[54] Ly. T. Gruyitch, *Galilean - Newtonean Rebuttal to Einsteins Relativity Theory*, Cambridge International Science Publishing, Cambridge UK, 2015.

[55] Ly. T. Gruyitch, "Gaussian generalisations of the relativity theory fundaments with applications", *Proceedings of the VII International Conference: Physical Interpretations of Relativity Theory*, Ed. M. C. Duffy, British Society for the Philosophy of Science, London, pp. 125 -136, September 15 - 18, 2000.

[56] Ly. T. Gruyitch, "Natural control of robots for fine tracking", *Proceedings of the 38th Conference on Decision and Control*, Phoenix, Arizona, USA, pp. 5102 - 5107, December 1999.

[57] Ly. T. Gruyitch, "Natural tracking control synthesis for lagrangian systems", *V International Seminar on Stability and Oscillations of Nonlinear Control Systems*, Russian Academy of Sciences, Moscow, pp. 115 - 120, June 3 - 5, 1998.

[58] Ly. T. Gruyitch, *Nonlinear Systems Tracking*, CRC Press/Taylor & Francis Group, Boca Raton, Florida, USA, 2016.

[59] Ly. T. Gruyitch, "On tracking theory with embedded stability: control duality resolution", *Proceedings of the 40th IEEE Conference on Decision and Control*, Orlando, Florida, USA, pp. 4003 - 4008, December 2001.

[60] Ly. T. Gruyitch, "Physical Continuity and Uniqueness Principle. Exponential Natural Tracking Control", *Neural, Parallel & Scientific Computations*, 6, pp. 143 - 170, 1998.

[61] Ly. T. Gruyitch, "Robot global tracking with finite vector reachability time", *Proceedings of the European Control Conference*, Karlsruhe, Germany, Paper # 132, pp. 1 - 6, 31 August - 3 September 1999.

[62] Ly. T. Gruyitch, "Robust prespecified quality tracking control synthesis for 2D systems", *Proc. International Conference on Advances in Systems, Signals, Control and Computers*, **3**, Durban, South Africa, pp. 171 - 175, 1998.

[63] Ly. T. Gruyitch, *Systèmes d'asservissement industriels*, Lecture Notes: Notes de cours SY 40, Belfort: Université de Technologie de Belfort - Montbéliard, 2001.

[64] Ly. T. Gruyitch, *Time and Consistent Relativity. Physical and Mathematical Fundamentals*, Apple Academic Press, Inc., Waretown N.J. and Oakville ON, 2015,

[65] Ly. T. Gruyitch, *TIME and TIME FIELDS. Modeling, Relativity, and Systems Control*, Trafford, Victoria, Canada, 2006.

[66] Ly. T. Gruyitch, *TIME. Fields, Relativity, and Systems*, Llumina, Coral Springs, Florida, USA, 2006.

[67] Ly. T. Gruyitch, "Time, Relativity and Physical Principle: Generalizations and Applications", *Proc. V International Conference: Physical Interpretations of Relativity Theory*, (Ed. M. C. Duffy), pp. 134 - 170, London, 11 - 14 September, 1998; (also in: *Nelinijni Koluvannya*, Vol. 2, No. 4, pp. 465 - 489, Kiev, Ukraine, 1999).

[68] Ly. T. Gruyitch, "Time, Systems, and Control: Qualitative Properties and Methods", Chapter 2 in *Stability and Control of Dynamical Systems with Applications*, Editors D. Liu and P. J. Antsaklis, pp. 23 - 46, Boston: Birkhâuser, pp. 23 - 46, 2003.

[69] Ly. T. Gruyitch, "Time and Uniform Relativity Theory Fundaments", *Problems of Nonlinear Analysis in Engineering Systems*, 7, N° 2(14), Kazan, Russia, pp. 1 - 29, 2001.

[70] Ly. T. Gruyitch, *Tracking Control of Linear Systems*, CRC Press/Taylor & Francis Group, Boca Raton, Florida, USA, 2013.

[71] Ly. T. Gruyitch and W. Pratt Mounfield, Jr., "Absolute output natural tracking control: MIMO Lurie systems", *Proceedings of the 14th Triennial World Congress*, Beijing, P. R. China, Pergamon - Elsevier Science, Vol. C, pp. 389 - 394, July 5 - 9, 1999.

[72] Ly. T. Gruyitch and W. Pratt Mounfield, Jr., "Constrained natural tracking control algorithms for bilinear DC shunt wound motors", *Proceedings of the 40th IEEE Conference on Decision and Control*, Orlando, Florida USA, pp. 4433 - 4438, December 2001.

[73] Ly. T. Gruyitch and W. Pratt Mounfield, Jr., "Elementwise stablewise tracking with finite reachability time: linear time-invariant continuous-time MIMO systems", *International Journal of Systems Science*, Vol. 33, No. 4, pp. 277 - 299, 2002.

[74] Ly. T. Gruyitch and W. P. Mounfield, Jr., "Robust elementwise exponential tracking control: IO linear systems", *Proceedings of the* 36th *IEEE Conference on Decision and Control*, San Diego, California, USA, pp. 3836 - 3841, December 1997.

[75] Ly. T. Gruyitch and W. Pratt Mounfield, Jr., "Stablewise absolute output natural tracking control with finite reachability time: MIMO Lurie systems", *CD ROM Proceedings of the 17th IMACS World Congress, Invited session IS-2 : Tracking theory and control of nonlinear systems*, Paris, France, pp. 1 - 17, July 11-15, 2005; *Mathematics and computers in simulation*, Vol. 76, pp. 330 - 344, 2008.

[76] W. Hahn, *Stability of Motion*, New York: Springer-Verlag, 1967.

[77] T. Kailath, *Linear Systems*, Englewood Cliffs: Prentice Hall, Inc., 1980.

[78] R. E. Kalman, "On the General Theory of Control Systems", *Proceedings of the First International Congress on Automatic Control*, pp. 481 - 491, London: Butterworth, 1960.

[79] R. E. Kalman, "Canonical structure of linear dynamical systems", *Proceedings of the National Academy of Science: Mathematics*, USA NAS, Vol. 48, pp. 596 - 600, 1962.

[80] R. E. Kalman, "Mathematical Description of Linear Dynamical Systems", *J. S. I. A. M. Control*, Ser. A, Vol. 1, No. 2, pp. 152 - 192, 1963.

[81] R. E. Kalman, "Algebraic Structure of Linear Dynamical Systems, I. The Module of Σ", *Proceedings of the National Academy of Science: Mathematics*, USA NAS, Vol. 54, pp. 1503 - 1508, 1965.

[82] R. E. Kalman, Y. C. Ho and K. S. Narendra, "Controllability of Linear Dynamical Systems", *Contributions to Differential Equations*, Vol.1, No. 2, pp. 189 - 213, 1963.

[83] N. N. Krasovskii, *Some Problems of the Theory of Stability of Motion*, in Russian, Moscow: FIZMATGIZ, 1959.

[84] N. N. Krasovskii, *Stability of Motion*, Stanford: Stanford University Press, 1963.

[85] B. C. Kuo, *Automatic Control Systems*, Englewood Cliffs: Prentice Hall, Inc., 1967.

[86] B. C. Kuo, *Automatic Control Systems*, Englewood Cliffs: Prentice Hall, Inc., 1987.

[87] H. Kwakernaak and R. Sivan, *Linear Optimal Control Systems*, New York: Wiley Interscience, 1972.

[88] A. M. Lyapunov, *The General Problem of Stability of Motion*, (in Russian), Kharkov Mathematical Society, Kharkov, 1892; in Academician A. M. Lyapunov: "Collected Papers", U.S.S.R. Academy of Science, Moscow, II, pp. 5 - 263, 1956. French translation: "Problème général de la stabilité du mouvement", *Ann. Fac. Toulouse*, 9, pp. 203 - 474; also in: *Annals of Mathematics Study*, No. 17, Princeton University Press, 1949. English translation: *Intern. J. Control*, 55, pp. 531 - 773, 1992; and Taylor & Francis, London, 1992.

[89] L. A. MacColl, *Fundamental Theory of Servomechanisms*, New York: D. Van Nostrand Company, Inc., 1945.

[90] J. M. Maciejowski, *Multivariable Feedback Systems, Wokingham: Addison-Wesley Publishing Company, 1989.*

[91] J. L. Melsa and D. G. Schultz, *Linear Control Systems*, New York: McGraw-Hill Book Company, 1969.

[92] A. N. Michel, L. Hou and D. Liu, *Stability of Dynamical Systems. Continuous, Discontinuous and Discrete Systems*, Boston-Basel-Berlin: Birkhäuser, 2008.

[93] B. R. Milojković and Lj. T. Grujić, *Automatic Control*, in Serbo-Croatian, Belgrade: Faculty of Mechanical Engineering, University of Belgrade, 1977.

[94] B. R. Milojković and L. T. Grujić, *Automatic control*, in Serbo-Croatian, Belgrade: Faculty of Mechanical Engineering, 1981.

[95] W. P. Mounfield, Jr. and Lj. T. Grujić, "High-Gain Natural Tracking Control of Linear Systems", *Proceedings of the 13th IMACS World*

Congress on Computation and Applied Mathematics, Eds. R. Vichnevetsky and J. J. H. Miller, Trinity College, Dublin, Ireland, Vol. 3, pp. 1271 - 1272, 1991.

[96] W. P. Mounfield, Jr. and Lj. T. Grujić, "High-Gain Natural Tracking Control of Time-Invariant Systems Described by IO Differential Equations", *Proceedings of the 30th Conference on Decision and Control*, Brighton, England, pp. 2447 - 2452, 1991.

[97] W. P. Mounfield, Jr. and Lj. T. Grujić, "High-Gain Natural Tracking Control of Linear Systems", *Proceedings of the 13th IMACS World Congress on Computation and Applied Mathematics*, Eds. R. Vichnevetsky and J. J. H. Miller, Trinity College, Dublin, Ireland, Vol. 3, pp. 1271 - 1272, July 22 - 26, 1991.

[98] W. P. Mounfield, Jr. and Lj. T. Grujić, "High-gain PI natural tracking control for exponential tracking of linear single-output systems with state-space description", *RAIRO- Automatique, Productique, Informatique Industrielle (APII)*, Vo. 26, pp. 125 - 146, 1992.

[99] W. P. Mounfield and Lj. T. Grujić, "High-Gain PI Control of an Aircraft Lateral Control System", *Proc. 1993 IEEE International Conference on Systems, Man and Cybernetics*, Le Touquet, Vol. 2, pp. 736-741, 1993.

[100] W. P. Mounfield, Jr. and Lj. T. Grujić, "High-gain PI natural tracking control for exponential tracking of linear MIMO systems with state-space description", *International Journal of Control*, Vol. 25, No. 11, pp. 1793 - 1817, 1994.

[101] W. P. Mounfield and Lj. T. Grujić, "Natural tracking control for exponential tracking: lateral high-gain PI control of an aircraft system with state-space description", *Neural, Parallel & Scientific Computations*, Vol. 1, No. 3, pp. 357-370, 1993.

[102] W. P. Mounfield and Lj. T. Grujić, "PID - Natural Tracking Control of a Robot: Application", *Proc. 1993 IEEE International Conference on Systems, Man and Cybernetics*, Le Touquet, Vol. 4, pp. 328-333, 1993.

[103] W. P. Mounfield and Lj. T. Grujić, "Robust Natural Tracking Control for Multi-Zone Space Heating Systems", *Proc. 14th IMACS World Congress*, Vol. 2, 841 - 843, 1994.

[104] W. P. Mounfield, Jr. and Ly. T. Gruyitch, "Control of aircrafts with redundant control surfaces: stablewise tracking control with finite reachability time", *Proceedings of the Second International Conference on Nonlinear Problems in Aviation and Aerospace*, European Conference Publishers, Cambridge, Vol. 2, 1999, pp. 547 - 554, 1999.

[105] W. P. Mounfield, Jr. and Ly. T. Gruyitch, "Elementwise stablewise finite reachability time natural tracking control of robots", *Proceedings of the 14th Triennial World Congress*, Beijing, P. R. China, Pergamon - Elsevier Science, Vol. B, pp. 31 - 36, July 5 - 9, 1999.

[106] K. Ogata, *State Space Analysis of Control Systems*, Englewood Cliffs: Prentice Hall, 1967.

[107] K. Ogata, *Modern Control Engineering*, Englewood Cliffs: Prentice Hall, 1970.

[108] D. H. Owens, *Feedback and Multivariable Systems*, Stevenage, Herts: Peter Peregrinus Ltd., 1978.

[109] H. M. Power and R. J. Simpson, *Introduction to Dynamics and Control*, London: McGraw-Hill Book Company (UK) Limited, 1978.

[110] H. H. Rosenbrock, *State-space and Multivariable Theory*, London: Thomas Nelson and Sons Ltd., 1970.

[111] D. D. Šiljak, *Feedback Systems Synthesis via Squared Error Conditional Optimization,* D. Sc. Dissertation (in Serbo-Croatian), Departement of Electrical Engineering, University of Belgrade, Belgrade (Yugoslavia - Serbia), 1963.

[112] D. D. Šiljak, *Large-scale dynamic systems: stability and structure*, New York: North Holland, 1978.

[113] D. D. Šiljak, *Nonlinear Systems*, John Wiley & Sons, New York, 1969.

[114] R. E. Skelton, *DYNAMIC SYSTEMS CONTROL. Linear Systems Analysis and Synthesis*, New York: John Wiley & Sons, 1988.

[115] E. D. Sontag, "Further facts about input to state stabilization", *IEEE Transactions on Automatic Control*, Vol. 35, No. 3, pp. 473 - 476, 1990.

[116] E. D. Sontag, *Mathematical Control Theory: Deterministic Finite Dimensional Systems*, New York: Springer, 1990.

[117] E. D. Sontag, "Input to State Stability: Basic Concepts and Results", pages 163-220 in P. Nistri and G. Stefani, editors, *Nonlinear and Optimal Control Theory*, Berlin: Springer-Verlag, 2007.

[118] M. Vidyasagar, *Nonlinear Systems Analysis*, Englewood Cliffs: Prentice Hall, 1978.

[119] J. C. West, *Textbook of Servomechanisms*, London: English Universities Press, 1953.

[120] D. M. Wiberg, *State Space and Linear Systems*, New York: McGraw-Hill Book Company, 1971.

[121] W. A. Wolovich, *Linear Multivariable Systems*, New York: Springer-Verlag, 1974.

[122] W. M. Wonham, *Linear Multivariable Control. A Geometric Approach*, Berlin: Springer-Verlag, 1974.

Part VI

APPENDICES

Appendix A

Notation

The meaning of the notation is explained in the text at its first use.

A.1 Abbreviations

This is the list of the main abbreviations.

GISO system *General Input-State-Output system (2.39)*
iff *if and only if*
I *Input*
II *Input-Internal*
IIO *Input-Internal and Output dynamical*
IIO system *the Input-Internal and Output dynamical system (2.22)*
IO *Input-Output*
IO system *the Input-Output system (2.1)*
IS *Input-State*
ISO *Input-State-Output*
ISO system *the Input-State-Output system (3.84), (3.85)*
LY *LY stability concept* means Lyapunov's stability concept
MIMO *Multiple-Input-Multiple-Output*
OR *Output response*
PCUP *Physical Continuity and Uniqueness Principle*
PMD *Polynomial Matrix Description (2.36)*
RS system *Rosenbrock system (2.37)*
SISO *Single-Input Single-Output*
System *Continuous-time time-invariant linear dynamical system*
TCUP *Time Continuity and Uniqueness Principle*

A.2 Indexes

We separate subscripts from superscripts.

A.2.1 Subscripts

d the subscript d denotes "desired"

e *equilibrium*

i the subscript i denotes "the i-th"

ir irreducible

j the subscript j denotes "the j-th"

nd *nondegenerate*

rd *reduced*

rnd *row nondegenerate*

$zero$ the subscript $zero$ denotes "the zero value"

0 the subscript 0 (zero) associated with a variable (.) denotes its initial value $(.)_0$; however, if $(.) \subset \mathfrak{T}$ then the subscript 0 (zero) associated with (.) denotes the *time* set $\mathfrak{T}_0$, $(.)_0 = \mathfrak{T}_0$.

A.2.2 Superscript

k the superscript k denotes "k-dimensional", $k \in \{1, 2, ..., n, ...\}$.

A.3 Letters

Lowercase block or italic letters are used for scalars. Lowercase bold block letters denote vectors. Uppercase block letters denote matrices, or points. Uppercase fraktur letters designate sets or spaces.

The notation "$; t_{(.)0}$" will be omitted as an argument of a variable if and only if a choice of the initial moment $t_{(.)0}$ does not have any influence on the value of the variable.

A.3.1 Blackboard bold letters

The blackboard letters are related to the complex domain.

$\mathbb{C}$ *set of complex numbers s*

$\mathbb{C}^k$ *k-dimensional complex vector space*

A.3.2 Calligraphic letters

The caligraphic letters are used for the Laplace transform and for the motion.

$\mathcal{L}^{\mp}\{\mathbf{i}(.)\}$ *the left (-), right (+), respectively, Laplace transform of a function* $\mathbf{i}(.)$, Section B

$$\mathcal{L}^{\mp}\{\mathbf{i}(t)\} = \mathbf{I}^{\mp}(s) = \int_{0^{\mp}}^{\infty} \mathbf{i}(t)e^{-st}dt = \lim\left[\int_{\mp\zeta}^{\infty} \mathbf{i}(t)e^{-st}dt : \zeta \longrightarrow 0^{+}\right].$$

$\mathcal{S}(,)$ *system motion*

A.3.3 Fraktur letters

Capital fraktur letters are used for spaces or sets.

$\mathfrak{A} \subseteq \mathfrak{R}^n$ *a nonempty subset of* $\mathfrak{R}^n$

$\mathfrak{B} \subseteq \mathfrak{R}^n$ *a nonempty subset of* $\mathfrak{R}^n$

$\mathfrak{B}_\xi(\mathbf{z})$ *an open hyperball with the radius* ξ *centered at the point* $\mathbf{z}$ in the corresponding space

$$\mathfrak{B}_\xi(\mathbf{z}) = \{\mathbf{w} : \ \|\mathbf{w} - \mathbf{z}\| < \xi\}$$

$\mathfrak{B}_\xi$ *an open hyperball with the radius* ξ *centered at the origin* of the corresponding space

$$\mathfrak{B}_\xi = \mathfrak{B}_\xi(\mathbf{0})$$

$\mathfrak{C}$ *the family of all continuous functions on* $\mathfrak{T}_0$

$\mathfrak{C}^{ki}(\mathfrak{S})$ *the family of all functions defined, continuous and k-times continuously differentiable on the set* $\mathfrak{S} \subseteq \mathfrak{R}^i$, $\mathfrak{C}^{ki}(\mathfrak{R}^i) = \mathfrak{C}^k(\mathfrak{R}^i) = \mathfrak{C}^{ki}$

$\mathfrak{C}^k$ *the family of all functions defined, continuous and k-times continuously differentiable on* $\mathfrak{T}_0$, $\mathfrak{C}^0 = \mathfrak{C}$

$\mathfrak{C}^0(\mathfrak{S})$ *the family of all functions defined and continuous on the set* $\mathfrak{S}$, $\mathfrak{C}^0(\mathfrak{R}^i) = \mathfrak{C}^{0,i}$

$\mathfrak{C}^{k-}(\mathfrak{R}^i)$ *the family of all functions defined everywhere and k-times continuously differentiable on* $\mathfrak{R}^i \backslash \{\mathbf{0}_i\}$, *which have defined and continuous derivatives at the origin* $\mathbf{0}_i$ of $\mathfrak{R}^i$ up to the order $(k-1)$, which are defined and continuous *at the origin* $\mathbf{0}_i$ and have defined the left and the right *k-th* order derivative *at the origin* $\mathbf{0}_i$

$\mathfrak{D}^k$ a given, or to be determined, *family of all bounded k-times continuously differentiable on* $\mathfrak{T}_0$ *permitted disturbance vector total functions* $\mathbf{D}(.)$, or *deviation functions* $\mathbf{d}(.)$, $\mathfrak{D}^k \subset \mathfrak{C}^k$, the Laplace transforms of which are strictly proper real rational complex functions

$$\mathfrak{D}^k = \left\{\mathbf{D}(.) : \mathbf{D}^{(k)}(\mathbf{t}) \in \mathfrak{C},\ \exists\zeta \in \mathfrak{R}^+ \Longrightarrow \left\|\mathbf{D}^k(\mathbf{t})\right\| < \zeta,\ \forall t \in \mathfrak{T}_0\right\},\ or$$
$$\mathfrak{D}^k = \left\{\mathbf{d}(.) : \mathbf{d}^{(k)}(\mathbf{t}) \in \mathfrak{C},\ \exists\xi \in \mathfrak{R}^+ \Longrightarrow \left\|\mathbf{d}^k(\mathbf{t})\right\| < \xi,\ \forall t \in \mathfrak{T}_0\right\}$$

$\mathfrak{D}^0 = \mathfrak{D}$ the *family of all bounded continuous permitted disturbance vector total functions* $\mathbf{D}(.)$ *or deviation functions* $\mathbf{d}(.)$, $\mathfrak{D} \subset \mathfrak{C}$, the Laplace transforms of which are strictly proper real rational complex functions,

$\mathfrak{I}^k$ a given, or to be determined, family of all bounded and k-times continuously differentiable permitted input vector functions $\mathbf{I}(.)$

$$\mathfrak{I}^k \subset \mathfrak{C}^k \cap \mathfrak{L}$$

$\mathfrak{I}^0 = \mathfrak{I}$ the family of all bounded continuous permitted input vector functions $\mathbf{I}(.)$

$$\mathfrak{I} \subset \mathfrak{C} \cap \mathfrak{L}$$

$\mathfrak{L}$ *the family of all strictly proper real rational complex functions, the original of which are bounded time-dependent functions*

$$\mathfrak{L} = \left\{ \mathbf{I}(.) : \left(\begin{array}{c} \exists\gamma(\mathbf{I}) \in \mathfrak{R}^+ \Longrightarrow \|\mathbf{I}(t)\| < \gamma(\mathbf{I}),\ \forall t \in \mathfrak{T}_0, \\ \mathcal{L}^{\mp}\{\mathbf{I}(t)\} = \mathbf{I}^{\mp}(s) = \left[I_1^{\mp}(s)\ \ I_2^{\mp}(s)\ \ \ldots\ I_M^{\mp}(s)\right]^T \Longrightarrow \\ \mathbf{I}^{\mp}(s) = \underbrace{blockdiag\left\{I_1^{\mp}(s)\ \ I_2^{\mp}(s)\ \ \ldots\ I_M^{\mp}(s)\right\}}_{I(s)}\mathbf{1}_M \Longrightarrow \\ \mathbf{1}_M = \left[1 \vdots 1 \vdots \ldots 1\right]^T \in \mathfrak{R}^M \Longrightarrow \mathbf{I}^{\mp}(s) = I(s)\mathbf{1}_M \\ I_k^{\mp}(s) = \dfrac{\sum_{j=0}^{j=\zeta_k} a_{kj}s^j}{\sum_{j=0}^{j=\psi_k} b_{kj}s^j}, 0 \leq \zeta_k < \psi_k,\ \forall k = 1, 2, ..., M, \end{array} \right) \right\}$$

$\mathfrak{R}$ *the set of all real numbers*

$\mathfrak{R}^+$ *the set of all positive real numbers*

$\mathfrak{R}_+$ *the set of all nonnegative real numbers*

$\mathfrak{R}^{\nu N}$ *the extended output space of the IO system*, which is simultaneously its *state space*

$\mathfrak{R}^n$ an *n-dimensional real vector space, the state space of the ISO system*

$\mathfrak{R}^{N\nu}\backslash\mathfrak{B}_\varepsilon$ *the set of all vectors* $\mathbf{y}^{\nu-1}$ *in* $\mathfrak{R}^{N\nu}$ *out of* $\mathfrak{B}_\varepsilon$

$$\mathfrak{R}^{N\nu}\backslash\mathfrak{B}_\varepsilon = \left\{\mathbf{y}^{\nu-1} : \mathbf{y}^{\nu-1} \in \mathfrak{R}^{N\nu}, \ \mathbf{y}^{\nu-1} \notin \mathfrak{B}_\varepsilon\right\}$$

.

$\mathfrak{T}$ *the accepted reference time set*, the arbitrary element of which is an arbitrary moment t and the *time* unit of which is second s, $1_t = s$, $t\langle s\rangle$

$$\mathfrak{T} = \{t : t[T]\langle s\rangle, \ numt \in \mathfrak{R}, \ dt > 0\}, \ \ inf\ \mathfrak{T} = -\infty, \ \ sup\ \mathfrak{T} = \infty.$$

$\mathfrak{T}_{0\mp}$ *the subset of* $\mathfrak{T}$, which has the minimal element $min\mathfrak{T}_{0\mp}$ that is the initial instant $t_{0\mp}$, $numt_{0\mp} = 0^{\mp}$

$$\mathfrak{T}_{0\mp} = \left\{t : t \in \mathfrak{T}, \ t \geq t_{0\mp}, \ numt_{0\mp} = 0^{\mp}\right\}, \mathfrak{T}_{0\mp} \subset \mathfrak{T},$$
$$min\mathfrak{T}_{0\mp} = t_{0\mp} \in \mathfrak{T}, \ \ sup\ \mathfrak{T}_{0\mp} = \infty.$$

$\mathfrak{Y}_d^k \subset \mathfrak{C}^{kN}$ a given, or to be determined, *family of all bounded k-times continuously differentiable realizable desired total output vector functions* $\mathbf{Y}_d(.)$, the Laplace transforms of which are strictly proper real rational complex functions,

$$\mathfrak{Y}_d^k = \left\{\mathbf{Y}_d(.) : \mathbf{Y}_d(t) \in \mathfrak{C}^{kN}, \ \exists\kappa \in \mathfrak{R}^+ \Longrightarrow \left\|\mathbf{Y}_d^k(t)\right\| < \kappa, \ \forall t \in \mathfrak{T}_0\right\}.$$

$\mathfrak{Y}_d = \mathfrak{Y}_d^0$ the *family of all bounded continuous realizable desired total output vector functions* $\mathbf{Y}_d(.)$, $\mathfrak{Y}_d = \mathfrak{Y}_d^0 \subset \mathfrak{C}^{0d}$, the Laplace transforms of which are strictly proper real rational complex functions

A.3.4 Greek letters

Greek letters are used in the mathematical context.

α *a positive integer*

β *a nonnegative integer*

$\gamma = \max\{\beta, \mu\}$

δ_{ij} *the Kronecker delta*, $\delta_{ij} = 1$ for $i = j$, and $\delta_{ij} = 0$ for $i \neq j$

ε *the output error vector* $\varepsilon \in R^N$, (3.65), (in Subsection 3.5.1*)*,

$$\varepsilon = \mathbf{Y}_d - \mathbf{Y} = -\mathbf{y}, \ \varepsilon = [\varepsilon_1 \ \varepsilon_2 \ ... \ \varepsilon_N\ ]^T$$

θ *a nonnegative integer*

$\lambda_m(H)$ *the minimal eigenvalue of the symmetric matrix* $H = H^T$

$\lambda_M(H)$ *the maximal eigenvalue of the symmetric matrix* $H = H^T$

μ *a nonnegative integer*

ν *a nonnegative integer*

τ a subsidiary notation for *time t*

ϕ *the empty set*

ρ *a natural number*

A.3.5 Roman letters

Roman letters are mostly used.

$A \in \mathfrak{R}^{nxn}$ *the matrix describing the internal dynamics of the ISO system*

$A_k \in \mathfrak{R}^{N\times N}$ *the matrix associated with the k-th derivative* $\mathbf{Y}^{(k)}$ of the output vector $\mathbf{Y}$ of the *IO* system

$A^{(\nu)} \in \mathfrak{R}^{N\mathrm{x}(\nu+1)N}$ *the extended matrix describing the IO system internal dynamics,* $A^{(\nu)} = \left[A_0 \vdots\ A_1 \vdots ... \vdots A_\nu\right]$

$B_k \in \mathfrak{R}^{N\times M}$ *the matrix associated with the k-th derivative* $\mathbf{I}^{(k)}$ of the input vector $\mathbf{I}$ of the *IO* system

$B^{(\mu)} \in \mathfrak{R}^{N\mathrm{x}(\mu+1)M}$ *the extended matrix describing the transmission of the influence of the input vector* $\mathbf{I}(t)$ *on the system dynamics,* $B^{(\mu)} = \left[B_0 \vdots\ B_1 \vdots ... \vdots B_\mu\right]$

$C \in \mathfrak{R}^{Nxn}$ *the matrix of the ISO system, which describes the transmission of the state vector action on the system output vector* $\mathbf{Y}$

$\mathbf{C}_0$ is *the vector of all initial conditions* acting on the system,

$$\mathbf{C}_0 = \left[I_0 \vdots I_0^{(1)} \vdots ... \vdots I_0^{(\mu-1)} \vdots X_0 \vdots Y_0 \vdots Y_0^{(1)} \vdots ... \vdots Y_0^{(\nu-1)}\right]^T \in \mathfrak{R}^{\mu M+n+\nu N}$$

d *a natural number*

$\mathbf{d} \in \mathfrak{R}^d$ *the disturbance deviation vector,* (3.69), (in Subsection 3.5.1)

$$\mathbf{d} = \mathbf{D} - \mathbf{D}_N$$

$\mathbf{D} \in \mathfrak{R}^d$ *the total disturbance vector*

$\mathbf{D}_N \in \mathfrak{R}^d$ *the nominal disturbance vector*

$D \in R^{Nxd}$ *the ISO system matrix describing the transmission of the influence of* $\mathbf{I}(t)$ *on the system output*

$F(.) : \mathfrak{T}_0 \longrightarrow \mathfrak{R}^{N\times N}$ *a matrix function associated with* $\mathbf{f}(.)$

$$\mathbf{f} = [f_1\ \ f_2\ \ldots\ f_N]^T \Longrightarrow F = diag\{f_1\ \ f_2\ \ldots\ f_N\},$$

$F(s)$ *the full (complete) transfer function matrix of a time-invariant continuous-time linear dynamical system*

$F_{IIO}(s) \in \mathfrak{C}^{N\times[(\gamma+1)M+\delta\rho+\nu N]}$ *the full transfer function matrix of the IIO system (Definition 140)*

$F_{IIOII}(s) \in \mathfrak{C}^{\rho\times[(\beta+1)M+\alpha\rho]}$ *the full (complete) II transfer function matrix of the IIO system (Definition 144)*

$F_{IO}(s) \in \mathfrak{C}^{N\times[(\gamma+1)M+\alpha\rho]}$ *the full transfer function matrix of the IO system (Definition 7.2)*

$F_{IOISO}(s)$ *the full transfer function matrix obtained from the IO mathematical model of the given ISO system* (Section 8.2, Note 164)

$F_{ISO}(s) \in \mathfrak{C}^{N\times(M+n)}$ *the full ISO transfer function matrix of the ISO system (Definition 131)*

$F_{ISOIS}(s) \in \mathfrak{C}^{n\times(M+n)}$ *the full (complete) IS transfer function matrix of the ISO system (Definition 138)*

$G = G^T \in R^{p\times p}$ *the symmetric matrix of the quadratic form* $v(\mathbf{w}) = \mathbf{w}^T G\mathbf{w}$

$G(s)$ *the transfer function matrix of a time-invariant continuous-time linear dynamical system*

$G_{IIO}(s) \in \mathfrak{C}^{N\times M}$ *e transfer function matrix of the IIO system (Definition 140)*

$G_{IIOII}(s) \in \mathfrak{C}^{\rho\times M}$ *the II transfer function matrix of the IIO system (Definition 144)*

$G_{IIOi_0}(s) \in \mathfrak{C}^{N\times(\gamma+1)M}$ *the transfer function matrix relative to* $\mathbf{i}_{0(\mp)}^{\gamma-1}$ *of the IIO system (Definition 140)*

$G_{IIOi_0II}(s) \in \mathfrak{C}^{\rho\times(\beta+1)M}$ *the II transfer function matrix relative to* $\mathbf{i}_{0(\mp)}^{\beta-1}$ *of the IIO system (Definition 144)*

$G_{IIOr_0}(s) \in \mathfrak{C}^{N\times\delta\rho}$ *the transfer function matrix relative to* $\mathbf{r}_{0(\mp)}^{\delta-1}$ *of the IIO system (Definition 140)*

$G_{IIOr_0II}(s) \in \mathfrak{C}^{\rho\times\alpha\rho}$ *the II transfer function matrix relative to* $\mathbf{r}_{0(\mp)}^{\alpha-1}$ *of the IIO system (Definition 144)*

$G_{IIOy_0}(s) \in \mathfrak{C}^{N\times\nu N}$ *the transfer function matrix relative to* $\mathbf{y}_{0(\mp)}^{\nu-1}$ *of the IIO system (Definition 140)*

$G_{IO}(s) \in \mathfrak{C}^{N\times M}$ *the transfer function matrix of the IO system (Definition 7.2)*

$G_{IO_0}(s) \in \mathfrak{C}^{N\times(\mu M+\nu N)}$ *the transfer function matrix relative to all initial conditions of the IO system (Definition 7.2)*

$G_{IOi_0}(s) \in \mathfrak{C}^{N\times\mu M}$ *the transfer function matrix relative to* $i_{0(\mp)}^{\mu-1}$ *of the IO system (Definition 7.2)*

$G_{IOy_0}(s) \in \mathfrak{C}^{N\text{x}\nu N}$ *the transfer function matrix relative to* $y_{0(\mp)}^{\nu-1}$ *of the IO system (Definition 7.2)*

$G_{IOISO}(s) \in \mathfrak{C}^{N\text{x}M}$ *the transfer function obtained from the IO mathematical model of the given ISO system* (Section 8.2, Note 164),

$G_{ISO}(s) \in \mathfrak{C}^{N\text{x}M}$ *the ISO transfer function matrix of the ISO system (Definition 131)*

$G_{ISOIS}(s) \in \mathfrak{C}^{n\text{x}M}$ *the IS transfer function matrix of the ISO system (Definition 138)*

$G_{ISOSS}(s) \in \mathfrak{C}^{n\text{x}n}$ *the SS transfer function matrix of the ISO system (Definition 138)*

$G_{ISOx_0}(s) \in \mathfrak{C}^{N\text{x}n}$ *the ISO transfer function matrix relative to* $\mathbf{x}_{0\mp}$ *of the ISO system (Definition 131)*

$h(.)$ *the Heaviside function*, i.e., *the unite step function,*

$$h(.) : \mathfrak{T} \rightarrow [0,1],\ h(t) \left\{ \begin{array}{c} = 0 \ for\ t < 0, \\ \in [0,1] \ for\ t = 0, \\ = 1 \ for\ t > 0, \end{array} \right\}$$

Fig. B.4 (Appendix B.2)

$H = H^T \in R^{p\text{x}p}$ *the symmetric matrix of the quadratic form* $v(\mathbf{w}) = \mathbf{w}^T H \mathbf{w}$

i *an arbitrary natural number*, or *the input deviation variable*

$\mathbf{i} \in \mathfrak{R}^M$ *the input deviation vector*, $\mathbf{i} = [i_1 \ \ i_2 \ ... \ i_M]^T$, (3.70), (in Subsection 3.5.1*)*,

$$\mathbf{i} = \mathbf{I} - \mathbf{I}_N.$$

$\mathbf{i}^{\mu}(t) \in \mathfrak{R}^{(\mu+1)M}$ *the extended input vector at a moment t,*

$$\mathbf{i}^{\mu}(t) = \left[\mathbf{i}^T(t) \vdots \left(\mathbf{i}^{(1)}(t)\right)^T \vdots ... \vdots \left(\mathbf{i}^{(\mu)}(t)\right)^T \right]^T.$$

$\mathbf{i}_{0\mp}^{\mu-1} \in \mathfrak{R}^{\mu M}$ *the initial extended input vector at the initial moment* $t_0 = 0$,

$$\mathbf{i}_{0\mp}^{\mu-1} = \mathbf{i}^{\mu-1}(0^{\mp}) = \left[\mathbf{i}_{0(\mp)}^T \vdots \left(\mathbf{i}_{0(\mp)}^{(1)}\right)^T \vdots ... \vdots \left(\mathbf{i}_{0(\mp)}^{(\mu-1)}\right)^T \right]^T \in \mathfrak{R}^{\mu M}.$$

I *the identity matrix of the n-th order,* $I = diag\{1 \ \ 1 \ ... \ 1\} \in \mathfrak{R}^{n\text{x}n}$, or *the total input variable*

I_k *the identity matrix of the k-th order,* $I_k = diag\{1 \ \ 1 \ ... \ 1\} \in \mathfrak{R}^{k\text{x}k}$, $I_n = I$

$\mathbf{I} \in \mathfrak{R}^M$ *the total input vector,* $\mathbf{I} = [I_1 \quad I_2 \ ... \ I_M]^T$

$\mathbf{I}_N \in \mathfrak{R}^M$ *the nominal total input vector,* $\mathbf{I}_N = [I_{N1} \quad I_{N2} \ ... \ I_{NM}]^T$

$Int\mathfrak{S}$ *the interior* of the set $\mathfrak{S}$

$Int\mathfrak{T}_0$ *the interior* of the set $\mathfrak{T}_0$, $Int\mathfrak{T}_0 = \{t: \ t \in \mathfrak{T}_0, \ t > 0\}$

$\operatorname{Im} s$ the imaginary part of $s = \sigma + j\omega$, $\operatorname{Im} s = j\omega$

j *an arbitrary natural number,* or $j = \sqrt{-1}$ is the imaginary unit

k *an arbitrary natural number*

m *a nonnegative integer*

n *a natural number*

N *a natural number,* if N is the dimension of the output vector and if n is the dimension of the state vector then $N \leq n$

O *is the origin of* $\mathfrak{R}^n$; or *the zero matrix of the appropriate order*

p *a natural number*

$P \in R^{n \times N}$ *a matrix*

$P_k \in \mathfrak{R}^{\rho \times M}$ *a matrix*

$P^{(\beta)} \in \mathfrak{R}^{\rho \times M(\beta+1)}$ *an extended matrix describing the transmission of the influence of* $\mathbf{i}^\beta(t)$ *on the internal dynamics of the IIO system,* $P^{(\beta)} = [P_0 \vdots P_1 \vdots ... \vdots P_\beta]$

q *a natural number*

$Q \in R^{N \times N}$ *a matrix*

$Q_k \in \mathfrak{R}^{\rho \times \rho}$ *a matrix*

$Q^{(\alpha)} \in \mathfrak{R}^{\rho \times \rho(\alpha+1)}$ *the extended matrix describing the internal dynamics of the IIO system,* $Q^{(\alpha)} = \left[Q_0 \vdots Q_1 \vdots ... \ Q_\alpha\right]$

$\mathbf{r} \in \mathfrak{R}^\rho$ *a subsidiary deviation vector,* which is *the substate deviation vector of the IIO system,* (3.71), (in Subsection 3.5.1*),*

$$\mathbf{r} = \mathbf{R} - \mathbf{R}_N.$$

$\mathbf{R} \in \mathfrak{R}^\rho$ *a subsidiary total vector,* which is *the total substate vector of the IIO system*

$\mathbf{R}_N \in \mathfrak{R}^\rho$ *a subsidiary nominal vector,* which is *the nominal substate vector of the IIO system*

$\mathfrak{R}_+$ *the set of all nonnegative real numbers*

$\mathfrak{R}^+$ *the set of all positive real numbers*

$R_k \in \mathfrak{R}^{N \times \rho}$ *a matrix*

$R^{(\theta)} \in \mathfrak{R}^{N \times \rho(\theta+1)}$ *the extended matrix describing the action of the extended internal dynamics vector* $\mathbf{r}^\theta$ *on the output dynamics of the IIO system,*

$$R^{(\theta)} = \left[R_0 \vdots R_1 \vdots ... \vdots R_\theta\right].$$

$\mathrm{Re}\, s$ the real part of $s = \sigma + j\omega$, $\mathrm{Re}\, s = \sigma$

s *the basic time unit*: *second*, or *a complex variable or a complex number* $s = \sigma + j\omega$

$sign(.) : \mathfrak{R} \to \{-1, 0, 1\}$ *the scalar sign function*

$$sign(x) = |x|^{-1}\, x \text{ if } x \neq 0, \text{ and } sign(0) = 0$$

$S_i^{(k)}(.) : \mathfrak{C} \longrightarrow \mathfrak{C}^{\,i(k+1)\text{x}i}$ *the matrix function of s defined by (8.4) in the in Subsection 8.1:*

$$S_i^{(k)}(s) = \left[s^0 I_i \vdots s^1 I_i \vdots s^2 I_i \vdots ... \vdots s^k I_i \right]^T \in \mathfrak{C}^{\,i(k+1)\text{x}i},$$
$$(k, i) \in \{(\mu, M),\ (\nu, N)\}.$$

t *time* (*temporal variable*), or *an arbitrary time value* (*an arbitrary moment, an arbitrary instant*); and formally mathematically t denotes for short also the numerical *time* value *numt* if it does not create a confusion,

$$t[\mathrm{T}]\, \langle s \rangle,\ numt \in \mathfrak{R},\ dt > 0\ ,\ or\ equivalently\text{: } t \in \mathfrak{T}.$$

It has been the common attitude to use the notation t of *time* and of its arbitrary temporal value also for its numerical value *numt*, e.g., $t = 0$ is used in the sense $numt = 0$. We do the same throughout the book if there is not any confusion because we can replace t everywhere by $t1_t^{-1}$, $(t1_t^{-1}) \in \mathfrak{R}$, that we denote again by t, $numt = num\left(t1_t^{-1}\right)$

t_0 a conventionally accepted *initial value of time* (*initial instant, initial moment*), $t_0 \in \mathfrak{T}$, $numt_0 = 0$, i.e., simply $t_0 = 0$ in the sense $numt_0 = 0$

$t_{\inf}$ *the first instant*, which has not happened, $t_{\inf} = -\infty$

$t_{\sup}$ *the last instant*, which will not occur, $t_{\sup} = \infty$

$t_{ZeroTotal}$ *the total zero value of time*, which has not existed and will not happen,

t_{zero} *a conventionally accepted relative zero value of time*

T *the temporal dimension*, "the *time* dimension", which is the physical dimension of *time*

$T \in \mathfrak{R}^+$ *the period of a periodic behavior*

$T_k \in \mathfrak{R}^{N\text{x}M}$ *a matrix*

$T^{(\mu)} \in \mathfrak{R}^{N\text{x}M(\mu+1)}$ *the extended matrix describing the action of the extended input vector* $\mathbf{i}^{\mu}$ *on the output dynamics of the IIO system,*

$$T^{(\mu)} = \left[T_0 \vdots T_1 \vdots ... \vdots T_\mu \right].$$

$v(.) : \mathfrak{R}^p \to \mathfrak{R}$ *a quadratic form,* $v(\mathbf{w}) = \mathbf{w}^T \mathbf{W} \mathbf{w}$

$\mathbf{V}(s)$ the *Laplace transform of all actions on the system*; it is composed of the Laplace transform $\mathbf{I}(s)$ of the input vector $\mathbf{I}(t)$ and of all (input and output) initial conditions,

$$\mathbf{V}(s) = \begin{bmatrix} \mathbf{I}(s) \\ \mathbf{C}_0 \end{bmatrix}.$$

$\mathbf{w} \in \mathfrak{R}^p$ *a subsidiary real-valued vector,*

$$\mathbf{w} = [w_1 \ \ w_2 \ \ ... \ wp]^T \in \left\{ \left[\left(\mathbf{r}^{\alpha-1}\right)^T \vdots \left(\mathbf{y}^{\nu-1}\right)^T \right]^T, \ \mathbf{x}, \ \mathbf{y}^{\nu-1} \right\},$$
$$p \in \{\rho, \ n, \ N\}.$$

$W = W^T \in R^{p \times p}$ *the symmetric matrix of the quadratic form* $v(\mathbf{w})$, $v(\mathbf{w}) = \mathbf{w}^T W \mathbf{w}$, $W \in \{G = G^T, \ H = H^T\}$

$x \in \mathfrak{R}$ *a real valued scalar state deviation variable*

$\mathbf{x} \in \mathfrak{R}^n$ *the state vector deviation of the ISO system,* (3.72), (in Subsection 3.5.1*),*

$$\mathbf{x} = [x_1 \ \ x_2 \ \ ... \ x_n]^T, \ \mathbf{x} = \mathbf{X} - \mathbf{X}_N = \mathbf{X} - \mathbf{X}_d.$$

$\mathbf{X} \in \mathfrak{R}^n$ *the total state vector of the ISO system,*

$$\mathbf{X} = [X_1 \ X_2 \ .. \ X_n]^T.$$

$\mathbf{X}_N \in \mathfrak{R}^n$ *the total nominal state vector of the ISO system,*

$$\mathbf{X}_N = [X_{N1} \ \ X_{N2} \ \ ... \ X_{Nn}]^T.$$

$y \in \mathfrak{R}$ *a real valued scalar output deviation variable*

$\mathbf{y} \in \mathfrak{R}^N$ *a real valued vector output deviation variable-the output deviation vector of the system,*

$$\mathbf{y} = [y_1 \ \ y_2 \ \ ... \ y_N]^T,$$

(3.64), (in Subsection 3.5.1*),*

$$\mathbf{y} = \mathbf{Y} - \mathbf{Y}_d = -\varepsilon.$$

$\mathbf{Y} \in \mathfrak{R}^N$ *a real total valued vector output-the total output vector of the system,*

$$\mathbf{Y} = [Y_1 \ \ Y_2 \ \ ... \ Y_N]^T.$$

$\mathbf{Y}_d \in \mathfrak{R}^N$ *a desired (a nominal) total valued vector output-the desired total output vector of the system,*

$$\mathbf{Y}_d = [Y_{d1} \ \ Y_{d2} \ \ ... \ Y_{dN}]^T ,$$

$\mathbf{y}_{0\mp}^{\nu-1} \in \mathfrak{R}^{\nu N}$ *the initial extended output vector at the initial moment* $t_0 = 0$,

$$\mathbf{y}_{0\mp}^{\nu-1} = \mathbf{y}^{\nu-1}(0^{\mp}) = \left[\mathbf{y}_{0(\mp)}^T \vdots \left(\mathbf{y}_{0(\mp)}^{(1)} \right)^T \vdots ... \vdots \left(\mathbf{y}_{0(\mp)}^{(\nu-1)} \right)^T \right]^T ,$$
$$\mathbf{y}_{0\mp}^0 = \mathbf{y}^0(0^{\mp}) = \mathbf{y}_{0\mp} = \mathbf{y}(0^{\mp}).$$

$Z_k^{(\varsigma-1)}(.) : \mathfrak{C} \rightarrow \mathfrak{C}^{(\varsigma+1)k\times\varsigma k}$ *the matrix function of* s *defined by (8.5) in Section 8.1:*

$$Z_k^{(\varsigma-1)}(s) = \begin{bmatrix} O_k & O_k & O_k & ... & O_k \\ s^0 I_k & O_k & O_k & ... & O_k \\ ... & ... & ... & ... & ... \\ s^{\varsigma-1} I_k & s^{\varsigma-2} I_k & s^{\varsigma-3} I_k & ... & s^0 I_k \end{bmatrix}, \ \varsigma \geq 1,$$
$$Z_k^{(\varsigma-1)}(s) \in \mathfrak{C}^{(\varsigma+1)k\times\varsigma k}, \ (\varsigma, k) \in \{(\mu, M), \ (\nu, N)\}.$$

See Note 55 (in Subsection 3.4.2) on $Z_k^{(\zeta-1)}(.)$ for $\zeta \leq 0$.

A.4 Names

This is the list of the characteristic names.

General Input-State-Output Systems (*GISO systems*) are described in (2.39) (Section 2.3)

Input-Internal and Output dynamical systems (IIO systems) are described in (2.22) (Section 2.3)

Input-Output (*IO*) *systems are described in* (2.1) *(Section 2.1)*

Input-State-Output (*ISO*) *systems* described by *the state space equation* (2.17) and by *the output equation* (2.18) *(Section 2.2,*

PMD systems are described by (2.36) (Section 2.3)

Rosenbrock systems (*RS*) are described in (2.37) (Section 2.3)

Stable (stability) matrix: a square matrix is *stable (stability) matrix* if and only if the real parts of all its eigenvalues are negative

A.5 Symbols and vectors

Some of the symbols are used in various senses in litterature.

$(.)$ *an arbitrary variable*, or *an index*

$|(.)|:\mathfrak{R}\rightarrow\mathfrak{R}_+$ *the absolute value (module) of a (complex valued) scalar variable* $(.)$

$||.||:\mathfrak{R}^n\rightarrow\mathfrak{R}_+$ *an accepted norm on* $\mathfrak{R}^n$, which is the *Euclidean norm on* $\mathfrak{R}^n$ iff not stated otherwise:

$$||\mathbf{x}|| = ||\mathbf{x}||_2 = \sqrt{\mathbf{x}^T\mathbf{x}} = \sqrt{\sum_{i=1}^{i=n} x_i^2}.$$

$\langle 1_{..}\rangle$ shows *the units 1... of a physical variable*

$[\,\alpha,\,\beta\,]\subset\mathfrak{R}$ *a compact interval,* $[\alpha,\beta]=\{x:\ x\in\mathfrak{R},\ \alpha\leq x\leq\beta\}$

$[\,\alpha,\,\beta\,[\,\subseteq\mathfrak{R}$ *a left closed, right open interval,* $[\alpha,\beta[=\{x: x\in\mathfrak{R},\ \alpha\leq x<\beta\}$

$]\,\alpha,\,\beta\,]\subseteq\mathfrak{R}$ *a left open, right closed interval,* $[\alpha,\beta[=\{x: x\in\mathfrak{R},\ \alpha< x\leq\beta\}$

$]\,\alpha,\,\beta\,[\,\subseteq\mathfrak{R}$ *an open interval,* $]\alpha,\beta[=\{x:\ x\in\mathfrak{R},\ \alpha<x<\beta\}$,

$$(\sigma,\infty[\in\{]\sigma,\infty[,\ [\sigma,\infty[\}\,.$$

$(\,\alpha,\,\beta\,)\subseteq\mathfrak{R}$ *a general interval,*

$$(\alpha,\beta)\in\{[\alpha,\beta],\ \ [\alpha,\beta[,\ \ ]\alpha,\beta],\ \ ]\alpha,\beta[\}\,.$$

$\mathfrak{A}\backslash\mathfrak{B}$ is the set difference between the set $\mathfrak{A}$ and the set $\mathfrak{B}$,

$$\mathfrak{A}\backslash\mathfrak{B}=\{x:\ x\in\mathfrak{A},\ x\notin\mathfrak{B}\}\,.$$

$\lambda_i(A)$ *the eigenvalue* $\lambda_i(A)$ *of the matrix* A

$[\,\mathrm{A..}\,]$ shows *the physical dimension* A... *of a physical variable*

$\left[A_1 \vdots A_2 \vdots ... \vdots A_\nu\right]$ *a structured matrix* composed of the submatrices A_1, A_2, ..., A_ν

$\mathbf{0}_k=[0\ \ 0\ ...0]^T\in\mathfrak{R}^k$ *the elementwise zero vector,* $\mathbf{0}_n=\mathbf{0}$

$\mathbf{1}_k=[1\ 1...1]^T\in\mathfrak{R}^k$ *the elementwise unity vector,* $\mathbf{1}_n=\mathbf{1}$

$\forall$ *for every*

$adjA$ *the adjoint matrix of the nonsingular square matrix* A

$$detA\neq 0\Longrightarrow\ A\,adjA=(detA)\,I.$$

detA *the determinant of the matrix A*, $detA = |A|$
A^{-1} *the inverse matrix of the nonsingular square matrix A*,

$$detA \neq 0 \Longrightarrow A^{-1} = adjA/detA.$$

$d(\mathbf{v}, \mathfrak{S})$ *the scalar distance of a vector* $\mathbf{v}$ *from a set* $\mathfrak{S}$,

$$d(\mathbf{v}, \mathfrak{S}) = \inf[\|\mathbf{v} - \mathbf{w}\| : \mathbf{w} \in \mathfrak{S}].$$

dim $\mathbf{z}$ *the mathematical dimension of a vector* $\mathbf{z}$, $\mathbf{z} \in \mathfrak{R}^n \Longrightarrow dim$ $\mathbf{z} = n$

ddim *the dynamical dimension of a system* composed of the system order and the system dimension

$DenF(s)$ *the denominator matrix polynomial of the real rational matrix* $F(s)$,

$$F(s) = [DenF(s)]^{-1} NumF(s), \text{ or } F(s) = NumF(s)\,[DenF(s)]^{-1}.$$

mddim *the minimal dynamical dimension of a system*

$deg\left[adj\left(\sum_{k=0}^{k=\nu} A_k s^k\right)\right]$ denotes *the greatest power of s over all elements of* $adj\left(\sum_{k=0}^{k=\nu} A_k s^k\right)$

$deg\left(\sum_{k=0}^{k=\mu} B_k s^k\right)$ denotes *the greatest power of s over all elements of* $\sum_{k=0}^{k=\mu} B_k s^k$

$deg\left[det\left(\sum_{k=0}^{k=\nu} A_k s^k\right)\right]$ denotes *the greatest power of s in*

$$det\left(\sum_{k=0}^{k=\nu} A_k s^k\right).,$$

$gradv\left(\mathbf{y}^{\nu-1}\right)$ the gradient of $v\left(\mathbf{y}^{\nu-1}\right)$

$$gradv\left(\mathbf{y}^{\nu-1}\right) = \left[\frac{\partial v\left(\mathbf{y}^{\nu-1}\right)}{\partial y_1} .. \frac{\partial v\left(\mathbf{y}^{\nu-1}\right)}{\partial y_N} .... \frac{\partial v\left(\mathbf{y}^{\nu-1}\right)}{\partial y_1^{(\nu-1)}} .. \frac{\partial v\left(\mathbf{y}^{\nu-1}\right)}{\partial y_N^{(\nu-1)}}\right]^T.$$

$Im\lambda_i(A)$ *the imaginary part* of the eigenvalue $\lambda_i(A)$ of the matrix A
$\min(\delta, \Delta)$ denotes *the smaller between* δ *and* Δ,

$$\min(\delta, \Delta) = \left\{ \begin{array}{c} \delta, \ \delta \leq \Delta, \\ \Delta, \ \Delta \leq \delta \end{array} \right\}.$$

$NumF(s)$ *the numerator matrix polynomial of the real rational matrix* $F(s)$,

$$F(s) = [DenF(s)]^{-1} NumF(s), \text{ or } NumF(s) [DenF(s)]^{-1}.$$

$Re\lambda_i(A)$ *the real part* of the eigenvalue $\lambda_i(A)$ of the matrix A
$\exists$ *there exist(s)*
$\exists!$ *there exists exactly one*
$\in$ *belong(s) to, are (is) members (a member) of*, respectively
$\subset$ *a proper subset of* (it cannot be equal to)
$\subseteq$ *a subset of* (it can be equal to)
$\simeq$ *equivalent*
$\sqrt{-1}$ *the imaginary unit* denoted by j, $j = \sqrt{-1}$
inf *infimum*
max *maximum*
min *minimum*
numx *the numerical value of* x, if $x = 50V$ then $numx = 50$,
phdim $x(.)$ *the physical dimension of a variable* $x(.)$,

$$x(.) = t \Longrightarrow phdim\ x(.) = phdim\ t = \mathrm{T}, \text{ but } dim\ t = 1.$$

sup *supremum*
$\otimes$ the Kronecker matrix product, (13.72), (in Section 13.3)

A.6 Units

Units are related to physical variables.
$1_{(.)}$ *the unit of a physical variable* (.)
1_t *the time unit of the reference time axis* T, $1_t = s$

Appendix B

Laplace transforms and Dirac impulses

B.1 Laplace transforms

If and only if the following integral exists,

$$\exists \int_{-\infty}^{\infty} x(t)e^{-st}dt, \tag{B.1}$$

then it is the *(two-sided) Laplace transform of the scalar function* $x(.) : \mathfrak{T} \longrightarrow \mathfrak{R}$. We accept that the values of all functions considered in the sequel are equal to zero for all $t < 0$, e.g. $x(t) = 0 \ \forall t < 0^-$, where $0^\mp$ denotes $\mp\varepsilon$, respectively, for an arbitrarily small $\varepsilon \in \mathfrak{R}^+$, in the limit as $\varepsilon \longrightarrow 0$. The set $\mathfrak{R}^+ = \{\alpha : 0 < \alpha < \infty\} =]0, \infty[$ is the set of all positive real numbers. The function $x(.)$ can be discontinuous at $t = 0$ so that $x(0^\mp)$ can be different from zero, $x(0^\mp) \neq 0$. For such functions, the integral (B.1) becomes

$$\int_{0^\mp}^{\infty} x(t)e^{-st}dt = \lim \left[\int_{\mp\zeta}^{\infty} x(t)e^{-st}dt : \zeta \longrightarrow 0^+ \right]. \tag{B.2}$$

If and only if it exists then it is, respectively, *the left* $\mathcal{L}^- \{.\}$ / *right* $\mathcal{L}^+ \{.\}$ *Laplace transform of the function* $x(.)$,

$$\mathcal{L}^\mp \{x(t)\} = X^\mp(s) = \int_{0^\mp}^{\infty} x(t)e^{-st}dt = \lim \left[\int_{\mp\zeta}^{\infty} x(t)e^{-st}dt : \zeta \longrightarrow 0^+ \right]. \tag{B.3}$$

The function $x(.)$ has *the Laplace transform* $\mathcal{L} \{x(t)\} = X(s)$ if and only if it has both the left Laplace transform $\mathcal{L}^- \{x(t)\} = X^-(s)$ and the right

Laplace transform $\mathcal{L}^{+}\{x(t)\} = X^{+}(s)$ and they are equal,

$$\begin{aligned}
&\exists\mathcal{L}\{x(t)\} = X(s) \Longleftrightarrow \exists\mathcal{L}^{\mp}\{x(t)\} = X^{\mp}(s) \text{ and} \\
&\mathcal{L}^{-}\{x(t)\} = X^{-}(s) = \mathcal{L}^{+}\{x(t)\} = X^{+}(s). \\
&\text{Then, } \mathcal{L}\{x(t)\} = X(s) = \int_0^\infty x(t)e^{-st}dt = \\
&= \mathcal{L}^{-}\{x(t)\} = X^{-}(s) = \mathcal{L}^{+}\{x(t)\} = X^{+}(s).
\end{aligned} \tag{B.4}$$

Definition 411 *A function $x(.)$ is of the exponential order if and only if there are real numbers α and K, and a moment $\tau \in \mathfrak{T}$ such that*

$$e^{-\alpha t}|x(t)| < K, \ \forall t \geq \tau. \tag{B.5}$$

The infimum γ of the set of all numbers α that obey (B.5) is the abscissa of the absolute convergence of the function $x(.)$,

$$\gamma = \inf\{\alpha : \exists K \in \mathfrak{R}, \exists\tau \in \mathfrak{T} \Longrightarrow e^{-\alpha t}|x(t)| < K, \forall t \geq \tau\}. \tag{B.6}$$

The following theorem results from Theorem 3.3.1 by Debnath [16, p. 87] and the preceding definition.

Theorem 412 ***Conditions for the existence of the Laplace transform***

If a function $x(.)$ is continuous or piecewise continuous in every finite time interval $]0, \tau[\subset \mathfrak{T}_0$, and of the exponential order, then the Laplace transform $X(s)$ of $x(t)$ exists for all $s \in \mathfrak{C}$ provided $\operatorname{Re} s > \gamma$.

The Inverse Laplace transform $\mathcal{L}^{-1}\{X(s)\}$ of $X(.)$ is defined by

$$\mathcal{L}^{-1}\{X(s)\} = \frac{1}{2\pi j}\int_{c-j\infty}^{c+j\infty} X(s)e^{st}ds = \frac{x(t^-) + x(t^+)}{2}. \tag{B.7}$$

If and only if $x(.)$ is continuous at the moment t, then

$$\mathcal{L}^{-1}\{X(s)\} = \frac{1}{2\pi j}\int_{c-j\infty}^{c+j\infty} X(s)e^{st}ds = x(t). \tag{B.8}$$

The real number c is greater than the abscissa γ of the absolute convergence of the function $x(.)$, $c \in]\gamma, \infty[$.

The left / right Laplace transform of a vector function $\mathbf{x}(.) : \mathfrak{T} \longrightarrow \mathfrak{R}^n$, $\mathbf{x}(.) = [x_1(.) \ \ x_2(.) \ \ ... \ x_n(.)]^T$, is the vector of the left / right Laplace transform of the entries of $\mathbf{x}(.)$, respectively,

$$\mathcal{L}^{\mp}\{\mathbf{x}(t)\} = \mathbf{X}^{\mp}(s) = [X_1^{\mp}(s) \ \ X_2^{\mp}(s) \ ...X_n^{\mp}(s)]^T. \tag{B.9}$$

The following two properties of the Laplace transform are crucial:
° *linearity of the Laplace transform*

$$\mathcal{L}^{\mp}\left\{\sum_{k=1}^{k=p} M_k \mathbf{x}_k(t)\right\} = \sum_{k=1}^{k=p} M_k \mathcal{L}^{\mp}\left\{\mathbf{x}_k(t)\right\} = \sum_{k=1}^{k=p} M_k \mathbf{X}_k^{\mp}(s), \tag{B.10}$$

° *the Laplace transforms of the k-th derivative of a k-times continuously differentiable function* $\mathbf{x}(.)$, which contains the influence of initial conditions, read

$$\mathcal{L}^{\mp}\left\{\mathbf{x}^{(k)}(t)\right\} = s^k \mathcal{L}^{\mp}\left\{\mathbf{x}(t)\right\} - \sum_{i=1}^{i=k} s^{i-1}\mathbf{x}^{(k-i)}\left(0^{\mp}\right) =$$

$$= s^k \mathbf{X}^{\mp}(s) - \sum_{i=1}^{i=k} s^{i-1}\mathbf{x}^{(k-i)}\left(0^{\mp}\right). \tag{B.11}$$

° *the Laplace transforms* $\mathcal{L}^{\mp}\left\{\mathbf{x}_d(t)\right\}$ *of the delayed function* $\mathbf{x}_d(.)$, $\mathbf{x}_d(.) = \mathbf{x}(. - \tau_d)$ expressed in terms of the Laplace transforms $\mathbf{X}^{\mp}(s)$ of the function $\mathbf{x}(.)$ without delay, read

$$\mathcal{L}^{\mp}\left\{\mathbf{x}_d(t)\right\} = \mathcal{L}^{\mp}\left\{\mathbf{x}(t-\tau_d)\right\} = e^{-\tau_d s}\mathbf{X}^{\mp}(s) \tag{B.12}$$

These features of the Laplace transform permit us to transfer time domain studies of properties of linear time-invariant continuous-time linear systems without delay to their complex domain studies.

Appendix B.2 presents the Laplace transforms of Dirac impulses and Heaviside functions.

B.2 Dirac impulses

Let $\delta^{\mp}(.)$ be the left/right unit impulse (unit Dirac impulse, unit Dirac delta distribution) defined symbolically by (B.13), respectively, (for the origin of what follows see [4, p. 73], [114, p. 494]),

$$\delta^{\mp}(t) = \lim[\varphi_{\varepsilon}^{\mp}(t) : \varepsilon = \frac{\zeta}{2} \longrightarrow 0^{+}], \tag{B.13}$$

where

$$\varphi_{\varepsilon}^{-}(t) = \left\{ \begin{array}{c} 0,\ t \in]-\infty, -2\varepsilon] \cup [0, +\infty[, \\ \frac{1}{\varepsilon^2}(t+2\varepsilon),\ t \in [-2\varepsilon, -\varepsilon], \\ -\frac{1}{\varepsilon^2}t,\ t \in [-\varepsilon, 0], \end{array} \right\}, \tag{B.14}$$

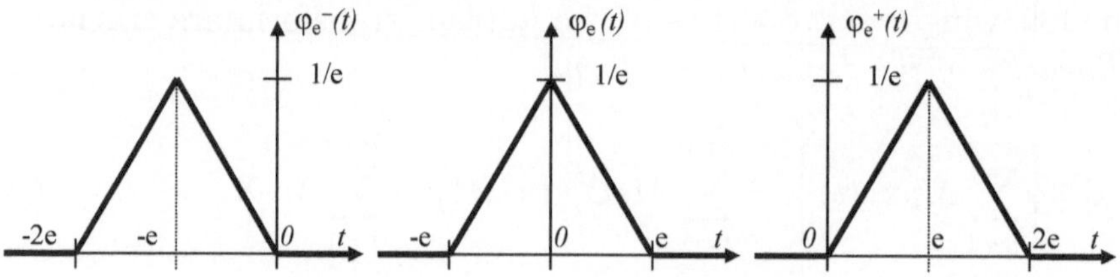

Figure B.1: Triangular functions generating Dirac unit impulses.

$$\varphi_\varepsilon(t) = \left\{ \begin{array}{c} 0,\ t \in]-\infty, -\varepsilon] \cup [\varepsilon, +\infty[, \\ \frac{1}{\varepsilon^2}(t+\varepsilon),\ t \in [-\varepsilon, 0], \\ \frac{1}{\varepsilon^2}\left(-t+\varepsilon\right),\ t \in [0, \varepsilon] \end{array} \right\}, \tag{B.15}$$

$$\varphi_\varepsilon^+(t) = \left\{ \begin{array}{c} 0,\ t \in]-\infty, 0] \cup [2\varepsilon, +\infty[, \\ \frac{1}{\varepsilon^2}t,\ t \in [0, \varepsilon], \\ \frac{1}{\varepsilon^2}\left(-t+2\varepsilon\right),\ t \in [\varepsilon, 2\varepsilon] \end{array} \right\}, \tag{B.16}$$

Fig. B.1 in which e stands for ε, $e = \varepsilon$.

Claim 413 *The Laplace transforms $\mathcal{L}^{\mp}\left\{\delta^{\mp}(t)\right\}$ of $\delta^{\mp}(t)$ are the following:*

$$\mathcal{L}^-\left\{\delta^-(t)\right\} = 1,\ \mathcal{L}\left\{\delta^-(t)\right\} = \mathcal{L}^+\left\{\delta^-(t)\right\} = 0, \tag{B.17}$$

$$\mathcal{L}^-\left\{\delta(t)\right\} = 1,\ \mathcal{L}\left\{\delta(t)\right\} = \frac{1}{2},\ \mathcal{L}^+\left\{\delta(t)\right\} = 0, \tag{B.18}$$

$$\mathcal{L}^-\left\{\delta^+(t)\right\} = \mathcal{L}\left\{\delta^+(t)\right\} = 1,\ \mathcal{L}^+\left\{\delta^+(t)\right\} = 0. \tag{B.19}$$

Proof. We begin with the definitions. Let us determine the Laplace transforms of $\delta^-(.)$, i.e., let us prove (B.17),

$$\mathcal{L}^-\left\{\delta^-(t)\right\} = \lim_{\zeta = 2\varepsilon \longrightarrow 0^+} \left[\int_{-\zeta}^{\infty} \delta^-(t)e^{-st}dt\right] =$$

$$= \lim_{\varepsilon \longrightarrow 0^+} \left[\int_{-2\varepsilon}^{\infty} \left\{ \begin{array}{c} 0,\ t \in]-\infty, -2\varepsilon] \cup [0, +\infty[, \\ \frac{1}{\varepsilon^2}(t+2\varepsilon),\ t \in [-2\varepsilon, -\varepsilon], \\ -\frac{1}{\varepsilon^2}t,\ t \in [-\varepsilon, 0], \end{array} \right\} \bullet \atop \bullet e^{-st}dt \right] =$$

$$= \lim_{\varepsilon \longrightarrow 0^+} \left[\int_{-2\varepsilon}^{-\varepsilon} \frac{1}{\varepsilon^2}(t+2\varepsilon)e^{-st}dt + \int_{-\varepsilon}^{0} \left(-\frac{1}{\varepsilon^2}t\right) e^{-st}dt\right],$$

$$\tau = t + 2\varepsilon,\ t = \tau - 2\varepsilon,\ d\tau = dt,$$
$$t = -2\varepsilon \Longrightarrow \tau = 0,\ t = -\varepsilon \Longrightarrow \tau = \varepsilon,$$

$$\Longrightarrow$$

$$\mathcal{L}^{-}\left\{\delta^{-}(t)\right\} =$$

$$= \lim_{\varepsilon\longrightarrow 0+}\frac{1}{\varepsilon^2}\left[\int_0^{\varepsilon}\tau e^{-s(\tau-2\varepsilon)}d\tau - \int_{-\varepsilon}^{0}te^{-st}dt\right] =$$

$$= \lim_{\varepsilon\longrightarrow 0+}\frac{1}{\varepsilon^2}\left(e^{2\varepsilon s}\int_0^{\varepsilon}\tau e^{-s\tau}d\tau - \int_{-\varepsilon}^{0}te^{-st}dt\right) =$$

$$= \lim_{\varepsilon\longrightarrow 0+}\frac{1}{\varepsilon^2}\left\{\begin{array}{c} e^{2\varepsilon s}\left[-\frac{1}{s}\left(\tau e^{-s\tau}\mid_0^{\varepsilon} - \int_0^{\varepsilon}e^{-s\tau}d\tau\right)\right] - \\ -\left[-\frac{1}{s}\left(\tau e^{-s\tau}\mid_{-\varepsilon}^{0} - \int_{-\varepsilon}^{0}e^{-st}dt\right)\right]\end{array}\right\} =$$

$$= \frac{1}{s}\lim_{\varepsilon\longrightarrow 0+}\frac{1}{\varepsilon^2}\left\{\begin{array}{c} -e^{2\varepsilon s}\left[\varepsilon e^{-\varepsilon s} - 0 - \left(-\frac{1}{s}e^{-s\tau}\mid_0^{\varepsilon}\right)\right] + \\ +\left[0 + \varepsilon e^{\varepsilon s} - \left(-\frac{1}{s}e^{-st}\mid_{-\varepsilon}^{0}\right)\right]\end{array}\right\} =$$

$$= \frac{1}{s}\lim_{\varepsilon\longrightarrow 0+}\frac{1}{\varepsilon^2}\left\{\begin{array}{c} -e^{2\varepsilon s}\left[\varepsilon e^{-\varepsilon s} + \left(\frac{1}{s}e^{-\varepsilon s} - \frac{1}{s}\right)\right] + \\ +\left[\varepsilon e^{\varepsilon s} + \left(\frac{1}{s} - \frac{1}{s}e^{\varepsilon s}\right)\right]\end{array}\right\} =$$

$$= \frac{1}{s}\lim_{\varepsilon\longrightarrow 0+}\frac{1}{\varepsilon^2}\left\{\begin{array}{c} -e^{2\varepsilon s}\left[\varepsilon e^{-\varepsilon s} + \frac{1}{s}\left(e^{-\varepsilon s} - 1\right)\right] + \\ +\left[\varepsilon e^{\varepsilon s} + \frac{1}{s}\left(1 - e^{\varepsilon s}\right)\right]\end{array}\right\} =$$

$$= \frac{1}{s}\lim_{\varepsilon\longrightarrow 0+}\frac{1}{\varepsilon^2}\left\{\begin{array}{c} -\varepsilon e^{\varepsilon s} - \frac{1}{s}e^{2\varepsilon s}\left(e^{-\varepsilon s} - 1\right) + \\ +\varepsilon e^{\varepsilon s} + \frac{1}{s}\left(1 - e^{\varepsilon s}\right)\end{array}\right\} =$$

$$= \frac{1}{s}\lim_{\varepsilon\longrightarrow 0+}\frac{1}{\varepsilon^2}\left\{\frac{1}{s}e^{2\varepsilon s}\left(-e^{-\varepsilon s} + 1\right) + \frac{1}{s}\left(1 - e^{\varepsilon s}\right)\right\} =$$

$$= \frac{1}{s^2}\lim_{\varepsilon\longrightarrow 0+}\frac{1}{\varepsilon^2}\left[e^{2\varepsilon s}\left(1 - e^{-\varepsilon s}\right) + \left(1 - e^{\varepsilon s}\right)\right] =$$

$$= \frac{1}{s^2}\lim_{\varepsilon\longrightarrow 0+}\frac{1}{\varepsilon^2}\left(e^{2\varepsilon s} - e^{\varepsilon s} + 1 - e^{\varepsilon s}\right) =$$

$$= \frac{1}{s^2}\lim_{\varepsilon\longrightarrow 0+}\frac{1}{\varepsilon^2}\left(e^{2\varepsilon s} + 1 - 2e^{\varepsilon s}\right) =$$

$$= \frac{1}{s^2}\lim_{\varepsilon\longrightarrow 0+}\frac{1}{2\varepsilon}\left(2se^{2\varepsilon s} - 2se^{\varepsilon s}\right) =$$

$$= \frac{1}{s}\lim_{\varepsilon\longrightarrow 0+}\frac{1}{\varepsilon}\left(e^{2\varepsilon s} - e^{\varepsilon s}\right) =$$

$$= \frac{1}{s}\lim_{\varepsilon\longrightarrow 0+}\frac{1}{1}\left(2se^{2\varepsilon s} - se^{\varepsilon s}\right) =$$

$$= \lim_{\varepsilon\longrightarrow 0^+}\left(2e^{2\varepsilon s} - e^{\varepsilon s}\right) = 2 - 1 = 1.$$

Or, equivalently,

$$\mathcal{L}^{-}\left\{\delta^{-}(t)\right\} = \lim_{\zeta=2\varepsilon\longrightarrow 0^{+}}\left[\int_{-\zeta}^{\infty}\delta^{-}(t)e^{-st}dt\right] =$$

$$= \lim_{\varepsilon\longrightarrow 0^{+}}\left[\begin{array}{c}\int_{-2\varepsilon}^{\infty}\left\{\begin{array}{c}0,\ t\in]-\infty,-2\varepsilon]\cup[0,+\infty[,\\ \frac{1}{\varepsilon^{2}}(t+2\varepsilon),\ t\in[-2\varepsilon,-\varepsilon],\\ -\frac{1}{\varepsilon^{2}}t,\ t\in[-\varepsilon,0],\end{array}\right\}\bullet\\ \bullet e^{-st}dt\end{array}\right] =$$

$$= \lim_{\varepsilon\longrightarrow 0^{+}}\left[\int_{-2\varepsilon}^{-\varepsilon}\frac{1}{\varepsilon^{2}}(t+2\varepsilon)e^{-st}dt + \int_{-\varepsilon}^{0}\left(-\frac{1}{\varepsilon^{2}}t\right)e^{-st}dt\right] =$$

$$= \lim_{\varepsilon\longrightarrow 0^{+}}\frac{1}{\varepsilon^{2}}\left[\int_{-2\varepsilon}^{-\varepsilon}te^{-st}dt + \int_{-2\varepsilon}^{-\varepsilon}2\varepsilon e^{-st}dt - \int_{-\varepsilon}^{0}te^{-st}dt\right] =$$

$$= \lim_{\varepsilon\longrightarrow 0^{+}}\frac{1}{\varepsilon^{2}}\left[\begin{array}{c}\int_{-2\varepsilon}^{-\varepsilon}te^{-st}dt+\\ +\int_{-2\varepsilon}^{-\varepsilon}2\varepsilon e^{-st}dt-\\ -\int_{-\varepsilon}^{0}te^{-st}dt\end{array}\right] =$$

$$= \lim_{\varepsilon\longrightarrow 0^{+}}\frac{1}{\varepsilon^{2}}\left[\begin{array}{c}-\frac{1}{s}\left[te^{-st}\mid_{-2\varepsilon}^{-\varepsilon} - \int_{-2\varepsilon}^{-\varepsilon}e^{-st}dt\right]+\\ +2\varepsilon\int_{-2\varepsilon}^{-\varepsilon}e^{-st}dt-\\ -\int_{-\varepsilon}^{0}te^{-st}dt\end{array}\right] =$$

$$= \lim_{\varepsilon\longrightarrow 0^{+}}\frac{1}{\varepsilon^{2}}\left[\begin{array}{c}\frac{1}{s}\left[-te^{-st}\mid_{-2\varepsilon}^{-\varepsilon} + \int_{-2\varepsilon}^{-\varepsilon}e^{-st}dt\right]+\\ +2\varepsilon\int_{-2\varepsilon}^{-\varepsilon}e^{-st}dt+\\ +\frac{1}{s}\left[te^{-st}\mid_{-\varepsilon}^{0} - \int_{-\varepsilon}^{0}e^{-st}dt\right]\end{array}\right] =$$

$$= \lim_{\varepsilon\longrightarrow 0^{+}}\frac{1}{\varepsilon^{2}}\left[\begin{array}{c}\frac{1}{s}\left[\varepsilon e^{\varepsilon s} - 2\varepsilon e^{2\varepsilon s} + \left(-\frac{1}{s}e^{\varepsilon s} + \frac{1}{s}e^{2\varepsilon s}\right)\right]+\\ +2\varepsilon\left(-\frac{1}{s}e^{-st}\mid_{-2\varepsilon}^{-\varepsilon}\right)+\\ +\frac{1}{s}\left[\varepsilon e^{\varepsilon s} - \left(-\frac{1}{s} + \frac{1}{s}e^{\varepsilon s}\right)\right]\end{array}\right] =$$

$$= \lim_{\varepsilon\longrightarrow 0^{+}}\frac{1}{\varepsilon^{2}}\left[\begin{array}{c}\frac{1}{s}\left[\varepsilon e^{\varepsilon s}\left(1 - 2e^{\varepsilon s}\right) + \frac{1}{s}e^{\varepsilon s}\left(-1 + e^{\varepsilon s}\right)\right]+\\ +2\varepsilon\left(-\frac{1}{s}e^{\varepsilon s} + \frac{1}{s}e^{2\varepsilon s}\right)+\\ +\frac{1}{s}\left[\varepsilon e^{\varepsilon s} - \left(-\frac{1}{s} + \frac{1}{s}e^{\varepsilon s}\right)\right]\end{array}\right] =$$

$$= \lim_{\varepsilon\longrightarrow 0^{+}}\frac{1}{\varepsilon^{2}}\left[\begin{array}{c}\frac{1}{s}\left[\varepsilon e^{\varepsilon s}\left(1 - 2e^{\varepsilon s}\right) - \frac{1}{s}e^{\varepsilon s}\left(1 - e^{\varepsilon s}\right)\right]-\\ -2\varepsilon\frac{1}{s}e^{\varepsilon s}\left(1 - e^{\varepsilon s}\right)+\\ +\frac{1}{s}\left[\varepsilon e^{\varepsilon s} + \frac{1}{s}\left(1 - e^{\varepsilon s}\right)\right]\end{array}\right] =$$

$$= \lim_{\varepsilon\longrightarrow 0^{+}}\frac{1}{\varepsilon^{2}}\frac{1}{s}\left[\begin{array}{c}\left[\varepsilon e^{\varepsilon s}\left(1 - 2e^{\varepsilon s}\right) - \frac{1}{s}e^{\varepsilon s}\left(1 - e^{\varepsilon s}\right)\right]-\\ -2\varepsilon e^{\varepsilon s}\left(1 - e^{\varepsilon s}\right)+\\ +\left[\varepsilon e^{\varepsilon s} + \frac{1}{s}\left(1 - e^{\varepsilon s}\right)\right]\end{array}\right] =$$

$$= \lim_{\varepsilon \longrightarrow 0+} \frac{1}{\varepsilon^2} \frac{1}{s} \begin{bmatrix} \varepsilon e^{\varepsilon s}\left(1 - 2e^{\varepsilon s}\right) - \frac{1}{s} e^{\varepsilon s}\left(1 - e^{\varepsilon s}\right) - \\ -2\varepsilon e^{\varepsilon s}\left(1 - e^{\varepsilon s}\right) + \\ +\varepsilon e^{\varepsilon s} + \frac{1}{s}\left(1 - e^{\varepsilon s}\right) \end{bmatrix} =$$

$$= \lim_{\varepsilon \longrightarrow 0+} \frac{1}{\varepsilon^2} \frac{1}{s} \begin{bmatrix} \varepsilon e^{\varepsilon s}\left(1 - 2e^{\varepsilon s} + 1\right) - \frac{1}{s} e^{\varepsilon s}\left(1 - e^{\varepsilon s}\right) - \\ -2\varepsilon e^{\varepsilon s}\left(1 - e^{\varepsilon s}\right) + \frac{1}{s}\left(1 - e^{\varepsilon s}\right) \end{bmatrix} =$$

$$= \lim_{\varepsilon \longrightarrow 0+} \frac{1}{\varepsilon^2} \frac{1}{s} \begin{bmatrix} 2\varepsilon e^{\varepsilon s}\left(1 - e^{\varepsilon s}\right) - 2\varepsilon e^{\varepsilon s}\left(1 - e^{\varepsilon s}\right) - \\ -\frac{1}{s} e^{\varepsilon s}\left(1 - e^{\varepsilon s}\right) + \frac{1}{s}\left(1 - e^{\varepsilon s}\right) \end{bmatrix} =$$

$$= \lim_{\varepsilon \longrightarrow 0+} \frac{1}{\varepsilon^2} \frac{1}{s} \left[-\frac{1}{s}\left(e^{\varepsilon s} - 1\right)\left(1 - e^{\varepsilon s}\right) \right] =$$

$$= \lim_{\varepsilon \longrightarrow 0+} \frac{1}{\varepsilon^2} \frac{1}{s} \left[\frac{1}{s}\left(1 - e^{\varepsilon s}\right)\left(1 - e^{\varepsilon s}\right) \right] =$$

$$= \lim_{\varepsilon \longrightarrow 0+} \frac{1}{\varepsilon^2} \frac{1}{s^2}\left(1 - e^{\varepsilon s}\right)^2 = \lim_{\varepsilon \longrightarrow 0+} \frac{1}{2\varepsilon} \frac{1}{s^2} 2(-s)\left(1 - e^{\varepsilon s}\right) =$$

$$= \lim_{\varepsilon \longrightarrow 0+} \frac{1}{\varepsilon} \frac{1}{s}(-1)\left(1 - e^{\varepsilon s}\right) = \lim_{\varepsilon \longrightarrow 0+} \frac{1}{1} \frac{1}{s}(-1)(-s e^{\varepsilon s}) =$$

$$= \lim_{\varepsilon \longrightarrow 0+} e^{\varepsilon s} = 1.$$

Further,

$$\mathcal{L}\left\{\delta^{-}(t)\right\} = \int_{0}^{\infty} \delta^{-}(t) e^{-st} dt = \int_{0}^{\infty} 0 e^{-st} dt = 0,$$

$$\mathcal{L}^{+}\left\{\delta^{-}(t)\right\} = \lim_{\zeta \longrightarrow 0+} \left[\int_{\zeta}^{\infty} \delta^{-}(t) e^{-st} dt \right] =$$

$$= \lim_{\zeta \longrightarrow 0+} \int_{\zeta}^{\infty} 0 e^{-st} dt = 0.$$

The preceding results prove (B.17).

We determine now the Laplace transforms of $\delta(.)$, i.e., let us prove (B.18).

$$\mathcal{L}^{-}\left\{\delta(t)\right\} = \lim_{\zeta = 2\varepsilon \longrightarrow 0+} \left[\int_{-\zeta}^{\infty} \delta(t) e^{-st} dt \right] =$$

$$= \lim_{\varepsilon \longrightarrow 0+} \left[\int_{-2\varepsilon}^{\infty} \left\{ \begin{array}{c} 0,\, t \in]-\infty, -\varepsilon] \cup [\varepsilon, +\infty[, \\ \frac{1}{\varepsilon^2}(t + \varepsilon),\, t \in [-\varepsilon, 0], \\ \frac{1}{\varepsilon^2}\left(-t + \varepsilon\right),\, t \in [0, \varepsilon] \end{array} \right\} e^{-st} dt \right] =$$

$$= \lim_{\varepsilon \longrightarrow 0+} \left[\int_{-\varepsilon}^{\varepsilon} \left\{ \begin{array}{c} \frac{1}{\varepsilon^2}(t + \varepsilon),\, t \in [-\varepsilon, 0], \\ \frac{1}{\varepsilon^2}\left(-t + \varepsilon\right),\, t \in [0, \varepsilon] \end{array} \right\} e^{-st} dt \right] =$$

$$= \lim_{\varepsilon \longrightarrow 0+} \frac{1}{\varepsilon^2} \left[\int_{-\varepsilon}^{0} (t+\varepsilon) e^{-st} dt + \int_{0}^{\varepsilon} (-t+\varepsilon) e^{-st} dt \right],$$

$$\tau = t + \varepsilon \Longrightarrow t = \tau - \varepsilon,\ dt = d\tau,$$

$$t = -\varepsilon \Longrightarrow \tau = 0,\ t = 0 \Longrightarrow \tau = \varepsilon,$$

$$\sigma = -t + \varepsilon \Longrightarrow t = -\sigma + \varepsilon,\ dt = -d\sigma,$$

$$t = 0 \Longrightarrow \sigma = \varepsilon,\ t = \varepsilon \Longrightarrow \sigma = 0,$$

$$\Longrightarrow \mathcal{L}^{-} \{\delta(t)\} =$$

$$= \lim_{\varepsilon \longrightarrow 0+} \frac{1}{\varepsilon^2} \left[\int_{0}^{\varepsilon} \tau e^{-s(\tau-\varepsilon)} d\tau + \int_{\varepsilon}^{0} \sigma e^{-s(-\sigma+\varepsilon)} (-d\sigma) \right] =$$

$$= \lim_{\varepsilon \longrightarrow 0+} \frac{1}{\varepsilon^2} \left\{ \begin{array}{c} e^{\varepsilon s} \left[-\frac{1}{s} \left(\tau e^{-s\tau} \mid_0^\varepsilon - \int_0^\varepsilon e^{-s\tau} d\tau \right) \right] - \\ - \int_\varepsilon^0 \sigma e^{-s(-\sigma+\varepsilon)} (d\sigma) \end{array} \right\} =$$

$$= \lim_{\varepsilon \longrightarrow 0+} \frac{1}{\varepsilon^2} \left\{ \begin{array}{c} e^{\varepsilon s} \left[-\frac{1}{s} \left(\tau e^{-s\tau} \mid_0^\varepsilon - \int_0^\varepsilon e^{-s\tau} d\tau \right) \right] + \\ + \int_0^\varepsilon \sigma e^{-s(-\sigma+\varepsilon)} (d\sigma) \end{array} \right\} =$$

$$= \lim_{\varepsilon \longrightarrow 0+} \frac{1}{\varepsilon^2} \left\{ \begin{array}{c} e^{\varepsilon s} \left[-\frac{1}{s} \left(\tau e^{-s\tau} \mid_0^\varepsilon - \int_0^\varepsilon e^{-s\tau} d\tau \right) \right] + \\ + e^{-\varepsilon s} \int_0^\varepsilon \sigma e^{\sigma s} d\sigma \end{array} \right\} =$$

$$= \lim_{\varepsilon \longrightarrow 0+} \frac{1}{\varepsilon^2} \left\{ \begin{array}{c} e^{\varepsilon s} \left[-\frac{1}{s} \left(\tau e^{-s\tau} \mid_0^\varepsilon - \int_0^\varepsilon e^{-s\tau} d\tau \right) \right] + \\ + e^{-\varepsilon s} \left[\frac{1}{s} \left(\sigma e^{\sigma s} \mid_0^\varepsilon - \int_0^\varepsilon \sigma e^{\sigma s} d\sigma \right) \right] \end{array} \right\} =$$

$$= \lim_{\varepsilon \longrightarrow 0+} \frac{1}{\varepsilon^2} \left\{ \begin{array}{c} e^{\varepsilon s} \left[-\frac{1}{s} \left(\varepsilon e^{-\varepsilon s} - 0 - \left(-\frac{1}{s} e^{-s\tau} \mid_0^\varepsilon \right) \right) \right] + \\ + e^{-\varepsilon s} \left[\frac{1}{s} \left(\varepsilon e^{\varepsilon s} - 0 - \left(\frac{1}{s} e^{\sigma s} \mid_0^\varepsilon \right) \right) \right] \end{array} \right\} =$$

$$= -\frac{1}{s} \lim_{\varepsilon \longrightarrow 0+} \frac{1}{\varepsilon^2} \left\{ \begin{array}{c} e^{\varepsilon s} \left[\left(\varepsilon e^{-\varepsilon s} + \left(\frac{1}{s} e^{-\varepsilon s} - \frac{1}{s} \right) \right) \right] - \\ - e^{-\varepsilon s} \left[\left(\varepsilon e^{\varepsilon s} - \left(\frac{1}{s} e^{\varepsilon s} - \frac{1}{s} \right) \right) \right] \end{array} \right\} =$$

$$= -\frac{1}{s} \lim_{\varepsilon \longrightarrow 0+} \frac{1}{\varepsilon^2} \left\{ \begin{array}{c} \varepsilon + \frac{1}{s} \left(1 - e^{\varepsilon s} \right) - \\ - \left[\varepsilon - \frac{1}{s} \left(1 - e^{-\varepsilon s} \right) \right] \end{array} \right\} =$$

$$= -\frac{1}{s} \lim_{\varepsilon \longrightarrow 0+} \frac{1}{\varepsilon^2} \left\{ \varepsilon - \varepsilon + \frac{1}{s} \left(1 - e^{\varepsilon s} + 1 - e^{-\varepsilon s} \right) \right\} =$$

$$= -\frac{1}{s^2} \lim_{\varepsilon \longrightarrow 0+} \frac{1}{\varepsilon^2} \left(2 - e^{\varepsilon s} - e^{-\varepsilon s} \right) =$$

$$= -\frac{1}{s^2} \lim_{\varepsilon \longrightarrow 0+} \frac{1}{2\varepsilon} \left(-s e^{\varepsilon s} + s e^{-\varepsilon s} \right) =$$

$$= -\frac{1}{s} \lim_{\varepsilon \longrightarrow 0+} \frac{1}{2\varepsilon} \left(-e^{\varepsilon s} + e^{-\varepsilon s} \right) =$$

$$= -\frac{1}{s} \lim_{\varepsilon \longrightarrow 0^+} \frac{1}{2}\left(-se^{\varepsilon s} - se^{-\varepsilon s}\right) =$$

$$= \lim_{\varepsilon \longrightarrow 0^+} \frac{1}{2}\left(e^{\varepsilon s} + e^{-\varepsilon s}\right) = \frac{2}{2} = 1.$$

Or equivalently,

$$\mathcal{L}^- \{\delta(t)\} = \lim_{\zeta = 2\varepsilon \longrightarrow 0^+} \left[\int_{-\zeta}^{\infty} \delta(t) e^{-st} dt\right] =$$

$$= \lim_{\varepsilon \longrightarrow 0^+} \left[\int_{-2\varepsilon}^{\infty} \left\{ \begin{array}{c} 0,\ t \in]-\infty, -\varepsilon] \cup [\varepsilon, +\infty[, \\ \frac{1}{\varepsilon^2}(t+\varepsilon),\ t \in [-\varepsilon, 0], \\ \frac{1}{\varepsilon^2}\left(-t+\varepsilon\right),\ t \in [0, \varepsilon] \end{array} \right\} e^{-st} dt\right] =$$

$$= \lim_{\varepsilon \longrightarrow 0^+} \left[\int_{-\varepsilon}^{\varepsilon} \left\{ \begin{array}{c} \frac{1}{\varepsilon^2}(t+\varepsilon),\ t \in [-\varepsilon, 0], \\ \frac{1}{\varepsilon^2}\left(-t+\varepsilon\right),\ t \in [0, \varepsilon] \end{array} \right\} e^{-st} dt\right] =$$

$$= \lim_{\varepsilon \longrightarrow 0^+} \frac{1}{\varepsilon^2} \left[\int_{-\varepsilon}^{0} (t+\varepsilon) e^{-st} dt + \int_{0}^{\varepsilon} \left(-t+\varepsilon\right) e^{-st} dt\right] =$$

$$= \lim_{\varepsilon \longrightarrow 0^+} \frac{1}{\varepsilon^2} \left[\begin{array}{c} \int_{-\varepsilon}^{0} te^{-st} dt + \int_{-\varepsilon}^{0} \varepsilon e^{-st} dt - \\ - \int_{0}^{\varepsilon} te^{-st} dt + \int_{0}^{\varepsilon} \varepsilon e^{-st} dt \end{array} \right] =$$

$$= \lim_{\varepsilon \longrightarrow 0^+} \frac{1}{\varepsilon^2} \left[\int_{-\varepsilon}^{0} te^{-st} dt + \int_{-\varepsilon}^{\varepsilon} \varepsilon e^{-st} dt - \int_{0}^{\varepsilon} te^{-st} dt\right] =$$

$$= \lim_{\varepsilon \longrightarrow 0^+} \frac{1}{\varepsilon^2} \left[\begin{array}{c} \left(-\frac{1}{s}\right)\left[te^{-st}\ |_{-\iota}^{0} - \int_{-\varepsilon}^{0} e^{-st} dt\right] + \\ +\varepsilon\left(-\frac{1}{s}\right)\left[e^{-st}\ |_{-\varepsilon}^{\varepsilon}\right] - \\ -\left(-\frac{1}{s}\right)\left[te^{-st}\ |_{0}^{\varepsilon} - \int_{0}^{\varepsilon} e^{-st} dt\right] \end{array} \right] =$$

$$= \lim_{\varepsilon \longrightarrow 0^+} \frac{1}{\varepsilon^2} \left[\begin{array}{c} \left(-\frac{1}{s}\right)\left[0 + \varepsilon e^{\varepsilon s} + \frac{1}{s}\left(e^{-st}\ |_{-\varepsilon}^{0}\right)\right] + \\ +\varepsilon\left(-\frac{1}{s}\right)\left(e^{-\varepsilon s} - e^{\varepsilon s}\right) + \\ +\frac{1}{s}\left[\varepsilon e^{-\varepsilon s} + \frac{1}{s}\left[e^{-st}\ |_{0}^{\varepsilon}\right]\right] \end{array} \right] =$$

$$= \lim_{\varepsilon \longrightarrow 0^+} \frac{1}{\varepsilon^2} \left[\begin{array}{c} \left(-\frac{1}{s}\right)\left[\varepsilon e^{\varepsilon s} + \frac{1}{s}\left(1 - e^{\varepsilon s}\right)\right] - \\ -\varepsilon\frac{1}{s}\left(e^{-\varepsilon s} - e^{\varepsilon t}\right) + \\ +\frac{1}{s}\left[\varepsilon e^{-\varepsilon s} + \frac{1}{s}\left(e^{-\varepsilon s} - 1\right)\right] \end{array} \right] =$$

$$= \lim_{\varepsilon \longrightarrow 0^+} \frac{1}{\varepsilon^2}\frac{1}{s} \left[\begin{array}{c} -\varepsilon e^{\varepsilon s} - \frac{1}{s}\left(1 - e^{\varepsilon s}\right) - \\ -\varepsilon\left(e^{-\varepsilon s} - e^{\varepsilon t}\right) + \\ +\left[\varepsilon e^{-\varepsilon s} + \frac{1}{s}\left(e^{-\varepsilon s} - 1\right)\right] \end{array} \right] =$$

$$= \lim_{\varepsilon \longrightarrow 0^+} \frac{1}{\varepsilon^2}\frac{1}{s} \left[\begin{array}{c} \varepsilon e^{-\varepsilon s} - \varepsilon e^{\varepsilon s} - \frac{1}{s}\left(1 - e^{\varepsilon s}\right) - \\ -\varepsilon\left(e^{-\varepsilon s} - e^{\varepsilon t}\right) + \frac{1}{s}\left(e^{-\varepsilon s} - 1\right) \end{array} \right] =$$

$$= \lim_{\varepsilon \longrightarrow 0+} \frac{1}{\varepsilon^2} \frac{1}{s} \left[\begin{array}{c} \varepsilon(e^{-\varepsilon s} - e^{\varepsilon s}) - \frac{1}{s}(1 - e^{\varepsilon s}) - \\ -\varepsilon\left(e^{-\varepsilon s} - e^{\varepsilon t}\right) + \frac{1}{s}\left(e^{-\varepsilon s} - 1\right) \end{array} \right] =$$

$$= \lim_{\varepsilon \longrightarrow 0+} \frac{1}{\varepsilon^2} \frac{1}{s} \left[-\frac{1}{s}\left(1 - e^{\varepsilon s}\right) + \frac{1}{s}\left(e^{-\varepsilon s} - 1\right) \right] =$$

$$= \frac{1}{s^2} \lim_{\varepsilon \longrightarrow 0+} \frac{1}{\varepsilon^2} \left(-2 + e^{\varepsilon s} + e^{-\varepsilon s}\right) =$$

$$= \frac{1}{s^2} \lim_{\varepsilon \longrightarrow 0+} \frac{1}{2\varepsilon} \left(se^{\varepsilon s} - se^{-\varepsilon s}\right) =$$

$$= \frac{1}{s} \lim_{\varepsilon \longrightarrow 0+} \frac{1}{2\varepsilon} \left(e^{\varepsilon s} - e^{-\varepsilon s}\right) =$$

$$= \frac{1}{s} \lim_{\varepsilon \longrightarrow 0+} \frac{1}{2} \left(se^{\varepsilon s} + se^{-\varepsilon s}\right) =$$

$$= \lim_{\varepsilon \longrightarrow 0+} \frac{1}{2} \left(e^{\varepsilon s} + e^{-\varepsilon s}\right) = \frac{2}{2} = 1.$$

By repeating the procedure,

$$\mathcal{L}\left\{\delta(t)\right\} = \int_0^\infty \delta^-(t) e^{-st} dt =$$

$$= \int_0^\infty \lim_{\varepsilon \longrightarrow 0+} \left\{ \begin{array}{c} 0,\, t \in]-\infty, -\varepsilon] \cup [\varepsilon, +\infty[, \\ \frac{1}{\varepsilon^2}(t+\varepsilon),\, t \in [-\varepsilon, 0], \\ \frac{1}{\varepsilon^2}\left(-t+\varepsilon\right),\, t \in [0, \varepsilon] \end{array} \right\} e^{-st} dt =$$

$$= \lim_{\varepsilon \longrightarrow 0+} \frac{1}{\varepsilon^2} \int_0^\varepsilon \left(-t + \varepsilon\right) e^{-st} dt,$$

$$\sigma = -t + \varepsilon \Longrightarrow t = -\sigma + \varepsilon,\, dt = -d\sigma,$$

$$t = 0 \Longrightarrow \sigma = \varepsilon,\, t = \varepsilon \Longrightarrow \sigma = 0,$$

$$\Longrightarrow$$

$$\mathcal{L}\left\{\delta^-(t)\right\} = \lim_{\varepsilon \longrightarrow 0+} \frac{1}{\varepsilon^2} \int_\varepsilon^0 \sigma e^{-s(-\sigma+\varepsilon)} \left(-d\sigma\right) =$$

$$= \lim_{\varepsilon \longrightarrow 0+} \frac{1}{\varepsilon^2} e^{-\varepsilon s} \int_0^\varepsilon \sigma e^{\sigma s} d\sigma =$$

$$= \lim_{\varepsilon \longrightarrow 0+} \frac{1}{\varepsilon^2} e^{-\varepsilon s} \left[\frac{1}{s} \left(\sigma e^{\sigma s} \mid_0^\varepsilon - \int_0^\varepsilon e^{\sigma s} d\sigma \right) \right] =$$

$$= \lim_{\varepsilon \longrightarrow 0+} \frac{1}{\varepsilon^2} e^{-\varepsilon s} \left[\frac{1}{s} \left(\varepsilon e^{\varepsilon s} - 0 - \frac{1}{s} e^{\sigma s} \mid_0^\varepsilon \right) \right] =$$

$$= \lim_{\varepsilon \longrightarrow 0^+} \frac{1}{\varepsilon^2} e^{-\varepsilon s} \frac{1}{s} \left(\varepsilon e^{\varepsilon s} - \frac{1}{s} e^{\varepsilon s} + \frac{1}{s} \right) =$$

$$= \lim_{\varepsilon \longrightarrow 0^+} \frac{1}{\varepsilon^2 s} \left(\varepsilon - \frac{1}{s} + \frac{1}{s} e^{-\varepsilon s} \right) =$$

$$= \lim_{\varepsilon \longrightarrow 0^+} \frac{1}{2\varepsilon s} \left(1 - s \frac{1}{s} e^{-\varepsilon s} \right) =$$

$$= \lim_{\varepsilon \longrightarrow 0^+} \frac{1}{2\varepsilon s} \left(1 - e^{-\varepsilon s} \right) =$$

$$= \lim_{\varepsilon \longrightarrow 0^+} \frac{1}{2s} s e^{-\varepsilon s} = \lim_{\varepsilon \longrightarrow 0^+} \frac{1}{2} e^{-\varepsilon s} = \frac{1}{2}.$$

Or equivalently,

$$\mathcal{L} \left\{ \delta(t) \right\} = \int_0^\infty \delta^-(t) e^{-st} dt =$$

$$= \int_0^\infty \lim_{\varepsilon \longrightarrow 0^+} \left\{ \begin{array}{c} 0,\ t \in]-\infty, -\varepsilon] \cup [\varepsilon, +\infty[, \\ \frac{1}{\varepsilon^2}(t+\varepsilon),\ t \in [-\varepsilon, 0], \\ \frac{1}{\varepsilon^2}\left(-t+\varepsilon\right),\ t \in [0, \varepsilon] \end{array} \right\} e^{-st} dt =$$

$$= \lim_{\varepsilon \longrightarrow 0^+} \frac{1}{\varepsilon^2} \int_0^\varepsilon \left(-t+\varepsilon\right) e^{-st} dt =$$

$$= \lim_{\varepsilon \longrightarrow 0^+} \frac{1}{\varepsilon^2} \left[-\int_0^\varepsilon t e^{-st} dt + \varepsilon \int_0^\varepsilon e^{-st} dt \right] =$$

$$= \lim_{\varepsilon \longrightarrow 0^+} \frac{1}{\varepsilon^2} \left\{ \begin{array}{c} -\left[-\frac{1}{s} \left(t e^{-st} \mid_0^\varepsilon - \int_0^\varepsilon e^{-st} dt \right) \right] + \\ +\varepsilon \left[-\frac{1}{s} \left(e^{-st} \mid_0^\varepsilon \right) \right] \end{array} \right\} =$$

$$= \lim_{\varepsilon \longrightarrow 0^+} \frac{1}{\varepsilon^2} \left\{ \begin{array}{c} \frac{1}{s} \left[\left(\varepsilon e^{-\varepsilon s} - 0 - \left(-\frac{1}{s} e^{-st} \mid_0^\varepsilon \right) \right) \right] + \\ +\varepsilon \left(-\frac{1}{s} e^{-st} \mid_0^\varepsilon \right) \end{array} \right\} =$$

$$= \lim_{\varepsilon \longrightarrow 0^+} \frac{1}{\varepsilon^2} \left\{ \begin{array}{c} \frac{1}{s} \left[\varepsilon e^{-\varepsilon s} + \left(\frac{1}{s} e^{-\varepsilon s} - \frac{1}{s} \right) \right] - \\ -\varepsilon \left(\frac{1}{s} e^{-\varepsilon s} - \frac{1}{s} \right) \end{array} \right\} =$$

$$= \lim_{\varepsilon \longrightarrow 0^+} \frac{1}{\varepsilon^2} \frac{1}{s} \left\{ \begin{array}{c} \varepsilon e^{-\varepsilon s} + \left(\frac{1}{s} e^{-\varepsilon s} - \frac{1}{s} \right) - \\ -\varepsilon \left(e^{-\varepsilon s} - 1 \right) \end{array} \right\} =$$

$$= \lim_{\varepsilon \longrightarrow 0^+} \frac{1}{\varepsilon^2} \frac{1}{s} \left\{ \varepsilon e^{-\varepsilon s} + \frac{1}{s} e^{-\varepsilon s} - \frac{1}{s} - \varepsilon e^{-\varepsilon s} + \varepsilon \right\} =$$

$$= \lim_{\varepsilon \longrightarrow 0^+} \frac{1}{2\varepsilon} \frac{1}{s} \left\{ e^{-\varepsilon s} - \varepsilon s e^{-\varepsilon s} + \frac{-s}{s} e^{-\varepsilon s} - e^{-\varepsilon s} + \varepsilon s e^{-\varepsilon s} + 1 \right\} =$$

$$= \lim_{\varepsilon \longrightarrow 0^+} \frac{1}{2\varepsilon} \frac{1}{s} \left\{ e^{-\varepsilon s} - e^{-\varepsilon s} - e^{-\varepsilon s} + 1 \right\} =$$

$$= \lim_{\varepsilon \longrightarrow 0+} \frac{1}{2\varepsilon} \frac{1}{s} \left\{ -e^{-\varepsilon s} + 1 \right\} =$$

$$= \lim_{\varepsilon \longrightarrow 0+} \frac{1}{2} \frac{1}{s} \left\{ s e^{-\varepsilon s} \right\} = \lim_{\varepsilon \longrightarrow 0+} \frac{1}{2} e^{-\varepsilon s} = \frac{1}{2}.$$

$$\mathcal{L}^{+} \{\delta(t)\} = \lim_{\zeta = 2\varepsilon \longrightarrow 0+} \left[\int_{\zeta}^{\infty} \delta(t) e^{-st} dt \right] = \lim_{\varepsilon \longrightarrow 0+} \int_{2\varepsilon}^{\infty} 0 e^{-st} dt = 0.$$

The preceding results prove (B.17).

Finally, we prove (B.17) as follows:

$$\mathcal{L}^{-} \left\{ \delta^{+}(t) \right\} = \lim_{\zeta = 2\varepsilon \longrightarrow 0+} \left[\int_{-\zeta}^{\infty} \delta^{+}(t) e^{-st} dt \right] =$$

$$= \lim_{\zeta = 2\varepsilon \longrightarrow 0+} \left[\int_{-\zeta}^{\infty} \left\{ \begin{array}{c} 0,\, t \in]-\infty, 0] \cup [2\varepsilon, +\infty[, \\ \frac{1}{\varepsilon^2} t,\, t \in [0, \varepsilon], \\ \frac{1}{\varepsilon^2} \left(-t + 2\varepsilon \right),\, t \in [\varepsilon, 2\varepsilon] \end{array} \right\} e^{-st} dt \right] =$$

$$= \lim_{\varepsilon \longrightarrow 0+} \left[\int_{-2\varepsilon}^{\infty} \left\{ \begin{array}{c} 0,\, t \in]-\infty, 0] \cup [2\varepsilon, +\infty[, \\ \frac{1}{\varepsilon^2} t,\, t \in [0, \varepsilon], \\ \frac{1}{\varepsilon^2} \left(-t + 2\varepsilon \right),\, t \in [\varepsilon, 2\varepsilon] \end{array} \right\} e^{-st} dt \right] =$$

$$= \lim_{\varepsilon \longrightarrow 0+} \left[\int_{0}^{2\varepsilon} \left\{ \begin{array}{c} \frac{1}{\varepsilon^2} t,\, t \in [0, \varepsilon], \\ \frac{1}{\varepsilon^2} \left(-t + 2\varepsilon \right),\, t \in [\varepsilon, 2\varepsilon] \end{array} \right\} e^{-st} dt \right] =$$

$$= \lim_{\varepsilon \longrightarrow 0+} \frac{1}{\varepsilon^2} \left[\int_{0}^{\varepsilon} t e^{-st} dt + \int_{\varepsilon}^{2\varepsilon} \left(-t + 2\varepsilon \right) e^{-st} dt \right],$$

$$\rho = -t + 2\varepsilon \Longrightarrow t = -\rho + 2\varepsilon,\, dt = -d\rho,$$

$$t = \varepsilon \Longrightarrow \rho = \varepsilon,\, t = 2\varepsilon \Longrightarrow \rho = 0,$$

$$\Longrightarrow \mathcal{L}^{-} \left\{ \delta^{+}(t) \right\} =$$

$$= \lim_{\varepsilon \longrightarrow 0+} \frac{1}{\varepsilon^2} \left[\int_{0}^{\varepsilon} t e^{-st} dt + \int_{\varepsilon}^{0} \rho e^{-s(-\rho + 2\varepsilon)} \left(-d\rho \right) \right] =$$

$$= \lim_{\varepsilon \longrightarrow 0+} \frac{1}{\varepsilon^2} \left[\int_{0}^{\varepsilon} t e^{-st} dt + e^{-2\varepsilon s} \int_{0}^{\varepsilon} \rho e^{\rho s} d\rho \right] =$$

$$= \lim_{\varepsilon \longrightarrow 0+} \frac{1}{\varepsilon^2} \left\{ \begin{array}{c} \left[-\frac{1}{s} \left(t e^{-st} \mid_0^{\varepsilon} - \int_0^{\varepsilon} e^{-st} dt \right) \right] + \\ + e^{-2\varepsilon s} \left[\frac{1}{s} \left(\rho e^{\rho s} \mid_0^{\varepsilon} - \int_0^{\varepsilon} e^{\rho s} d\rho \right) \right] \end{array} \right\} =$$

$$= \lim_{\varepsilon \longrightarrow 0+} \frac{1}{\varepsilon^2} \left\{ \begin{array}{c} -\frac{1}{s} \left[\varepsilon e^{-\varepsilon s} - 0 - \left(-\frac{1}{s} e^{-st} \mid_0^{\varepsilon} \right) \right] + \\ + e^{-2\varepsilon s} \frac{1}{s} \left[\varepsilon e^{\varepsilon s} - 0 - \left(\frac{1}{s} e^{\rho s} \mid_0^{\varepsilon} \right) \right] \end{array} \right\} =$$

$$= \lim_{\varepsilon \longrightarrow 0+} \frac{1}{\varepsilon^2} \left\{ \begin{array}{c} -\frac{1}{s}\left[\varepsilon e^{-\varepsilon s} + \left(\frac{1}{s} e^{-\varepsilon s} - \frac{1}{s}\right)\right] + \\ +e^{-2\varepsilon s}\frac{1}{s}\left[\varepsilon e^{\varepsilon s} - \left(\frac{1}{s} e^{\varepsilon s} - \frac{1}{s}\right)\right] \end{array} \right\} =$$

$$= \frac{1}{s} \lim_{\varepsilon \longrightarrow 0+} \frac{1}{\varepsilon^2} \left\{ \begin{array}{c} -\varepsilon e^{-\varepsilon s} - \left(\frac{1}{s} e^{-\varepsilon s} - \frac{1}{s}\right) + \\ +\varepsilon e^{-\varepsilon s} - \left(\frac{1}{s} e^{-\varepsilon s} - \frac{1}{s} e^{-2\varepsilon s}\right) \end{array} \right\} =$$

$$= \frac{1}{s^2} \lim_{\varepsilon \longrightarrow 0+} \frac{1}{\varepsilon^2} \left\{ -2e^{-\varepsilon s} + 1 + e^{-2\varepsilon s} \right\} =$$

$$= \frac{1}{s^2} \lim_{\varepsilon \longrightarrow 0+} \frac{1}{2\varepsilon} \left\{ 2se^{-\varepsilon s} - 2se^{-2\varepsilon s} \right\} =$$

$$= \frac{1}{s} \lim_{\varepsilon \longrightarrow 0+} \frac{1}{\varepsilon} \left\{ e^{-\varepsilon s} - e^{-2\varepsilon s} \right\} =$$

$$= \frac{1}{s} \lim_{\varepsilon \longrightarrow 0+} \frac{1}{1} \left\{ -se^{-\varepsilon s} + 2se^{-2\varepsilon s} \right\} =$$

$$= \lim_{\varepsilon \longrightarrow 0+} \left\{ -e^{-\varepsilon s} + 2e^{-2\varepsilon s} \right\} = -1 + 2 = 1.$$

Or equivalently,

$$\mathcal{L}^- \left\{ \delta^+(t) \right\} = \lim_{\zeta = 2\varepsilon \longrightarrow 0+} \left[\int_{-\zeta}^{\infty} \delta^+(t) e^{-st} dt \right] =$$

$$= \lim_{\zeta = 2\varepsilon \longrightarrow 0+} \left[\int_{-\zeta}^{\infty} \left\{ \begin{array}{c} 0,\ t \in]-\infty, 0] \cup [2\varepsilon, +\infty[, \\ \frac{1}{\varepsilon^2} t,\ t \in [0, \varepsilon], \\ \frac{1}{\varepsilon^2}\left(-t + 2\varepsilon\right),\ t \in [\varepsilon, 2\varepsilon] \end{array} \right\} e^{-st} dt \right] =$$

$$= \lim_{\varepsilon \longrightarrow 0+} \left[\int_{-2\varepsilon}^{\infty} \left\{ \begin{array}{c} 0,\ t \in]-\infty, 0] \cup [2\varepsilon, +\infty[, \\ \frac{1}{\varepsilon^2} t,\ t \in [0, \varepsilon], \\ \frac{1}{\varepsilon^2}\left(-t + 2\varepsilon\right),\ t \in [\varepsilon, 2\varepsilon] \end{array} \right\} e^{-st} dt \right] =$$

$$= \lim_{\varepsilon \longrightarrow 0+} \left[\int_{0}^{2\varepsilon} \left\{ \begin{array}{c} \frac{1}{\varepsilon^2} t,\ t \in [0, \varepsilon], \\ \frac{1}{\varepsilon^2}\left(-t + 2\varepsilon\right),\ t \in [\varepsilon, 2\varepsilon] \end{array} \right\} e^{-st} dt \right] =$$

$$= \lim_{\varepsilon \longrightarrow 0+} \frac{1}{\varepsilon^2} \left[\int_0^{\varepsilon} te^{-st} dt - \int_{\varepsilon}^{2\varepsilon} te^{-st} dt + 2\varepsilon \int_{\varepsilon}^{2\varepsilon} e^{-st} dt \right] =$$

$$= \lim_{\varepsilon \longrightarrow 0+} \frac{1}{\varepsilon^2} \left\{ \begin{array}{c} \left[-\frac{1}{s}\left(te^{-st} \mid_0^{\varepsilon} - \int_0^{\varepsilon} e^{-st} dt\right)\right] - \\ -\left[-\frac{1}{s}\left(te^{-st} \mid_{\varepsilon}^{2\varepsilon} - \int_{\varepsilon}^{2\varepsilon} e^{-st} dt\right)\right] + \\ +2\varepsilon \int_{\varepsilon}^{2\varepsilon} e^{-st} dt \end{array} \right\} =$$

$$= \lim_{\varepsilon \longrightarrow 0+} \frac{1}{\varepsilon^2} \left\{ \begin{array}{c} -\frac{1}{s}\left[\varepsilon e^{-\varepsilon s} - 0 - \left(-\frac{1}{s} e^{-st} \mid_0^{\varepsilon}\right)\right] + \\ +\frac{1}{s}\left[2\varepsilon e^{-2\varepsilon s} - \varepsilon e^{-\varepsilon s} - \left(-\frac{1}{s} e^{-st} \mid_{\varepsilon}^{2\varepsilon}\right)\right] + \\ +2\varepsilon\left(-\frac{1}{s} e^{-st} \mid_{\varepsilon}^{2\varepsilon}\right) \end{array} \right\} =$$

$$= \frac{1}{s} \lim_{\varepsilon \longrightarrow 0+} \frac{1}{\varepsilon^2} \left\{ \begin{array}{c} -\left[\varepsilon e^{-\varepsilon s} + \left(\frac{1}{s} e^{-\varepsilon s} - \frac{1}{s}\right)\right] + \\ + \left[2\varepsilon e^{-2\varepsilon s} - \varepsilon e^{-\varepsilon s} - \left(-\frac{1}{s} e^{-2\varepsilon s} + \frac{1}{s} e^{-\varepsilon s}\right)\right] + \\ +2\varepsilon \left(-e^{-2\varepsilon s} + e^{-\varepsilon s}\right) \end{array} \right\} =$$

$$= \frac{1}{s} \lim_{\varepsilon \longrightarrow 0+} \frac{1}{\varepsilon^2} \left\{ \begin{array}{c} -\varepsilon e^{-\varepsilon s} - \frac{1}{s} e^{-\varepsilon s} + \frac{1}{s} + \\ +2\varepsilon e^{-2\varepsilon s} - \varepsilon e^{-\varepsilon s} + \frac{1}{s} e^{-2\varepsilon s} - \frac{1}{s} e^{-\varepsilon s} - \\ -2\varepsilon e^{-2\varepsilon s} + 2\varepsilon e^{-\varepsilon s} \end{array} \right\} =$$

$$= \frac{1}{s^2} \lim_{\varepsilon \longrightarrow 0+} \frac{1}{\varepsilon^2} \left\{-2e^{-\varepsilon s} + 1 + e^{-2\varepsilon s}\right\} =$$

$$= \frac{1}{s^2} \lim_{\varepsilon \longrightarrow 0+} \frac{1}{2\varepsilon} \left\{2se^{-\varepsilon s} - 2se^{-2\varepsilon s}\right\} =$$

$$= \frac{1}{s} \lim_{\varepsilon \longrightarrow 0+} \frac{1}{\varepsilon} \left\{e^{-\varepsilon s} - e^{-2\varepsilon s}\right\} =$$

$$= \frac{1}{s} \lim_{\varepsilon \longrightarrow 0+} \frac{1}{1} \left\{-se^{-\varepsilon s} + 2se^{-2\varepsilon s}\right\} =$$

$$= \lim_{\varepsilon \longrightarrow 0+} \frac{1}{1} \left\{-e^{-\varepsilon s} + 2e^{-2\varepsilon s}\right\} = -1 + 2 = 1.$$

Further,

$$\mathcal{L}\left\{\delta^+(t)\right\} = \left[\int_0^\infty \delta^+(t) e^{-st} dt\right] =$$

$$= \lim_{\varepsilon \longrightarrow 0+} \left[\int_0^\infty \left\{ \begin{array}{c} 0,\, t \in]-\infty, 0] \cup [2\varepsilon, +\infty[, \\ \frac{1}{\varepsilon^2} t,\, t \in [0, \varepsilon], \\ \frac{1}{\varepsilon^2}(-t + 2\varepsilon),\, t \in [\varepsilon, 2\varepsilon] \end{array} \right\} e^{-st} dt\right] =$$

$$= \lim_{\varepsilon \longrightarrow 0+} \left[\int_0^{2\varepsilon} \left\{ \begin{array}{c} \frac{1}{\varepsilon^2} t,\, t \in [0, \varepsilon], \\ \frac{1}{\varepsilon^2}(-t + 2\varepsilon),\, t \in [\varepsilon, 2\varepsilon] \end{array} \right\} e^{-st} dt\right] =$$

$$= \lim_{\varepsilon \longrightarrow 0+} \frac{1}{\varepsilon^2} \left[\int_0^\varepsilon te^{-st} dt - \int_\varepsilon^{2\varepsilon} te^{-st} dt + 2\varepsilon \int_\varepsilon^{2\varepsilon} e^{-st} dt\right] =$$
$$= \mathcal{L}^-\left\{\delta^+(t)\right\} = 1.$$

Finally,

$$\mathcal{L}^+\left\{\delta^+(t)\right\} = \lim_{\zeta = 2\varepsilon \longrightarrow 0+} \left[\int_\zeta^\infty \delta^+(t) e^{-st} dt\right] =$$

$$= \lim_{\varepsilon \longrightarrow 0+} \left[\int_{2\varepsilon}^\infty \left\{ \begin{array}{c} 0,\, t \in]-\infty, 0] \cup [2\varepsilon, +\infty[, \\ \frac{1}{\varepsilon^2} t,\, t \in [0, \varepsilon], \\ \frac{1}{\varepsilon^2}(-t + 2\varepsilon),\, t \in [\varepsilon, 2\varepsilon] \end{array} \right\} e^{-st} dt\right] =$$

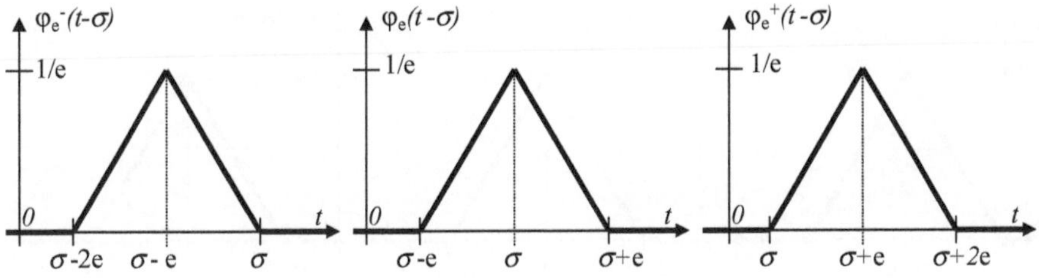

Figure B.2: Translated, i.e., delayed, triangular functions generating unit Dirac impulses.

$$= \lim_{\varepsilon \longrightarrow 0^+} \left[\int_{2\varepsilon}^{\infty} 0 e^{-st} dt \right] = 0.$$

The preceding results prove (B.17) ■

Let $\delta^{\mp}(. - \sigma)$ be the delayed left/right unit Dirac impulse (delayed unit Dirac delta distribution) defined by, respectively,

$$\delta^{\mp}(t-\sigma) = \lim[\varphi_{\varepsilon}^{\mp}(t-\sigma) : \varepsilon \longrightarrow 0^+] \tag{B.20}$$

where

$$\varphi_{\varepsilon}^{-}(t-\sigma) = \left\{ \begin{array}{c} 0,\ t \in]-\infty, \sigma - 2\varepsilon] \cup [0, +\infty[, \\ \frac{1}{\varepsilon^2}(t - \sigma + 2\varepsilon),\ t \in [\sigma - 2\varepsilon, \sigma - \varepsilon], \\ -\frac{1}{\varepsilon^2}(t-\sigma),\ t \in [\sigma - \varepsilon, \sigma], \\ 0,\ \ t \in [\sigma, \infty[\end{array} \right\}, \tag{B.21}$$

$$\varphi_{\varepsilon}(t-\sigma) = \left\{ \begin{array}{c} 0,\ t \in]-\infty, \sigma - \varepsilon] \cup [\sigma + \varepsilon, +\infty[, \\ \frac{1}{\varepsilon^2}(t - \sigma + \varepsilon),\ t \in [\sigma - \varepsilon, 0], \\ \frac{1}{\varepsilon^2}(-t + \sigma + \varepsilon),\ t \in [0, \sigma + \varepsilon] \end{array} \right\}, \tag{B.22}$$

$$\varphi_{\varepsilon}^{+}(t-\sigma) = \left\{ \begin{array}{c} 0,\ t \in]-\infty, \sigma] \cup [\sigma + 2\varepsilon, +\infty[, \\ \frac{1}{\varepsilon^2}(t-\sigma),\ t \in [\sigma, \sigma + \varepsilon], \\ \frac{1}{\varepsilon^2}(-t + \sigma + 2\varepsilon),\ t \in [\sigma + \varepsilon, \sigma + 2\varepsilon], \end{array} \right\}, \tag{B.23}$$

Fig. B.2 in which e stands for ε, $e = \varepsilon$. Besides,

$$\delta^{\mp}(\sigma - t) = \lim[\varphi_{\varepsilon}^{\mp}(\sigma - t) : \varepsilon \longrightarrow 0^+] \tag{B.24}$$

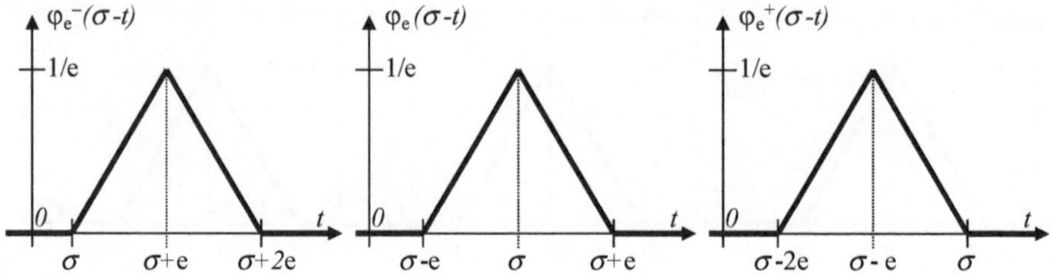

Figure B.3: Delayed triangular functions generating unit Dirac impulses.

where

$$\varphi_{\varepsilon}^{-}(\sigma-t)=$$
$$=\left\{\begin{array}{c} 0,\ t\in]-\infty,\sigma]\cup[\sigma+2\varepsilon,+\infty[, \\ \frac{1}{\varepsilon^2}\left(t-\sigma\right),\ t\in[\sigma,\sigma+\varepsilon], \\ \frac{1}{\varepsilon^2}\left(-t+\sigma+2\varepsilon\right),\ t\in[\sigma+\varepsilon,\sigma+2\varepsilon], \end{array}\right\}=$$
$$=\varphi_{\varepsilon}^{+}(t-\sigma), \tag{B.25}$$

$$\varphi_{\varepsilon}(\sigma-t)=$$
$$=\left\{\begin{array}{c} 0,\ t\in]-\infty,\sigma-\varepsilon]\cup[\sigma+\varepsilon,+\infty[, \\ \frac{1}{\varepsilon^2}(t-\sigma+\varepsilon),\ t\in[\sigma-\varepsilon,0], \\ \frac{1}{\varepsilon^2}\left(-t+\sigma+\varepsilon\right),\ t\in[0,\sigma+\varepsilon] \end{array}\right\}=$$
$$=\varphi_{\varepsilon}(t-\sigma), \tag{B.26}$$

$$\varphi_{\varepsilon}^{+}(\sigma-t)=$$
$$=\left\{\begin{array}{c} 0,\ t\in]-\infty,\sigma-2\varepsilon]\cup[0,+\infty[, \\ \frac{1}{\varepsilon^2}(t-\sigma+2\varepsilon),\ t\in[\sigma-2\varepsilon,\sigma-\varepsilon], \\ -\frac{1}{\varepsilon^2}\left(t-\sigma\right),\ t\in[\sigma-\varepsilon,\sigma], \\ 0,\ t\in[\sigma,\infty[\end{array}\right\}=$$
$$=\varphi_{\varepsilon}^{-}(t-\sigma), \tag{B.27}$$

Fig. B.3 in which e stands for ε, $e=\varepsilon$.

We will use (*left* for -, *right* for +) *Heaviside functions*, called also *the (left for -, right for +) unit step function*, $h^{\mp}(.):\mathfrak{T}\longrightarrow[0,1]$,

$$h^{\mp}(t)\left\{\begin{array}{c} =0,\ t<0^{\mp}, \\ \in[0,1],\ t=0^{\mp}, \\ =1,\ t>0^{\mp} \end{array}\right\}, \tag{B.28}$$

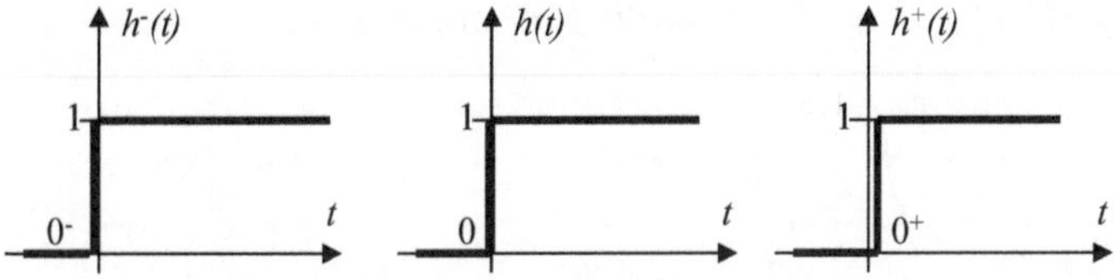

Figure B.4: Heaviside functions.

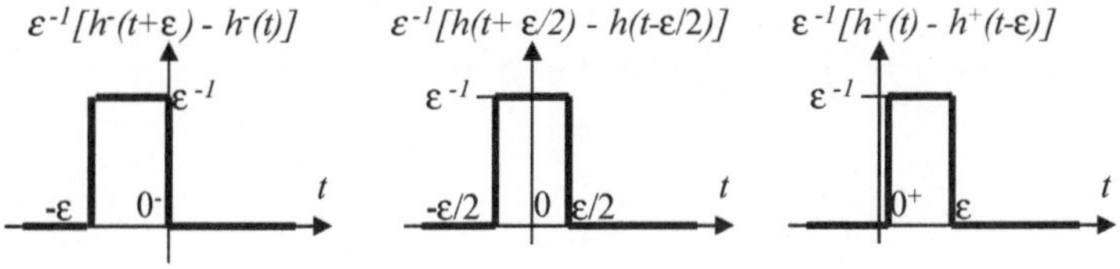

Figure B.5: Unit Dirac pulses that become unit Dirac impulses for $\varepsilon \longrightarrow 0^+$.

See Fig. B.4 Although they are not physically exactly realizable due to $PCUP$ (Principle 5 to Principle 7, Corollary 8) and $TCUP$ (Principle 9), [65], [66] they are very useful for mathematical treatments of mathematical models of physical systems. They permit us to determine symbolically $\delta^{\mp}(.)$ in terms of them,

$$\begin{aligned}
\delta^-(t) &= \lim \left[\frac{h^-(t+\varepsilon) - h^-(t)}{\varepsilon} : \varepsilon \longrightarrow 0^+ \right], \\
\delta(t) &= \lim \left[\frac{h(t+\varepsilon/2) - h(t-\varepsilon/2)}{\varepsilon} : \varepsilon \longrightarrow 0^+ \right], \\
\delta^+(t) &= \lim \left[\frac{h^+(t) - h^+(t-\varepsilon)}{\varepsilon} : \varepsilon \longrightarrow 0^+ \right], \qquad \text{(B.29)}
\end{aligned}$$

See Fig. B.5.

We can summarize several results of P. J. Antsaklis and A. N. Michel [4, pp. 72-75, Lemma 16.1] as follows:

Lemma 414 *Antsaklis and Michel lemma*

Let $f(.)$ be a continuous real-valued function defined on $\mathfrak{R}$ and let $\varphi_\varepsilon(t)$

be defined by (B.15), Fig. B.1. Then for any $\sigma \in \mathfrak{T}$,

$$\lim \left[\int_{-\infty}^{+\infty} f(\tau)\varphi_{\varepsilon}(\sigma - \tau)d\tau : \varepsilon \longrightarrow 0^{+} \right] =$$
$$= \int_{-\infty}^{+\infty} f(\tau)\delta(\sigma - \tau)d\tau =$$
$$= \lim \left[\int_{-\infty}^{+\infty} f(\tau)\varphi_{\varepsilon}(\tau - \sigma)d\tau : \varepsilon \longrightarrow 0^{+} \right] =$$
$$= \int_{-\infty}^{+\infty} f(\tau)\delta(\tau - \sigma)d\tau = f(\sigma). \tag{B.30}$$

Appendix C

System forms

C.1 From *IO* system to *ISO* system

In order to transform the *IO* system (2.1), (in Section 2.1),

$$\sum_{k=0}^{k=\nu} A_k \mathbf{Y}^{(k)}(t) = \sum_{k=0}^{k=\mu} B_k \mathbf{I}^{(k)}(t),\ detA_\nu \neq 0,\ \forall t \in \mathfrak{T}_0,\ \nu \geq 1,\ 0 \leq \mu \leq \nu,$$

$$\mu < \nu \Longrightarrow B_k = O_{N,M},\ \forall k = \mu+1,\ \mu+2,\ \dots,\ \nu, \tag{C.1}$$

into the equivalent *ISO* system (2.17) and (2.18), (in Section 2.2), i.e., into

$$\frac{d\mathbf{X}(t)}{dt} = A\mathbf{X}(t) + B\mathbf{I}(t),\ ,\ \forall t \in \mathfrak{T}_0, \tag{C.2}$$

$$\mathbf{Y}(t) = C\mathbf{X}(t) + D\mathbf{I}(t),\ ,\ \forall t \in \mathfrak{T}_0, \tag{C.3}$$

we define mathematically, without any physical justification or meaning, subsidiary vector variables $\mathbf{X}_1$, $\mathbf{X}_2$, ... $\mathbf{X}_\nu$ by

$$\mathbf{X}_1 = \mathbf{Y} - B_\nu \mathbf{I}, \tag{C.4}$$

$$\mathbf{X}_2 = \dot{\mathbf{X}}_1 + A_{\nu-1}\ \mathbf{Y} - B_{\nu-1}\mathbf{I}, \tag{C.5}$$

$$\mathbf{X}_3 = \dot{\mathbf{X}}_2 + A_{\nu-2}\ \mathbf{Y} - B_{\nu-2}\mathbf{I}, \tag{C.6}$$

$$\dots. \tag{C.7}$$

$$\mathbf{X}_{\nu-2} = \dot{\mathbf{X}}_{\nu-3} + A_3\ \mathbf{Y} - B_3\mathbf{I} \tag{C.8}$$

$$\mathbf{X}_{\nu-1} = \dot{\mathbf{X}}_{\nu-2} + A_2\ \mathbf{Y} - B_2\mathbf{I} \tag{C.9}$$

$$\mathbf{X}_{\nu} = \dot{\mathbf{X}}_{\nu-1} + A_1\ \mathbf{Y} - B_1\mathbf{I} \tag{C.10}$$

which we can rewrite as

$$\begin{aligned}
\mathbf{X}_1 &= \mathbf{Y} - B_\nu \mathbf{I}, \\
\dot{\mathbf{X}}_1 &= \mathbf{X}_2 - A_{\nu-1}\ \mathbf{Y} + B_{\nu-1}\mathbf{I}, \\
\dot{\mathbf{X}}_2 &= \mathbf{X}_3 - A_{\nu-2}\ \mathbf{Y} + B_{\nu-2}\mathbf{I}, \\
&\cdots\cdot \\
\dot{\mathbf{X}}_{\nu-3} &= \mathbf{X}_{\nu-2} - A_3\ \mathbf{Y} + B_3\mathbf{I} \\
\dot{\mathbf{X}}_{\nu-2} &= \mathbf{X}_{\nu-1} - A_2\ \mathbf{Y} + B_2\mathbf{I} \\
\dot{\mathbf{X}}_{\nu-1} &= \mathbf{X}_\nu - A_1\ \mathbf{Y} + B_1\mathbf{I}
\end{aligned} \tag{C.11}$$

We solve the first Equation (C.11) for $\mathbf{Y}$,

$$\mathbf{Y} = \mathbf{X}_1 + B_\nu \mathbf{I}. \tag{C.12}$$

and then we replace $\mathbf{Y}$ by $\mathbf{X}_1 + B_\nu \mathbf{I}$ in all other Equations (C.11):

$$\begin{aligned}
\dot{\mathbf{X}}_1 &= \mathbf{X}_2 - A_{\nu-1}\ (\mathbf{X}_1 + B_\nu \mathbf{I}) + B_{\nu-1}\mathbf{I}, \\
\dot{\mathbf{X}}_2 &= \mathbf{X}_3 - A_{\nu-2}\ (\mathbf{X}_1 + B_\nu \mathbf{I}) + B_{\nu-2}\mathbf{I}, \\
&\cdots\cdot \\
\dot{\mathbf{X}}_{\nu-3} &= \mathbf{X}_{\nu-2} - A_3\ (\mathbf{X}_1 + B_\nu \mathbf{I}) + B_3\mathbf{I} \\
\dot{\mathbf{X}}_{\nu-2} &= \mathbf{X}_{\nu-1} - A_2\ (\mathbf{X}_1 + B_\nu \mathbf{I}) + B_2\mathbf{I} \\
\dot{\mathbf{X}}_{\nu-1} &= \mathbf{X}_\nu - A_1\ (\mathbf{X}_1 + B_\nu \mathbf{I}) + B_1\mathbf{I}
\end{aligned}$$

which we set in the following form

$$\begin{aligned}
\dot{\mathbf{X}}_1 &= -A_{\nu-1}\mathbf{X}_1 + \mathbf{X}_2 + (B_{\nu-1} - A_{\nu-1}B_\nu)\,\mathbf{I}, \\
\dot{\mathbf{X}}_2 &= -A_{\nu-2}\ \mathbf{X}_1 + \mathbf{X}_3 + (B_{\nu-2} - A_{\nu-2}B_\nu)\,\mathbf{I} \\
&\cdots\cdot\cdot \\
\dot{\mathbf{X}}_{\nu-3} &= -A_3\ \mathbf{X}_1 + \mathbf{X}_{\nu-2} + (B_3 - A_3B_\nu)\,\mathbf{I} \\
\dot{\mathbf{X}}_{\nu-2} &= -A_2\ \mathbf{X}_1 + \mathbf{X}_{\nu-1} + (B_2 - A_2B_\nu)\,\mathbf{I} \\
\dot{\mathbf{X}}_{\nu-1} &= -A_1\ \mathbf{X}_1 + \mathbf{X}_\nu + (B_1 - A_1B_\nu)\,\mathbf{I}.
\end{aligned} \tag{C.13}$$

We replace the derivatives $\overset{\bullet}{\mathbf{X}}_1$, $\overset{\bullet}{\mathbf{X}}_2$, ... $\overset{\bullet}{\mathbf{X}}_{\nu-1}$into (C.5) through (C.10):

$$\mathbf{X}_1 = \mathbf{Y} - B_\nu \mathbf{I},$$
$$\mathbf{X}_2 = \mathbf{Y}^{(1)} - B_\nu \mathbf{I}^{(1)} + A_{\nu-1}\ \mathbf{Y} - B_{\nu-1}\mathbf{I},$$
$$\mathbf{X}_3 = \mathbf{Y}^{(2)} - B_\nu \mathbf{I}^{(2)} + A_{\nu-1}\ \mathbf{Y}^{(\mathbf{1})} - B_{\nu-1}\mathbf{I}^{(1)} + A_{\nu-2}\ \mathbf{Y} - B_{\nu-2}\mathbf{I},$$
$$....$$
$$\mathbf{X}_{\nu-2} = \begin{pmatrix} \mathbf{Y}^{(\nu-3)} - B_\nu \mathbf{I}^{(\nu-3)} + A_{\nu-1}\ \mathbf{Y}^{(\nu-\mathbf{4})} - B_{\nu-1}\mathbf{I}^{(\nu-4)} + . \\ ... + A_{\nu-2}\ \mathbf{Y}^{(\mathbf{1})} - B_{\nu-2}\mathbf{I}^{(1)} + A_3\ \mathbf{Y} - B_3\mathbf{I} \end{pmatrix}$$
$$\mathbf{X}_{\nu-1} = \begin{pmatrix} \mathbf{Y}^{(\nu-2)} - B_\nu \mathbf{I}^{(\nu-2)} + A_{\nu-1}\ \mathbf{Y}^{(\nu-\mathbf{3})} - B_{\nu-1}\mathbf{I}^{(\nu-3)} + \\ +A_{\nu-2}\ \mathbf{Y}^{(\nu-\mathbf{4})} - B_{\nu-2}\mathbf{I}^{(\nu-4)} + ... \\ +A_3\ \mathbf{Y}^{(\mathbf{1})} - B_3\mathbf{I}^{(1)} + A_2\ \mathbf{Y} - B_2\mathbf{I} \end{pmatrix}$$
$$\mathbf{X}_\nu = \begin{pmatrix} \mathbf{Y}^{(\nu-1)} - B_\nu \mathbf{I}^{(\nu-1)} + A_{\nu-1}\ \mathbf{Y}^{(\nu-\mathbf{2})} - B_{\nu-1}\mathbf{I}^{(\nu-2)} + \\ +A_{\nu-2}\ \mathbf{Y}^{(\nu-\mathbf{3})} - B_{\nu-2}\mathbf{I}^{(\nu-3)} + ... + A_3\ \mathbf{Y}^{(\mathbf{2})} - B_3\mathbf{I}^{(2)} + \\ +A_2\ \mathbf{Y}^{(\mathbf{1})} - B_2\mathbf{I}^{(1)} + A_1\ \mathbf{Y} - B_1\mathbf{I} \end{pmatrix}..$$

We differentiate once the last equation

$$\overset{\bullet}{\mathbf{X}}_\nu = \mathbf{Y}^{(\nu)} + A_{\nu-1}\ \mathbf{Y}^{(\nu-\mathbf{1})} + A_{\nu-2}\ \mathbf{Y}^{(\mathbf{2})} + ... + A_3\ \mathbf{Y}^{(\mathbf{3})} +$$
$$+A_2\ \mathbf{Y}^{(\mathbf{2})} + A_1\ \mathbf{Y}^{(1)} - B_\nu \mathbf{I}^{(\nu)} - B_{\nu-1}\mathbf{I}^{(\nu-1)} - B_{\nu-2}\mathbf{I}^{(\nu-2)}..$$
$$.. - B_3\mathbf{I}^{(3)} - B_2\mathbf{I}^{(2)} - B_1\mathbf{I}^{(1)} = \sum_{k=1}^{k=\nu} A_k \mathbf{Y}^{(k)}(t) - \sum_{k=1}^{k=\mu} B_k \mathbf{I}^{(k)}(t),$$

and we use (C.1)

$$\sum_{k=1}^{k=\nu} A_k \mathbf{Y}^{(k)}(t) - \sum_{k=1}^{k=\mu} B_k \mathbf{I}^{(k)}(t) = B_0 \mathbf{I}(t) - A_0 \mathbf{Y}(t) =$$
$$= B_0 \mathbf{I}(t) - A_0 (\mathbf{X}_1\ + B_\nu \mathbf{I})\ = -A_0 \mathbf{X}_1 + (B_0 - A_0 B_\nu)\,\mathbf{I}.$$

Hence,

$$\overset{\bullet}{\mathbf{X}}_\nu = -A_0 \mathbf{X}_1 + (B_0 - A_0 B_\nu)\,\mathbf{I}.$$

We gather this and (C.13),

$$\underbrace{\begin{bmatrix} \overset{\bullet}{\mathbf{X}}_1 \\ \overset{\bullet}{\mathbf{X}}_2 \\ \cdots \\ \overset{\bullet}{\mathbf{X}}_{\nu-3} \\ \overset{\bullet}{\mathbf{X}}_{\nu-2} \\ \overset{\bullet}{\mathbf{X}}_{\nu-1} \\ \overset{\bullet}{\mathbf{X}}_{\nu} \end{bmatrix}}_{\overset{\bullet}{\mathbf{X}}} = \begin{bmatrix} -A_{\nu-1}\mathbf{X}_1 + \mathbf{X}_2 + (B_{\nu-1} - A_{\nu-1}B_\nu)\,\mathbf{I} \\ -A_{\nu-2}\,\mathbf{X}_1 + \mathbf{X}_3 + (B_{\nu-2} - A_{\nu-2}B_\nu)\,\mathbf{I} \\ \\ -A_3\,\mathbf{X}_1 + \mathbf{X}_{\nu-2} + (B_3 - A_3B_\nu)\,\mathbf{I} \\ -A_2\,\mathbf{X}_1 + \mathbf{X}_{\nu-1} + (B_2 - A_2B_\nu)\,\mathbf{I} \\ -A_1\,\mathbf{X}_1 + \mathbf{X}_\nu + (B_1 - A_1B_\nu)\,\mathbf{I}. \\ -A_0\mathbf{X}_1 + (B_0 - A_0B_\nu)\,\mathbf{I} \end{bmatrix} =$$

$$= \underbrace{\begin{bmatrix} -A_{\nu-1} & I_N & O_N & \dots & O_N & O_N & O_N \\ -A_{\nu-2} & O_N & I_N & \dots & O_N & O_N & O_N \\ \dots & \dots & \dots & \dots & \dots & \dots & \dots \\ -A_3 & O_N & O_N & \dots & I_N & O_N & O_N \\ -A_2 & O_N & O_N & \dots & O_N & I_N & O_N \\ -A_1 & O_N & O_N & \dots & O_N & O_N & I_N \\ -A_0 & O_N & O_N & \dots & O_N & O_N & O_N \end{bmatrix}}_{A} \underbrace{\begin{bmatrix} \mathbf{X}_1 \\ \mathbf{X}_2 \\ \\ \mathbf{X}_{\nu-3} \\ \mathbf{X}_{\nu-2} \\ \mathbf{X}_{\nu-1} \\ \mathbf{X}_\nu \end{bmatrix}}_{\mathbf{X}} +$$

$$+ \underbrace{\begin{bmatrix} B_{\nu-1} - A_{\nu-1}B_\nu \\ B_{\nu-2} - A_{\nu-2}B_\nu \\ \dots. \\ B_3 - A_3B_\nu \\ B_2 - A_2B_\nu \\ B_1 - A_1B_\nu \\ B_0 - A_0B_\nu \end{bmatrix}}_{B} \mathbf{I}. \tag{C.14}$$

We set (C.12) into the following form:

$$\mathbf{Y} = \mathbf{X}_1 + B_\nu\mathbf{I}. = \underbrace{\begin{bmatrix} I_N & O_N & O_N & \dots & O_N & O_N & O_N \end{bmatrix}}_{C}\mathbf{X} + \underbrace{[B_\nu]}_{D}\mathbf{I},$$

which, together with (C.14), yields

$$\frac{d\mathbf{X}(t)}{dt} = A\mathbf{X}(t) + B\mathbf{I}(t),\ \mathbf{Y}(t) = C\mathbf{X}(t) + D\mathbf{I}(t),\ \forall t \in \mathfrak{T}_0. \tag{C.15}$$

The quadruple (A, B, C, D) is the *ISO* realization of the *IO* system (2.7).

C.2 From *ISO* system to *IO* system

We can transform the *ISO* system (2.17) and (2.18), (in Section 2.2),

$$\frac{d\mathbf{X}(t)}{dt} = A\mathbf{X}(t) + B\mathbf{I}(t),\ \forall t \in \mathfrak{T}_0, \tag{C.16}$$

$$\mathbf{Y}(t) = C\mathbf{X}(t) + D\mathbf{I}(t),\ \forall t \in \mathfrak{T}_0, \tag{C.17}$$

into the *IO* system by applying the Laplace transform to (C.16) and (C.17) for all zero initial conditions,

$$(sI - A)\,\mathbf{X}(s) = B\mathbf{I}(s) \Longrightarrow$$
$$\mathbf{X}(s) = (sI - A)^{-1}\,B\mathbf{I}(s)$$
$$\mathbf{Y}(s) = \left[C\,(sI - A)^{-1}\,B + D\right]\mathbf{I}(s) \Longrightarrow$$
$$[\det(sI - A)]\,\mathbf{Y}(s) = [C adj\,(sI - A)\,B + D\det(sI - A)]\,\mathbf{I}(s).$$

Let

$$f(s) = \det(sI - A) = \sum_{i=0}^{i=n} c_i s^i,\ c_n = 1,$$

$$C adj\,(sI - A)\,B + D\det(sI - A) = \sum_{i=0}^{i=n} K_i s^i,\ K_i \in \mathfrak{R}^{N\times M},$$

so that

$$\sum_{i=0}^{i=n} c_i\left[s^i\mathbf{Y}(s)\right] = \sum_{i=0}^{i=n} K_i\left[s^i\mathbf{I}(s)\right].$$

The application of the Inverse Laplace transform for all zero initial conditions results into

$$\sum_{i=0}^{i=n} c_i\mathbf{Y}^{(i)}(t) = \sum_{i=0}^{i=n} K_i\mathbf{I}^{(i)}(t),\ \forall t \in \mathfrak{T}_0.$$

This is the *IO* description of the *ISO* system (C.16) and (C.17).

Let

$$C^{(n)} = [c_0 I\ \ c_1 I\ ...\ c_n I] \in \mathfrak{R}^{N\times(n+1)N},$$
$$K^{(n)} = [K_0\ \ K_1\ ...\ K_n] \in \mathfrak{R}^{N\times(n+1)M}.$$

The quadruple $(n,\ n,\ C^{(n)},\ K^{(n)})$ is the *IO* realization of the *ISO* system (C.16) and (C.17).

C.3 Relationships among system descriptions

If $Q_k = A_k$, for $k = 0,1,2,...,\alpha$, $P_k = B_k$, for $k = 0,1,2,...,\beta$, $E_k = O_N$, for $k = 1,2,...,\nu$, i.e.,$\nu = 0$, $E_0 = I_N$, and $T_k = O_{N,M}$, for $k = 0,1,2,...,\mu$, then (2.22) (in Section 2.3) reduces to (2.1) (in Section 2.1), and (3.89), (in Subsection 3.5.4), reduces to (3.76), (in Subsection 3.5.2), i.e., the *IIO* system becomes the *IO* system. Then the *IIO* systems incorporate the *IO* systems. The *IO* system (2.7),

$$A^{(\nu)}\mathbf{Y}^{\nu}(t) = B^{(\mu)}\mathbf{I}^{\mu}(t),\ \forall t \in \mathfrak{T}_0.$$

can be set in the special *IIO* system (2.35) form, i.e., in the *GISO system form (2.39)*, by formally defining $\mathbf{R} = \mathbf{Y}$, $\alpha = \nu$, $\beta = \mu$, and $\rho = N$,

$$\begin{aligned} A^{(\alpha)}\mathbf{R}^{\alpha}(t) &= B^{(\beta)}\mathbf{I}^{\beta}(t),\ \forall t \in \mathfrak{T}_0, \\ \mathbf{Y}(t) &= \mathbf{R}(t),\ \forall t \in \mathfrak{T}_0. \end{aligned}$$

Hence, we can formally consider the *IO* systems as a subclass of the *IIO* systems. These two classes of the systems are equivalent if $\nu = 0$. However, if $\nu > 0$, then the *IIO* system (2.35) can be put into the following form:

$$\underbrace{\begin{bmatrix} Q^{(\alpha)} & O_{\rho,(\nu+1)N} \\ -R^{(\eta)} & E^{(\nu)} \end{bmatrix}}_{A^{(\alpha+\nu)}} \underbrace{\begin{bmatrix} \mathbf{R}^{\alpha}(t) \\ \mathbf{Y}^{\nu}(t) \end{bmatrix}}_{\mathbf{Y}^{*\alpha+\nu}(t)} = \underbrace{\begin{bmatrix} P^{(\beta)} & O_{\rho,(\beta+1)M} \\ O_{N,(\mu+1)M} & T^{(\mu)} \end{bmatrix}}_{B^{(\beta+\mu)}} \underbrace{\begin{bmatrix} \mathbf{I}^{\beta}(t) \\ \mathbf{I}^{\mu}(t) \end{bmatrix}}_{\mathbf{I}^{*\gamma}(t)},$$
$$\forall t \in \mathfrak{T}_0. \tag{C.18}$$

so that

$$A^{(\alpha+\nu)}\mathbf{Y}^{*\alpha+\nu}(t) = B^{(\beta+\mu)}\mathbf{I}^{*\gamma}(t),\ \forall t \in \mathfrak{T}_0, \tag{C.19}$$

$$\mathbf{Y}(t) = \left[O_{N,(\alpha+1)\rho} \vdots I_{\nu+1}\right] \mathbf{Y}^{*\alpha+\nu}(t),\ \forall t \in \mathfrak{T}_0. \tag{C.20}$$

This is the *IO* form relative to the vector $\mathbf{Y}^*$, but it rests the *IIO* form relative to the real output vector $\mathbf{Y}$. The *IO* system (2.7), (in Section 2.1), and the *IIO* system (2.35), are different systems if $\nu > 0$. These two classes of the systems are not equivalent if $\nu > 0$.

If $\rho = n$, $\mathbf{r} = \mathbf{x}$, $\alpha = 1$, $Q_1 = I_n$, $Q_0 = -A$, $\beta = 0$, $P_0 = B$, $\nu = 0$, $E_0 = I_N$, $\eta = 0$, $R_0 = C$, and $\mu = 0$, $T_0 = D$, then (2.22) reduces to (2.17) and (2.18) (in Section 2.2), and (3.89) reduces to (3.84) and (3.85), (in Subsection 3.5.3). The *IIO* becomes the *ISO* system. The latter is a special case of the former.

If $\nu = 0$, $E_0 = I_N$, then (2.22) (in Section 2.3) corresponds to the PMD system (2.36), (in Section 2.3), and (3.89) reduces to (3.91) (in Subsection 3.5.4). The IIO is then a PMD system. The family of the IIO systems incorporates the family of the PMD systems described by (2.36), (in Section 2.3).

If $\rho = n$, $\mathbf{r} = \mathbf{x}$, $\alpha = 1$, $Q_1 = I_n$, $Q_0 = -A$, $\beta = 0$, $P_0 = B$, $\nu = 0$, $E_0 = I_N$, $R_0 = C$ and $R_1 = O_{N \times \rho}$, then the IIO system (2.22) (in Section 2.3) reduces to the Rosenbrock system (2.37) (in Section 2.3), and (3.90), (in Subsection 3.5.4), represents the Rosenbrock system RS (3.92), (in Subsection 3.5.4), which is the special case of the former and the PMD system (2.36).

If $\nu = 0$, $E_0 = I_N$, $R_0 = R$, $R_k = O_{N \times \rho}$, $k = 1, 2, ..., \eta \leq \alpha - 1$, and $T_0 = T$, $T_k = O_{N \times M}$, $k = 1, 2, ..., \mu$, then (2.22) reduces to the $GISO$ system (2.39) (in Section 2.3), and (3.89), i.e., (3.90), becomes (3.94) (in Subsection 3.5.4). The $GISO$ systems are a special subclass of the PMD systems, which incorporates Rosenbrock systems.

Appendix C.1 presents the transformation of the IO system (2.1), in Section 2.1, into the equivalent ISO system (2.17) and (2.18). Appendix C.2 shows how the ISO system (2.17) and (2.18) can be transformed into the IO system (2.1).

We can summarize this by using notation $\{.\}$ for a class of systems, e.g., $\{IIO\}$ is the family of all IIO systems (3.89),

$$\{IO\} \simeq \{ISO\}, \ \{ISO\} \subset \{GISO\} \subset \{RS\} \subset \{PMD\} \subset \{IIO\},$$
$$\nu = \eta = \mu = 0 \Longrightarrow \{IO\} \simeq \{IIO\}, \ \nu > 0 \Longrightarrow \{IO\} \subset \{IIO\},$$

where $\simeq$ means *equivalent*.

Appendix D

Proof of Theorem 142

Proof. [47]

a) The application of the (left, right) Laplace transform $\mathcal{L}^{\mp}\{.\}$ (Appendix B) of the left-hand side of (3.76) or of (3.77), (in Subsection 3.5.2), i.e., (8.1), (in Subsection 8.1), yields the following:

$$\mathcal{L}^{\mp}\left\{\sum_{k=0}^{k=\nu} A_k \mathbf{y}^{(k)}(t)\right\} = \mathcal{L}^{\mp}\left\{A^{(\nu)}\mathbf{y}^{\nu}(t)\right\} =$$

$$= \overbrace{\left(A_0 s^0 + A_1 s^1 + ... + A_\nu s^\nu\right)}^{A^{(\nu)}S_N^{(\nu)}(s)} \mathbf{Y}^{\mp}(s) -$$

$$-A_0 O_N -$$

$$-A_1\left(s^{1-1}\mathbf{y}^{(1-1)}(0^{\mp})\right) -$$

$$-A_2\left(s^{2-1}\mathbf{y}^{(1-1)}(0^{\mp}) + s^{2-2}\mathbf{y}^{(2-1)}(0^{\mp})\right) -$$

$$\cdots\cdots\cdots\cdots\cdots\cdots\cdots\cdots\cdots$$

$$-A_k\left(\begin{array}{c} s^{k-1}\mathbf{y}^{(1-1)}(0^{\mp}) + s^{k-2}\mathbf{y}^{(2-1)}(0^{\mp}) + . \\ . + s^{k-i}\mathbf{y}^{(i-1)}(0^{\mp}) + .. + s^{k-k}\mathbf{y}^{(k-1)}(0^{\mp}) \end{array}\right) -$$

$$\cdots\cdots\cdots\cdots\cdots\cdots\cdots\cdots\cdots$$

$$-A_\nu\left(\begin{array}{c} s^{\nu-1}\mathbf{y}^{(1-1)}(0^{\mp}) + s^{\nu-2}\mathbf{y}^{(2-1)}(0^{\mp}) + .. \\ .. + s^{\nu-\nu}\mathbf{y}^{(\nu-1)}(0^{\mp}) \end{array}\right) =$$

$$= A^{(\nu)} S_N^{(\nu)}(s)\mathbf{Y}^{\mp}(s) - \overbrace{\begin{bmatrix} A_0 \vdots A_1 \vdots \ldots \vdots A_k \vdots \ldots \vdots A_\nu \end{bmatrix}}^{A^{(\nu)}} \bullet$$

$$\bullet \begin{bmatrix} O_N \\ s^{1-1}\mathbf{y}^{(1-1)}(0^{\mp}) \\ \ldots \\ s^{k-1}\mathbf{y}^{(1-1)}(0^{\mp}) + s^{k-2}\mathbf{y}^{(2-1)}(0^{\mp}) + \ldots + s^{k-k}\mathbf{y}^{(k-1)}(0^{\mp}) \\ \ldots \\ s^{\nu-1}\mathbf{y}^{(1-1)}(0^{\mp}) + s^{\nu-2}\mathbf{y}^{(2-1)}(0^{\mp}) + \ldots + s^{\nu-\nu}\mathbf{y}^{(\nu-1)}(0^{\mp}) \end{bmatrix} =$$

$$= \overbrace{\begin{bmatrix} A_0 \vdots A_1 \vdots \ldots \vdots A_k \vdots \ldots \vdots A_\nu \end{bmatrix}}^{A^{(\nu)},\ (8.4)} \bullet$$

$$\underbrace{\bullet \begin{bmatrix} O_N & O_N & \ldots & O_N & \ldots & O_N \\ s^{1-1}I_N & O_N & \ldots & O_N & \ldots & O_N \\ \ldots & \ldots & \ldots & O_N & \ldots & O_N \\ s^{k-1}I_N & s^{k-2}I_N & \ldots & s^{k-k}I_N & \ldots & O_N \\ \ldots & \ldots & \ldots & \ldots & \ldots & \ldots \\ s^{\nu-1}I_N & s^{\nu-2}I_N & \ldots & s^{\nu-2}I_N & \ldots & s^{\nu-\nu}I_N \end{bmatrix}}_{Z_N^{(\nu-1)}(s),\ (8.5)} \bullet$$

$$\underbrace{\bullet \begin{bmatrix} \mathbf{y}^{(1-1)}(0^{\mp}) \\ \mathbf{y}^{(2-1)}(0^{\mp}) \\ \ldots \\ \mathbf{y}^{(k-1)}(0^{\mp}) \\ \ldots \\ \mathbf{y}^{(\nu-1)}(0^{\mp}) \end{bmatrix}}_{\mathbf{y}_{0^{\mp}}^{\nu-1}(0^{\mp}),\ (2.6)} = A^{(\nu)} S_N^{(\nu)}(s)\mathbf{Y}^{\mp}(s) - A^{(\nu)} Z_N^{(\nu-1)}(s)\mathbf{y}_{0^{\mp}}^{\nu-1}. \tag{D.1}$$

By repeating the above procedure applied to

$$\mathcal{L}^{\mp}\left\{\sum_{k=0}^{k=\nu} B_k \mathbf{i}^{(k)}(t)\right\} = \mathcal{L}^{\mp}\left\{B^{(\mu)}\mathbf{i}^{\mu}(t)\right\},$$

and in view of (8.5),

$$\mathcal{L}^{\mp}\left\{\sum_{k=0}^{k=\nu} B_k \mathbf{i}^{(k)}(t)\right\} =$$
$$= \left\{\begin{array}{c} B^{(\mu)}S_M^{(\mu)}(s)\mathbf{I}^{\mp}(s) - B^{(\mu)}Z_M^{(\mu-1)}(s)\mathbf{i}^{\mu-1}(0^{\mp}),\ \mu \geq 1, \\ B^{(\mu)}S_M^{(\mu)}(s)\mathbf{I}^{\mp}(s) = B_0\mathbf{I}^{\mp}(s),\ \mu = 0. \end{array}\right\} \tag{D.2}$$

These results imply the following compact form of (left, right) Laplace transform $\mathcal{L}^{\mp}\{.\}$ of both (3.76) and (3.77):

$$A^{(\nu)}S_N^{(\nu)}(s)\mathbf{Y}^{\mp}(s) - A^{(\nu)}Z_N^{(\nu-1)}(s)\mathbf{y}_{0^{\mp}}^{\nu-1}(0^{\mp}) =$$
$$= \left\{\begin{array}{c} B^{(\mu)}S_M^{(\mu)}(s)\mathbf{I}^{\mp}(s) - B^{(\mu)}Z_M^{(\mu-1)}(s)\mathbf{i}^{\mu-1}(0^{\mp}),\ \mu \geq 1, \\ B^{(\mu)}S_M^{(\mu)}(s)\mathbf{I}^{\mp}(s) = B_0\mathbf{I}^{\mp}(s),\ \mu = 0. \end{array}\right\}. \tag{D.3}$$

This determines $\mathbf{Y}^{\mp}(s)$ linearly in terms of the vector function

$$\left[\left(\mathbf{I}^{(\mp)}(s)\right)^T \vdots \left(\mathbf{i}_{0^{\mp}}^{\mu-1}\right)^T \vdots \left(\mathbf{y}_{0^{\mp}}^{\nu-1}\right)^T\right]^T$$

as follows:

$$\mathbf{Y}^{\mp}(s) =$$
$$= \underbrace{\left[\begin{array}{c} \left(A^{(\nu)}S_N^{(\nu)}(s)\right)^{-1}\cdot \\ \cdot\left\{\begin{array}{c} \left[B^{(\mu)}S_M^{(\mu)}(s)\vdots - B^{(\mu)}Z_M^{(\mu-1)}(s)\vdots A^{(\nu)}Z_N^{(\nu-1)}(s)\right],\ \mu \geq 1 \\ \left[B_0 \vdots A^{(\nu)}Z_N^{(\nu-1)}(s)\right],\ \mu = 0 \end{array}\right\} \end{array}\right]}_{F_{IO}(s)}\cdot$$
$$\cdot\left\{\begin{array}{c} \left[\left(\mathbf{I}^{\mp}(s)\right)^T \vdots \left(\mathbf{i}_{0^{\mp}}^{\mu-1}\right)^T \vdots \left(\mathbf{y}_{0^{\mp}}^{\nu-1}\right)^T\right]^T,\ \mu \geq 1, \\ \left[\left(\mathbf{I}^{\mp}(s)\right)^T \vdots \left(\mathbf{y}_{0^{\mp}}^{\nu-1}\right)^T\right]^T,\ \mu = 0 \end{array}\right\}. \tag{D.4}$$

The definition of $F_{IO}(s)$ (Definition 7.2, in Section 7.2) and this equation prove the statement under a) of the theorem.

b) The statement under b) results directly from a), and the definition of $G_{IO}(s)$ (7.15). The formulae under c) through e) result directly from (7.16) through (7.18) linked with (D.1) through (D.4). Q. E. D. ■

Appendix E

Example: $\boldsymbol{F}(\mathrm{s})$ of a MIMO system

Example 415 *The MIMO IO system is given by*

$$\mathbf{y}^{(2)}(t) + \overbrace{\begin{bmatrix} 1 & 2 \\ 0 & 3 \end{bmatrix}}^{A_1} \mathbf{y}^{(1)}(t) = \underbrace{\begin{bmatrix} -1 & 2 & 2 \\ 4 & 4 & 7 \end{bmatrix}}_{B_0} \mathbf{i}(t) + \underbrace{\begin{bmatrix} 2 & 1 & -5 \\ 3 & 1 & 6 \end{bmatrix}}_{B_1} \mathbf{i}^{(1)}(t).$$

It follows that $M = 3$, $N = 2$, $\nu = 2$, $\mu = 1$, $A_0 = O_2$ and $A_2 = I_2$. Equations (2.2), (8.4), and (8.5) become

$$A^{(\nu)} = A^{(2)} = \left[A_0 \vdots A_1 \vdots A_2 \right] = \begin{bmatrix} 0 & 0 & 1 & 2 & 1 & 0 \\ 0 & 0 & 0 & 3 & 0 & 1 \end{bmatrix},$$

$$B^{(\mu)} = B^{(1)} = \left(B_0 \vdots B_1 \right) = \begin{bmatrix} -1 & 2 & 2 & 2 & 1 & -5 \\ 4 & 4 & 7 & 3 & 1 & 6 \end{bmatrix},$$

$$S_M^{(\mu)}(s) = S_3^{(1)}(s) = \left(s^0 I_3 \vdots s^1 I_3 \right)^T = \begin{bmatrix} 1 & 0 & 0 & s & 0 & 0 \\ 0 & 1 & 0 & 0 & s & 0 \\ 0 & 0 & 1 & 0 & 0 & s \end{bmatrix}^T,$$

$$S_N^{(\nu)}(s) = S_2^{(2)}(s) = \left(s^0 I_2 \vdots s^1 I_2 \vdots s^2 I_2 \right)^T = \begin{bmatrix} 1 & 0 & s & 0 & s^2 & 0 \\ 0 & 1 & 0 & s & 0 & s^2 \end{bmatrix}^T,$$

$$A^{(2)}S_2^{(2)}(s) = \begin{bmatrix} 0 & 0 & 1 & 2 & 1 & 0 \\ 0 & 0 & 0 & 3 & 0 & 1 \end{bmatrix} \begin{bmatrix} 1 & 0 \\ 0 & 1 \\ s & 0 \\ 0 & s \\ s^2 & 0 \\ 0 & s^2 \end{bmatrix} =$$

$$= \begin{bmatrix} s^2+s & 2s \\ 0 & s^2+3s \end{bmatrix} \Longrightarrow$$

$$\left(A^{(2)}S_2^{(2)}(s)\right)^{-1} = \frac{1}{(s^2+s)(s^2+3s)} \begin{bmatrix} s^2+3s & -2s \\ 0 & s^2+s \end{bmatrix} =$$

$$= \begin{bmatrix} \frac{s^2+3s}{(s^2+s)(s^2+3s)} & -\frac{2s}{(s^2+s)(s^2+3s)} \\ 0 & \frac{s^2+s}{(s^2+s)(s^2+3s)} \end{bmatrix} =$$

$$= \frac{s}{s^2(s+1)(s+3)} \begin{bmatrix} s+3 & -2 \\ 0 & s+1 \end{bmatrix} = \begin{bmatrix} \frac{s+3}{s(s+1)(s+3)} & -\frac{2}{s(s+1)(s+3)} \\ 0 & \frac{s+1}{s(s+1)(s+3)} \end{bmatrix},$$

$$B^{(1)}S_3^{(1)}(s) = \begin{bmatrix} -1 & 2 & 2 & 2 & 1 & -5 \\ 4 & 4 & 7 & 3 & 1 & 6 \end{bmatrix} \begin{bmatrix} 1 & 0 & 0 \\ 0 & 1 & 0 \\ 0 & 0 & 1 \\ s & 0 & 0 \\ 0 & s & 0 \\ 0 & 0 & s \end{bmatrix} =$$

$$= \begin{bmatrix} -1+2s & 2+s & 2-5s \\ 4+3s & 4+s & 7+6s \end{bmatrix},$$

$$\left(A^{(2)}S_2^{(2)}(s)\right)^{-1}\left(B^{(1)}S_3^{(1)}(s)\right) = \frac{s}{(s^2+s)(s^2+3s)} \begin{bmatrix} s+3 & -2 \\ 0 & s+1 \end{bmatrix} \bullet$$

$$\bullet \begin{bmatrix} -1+2s & 2+s & 2-5s \\ 4+3s & 4+s & 7+6s \end{bmatrix} =$$

$$= \frac{s}{(s^2+s)(s^2+3s)} \bullet$$

$$\bullet \begin{bmatrix} 2s^2-s-11 & s^2+3s-2 & -5s^2-25s-8 \\ 3s^2+7s+4 & s^2+5s+4 & 6s^2+13s+7 \end{bmatrix},$$

$$Z_M^{(\mu-1)}(s) = Z_3^{(0)}(s) = \begin{pmatrix} 0 & 0 & 0 \\ 0 & 0 & 0 \\ 0 & 0 & 0 \\ 1 & 0 & 0 \\ 0 & 1 & 0 \\ 0 & 0 & 1 \end{pmatrix},$$

$$Z_N^{(\nu-1)}(s) = Z_2^{(1)}(s) = \begin{pmatrix} 0 & 0 & 0 & 0 \\ 0 & 0 & 0 & 0 \\ 1 & 0 & 0 & 0 \\ 0 & 1 & 0 & 0 \\ s & 0 & 1 & 0 \\ 0 & s & 0 & 1 \end{pmatrix},$$

$$A^{(2)} Z_2^{(1)}(s) = \begin{bmatrix} 0 & 0 & 1 & 2 & 1 & 0 \\ 0 & 0 & 0 & 3 & 0 & 1 \end{bmatrix} \begin{pmatrix} 0 & 0 & 0 & 0 \\ 0 & 0 & 0 & 0 \\ 1 & 0 & 0 & 0 \\ 0 & 1 & 0 & 0 \\ s & 0 & 1 & 0 \\ 0 & s & 0 & 1 \end{pmatrix} =$$

$$= \begin{bmatrix} 1+s & 2 & 1 & 0 \\ 0 & 3+s & 0 & 1 \end{bmatrix} \Longrightarrow$$

$$\left(A^{(2)} S_2^{(2)}(s)\right)^{-1} A^{(2)} Z_2^{(1)}(s) =$$
$$= \frac{s}{(s^2+s)(s^2+3s)} \begin{bmatrix} s+3 & -2 \\ 0 & s+1 \end{bmatrix} \begin{bmatrix} 1+s & 2 & 1 & 0 \\ 0 & 3+s & 0 & 1 \end{bmatrix} =$$
$$= \frac{s}{(s^2+s)(s^2+3s)} \begin{bmatrix} s^2+4s+3 & 0 & s+3 & -2 \\ 0 & s^2+4s+3 & 0 & s+1 \end{bmatrix},$$

$$B^{(1)} Z_3^{(0)}(s) = \begin{bmatrix} -1 & 2 & 2 & 2 & 1 & -5 \\ 4 & 4 & 7 & 3 & 1 & 6 \end{bmatrix} \begin{pmatrix} 0 & 0 & 0 \\ 0 & 0 & 0 \\ 0 & 0 & 0 \\ 1 & 0 & 0 \\ 0 & 1 & 0 \\ 0 & 0 & 1 \end{pmatrix} =$$

$$= \begin{bmatrix} 2 & 1 & -5 \\ 3 & 1 & 6 \end{bmatrix} \Longrightarrow$$

$$-\left(A^{(2)}S_2^{(2)}(s)\right)^{-1}B^{(1)}Z_3^{(0)}(s)=$$
$$=-\frac{s}{\left(s^2+s\right)\left(s^2+3s\right)}\begin{bmatrix} s+3 & -2 \\ 0 & s+1 \end{bmatrix}\begin{bmatrix} 2 & 1 & -5 \\ 3 & 1 & 6 \end{bmatrix}=$$
$$=-\frac{s}{\left(s^2+s\right)\left(s^2+3s\right)}\begin{bmatrix} 2s & s+1 & -5s-27 \\ 3s+3 & s+1 & 6s+6 \end{bmatrix}.$$

We can now determine $G_{IO}(s)$ in view of (8.2),

$$G_{IO}(s)=\left(A^{(2)}S_2^{(2)}(s)\right)^{-1}\left(B^{(1)}S_3^{(1)}(s)\right)=\frac{s}{\left(s^2+s\right)\left(s^2+3s\right)}\bullet$$
$$\bullet\begin{bmatrix} 2s^2-s-11 & s^2+3s-2 & -5s^2-25s-8 \\ 3s^2+7s+4 & s^2+5s+4 & 6s^2+13s+7 \end{bmatrix}. \tag{E.1}$$

The system full transfer function matrix $F_{IO}(s)$ has the following form due to both (8.6) and the above equations of this example:

$$F_{IO}(s)=\frac{s}{\left(s^2+s\right)\left(s^2+3s\right)}\bullet$$
$$\bullet\begin{bmatrix} \underbrace{\begin{bmatrix} 2s^2-s-11 & s^2+3s-2 & -5s^2-25s-8 \\ 3s^2+7s+4 & s^2+5s+4 & 6s^2+13s+7 \end{bmatrix}^T}_{G_{IO}(s)} \\ \underbrace{-\begin{bmatrix} 2s & s+1 & 5s-27 \\ 3s+3 & s+1 & 6s+6 \end{bmatrix}^T}_{G_{IOi_0}(s)} \\ \underbrace{\begin{bmatrix} s^2+4s+3 & 0 & s+3 & -2 \\ 0 & s^2+4s+3 & 0 & s+1 \end{bmatrix}^T}_{G_{IOy_0}(s)} \end{bmatrix}^T$$
$$=\left[G_{IO}(s)\,\vdots\,G_{IOi_0}(s)\,\vdots\,G_{IOy_0}(s)\right], \tag{E.2}$$

i.e.

$$F_{IO}(s) = \frac{s}{(s^2+s)(s^2+3s)} \bullet$$

$$\bullet \begin{bmatrix} 2s^2-s-11 & 3s^2+7s+4 \\ s^2+3s-2 & s^2+5s+4 \\ -5s^2-25s-8 & 6s^2+13s+7 \\ -2s & -3s-3 \\ -s-1 & -s-1 \\ -5s+27 & -6s-6 \\ s^2+4s+3 & 0 \\ 0 & s^2+4s+3 \\ s+3 & 0 \\ -2 & s+1 \end{bmatrix}^T \in \mathfrak{C}^{2\text{x}10}. \quad \text{(E.3)}$$

This is the degenerate form of $F_{IO}(s)$.. The nondegenerate form $F_{IOnd}(s)$ of $F_{IO}(s)$ reads:

$$F_{IOnd}(s) = \frac{1}{s(s+1)(s+3)} \bullet$$

$$\bullet \begin{bmatrix} 2s^2-s-11 & 3s^2+7s+4 \\ s^2+3s-2 & s^2+5s+4 \\ -5s^2-25s-8 & 6s^2+13s+7 \\ -2s & -3s+3 \\ -s-1 & -s-1 \\ -5s+27 & -6s-6 \\ s^2+4s+3 & 0 \\ 0 & s^2+4s+3 \\ s+3 & 0 \\ -2 & s+1 \end{bmatrix}^T \in \mathfrak{C}^{2\text{x}10}.$$

Appendix F

Proof of Theorem 165

Proof. The left/right Laplace transform of the first equation of the compact form (3.90) of (3.89) (in Subsection 3.5.4) reads

$$\left(Q^{(\alpha)}S_{\rho}^{(\alpha)}(s)\right)\mathbf{R}^{\mp}(s) =$$

$$= \left[P^{(\beta)}S_{M}^{(\beta)}(s) \vdots -P^{(\beta)}Z_{M}^{(\beta-1)}(s) \vdots Q^{(\alpha)}Z_{\rho}^{(\alpha-1)}(s)\right]\begin{bmatrix}\mathbf{I}^{\mp}(s)\\ \mathbf{i}_{0^{\mp}}^{\beta-1}\\ \mathbf{r}_{0^{\mp}}^{\alpha-1}\end{bmatrix} \Longrightarrow$$

$$\mathbf{R}^{\mp}(s) = \left(Q^{(\alpha)}S_{\rho}^{(\alpha)}(s)\right)^{-1}\bullet$$

$$\bullet\left[P^{(\beta)}S_{M}^{(\beta)}(s) \vdots -P^{(\beta)}Z_{M}^{(\beta-1)}(s) \vdots Q^{(\alpha)}Z_{\rho}^{(\alpha-1)}(s)\right]\begin{bmatrix}\mathbf{I}^{\mp}(s)\\ \mathbf{i}_{0^{\mp}}^{\beta-1}\\ \mathbf{r}_{0^{\mp}}^{\alpha-1}\end{bmatrix},$$

i.e.,

$$\mathbf{R}^{\mp}(s) = F_{IIOIS}(s)\left[\mathbf{I}^{\mp}(s) \vdots \mathbf{i}_{0^{\mp}}^{\beta-1} \vdots \mathbf{r}_{0^{\mp}}^{\alpha-1}\right]^{T},$$

where

$$F_{IIOIS}(s) = \left(Q^{(\alpha)}S_{\rho}^{(\alpha)}(s)\right)^{-1}\left[P^{(\beta)}S_{M}^{(\beta)}(s) \vdots -P^{(\beta)}Z_{M}^{(\beta-1)}(s) \vdots Q^{(\alpha)}Z_{\rho}^{(\alpha-1)}(s)\right]. \tag{F.1}$$

Hence, in view of Definition 144:

- IRIS transfer function matrix $G_{IIOIS}(s)$,

$$G_{IIOIS}(s) = \left(Q^{(\alpha)}S_{\rho}^{(\alpha)}(s)\right)^{-1}P^{(\beta)}S_{M}^{(\beta)}(s) =$$
$$= G_{IIOISD}^{-1}(s)G_{IIOISN}(s), \tag{F.2}$$

- IRII transfer function matrix $G_{IIOi_0}(s)$,

$$G_{IIOi_0IS}(s) = -\left(Q^{(\alpha)}S_{\rho}^{(\alpha)}(s)\right)^{-1} P^{(\beta)}Z_{M}^{(\beta-1)}(s) = \\ = G_{IIOi_0ISD}^{-1}(s)G_{IIOi_0ISN}(s), \tag{F.3}$$

- IRIR transfer function matrix $G_{IIOr_0}(s)$,

$$G_{IIOr_0IS}(s) = \left(Q^{(\alpha)}S_{\rho}^{(\alpha)}(s)\right)^{-1} Q^{(\alpha)}Z_{\rho}^{(\alpha-1)}(s) = \\ = G_{IIOr_0ISD}^{-1}(s)G_{IIOr_0ISN}(s), \tag{F.4}$$

so that the denominator matrix polynomial $F_{IIOISD}(s)$ and the numerator matrix polynomial $F_{IIOISN}(s)$ of $F_{IIOIS}(s) = F_{IIOISD}^{-1}(s)F_{IIOISN}(s)$, read:

$$F_{IIOISD}(s) = \left(Q^{(\alpha)}S_{\rho}^{(\alpha)}(s)\right), \\ F_{IIOISN}(s) = \\ \left[P^{(\beta)}S_{M}^{(\beta)}(s) \vdots - P^{(\beta)}Z_{M}^{(\beta-1)}(s) \vdots Q^{(\alpha)}Z_{\rho}^{(\alpha-1)}(s)\right]. \tag{F.5}$$

Equations (F.1)-(F.5) prove Equations (8.56)-(8.61).

The left/right Laplace transform of the second equation of the compact form (3.90) of (3.89) (in Subsection 3.5.4) reads

$$\left(E^{(\nu)}S_{N}^{(\nu)}(s)\right)\mathbf{Y}^{\mp}(s) = \\ = \begin{bmatrix} \left(R^{(\alpha)}S_{\rho}^{(\alpha)}(s)\right)^{T} \\ \left(T^{(\mu)}S_{M}^{(\mu)}(s)\right)^{T} \\ \left(-R^{(\alpha)}Z_{\rho}^{(\alpha-1)}(s)\right)^{T} \\ \left(-T^{(\mu)}Z_{M}^{(\mu-1)}(s)\right)^{T} \\ \left(E^{(\nu)}Z_{N}^{(\nu-1)}(s)\right)^{T} \end{bmatrix}^{T} \begin{bmatrix} \mathbf{R}^{\mp}(s) \\ \mathbf{I}^{\mp}(s) \\ \mathbf{r}_{0^{\mp}}^{\alpha-1} \\ \mathbf{i}_{0^{\mp}}^{\mu-1} \\ \mathbf{y}_{0^{\mp}}^{\nu-1} \end{bmatrix} \Longrightarrow$$

$$\left(E^{(\nu)}S_N^{(\nu)}(s)\right)\mathbf{Y}^{\mp}(s)=\begin{bmatrix}\left(R^{(\alpha)}S_\rho^{(\alpha)}(s)\right)^T\\ \left(T^{(\mu)}S_M^{(\mu)}(s)\right)^T\\ \left(-R^{(\alpha)}Z_\rho^{(\alpha-1)}(s)\right)^T\\ \left(-T^{(\mu)}Z_M^{(\mu-1)}(s)\right)^T\\ \left(E^{(\nu)}Z_N^{(\nu-1)}(s)\right)^T\end{bmatrix}^T\bullet$$

$$\bullet\begin{bmatrix}\left(Q^{(\alpha)}S_\rho^{(\alpha)}(s)\right)^{-1}\bullet\left[P^{(\beta)}S_M^{(\beta)}(s)\,\vdots\,-P^{(\beta)}Z_M^{(\beta-1)}(s)\,\vdots\,Q^{(\alpha)}Z_\rho^{(\alpha-1)}(s)\right]\bullet\\ \bullet\begin{bmatrix}\mathbf{I}^{\mp}(s)\\ \mathbf{i}_{0^{\mp}}^{\beta-1}\\ \mathbf{r}_{0^{\mp}}^{\alpha-1}\end{bmatrix}\\ \mathbf{I}^{\mp}(s)\\ \mathbf{r}_{0^{\mp}}^{\alpha-1}\\ \mathbf{i}_{0^{\mp}}^{\mu-1}\\ \mathbf{y}_{0^{\mp}}^{\nu-1}\end{bmatrix}$$

$$\Longrightarrow$$

$$\left(E^{(\nu)}S_N^{(\nu)}(s)\right)\mathbf{Y}^{\mp}(s)=$$

$$=\begin{bmatrix}\left(R^{(\alpha)}S_\rho^{(\alpha)}(s)\right)\left(Q^{(\alpha)}S_\rho^{(\alpha)}(s)\right)^{-1}\bullet\\ \bullet\left[P^{(\beta)}S_M^{(\beta)}(s)\,\vdots\,-P^{(\beta)}Z_M^{(\beta-1)}(s)\,\vdots\,Q^{(\alpha)}Z_\rho^{(\alpha-1)}(s)\right]\bullet\\ \bullet\begin{bmatrix}\mathbf{I}^{\mp}(s)\\ \mathbf{i}_{0^{\mp}}^{\beta-1}\\ \mathbf{r}_{0^{\mp}}^{\alpha-1}\end{bmatrix}+\\ +\left(T^{(\mu)}S_M^{(\mu)}(s)\right)\mathbf{I}^{\mp}(s)+\left(-R^{(\alpha)}Z_\rho^{(\alpha-1)}(s)\right)\mathbf{r}_{0^{\mp}}^{\alpha-1}+\\ +\left(-T^{(\mu)}Z_M^{(\mu-1)}(s)\right)\mathbf{i}_{0^{\mp}}^{\mu-1}+\left(E^{(\nu)}Z_N^{(\nu-1)}(s)\right)\mathbf{y}_{0^{\mp}}^{\nu-1}\end{bmatrix}\Longrightarrow$$

$$\left(E^{(\nu)}S_N^{(\nu)}(s)\right)\mathbf{Y}^{\mp}(s) =$$
$$= \left[\begin{array}{c} \left(R^{(\alpha)}S_\rho^{(\alpha)}(s)\right)\left(Q^{(\alpha)}S_\rho^{(\alpha)}(s)\right)^{-1}\bullet \\ \bullet\left[P^{(\beta)}S_M^{(\beta)}(s)\mathbf{I}^{\mp}(s) - P^{(\beta)}Z_M^{(\beta-1)}(s)\mathbf{i}_{0^{\mp}}^{\beta-1} + Q^{(\alpha)}Z_\rho^{(\alpha-1)}(s)\mathbf{r}_{0^{\mp}}^{\alpha-1}\right]+ \\ +\left(T^{(\mu)}S_M^{(\mu)}(s)\right)\mathbf{I}^{\mp}(s) + \left(-R^{(\alpha)}Z_\rho^{(\alpha-1)}(s)\right)\mathbf{r}_{0^{\mp}}^{\alpha-1}+ \\ +\left(-T^{(\mu)}Z_M^{(\mu-1)}(s)\right)\mathbf{i}_{0^{\mp}}^{\mu-1} + \left(E^{(\nu)}Z_N^{(\nu-1)}(s)\right)\mathbf{y}_{0^{\mp}}^{\nu-1} \end{array}\right],$$

$$\mathbf{Y}^{\mp}(s) = \left(E^{(\nu)}S_N^{(\nu)}(s)\right)^{-1}\bullet$$
$$\bullet\left[\begin{array}{c} \left[\left(R^{(\alpha)}S_\rho^{(\alpha)}(s)\right)\left(Q^{(\alpha)}S_\rho^{(\alpha)}(s)\right)^{-1}P^{(\beta)}S_M^{(\beta)}(s) + \left(T^{(\mu)}S_M^{(\mu)}(s)\right)\right]\mathbf{I}^{\mp}(s)+ \\ \left[\begin{array}{c} \left[\begin{array}{c} \left(R^{(\alpha)}S_\rho^{(\alpha)}(s)\right)\left(Q^{(\alpha)}S_\rho^{(\alpha)}(s)\right)^{-1}\left(-P^{(\beta)}Z_M^{(\beta-1)}(s)\right)\vdots \\ \vdots O_{N,(\gamma-\beta)M} \end{array}\right]+ \\ +\left(-T^{(\mu)}Z_M^{(\mu-1)}(s)\ \vdots\ O_{N,(\gamma-\mu)M}\right) \end{array}\right]\mathbf{i}_{0^{\mp}}^{\gamma-1}+ \\ +\left[\begin{array}{c} \left[\left(R^{(\alpha)}S_\rho^{(\alpha)}(s)\right)\left(Q^{(\alpha)}S_\rho^{(\alpha)}(s)\right)^{-1}Q^{(\alpha)}Z_\rho^{(\alpha-1)}(s)\right]- \\ -R^{(\alpha)}Z_\rho^{(\alpha-1)}(s) \end{array}\right]\mathbf{r}_{0^{\mp}}^{\alpha-1}+ \\ +\left(E^{(\nu)}Z_N^{(\nu-1)}(s)\right)\mathbf{y}_{0^{\mp}}^{\nu-1}. \end{array}\right]$$

These equations and (7.35) (Definition 140, in Section 7.4) yield

$$F_{IIO}(s) = \left[G_{IIO}(s)\ \vdots\ G_{IIOi_0}(s)\ \vdots\ G_{IIOr_0}(s)\ \vdots\ G_{IIOy_0}(s)\right],$$

$$G_{IIO}(s) = \left(E^{(\nu)}S_N^{(\nu)}(s)\right)^{-1}\bullet$$
$$\bullet\left(R^{(\alpha)}S_\rho^{(\alpha)}(s)\left(Q^{(\alpha)}S_\rho^{(\alpha)}(s)\right)^{-1}P^{(\beta)}S_M^{(\beta)}(s) + T^{(\mu)}S_M^{(\mu)}(s)\right) =$$
$$= G_{IIOD}^{-1}(s)G_{IION}(s),$$

$$G_{IIOi_0}(s) = -\left(E^{(\nu)}S_N^{(\nu)}(s)\right)^{-1}\bullet$$
$$\bullet\left(\begin{array}{c} \left[R^{(\alpha)}S_\rho^{(\alpha)}(s)\left(Q^{(\alpha)}S_\rho^{(\alpha)}(s)\right)^{-1}P^{(\beta)}Z_M^{(\beta-1)}(s)\vdots O_{N,(\gamma-\beta)M}\right]+ \\ +\left[T^{(\mu)}Z_M^{(\mu-1)}(s)\vdots O_{N,(\gamma-\mu)M}\right] \end{array}\right) =$$
$$= G_{IIOi_0D}^{-1}(s)G_{IIOi_0N}(s),$$

$$G_{IIOr_0}(s) = \left(E^{(\nu)} S_N^{(\nu)}(s)\right)^{-1} \bullet$$
$$\bullet \left(\begin{array}{c} \left[R^{(\alpha)} S_\rho^{(\alpha)}(s) \left(Q^{(\alpha)} S_\rho^{(\alpha)}(s)\right)^{-1} Q^{(\alpha)} Z_\rho^{(\alpha-1)}(s) \right] - \\ -R^{(\alpha)} Z_\rho^{(\alpha-1)}(s) \end{array} \right) =$$
$$= G_{IIOr_0D}^{-1}(s) G_{IIOr_0N}(s),$$

$$G_{IIOy_0}(s) = \left(E^{(\nu)} S_N^{(\nu)}(s)\right)^{-1} E^{(\nu)} Z_N^{(\nu-1)}(s) =$$
$$= G_{IIOy_0D}^{-1}(s) G_{IIOy_0N}(s). \tag{F.6}$$

Let

$$\mathbf{V}_{IIO}^{\mp}(s; \mathbf{i}_{0\mp}^{\gamma-1};\ \mathbf{r}_{0\mp}^{\alpha-1};\ \mathbf{y}_{0\mp}^{\nu-1}) = \left[\mathbf{I}^{\mp}(s) \vdots \mathbf{i}_{0\mp}^{\gamma-1} \vdots \mathbf{r}_{0\mp}^{\alpha-1} \vdots \mathbf{y}_{0\mp}^{\nu-1} \right]^T. \tag{F.7}$$

Equations (F.6), (F.7), and (7.35), (in Section 7.4) enable the following:

$$\mathbf{Y}^{\mp}(s) = F_{IIO}(s) \mathbf{V}_{IIO}^{\mp}(s; \mathbf{i}_{0\mp}^{\gamma-1} \vdots \mathbf{r}_{0\mp}^{\alpha-1} \vdots \mathbf{y}_{0\mp}^{\nu-1}), \tag{F.8}$$
$$F_{IIO}(s) = F_{IIOD}^{-1}(s) F_{IION}(s),$$
$$F_{IIOD}(s) = \left(E^{(\nu)} S_N^{(\nu)}(s)\right),$$

$$F_{IION}(s) =$$
$$\left[\begin{array}{c} \left[R^{(\alpha)} S_\rho^{(\alpha)}(s) \left(Q^{(\alpha)} S_\rho^{(\alpha)}(s)\right)^{-1} P^{(\beta)} S_M^{(\beta)}(s) + T^{(\mu)} S_M^{(\mu)}(s) \right]^T \vdots \\ \vdots \left[\begin{array}{c} -\left[R^{(\alpha)} S_\rho^{(\alpha)}(s) \left(Q^{(\alpha)} S_\rho^{(\alpha)}(s)\right)^{-1} P^{(\beta)} Z_M^{(\beta-1)}(s) \vdots O_{N,(\gamma-\beta)M} \right] - \\ -\left[T^{(\mu)} Z_M^{(\mu-1)}(s) \vdots O_{N,(\gamma-\mu)M} \right] \end{array} \right]^T \vdots \\ \vdots \left[\begin{array}{c} \left[R^{(\alpha)} S_\rho^{(\alpha)}(s) \left(Q^{(\alpha)} S_\rho^{(\alpha)}(s)\right)^{-1} Q^{(\alpha)} Z_\rho^{(\alpha-1)}(s) \right] - \\ -R^{(\alpha)} Z_\rho^{(\alpha-1)}(s) \end{array} \right]^T \vdots \\ \vdots \left(E^{(\nu)} Z_N^{(\nu-1)}(s)\right)^T. \end{array} \right]^T \tag{F.9}$$

Equations (F.8)-(F.9) prove (8.62) through (8.67). ■

Appendix G

Proof for Example 167

Proof. We apply the left/right Laplace transform to the *IIO* system described by (8.69), i.e., by

$$\begin{gathered} r^{(2)}(t)+4r^{(1)}(t)-5r(t)=3i^{(2)}(t)-21i^{(1)}(t)+18i(t), \\ y^{(2)}(t)+y^{(1)}(t)-2y\left(t\right)= \\ =5r^{(2)}(t)-5r^{(1)}(t)+2i^{(2)}(t)-14i^{(1)}(t)+12i(t). \end{gathered} \tag{G.1}$$

Its description and characteristics in the complex domain follow:

$$\begin{gathered} (s^2+4s-5)R^{\mp}(s)-(s+4)r_{0\mp}-r_{0\mp}^{(1)}= \\ =(3s^2-21s+18)I^{\mp}(s)-(3s-21)i_{0\mp}-3i_{0\mp}^{(1)}, \end{gathered}$$

$$\begin{gathered} (s^2+s-2)Y^{\mp}(s)-(s+1)y_{0\mp}-y_{0\mp}^{(1)}= \\ =(5s^2-5s)R^{\mp}(s)-(5s-5)r_{0\mp}-5r_{0\mp}^{(1)}+ \\ +(2s^2-14s+12)I^{\mp}(s)-(2s-14)i_{0\mp}-2i_{0\mp}^{(1)}\Longrightarrow \end{gathered}$$

$$\begin{gathered} (s^2+s-2)Y^{\mp}(s)-(s+1)y_{0\mp}-y_{0\mp}^{(1)}= \\ =(5s^2-5s)(s^2+4s-5)^{-1}\left[\begin{array}{c} (3s^2-21s+18)I^{\mp}(s)- \\ -(3s-21)i_{0\mp}-3i_{0\mp}^{(1)}+ \\ +(s+4)r_{0\mp}+r_{0\mp}^{(1)} \end{array}\right]- \\ -(5s-5)r_{0\mp}-5r_{0\mp}^{(1)}+(2s^2-14s+12)I^{\mp}(s)- \\ -(2s-14)i_{0\mp}-2i_{0\mp}^{(1)}\Longrightarrow \end{gathered}$$

$$Y^{\mp}(s) = (s^2+s-2)^{-1} \bullet$$
$$\bullet \left\{ \begin{array}{c} (5s^2-5s)(s^2+4s-5)^{-1} \bullet \\ \bullet \left[\begin{array}{c} (3s^2-21s+18)I^{\mp}(s)- \\ -(3s-21)i_{0\mp} - 3i_{0\mp}^{(1)}+ \\ +(s+4)r_{0\mp} + r_{0\mp}^{(1)} \end{array} \right] - \\ -(5s-5)r_{0\mp} - 5r_{0\mp}^{(1)} + (2s^2-14s+12)I^{\mp}(s)- \\ -(2s-14)i_{0\mp} - 2i_{0\mp}^{(1)} + (s+1)y_{0\mp} + y_{0\mp}^{(1)} \end{array} \right\} \Longrightarrow$$

$$Y^{\mp}(s) = (s^2+s-2)^{-1} \bullet$$
$$\bullet \left\{ \begin{array}{c} (5s^2-5s)(s^2+4s-5)^{-1}(3s^2-21s+18)I^{\mp}(s)- \\ -(5s^2-5s)(s^2+4s-5)^{-1}(3s-21)i_{0\mp}- \\ -(5s^2-5s)(s^2+4s-5)^{-1}3i_{0\mp}^{(1)}+ \\ +(5s^2-5s)(s^2+4s-5)^{-1}(s+4)r_{0\mp}+ \\ +(5s^2-5s)(s^2+4s-5)^{-1}r_{0\mp}^{(1)}- \\ -(5s-5)r_{0\mp} - 5r_{0\mp}^{(1)} + (2s^2-14s+12)I^{\mp}(s)- \\ -(2s-14)i_{0\mp} - 2i_{0\mp}^{(1)} + (s+1)y_{0\mp} + y_{0\mp}^{(1)} \end{array} \right\} \Longrightarrow$$

$$Y^{\mp}(s) = (s^2+s-2)^{-1} \bullet$$
$$\bullet \left\{ \begin{array}{c} \left[\begin{array}{c} (5s^2-5s)(s^2+4s-5)^{-1}(3s^2-21s+18)+ \\ +(2s^2-14s+12) \end{array} \right] I^{\mp}(s)- \\ -\left[(5s^2-5s)(s^2+4s-5)^{-1}(3s-21)+(2s-14)\right] i_{0\mp}- \\ -\left[(5s^2-5s)(s^2+4s-5)^{-1}3+2\right] i_{0\mp}^{(1)}+ \\ +\left[(5s^2-5s)(s^2+4s-5)^{-1}(s+4)-(5s-5)\right] r_{0\mp}+ \\ +\left[(5s^2-5s)(s^2+4s-5)^{-1}-5\right] r_{0\mp}^{(1)} + (s+1)y_{0\mp} + y_{0\mp}^{(1)} \end{array} \right\} \Longrightarrow$$

$$Y^{\mp}(s) = (s^2 + s - 2)^{-1} \bullet$$

$$\bullet \begin{bmatrix} \begin{pmatrix} (5s^2 - 5s)(s^2 + 4s - 5)^{-1}(3s^2 - 21s + 18) + \\ +(2s^2 - 14s + 12) \end{pmatrix} \\ -\left[(5s^2 - 5s)(s^2 + 4s - 5)^{-1}(3s - 21) + (2s - 14)\right] \\ (5s^2 - 5s)(s^2 + 4s - 5)^{-1}3 + 2 \\ (5s^2 - 5s)(s^2 + 4s - 5)^{-1}(s + 4) - (5s - 5) \\ -\left[(5s^2 - 5s)(s^2 + 4s - 5)^{-1} - 5\right] \\ (s + 1) \\ 1 \end{bmatrix}^T \bullet$$

$$\bullet \left[I^{\mp^T}(s) \vdots\, i_{0\mp} \vdots\, i_{0\mp}^{(1)} \vdots\, r_{0\mp} \vdots\, r_{0\mp}^{(1)} \vdots\, y_{0\mp} \vdots\, y_{0\mp}^{(1)} \right]^T \Longrightarrow \tag{G.2}$$

$$F_{IIO}(s) = (s^2 + s - 2)^{-1}(s^2 + 4s - 5)^{-1} \bullet$$

$$\bullet \begin{bmatrix} \begin{pmatrix} (5s^2 - 5s)(3s^2 - 21s + 18) + \\ +(s^2 + 4s - 5)(2s^2 - 14s + 12) \end{pmatrix} \\ -\left[(5s^2 - 5s)(3s - 21) + (s^2 + 4s - 5)(2s - 14)\right] \\ -\left[(5s^2 - 5s)3 + (s^2 + 4s - 5)2\right] \\ \left[(5s^2 - 5s)(s + 4) - (s^2 + 4s - 5)(5s - 5)\right] \\ (5s^2 - 5s) - (s^2 + 4s - 5)5 \\ \vdots(s^2 + 4s - 5)(s + 1) \\ (s^2 + 4s - 5) \end{bmatrix}^T \Longrightarrow$$

$$F_{IIO}(s) = \left[(s-1)^2(s+2)(s+5)\right]^{-1} \bullet$$
$$\bullet \begin{bmatrix} (s-1)^2(s-6)(17s+10) \\ -(s-1)(s-7)(17s+10) \\ -(s-1)(17s+10) \\ 25(s-1) \\ -25s(s-1) \\ (s-1)(s^2+6s+5) \\ (s-1)(s+5) \end{bmatrix}^T . \tag{G.3}$$

We can cancel $(s-1)$ because it is common factor to the denominator polynomial and to all entries of the numerator polynomial matrix,

$$F_{IIOnd}(s) = [(s-1)(s+2)(s+5)]^{-1} \bullet$$
$$\bullet \begin{bmatrix} (s-1)(s-6)(17s+10) \\ -(s-7)(17s+10) \\ -17s-10 \\ 10 \\ -25s \\ s^2+6s+5 \\ s+5 \end{bmatrix}^T . \tag{G.4}$$

The system transfer function $G_{IIO}(.)$ follows now from (G.3),

$$G_{IIO}(s) = \left[(s-1)^2(s+2)(s+5)\right]^{-1} (s-1)^2(s-6)(17s+10) . \tag{G.5}$$

We can cancel now $(s-1)^2$, which is not possible in $F_{IIOnd}(s)$,

$$G_{IIO}(s) = [(s+2)(s+5)]^{-1} (s-6)(17s+10) . \tag{G.6}$$

The left/right Laplace transform $\mathbf{V}_{IIO}^{\mp}(s)$ of the action vector function $\mathbf{v}_{IIO}(.)$ and the action vector function itself result from (G.2)

$$\mathbf{V}_{IIO}^{\mp}(s;\ \mathbf{i}_{0\mp}^{1}\ ;\mathbf{r}_{0\mp}^{1}\ ;\ \mathbf{y}_{0\mp}^{1}) = \left[I^{\mp}(s) \vdots \underbrace{i_{0\mp} \vdots i_{0\mp}^{(1)}}_{\mathbf{i}_{0\mp}^{1T}} \vdots \underbrace{r_{0\mp} \vdots r_{0\mp}^{(1)}}_{\mathbf{r}_{0\mp}^{1T}} \vdots \underbrace{y_{0\mp} \vdots y_{0\mp}^{(1)}}_{\mathbf{y}_{0\mp}^{1T}} \right]^{T},$$

$$\mathbf{v}_{IIO}(t;\ \mathbf{i}_{0\mp}^{1}\ ;\mathbf{r}_{0\mp}^{1}\ ;\ \mathbf{y}_{0\mp}^{1}) = \begin{bmatrix} i(t) \\ \delta(t)\mathbf{i}_{0\mp}^{1} \\ \delta(t)\mathbf{r}_{0\mp}^{1} \\ \delta(t)\mathbf{y}_{0\mp}^{1} \end{bmatrix}. \tag{G.7}$$

Equations (G.2), (G.3), and (G.7) yield

$$\mathbf{Y}^{-}(s) = F_{IIO}(s)\mathbf{V}_{IIO}^{-}(s;\ \mathbf{i}_{0-}^{1}\ ;\mathbf{r}_{0-}^{1}\ ;\ \mathbf{y}_{0-}^{1}). \tag{G.8}$$

This ends the proof. ∎

Appendix H

Proof of Theorem 168

Proof. The left/right Laplace transform of the compact form (3.90) of (3.89) (in Section 2.3) reads

$$\left(Q^{(\alpha)}S_{\rho}^{(\alpha)}(s)\right)\mathbf{R}^{\mp}(s)=$$

$$=\left[P^{(\beta)}S_{M}^{(\beta)}(s)\vdots -P^{(\beta)}Z_{M}^{(\beta-1)}(s)\vdots Q^{(\alpha)}Z_{\rho}^{(\alpha-1)}(s)\right]\begin{bmatrix}\mathbf{I}^{\mp}(s)\\ \mathbf{i}_{0^{\mp}}^{\beta-1}\\ \mathbf{r}_{0^{\mp}}^{\alpha-1}\end{bmatrix}\Longrightarrow$$

$$\mathbf{R}^{\mp}(s)=\left(Q^{(\alpha)}S_{\rho}^{(\alpha)}(s)\right)^{-1}\bullet$$

$$\bullet\left[P^{(\beta)}S_{M}^{(\beta)}(s)\vdots -P^{(\beta)}Z_{M}^{(\beta-1)}(s)\vdots Q^{(\alpha)}Z_{\rho}^{(\alpha-1)}(s)\right]\begin{bmatrix}\mathbf{I}^{\mp}(s)\\ \mathbf{i}_{0^{\mp}}^{\beta-1}\\ \mathbf{r}_{0^{\mp}}^{\alpha-1}\end{bmatrix},$$

These equations and (7.46) (Definition 144, in Section 7.4) yield

$$F_{IIOII}(s)=\left[G_{IIOII}(s)\vdots G_{IIOi_0II}(s)\vdots G_{IIOr_0II}(s)\right],$$

$$G_{IIOII}(s)=\left(Q^{(\alpha)}S_{\rho}^{(\alpha)}(s)\right)^{-1}P^{(\beta)}S_{M}^{(\beta)}(s)=$$
$$=G_{IIOIID}^{-1}(s)G_{IIOIIN}(s),$$

$$G_{IIOi_0II}(s)=-\left(Q^{(\alpha)}S_{\rho}^{(\alpha)}(s)\right)^{-1}P^{(\beta)}Z_{M}^{(\beta-1)}(s)=$$
$$=G_{IIOi_0IID}^{-1}(s)G_{IIOi_0IIN}(s),$$

$$G_{IIOr_0II}(s) = \left(Q^{(\alpha)} S_\rho^{(\alpha)}(s)\right)^{-1} Q^{(\alpha)} Z_\rho^{(\alpha-1)}(s) =$$
$$= G_{IIOr_0IID}^{-1}(s) G_{IIOr_0IIN}(s), \tag{H.1}$$

Let

$$\mathbf{V}_{IIOII}^{\mp}(s; \mathbf{i}_{0\mp}^{\beta-1};\ \mathbf{r}_{0\mp}^{\alpha-1}) = \left[\left(\mathbf{I}^{\mp}(s)\right)^T \vdots \left(\mathbf{i}_{0\mp}^{\beta-1}\right)^T \vdots \left(\mathbf{r}_{0\mp}^{\alpha-1}\right)^T\right]^T. \tag{H.2}$$

Equations (H.1), (H.2), and (7.46), (in Section 7.4) enable the following:

$$\mathbf{R}^{\mp}(s) = F_{IIOII}(s) \mathbf{V}_{IIO}^{\mp}(s; \mathbf{i}_{0\mp}^{\gamma-1} \vdots \mathbf{r}_{0\mp}^{\alpha-1}), \tag{H.3}$$

$$F_{IIOII}(s) = F_{IIOIID}^{-1}(s) F_{IIOIIN}(s),$$

$$F_{IIOIID}(s) = \left(Q^{(\alpha)} S_\rho^{(\alpha)}(s)\right),$$

$$F_{IION}(s) =$$
$$= \left[\ P^{(\beta)} S_M^{(\beta)}(s) \vdots -P^{(\beta)} Z_M^{(\beta-1)}(s) \vdots Q^{(\alpha)} Z_\rho^{(\alpha-1)}(s)\ \right]. \tag{H.4}$$

This completes the proof. ■

Appendix I

Proof of Theorem 176

Proof. (i) The inverse of the left Laplace transform of $\mathbf{Y}^{-}(s)$ leads to the inverse of the left Laplace transform of (7.11) (in Section 7.2):

$$\mathbf{y}(t;\mathbf{y}_0^{\nu-1};\mathbf{i}) = \mathcal{L}^{-1}\left\{\mathbf{Y}^{-}(s)\right\} = \frac{1}{2\pi j}\int_{c-j\infty}^{c+j\infty}\mathbf{Y}^{-}(s)e^{st}ds =$$

$$= \frac{1}{2\pi j}\int_{c-j\infty}^{c+j\infty}\left\{F_{IO}(s)\left(\left(\mathbf{I}^{-}(s)\right)^T \vdots \left(\mathbf{i}_{0^-}^{\mu-1}\right)^T \vdots \left(\mathbf{y}_{0^-}^{\nu-1}\right)^T\right)^T\right\}e^{st}ds,$$

where $j = \sqrt{-1}$. Let

$$\Xi(t) = \left[\Xi_I(t) \vdots \Xi_{io}(t) \vdots \Xi_{yo}(t)\right] = \mathcal{L}^{-1}\left\{F_{IO}(s)\right\}, \tag{I.1}$$

$$\left[\Xi_I(t) \vdots \Xi_{io}(t) \vdots \Xi_{yo}(t)\right] =$$
$$= \mathcal{L}^{-1}\left\{\left[G_{IO}(s) \vdots G_{IOio}(s) \vdots G_{IOyo}(s)\right]\right\},$$
$$\mathcal{L}\left\{\left[\Xi_I(t) \vdots \Xi_{io}(t) \vdots \Xi_{yo}(t)\right]\right\} =$$
$$= \left[G_{IO}(s) \vdots G_{IOio}(s) \vdots G_{IOyo}(s)\right], \tag{I.2}$$

which transforms the preceding result as follows (see Appendix B.2):

$$\mathbf{y}(t;\mathbf{y}_0^{\nu-1};\mathbf{i}) =$$

$$= \frac{1}{2\pi j}\int_{c-j\infty}^{c+j\infty} \left\{ \begin{array}{c} \left[G_{IO}(s) \vdots G_{IOio}(s) \vdots G_{IOyo}(s) \right] \\ \bullet \left[\left(\mathbf{I}^{-}(s)\right)^T \vdots \left(\mathbf{i}_{0-}^{\mu-1}\right)^T \vdots \left(\mathbf{y}_{0-}^{\nu-1}\right)^T \right]^T \end{array} \right\} e^{st} ds =$$

$$= \frac{1}{2\pi j}\int_{c-j\infty}^{c+j\infty} \left[G_{IO}(s)\mathbf{I}^{-}(s) + G_{IOio}(s)\mathbf{i}_{0-}^{\mu-1} + G_{IOyo}(s)\mathbf{y}_{0-}^{\nu-1} \right] e^{st} ds =$$

$$= \frac{1}{2\pi j}\int_{c-j\infty}^{c+j\infty} G_{IO}(s)\mathbf{I}^{-}(s) e^{st} ds +$$

$$+ \frac{1}{2\pi j}\int_{c-j\infty}^{c+j\infty} \left[G_{IOio}(s)\mathbf{i}_{0-}^{\mu-1} + G_{IOyo}(s)\mathbf{y}_{0-}^{\nu-1} \right] e^{st} ds \Longrightarrow$$

$$\mathbf{y}(t;\mathbf{y}_0^{\nu-1};\mathbf{i}) = \int_0^t \Xi_I(\tau)\mathbf{i}(t-\tau)d\tau + \Xi_{io}(t)\mathbf{i}_{0-}^{\mu-1} + \Xi_{yo}(t)\mathbf{y}_{0-}^{\nu-1}.$$

This and (10.4) (in Section 10.1) prove

$$\Xi(t) = \left[\Xi_I(t) \vdots \Xi_{io}(t) \vdots \Xi_{yo}(t) \right] =$$

$$= \Phi_{IO}(t) = \left[\Gamma_{IO}(t) \vdots \Gamma_{IOi_0}(t) \vdots \Gamma_{IOy_0}(t) \right],$$

$$\Phi_{IO}(t) = \mathcal{L}^{-1}\left\{ F_{IO}(s) \right\} =$$

$$= \left\{ \left[\Gamma_{IO}(t) \vdots \Gamma_{IOi_0}(t) \vdots \Gamma_{IOy_0}(t) \right] \right\}.$$

This proves (10.5) (in Section 10.1).

(ii) The Laplace transform of the preceding equations proves (10.6) (in Section 10.1).

(iii) Equations (8.6)-(8.13), (10.1)-(10.6) imply Equations (10.7)-(10.10), which completes the proof. ■

Appendix J

Proof of Theorem 179

Proof. (i) The inverse of the left Laplace transform of (7.29) (in Section 7.3) gives the following:

$$\mathbf{x}(t) = \mathcal{L}^{-1}\left\{\mathbf{X}^{-}(s)\right\} = \mathcal{L}^{-1}\left\{F_{ISOIS}(s)\left[\left(\mathbf{I}^{-}(s)\right)^{T} \vdots \mathbf{x}_{0^-}^{T}\right]^{T}\right\} =$$

$$= \frac{1}{2\pi j}\int_{c-j\infty}^{c+j\infty}\left\{F_{ISOIS}(s)\left[\left(\mathbf{I}^{-}(s)\right)^{T} \vdots \mathbf{x}_{0^-}^{T}\right]^{T}\right\}e^{st}ds =$$

$$= \frac{1}{2\pi j}\int_{c-j\infty}^{c+j\infty}\left\{\int_{0^-}^{\infty}\mathcal{L}^{-1}\left\{F_{ISOIS}(s)\right\}e^{-s\tau}d\tau\right\}\left[\left(\mathbf{I}^{-}(s)\right)^{T} \vdots \mathbf{x}_{0^-}^{T}\right]^{T}e^{st}ds =$$

$$= \int_{0^-}^{\infty}\mathcal{L}^{-1}\left\{F_{ISOIS}(s)\right\}\left[\frac{1}{2\pi j}\int_{c-j\infty}^{c+j\infty}\left[\left(\mathbf{I}^{-}(s)\right)^{T} \vdots \mathbf{x}_{0^-}^{T}\right]^{T}e^{s(t-\tau)}ds\right]d\tau =$$

$$= \int_{0^-}^{\infty}\mathcal{L}^{-1}\left\{F_{ISOIS}(s)\right\}\left[\mathbf{i}^{T}(t-\tau) \vdots \delta(t-\tau)\mathbf{x}_{0^-}^{T}\right]^{T}d\tau.$$

This, and (i) of Definition 193 (in Section 10.2) imply $\Phi_{ISOIS}(t) = \mathcal{L}^{-1}\left\{F_{ISOIS}(s)\right\}$ that is Equation (10.16).

(ii) The left Laplace transform of (10.16) is Equation (10.17). ■

Appendix K

Proof of Theorem 183

Proof. (i) The inverse of the left Laplace transform of (8.29) (in Section 8.2) gives the following:

$$\mathbf{y}(t)=\mathcal{L}^{-1}\left\{\mathbf{Y}^{-}(s)\right\}=\mathcal{L}^{-1}\left\{F_{ISO}(s)\left[\left(\mathbf{I}^{-}(s)\right)^{T}\vdots\,\mathbf{x}_{0^-}^{T}\right]^{T}\right\}=$$

$$\begin{aligned}&=\frac{1}{2\pi j}\int_{c-j\infty}^{c+j\infty}\mathbf{Y}^{-}(s)e^{st}ds=\\&=\frac{1}{2\pi j}\int_{c-j\infty}^{c+j\infty}\left\{F_{ISO}(s)\left[\left(\mathbf{I}^{-}(s)\right)^{T}\vdots\,\mathbf{x}_{0^-}^{T}\right]^{T}\right\}e^{st}ds=\end{aligned}$$

$$\begin{aligned}&=\frac{1}{2\pi j}\int_{c-j\infty}^{c+j\infty}\left\{\int_{0^-}^{\infty}\mathcal{L}^{-1}\left\{F_{ISO}(s)\right\}e^{-s\tau}d\tau\right\}\left[\left(\mathbf{I}^{-}(s)\right)^{T}\vdots\,\mathbf{x}_{0^-}^{T}\right]^{T}e^{st}ds=\\&=\int_{0^-}^{\infty}\mathcal{L}^{-1}\left\{F_{ISO}(s)\right\}\left[\frac{1}{2\pi j}\int_{c-j\infty}^{c+j\infty}\left[\left(\mathbf{I}^{-}(s)\right)^{T}\vdots\,\mathbf{x}_{0^-}^{T}\right]^{T}e^{s(t-\tau)}ds\right]d\tau=\\&=\int_{0^-}^{\infty}\mathcal{L}^{-1}\left\{F_{ISO}(s)\right\}\left[\mathbf{i}^{T}(t-\tau)\,\vdots\,\delta(t-\tau)\mathbf{x}_{0^-}^{T}\right]^{T}d\tau.\end{aligned}$$

This, and (i) of Definition 187 (in Section 10.2) imply $\Phi_{ISO}(t)=\mathcal{L}^{-1}\left\{F_{ISO}(s)\right\}$ that is Equation (10.22).

(ii) The left Laplace transform of (10.22) is Equation (10.23).

(iii) Equations (10.13)-(10.15), (10.19) and (10.21) imply (10.24), and (10.25). ■

Part VII

INDEX

Author Index

Subject Index